Lecture Notes in Computer Science   12755

More information about this subseries at http://www.springer.com/series/7407

Panos Pardalos · Michael Khachay ·
Alexander Kazakov (Eds.)

Mathematical Optimization Theory and Operations Research

20th International Conference, MOTOR 2021
Irkutsk, Russia, July 5–10, 2021
Proceedings

 Springer

Editors
Panos Pardalos (iD)
University of Florida
Gainesville, FL, USA

Alexander Kazakov (iD)
Matrosov Institute for System Dynamics
and Control Theory
Irkutsk, Russia

Michael Khachay (iD)
Krasovsky Institute of Mathematics
and Mechanics
Ekaterinburg, Russia

ISSN 0302-9743   ISSN 1611-3349 (electronic)
Lecture Notes in Computer Science
ISBN 978-3-030-77875-0   ISBN 978-3-030-77876-7 (eBook)
https://doi.org/10.1007/978-3-030-77876-7

LNCS Sublibrary: SL1 – Theoretical Computer Science and General Issues

This Springer imprint is published by the registered company Springer Nature Switzerland AG
The registered company address is: Gewerbestrasse 11, 6330 Cham, Switzerland

Preface

This volume contains the refereed proceedings of the 20th International Conference on Mathematical Optimization Theory and Operations Research (MOTOR 2021)[1] held during July 5–10, 2021, at Lake Baikal, near Irkutsk, Russia.

MOTOR 2021 was the third joint scientific event unifying a number of well-known international and Russian conferences that had been held in Ural, Siberia, and the Far East for a long time:

- The Baikal International Triennial School Seminar on Methods of Optimization and Their Applications (BITSS MOPT) established in 1969 by academician N. N. Moiseev; the 17th event[2] in this series was held in 2017 in Buryatia
- The All-Russian Conference on Mathematical Programming and Applications (MPA) established in 1972 by academician I. I. Eremin; was the 15th conference[3] in this series was held in 2015 near Ekaterinburg
- The International on Discrete Optimization and Operations Research (DOOR) was organized nine times since 1996 and the last event[4] was held in 2016 in Vladivostok
- The International Conference on Optimization Problems and Their Applications (OPTA) was organized regularly in Omsk since 1997 and the 7th event[5] was held in 2018

First two events of this series, MOTOR 2019[6] and MOTOR 2020[7], were held in Ekaterinburg and Novosibirsk, Russia, respectively.

As per tradition, the main conference scope included, but was not limited to, mathematical programming, bi-level and global optimization, integer programming and combinatorial optimization, approximation algorithms with theoretical guarantees and approximation schemes, heuristics and meta-heuristics, game theory, optimal control, optimization in machine learning and data analysis, and their valuable applications in operations research and economics.

In response to the call for papers, MOTOR 2021 received 181 submissions. Out of 102 full papers considered for reviewing (79 abstracts and short communications were excluded for formal reasons) only 30 papers were selected by the Program Committee (PC) for publication in this volume. Each submission was reviewed by at least three PC members or invited reviewers, experts in their fields, in order to supply detailed and helpful comments. In addition, the PC recommended 34 papers for inclusion in the

---

[1] https://conference.icc.ru/event/3/.

[2] http://isem.irk.ru/conferences/mopt2017/en/index.html.

[3] http://mpa.imm.uran.ru/96/en.

[4] http://www.math.nsc.ru/conference/door/2016/.

[5] http://opta18.oscsbras.ru/en/.

[6] http://motor2019.uran.ru.

[7] http://math.nsc.ru/conference/motor/2020/.

supplementary volume after their presentation and discussion during the conference and subsequent revision with respect to the reviewers' comments.

The conference featured nine invited lectures:

- Dr. Christian Blum (Artificial Intelligence Research Institute, Spain), "On the Design of Matheuristics that make Use of Learning"
- Prof. Emilio Carrizosa (Institute of Mathematics, University of Seville, Spain), "Optimal Classification and Regression Trees"
- Prof. François Clautiaux (Université de Bordeaux, France), "Integer Programming Formulations Based on Exponentially Large Networks: Algorithms and Applications"
- Prof. Andreas Griewank (Institute of Mathematics, Humboldt University, Germany), "Beyond Heuristic Gradient Descent in Machine Learning"
- Prof. Klaus Jansen (Christian-Albrechts-Universität, Germany) "Integer Programming and Convolution, with Applications"
- Prof. Sergey Kabanikhin (Institute of Numerical Mathematics and Mathematical Geophysics, Russia) "Optimization and Inverse Problems"
- Prof. Nenad Mladenovic (Khalifa University, United Arab Emirates), "Minimum Sum of Squares Clustering for Big Data – Heuristic Approach"
- Prof. Claudia Sagastizábal (IMECC - University of Campinas, Brazil), "Exploiting Structure in Nonsmooth Optimization"
- Prof. Mikhail Solodov (Institute for Pure and Applied Mathematics, Brazil), "State-of-the-art on Rates of Convergence and Cost of Iterations of Augmented Lagrangian Methods"

The following three tutorials were given by outstanding scientists:

- Prof. Alexander Gasnikov (Moscow Institute of Physics and Technology, Russia), "Reinforcement Learning from the Stochastic Optimization Point of View"
- Prof. Alexander Krylatov (Saint Petersburg State University, Russia), "Equilibrium Traffic Flow Assignment in a Multi-Subnet Urban Road Network"
- Prof. Alexander Strekalovsky (Matrosov Institute for System Dynamics and Control Theory, Irkutsk, Russia), "A Local Search Scheme for the Inequality-Constrained Optimal Control Problem"

We thank the authors for their submissions, the members of the Program Committee (PC), and the external reviewers for their efforts in providing exhaustive reviews. We thank our sponsors and partners: the Mathematical Center in Akademgorodok, Huawei Technologies Co., Ltd., the Sobolev Institute of Mathematics, the Krasovsky Institute of Mathematics and Mechanics, the Ural Mathematical Center, the Center for Research and Education in Mathematics, the Higher School of Economics (Campus Nizhny Novgorod), and the Matrosov Institute for System Dynamics and Control Theory. We are grateful to the colleagues from the Springer LNCS and CCIS editorial boards for their kind and helpful support.

July 2021   Panos Pardalos
                                                                Michael Khachay
                                                              Alexander Kazakov

Organization

Program Committee Chairs

Panos Pardalos	University of Florida, USA
Michael Khachay	Krasovsky Institute of Mathematics and Mechanics, Russia
Oleg Khamisov	Melentiev Energy Systems Institute, Russia
Yury Kochetov	Sobolev Institute of Mathematics, Russia
Alexander Strekalovsky	Matrosov Institute for System Dynamics and ControlTheory, Russia

Program Committee

Anatoly Antipin	Dorodnicyn Computing Centre of RAS, Russia
Alexander Arguchintscv	Irkutsk State University, Russia
Pasquale Avella	University of Sannio, Italy
Evripidis Bampis	Sorbonne Université, France
Olga Battaïa	ISAE-SUPAERO, France
René van Bevern	Novosibirsk State University, Russia
Maurizio Boccia	University of Naples Federico II, Italy
Scrgiy Butenko	Texas A&M University, USA
Igor Bychkov	Matrosov Institute for System Dynamics and Control Theory, Russia
Igor Bykadorov	Sobolev Institute of Mathematics, Russia
Tatjana Davidović	Mathematical Institute of the Serbian Academy of Sciences and Arts, Serbia
Stephan Dempe	Freiberg University, Germany
Gianni Di Pillo	University of Rome "La Sapienza", Italy
Alexandre Dolgui	IMT Atlantique, France
Mirjam Duer	University of Augsburg, Germany
Vladimir Dykhta	Matrosov Institute for System Dynamics and Control Theory, Russia
Rentsen Enkhbat	Institute of Mathematics and Digital Technology, Mongolia
Anton Eremeev	Sobolev Institute of Mathematics, Russia
Adil Erzin	Novosibirsk State University, Russia
Yuri Evtushenko	Dorodnicyn Computing Centre of RAS, Russia
Alexander Filatov	Far Eastern Federal University, Russia
Mikhail Falaleev	Irkutsk State University, Russia
Fedor Fomin	University of Bergen, Norway
Alexander Gasnikov	Moscow Institute of Physics and Technology, Russia
Victor Gergel	University of Nizhni Novgorod, Russia

Edward Gimadi	Sobolev Institute of Mathematics, Russia
Aleksander Gornov	Matrosov Institute for System Dynamics and Control Theory, Russia
Alexander Grigoriev	Maastricht University, Netherlands
Feng-Jang Hwang	University of Technology Sydney, Australia
Alexey Izmailov	Lomonosov Moscow State University, Russia
Milojica Jacimovic	University of Montenegro, Montenegro
Klaus Jansen	Kiel University, Germany
Sergey Kabanikhin	Institute of Numerical Mathematics and Mathematical Geophysics, Russia
Valeriy Kalyagin	Higher School of Economics, Russia
Vadim Kartak	Ufa State Aviation Technical University, Russia
Alexander Kazakov	Matrosov Institute of System Dynamics and ControlTheory, Russia
Lev Kazakovtsev	Siberian State Aerospace University, Russia
Andrey Kibzun	Moscow Aviation Institute, Russia
Donghyun (David) Kim	Kennesaw State University, USA
Igor Konnov	Kazan Federal University, Russia
Alexander Kononov	Sobolev Institute of Mathematics, Russia
Alexander Kruger	Federation University, Australia
Dmitri Kvasov	University of Calabria, Italy
Tatyana Levanova	Dostoevsky Omsk State University, Russia
Vadim Levit	Ariel University, Israel
Frank Lewis	The University of Texas at Arlington, USA
Leo Liberti	CNRS, France
Bertrand M. T. Lin	National Chiao Tung University, Taiwan
Marko Makela	University of Turku, Finland
Vittorio Maniezzo	University of Bologna, Italy
Pierre Marechal	Paul Sabatier University, France
Vladimir Mazalov	Institute of Applied Mathematical Research, Russia
Boris Mordukhovich	Wayne State University, USA
Yury Nikulin	University of Turku, Finland
Ivo Nowak	Hamburg University of Applied Sciences, Germany
Evgeni Nurminski	Far Eastern Federal University, Russia
Leon Petrosyan	Saint Petersburg State University, Russia
Alex Petunin	Ural Federal University, Russia
Boris Polyak	Trapeznikov Institute of Control Science, Russia
Leonid Popov	Krasovsky Institute of Mathematics and Mechanics, Russia
Mikhail Posypkin	Dorodnicyn Computing Centre of RAS, Russia
Oleg Prokopyev	University of Pittsburgh, USA
Artem Pyatkin	Sobolev Institute of Mathematics, Russia
Soumyendu Raha	Indian Institute of Science, India
Alexander Razgulin	Lomonosov Moscow State University, Russia
Jie Ren	Huawei Russian Research Institute, Russia
Anna N. Rettieva	Institute of Applied Mathematical Research, Russia

Claudia Sagastizabal	Unicamp, Brazil
Yaroslav Sergeyev	University of Calabria, Italy
Natalia Shakhlevich	University of Leeds, UK
Alexander Shananin	Moscow Institute of Physics and Technology, Russia
Vladimir Shikhman	Catholic University of Louvain, Belgium
Angelo Sifaleras	University of Macedonia, Greece
Vladimir Skarin	Krasovsky Institute of Mathematics and Mechanics, Russia
Vladimir Srochko	Irkutsk State University, Russia
Claudio Sterle	University of Naples Federico II, Italy
Petro Stetsyuk	Glushkov Institute of Cybernetics, Ukraine
Roman Strongin	University of Nizhni Novgorod, Russia
Nadia Sukhorukova	Swinburne University of Technology, Australia
Tatiana Tchemisova	University of Aveiro, Portugal
Alexander Tolstonogov	Matrosov Institute for System Dynamics and Control Theory, Russia
Ider Tseveendorj	University of Versailles, France
Vladimir Ushakov	Krasovsky Institute of Mathematics and Mechanics, Russia
Olga Vasilieva	Universidad del Valle, Colombia
Alexander Vasin	Lomonosov Moscow State University, Russia
Vitaly Zhadan	Dorodnicyn Computing Centre of RAS, Russia
Dong Zhang	Huawei Technologies, Co., Ltd., China
Anatoly Zhigljavsky	Cardiff University, UK
Yakov Zinder	Sydney Technical University, Australia

Additional Reviewers

Abbasov, Majid	Iljev, Victor	Neznakhina, Ekaterina
Berikov, Vladimir	Jaksic Kruger, Tatjana	Ogorodnikov, Yuri
Berndt, Sebastian	Khachay, Daniel	Orlov, Andrei
Brinkop, Hauke	Khoroshilova, Elena	Pinyagina, Olga
Buchem, Moritz	Khutoretskii, Alexandr	Plotnikov, Roman
Buldaev, Alexander	Kononov, Alexander	Plyasunov, Alexander
Buzdalov, Maxim	Kononova, Polina	Rohwedder, Lars
Chernykh, Ilya	Kovalenko, Yulia	Sandomirskaya, Marina
Dang, Duc-Cuong	Kulachenko, Igor	Semenov, Alexander
Davydov, Ivan	Kumacheva, Suriya	Servakh, Vladimir
Deineko, Vladimir	Kuzyutin, Denis	Sevastyanov, Sergey
Deppert, Max	Lassota, Alexandra	Shenmaier, Vladimir
Gluschenko, Konstantin	Lavlinskii, Sergey	Shkaberina, Guzel
Golak, Julian	Lee, Hunmin	Simanchev, Ruslan
Gonen, Rica	Lempert, Anna	Srochko, Vladimir
Grage, Kilian	Melnikov, Andrey	Stanimirovic, Zorica
Gromova, Ekaterina	Morshinin, Alexander	Stanovov, Vladimir

Staritsyn, Maxim	Tur, Anna	Vasin, Alexandr
Sukhoroslov, Oleg	Tyunin, Nikolay	Veremchuk, Natalia
Tovbis, Elena	Urazova, Inna	Yanovskaya, Elena
Tsidulko, Oxana	Urosevic, Dragan	Zalyubovskiy, Vyacheslav
Tsoy, Yury	van Lent, Freija	Zolotykh, Nikolai

Industry Section Chair

Vasilyev Igor   Matrosov Institute for System Dynamics and Control
                                  Theory, Russia

Organizing Committee

Alexander Kazakov (Chair)	ISDCT SB RAS
Andrei Orlov (Deputy Chair)	ISDCT SB RAS
Tatiana Gruzdeva (Scientific Secretary)	ISDCT SB RAS
Vladimir Antonik	IMIT ISU
Maria Barkova	ISDCT SB RAS
Oleg Khamisov	ESI SB RAS
Stepan Kochemazov	ISDCT SB RAS
Polina Kononova	IM SB RAS
Alexey Kumachev	ISDCT SB RAS
Pavel Kuznetsov	ISDCT SB RAS
Anna Lempert	ISDCT SB RAS
Taras Madzhara	ISDCT SB RAS
Nadezhda Maltugueva	ISDCT SB RAS
Ilya Minarchenko	ESI SB RAS
Ekaterina Neznakhina	IMM UB RAS
Yuri Ogorodnikov	IMM UB RAS
Nikolay Pogodaev	ISDCT SB RAS
Stepan Sorokin	ISDCT SB RAS
Pavel Sorokovikov	ISDCT SB RAS
Maxim Staritsyn	ISDCT SB RAS
Alexander Stolbov	ISDCT SB RAS
Anton Ushakov	ISDCT SB RAS
Igor Vasiliev	ISDCT SB RAS
Tatiana Zarodnyuk	ISDCT SB RAS
Maxim Zharkov	ISDCT SB RAS

Organizers

Matrosov Institute for System Dynamics and Control Theory, Russia
Sobolev Institute of Mathematics, Russia
Krasovsky Institute of Mathematics and Mechanics, Russia
Higher School of Economics (Campus Nizhny Novgorod), Russia

Sponsors

Center for Research and Education in Mathematics, Russia
Huawei Technologies Co., Ltd.
Mathematical Center in Akademgorodok, Russia
Ural Mathematical Center, Russia

Abstracts of Invited Talks

On the Design of Matheuristics that make Use of Learning

Christian Blum ⓘ

Artificial Intelligence Research Institute (IIIA-CSIC), Barcelona, Spain
christian.blum@iiia.csic.es .

Abstract. Approximation techniques for solving combinatorial optimization problems, such as metaheuristics, often make use of learning. Examples include evolutionary algorithms and ant colony optimization. On the other side, matheuristics — that is, heuristic techniques making use of mathematical programming - rarely include a learning component. Most variants of large neighbourhood search, for instance, do not take profit from learning. In this talk I will present examples of our recent work in which we design matheuristics that make successful use of learning, considering both positive and negative feedback.

Keywords: Combinatorial optimization · Approximation · Learning

Optimal Classification and Regression Trees

Emilio Carrizosa (iD)

Institute of Mathematics, University of Seville, Spain
ecarrizosa@us.es

Abstract. Classification and Regression Trees are very powerful Machine Learning tools. Their design expressed as an optimization problem enables us to obtain excellent accuracy performance, and, at the same time, to have some control on important issues such as sparsity, explainability or fairness. In this talk, some recent advances in the field and future research lines will be discussed.

Keywords: Classification · Regression trees

Integer Programming Formulations Based on Exponentially Large Networks: Algorithms and Applications

François Clautiaux (ID)

University of Bordeaux, France
francois.clautiaux@math.u-bordeaux.fr

Abstract. The last ten years have seen much progress in the field of so-called extended formulations, which aims at reformulating effectively a problem/polyhedron with the help of (exponentially many) additional variables. In particular, network-flow formulations have received an increasing interest from the community. A considerable difficulty to overcome when dealing with such a formulation is to handle its size. In this talk we recall some key results concerning these formulations, and present several recent successful applications that have been obtained using innovative aggregation/disaggregation techniques.

Keywords: Exponential size MIP models · Integer programming

Beyond Heuristic Gradient Descent in Machine Learning

Andreas Griewank [ID]

Institute of Mathematics, Humboldt University, Germany
griewank@mathematik.hu-berlin.de

Abstract. In neural network training and other large scale applications, deterministic and stochastic variants of Cauchy's steepest descent method are widely used for the minimization of objectives that are only piecewise smooth. From the classical optimization point of view the roaring and almost exclusive success of this most basic approach is somewhat puzzling. The lack of convergence analysis typically goes along with a large number of method parameters that have to be adjusted by trial and error. We explore several ideas to derive more rigorous but still efficient methods for classification problems. One is a Newton adaptation that can exploit the internal structure of the so-called sparse-max potential, the other a generalization of Wolfe's conjugate gradient method to nonsmooth and nonconvex problems. Our observations and results are demonstrated on the well known MNIST and CIFAR problems with one- and multilayer prediction functions.

Keywords: Neural networks · Conjugate gradient method

Integer Programming and Convolution, with Applications

Klaus Jansen (iD)

Christian-Albrechts-Universität, Kiel, Germany
kj@informatik.uni-kiel.de

Abstract. Integer programs (IP) with m constraints are solvable in pseudo-polynomial time. We give a new algorithm based on the Steinitz Lemma and dynamic programming with a better pseudo-polynomial running time than previous results. Moreover, we establish a strong connection to the problem (min,+) - convolution. (min,+) - convolution has a trivial quadratic time algorithm and it has been conjectured that this cannot be improved significantly. Finally we show for the feasibility problem also a tight lower bound, which is based on the Strong Exponential Time Hypothesis (SETH), and give some applications for knapsack and scheduling problems. This is joint work with Lars Rohwedder.

Keywords: Integer program · Strong exponential time hypothesis · Pseudo-polynomial time

Optimization and Inverse Problems

Sergey Kabanikhin [ID]

Institute of Numerical Mathematics and Mathematical Geophysics,
Novosibirsk, Russia
Ksi52@mai.ru

Abstract. Inverse problems arise in many applications in science and engineering. The term "inverse problem" is generally understood as the problem of finding a specific physical property, or properties, of the medium under investigation, using indirect measurements. In general, an inverse problem aims at recovering the unknown parameters of a physical system which produces the observations and/or measurements. Such problems are usually ill-posed. This is often solved via two approaches: a Bayesian approach which computes a posterior distribution of the models given prior knowledge and the regularized data fitting approach which chooses an optimal model by minimizing an objective taking into account both fitness to data and prior knowledge. Optimization plays an important role in solving many inverse problems. Indeed, the task of inversion often either involves or is fully cast as a solution to an optimization problem. In this talk, we discuss current state-of-the-art optimization methods widely used in inverse problems. We then survey recent related advances in addressing similar challenges in problems faced by the machine learning community and discuss their potential advantages for solving inverse problems.

Keywords: Inverse problem • Optimization • Machine learning

Minimum Sum of Squares Clustering for Big Data – Heuristic Approach

Nenad Mladenovic ⓘ

Khalifa University, United Arab Emirates
nenadmladenovic12@gmail.com

Abstract. We first present a review of local search methods that are usually used to solve the minimum sum-of-square clustering (MSSC) problem. We then present some their combinations within Variable neighbourhood descent (VND) scheme. They combine k-means, h-means and j-means heuristics in a nested and sequential way. To show how these local searches can be implemented within a metaheuristic framework, we apply the VND heuristics in the local improvement step of variable neighbourhood search (VNS) procedure. Computational experiments are carried out which suggest that this new and simple application of VNS is comparable to the state of the art. Then we discuss some decomposition and aggregation strategies for solving MSSC problem with huge data sets. Following the recent Less is more approach, the data set is divided randomly into a few smaller subproblems and after solving, the centroids of each subproblem is chosen to represent its cluster for a new aggregation stage. Encouraging computational results on instances of several million entities are presented.

Keywords: Minimum sum-of-square clustering · Variable neighbourhood search · Decomposition

Exploiting Structure in Nonsmooth Optimization

Claudia Sagastizábal (iD)

IMECC - University of Campinas, Brazil
sagastiz@unicamp.br

Abstract. In many optimization problems nonsmoothness appears in a structured manner. Composite structures are found in LASSO-type problems arising in machine-learning. Separable structures result from applying some decomposition technique to problems that cannot be solved directly. This context is frequent in stochastic programming, bilevel optimization, equilibrium problems. The talk will give a panorama of techniques that have proven successful in exploiting structural properties that are somewhat hidden behind nonsmoothness. Throughout the presentation the emphasis is put on transmitting the main ideas and concepts, illustrating with examples the presented material.

Keywords: Optimization • Structural properties

State-of-the-Art on Rates of Convergence and Cost of Iterations of Augmented Lagrangian Methods

Mikhail Solodov

Institute for Pure and Applied Mathematics, Brazil
solodov@impa.br

Abstract. We discuss state-of-the-art results on local convergence and rate of convergence of the classical augmented Lagrangian algorithm. The local primal-dual linear/superlinear convergence is obtained under the sole assumption that the dual starting point is close to a multiplier satisfying the second-order sufficient optimality condition. In fact, in the equality-constrained case, even the weaker noncriticality assumption is enough. In particular, no constraint qualifications of any kind are needed. Classical literature on the subject required the linear independence constraint qualification (in addition to other things). In addition to the most standard form of the augmented Lagrangian algorithm, the general lines of analysis apply also to its variant with partial penalization of the constraints, to the proximal-point version, and to the modification with smoothing of the max-function. Moreover, we show that to compute suitable approximate solutions of augmented Lagrangian subproblems which ensure the superlinear convergence of the algorithm, it is enough to make just two Newtonian steps (i.e., solve two quadratic programs, or two linear systems in the equality-constrained case). The two quadratic programs are related to stabilized sequential quadratic programming, and to second-order corrections, respectively.

Keywords: Convex programming · Augmented Lagrangian methods · Convergence rates

Abstracts of Tutorials

Reinforcement Learning from the Stochastic Optimization Point of View

Alexander Gasnikov

Moscow Institute of Physics and Technology, Russia
gasnikov@yandex.ru

Abstract. We consider the problem of learning the optimal policy for infinite-horizon Markov decision processes (MDPs). We discuss lower bounds and optimal algorithms for discount and average-reward MDPs with a generative model. We also pay attention to parallelization aspects. In the core of the described approaches lies the idea to relate the problem of learning the optimal policy for MDP with the stochastic optimization algorithms (Mirror Descent type) for optimization reformulations, based on Bellmans' equations (D. Bertsekas).

Keywords: Markov decision process • Reinforcement learning • Stochastic optimization

Equilibrium Traffic Flow Assignment in a Multi-subnet Urban Road Network

Alexander Krylatov (ID)

Saint-Petersburg State University, Russia
aykrylatov@yandex.ru

Abstract. Today urban road network of a modern city can include several subnets. Indeed, bus lanes form a transit subnet available only for public vehicles. Toll roads form a subnet, available only for drivers who ready to pay fees for passage. The common aim of developing such subnets is to provide better urban travel conditions for public vehicles and toll-paying drivers. The present paper is devoted to the equilibrium traffic flow assignment problem in a multi-subnet urban road network. We formulate this problem as a non-linear optimization program and prove that its solution corresponds to the equilibrium traffic assignment pattern in a multi-subnet road network. Moreover, we prove that obtained equilibrium traffic assignment pattern guarantees less or equal travel time for public vehicles and toll-paying drivers than experienced by all other vehicles. The findings of the paper contribute to the traffic theory and give fresh managerial insights for traffic engineers.

Keywords: Non-linear optimization • Traffic assignment problem • Multi-subnet urban road network

A Local Search
Scheme for the Inequality-Constrained
Optimal Control Problem

Alexander Strekalovsky ⓘ

Matrosov Institute for System Dynamics and Control Theory,
Irkutsk, Russiaa
strekal@icc.ru

Abstract. This paper addresses the nonconvex optimal control (OC) problem with the cost functional and inequality constraint given by the functionals of Bolza. All the functions in the statement of the problem are state-DC, i.e. presented by a difference of the state-convex functions. Meanwhile, the control system is state-linear. Further, with the help of the Exact Penalization Theory we propose the state-DC form of the penalized cost functional and, using the linearization with respect to the basic nonconvexity of the penalized problem, we study the linearized OC problem. On this basis, we develop a general scheme of the special Local Search Method with a varying penalty parameter. Finally, we address the convergence of the proposed scheme.

Keywords: Nonconvex optimal control · DC-functions · Local search

Contents

Game Theory and Optimal Control

Operational Research and Mathematical Economics

Data Analysis

Invited Talks

Equilibrium Traffic Flow Assignment in a Multi-subnet Urban Road Network

Alexander Krylatov[1,2(✉)] 🆔

[1] Saint Petersburg State University, Saint Petersburg, Russia
a.krylatov@spbu.ru
[2] Institute of Transport Problems RAS, Saint Petersburg, Russia

Abstract. Today urban road network of a modern city can include several subnets. Indeed, bus lanes form a transit subnet available only for public vehicles. Toll roads form a subnet, available only for drivers who ready to pay fees for passage. The common aim of developing such subnets is to provide better urban travel conditions for public vehicles and toll-paying drivers. The present paper is devoted to the equilibrium traffic flow assignment problem in a multi-subnet urban road network. We formulate this problem as a non-linear optimization program and prove that its solution corresponds to the equilibrium traffic assignment pattern in a multi-subnet road network. Moreover, we prove that obtained equilibrium traffic assignment pattern guarantees less or equal travel time for public vehicles and toll-paying drivers than experienced by all other vehicles. The findings of the paper contribute to the traffic theory and give fresh managerial insights for traffic engineers.

Keywords: Non-linear optimization · Traffic assignment problem · Multi-subnet urban road network

1 Introduction

An urban road area of a modern city is a multi-subnet complex composited network, which has been permanently growing over the past 40 years due to the worldwide urbanization process [8]. Indeed, the scale of many actual urban road networks today is truly incredible [20]. The continuing growth of large cities challenges authorities, civil engineers, and researchers to face a lot of complicated problems at all levels of management [15]. The service of huge urban networks requires a large budget which takes a significant part in the budget of a city [12]. Thus, errors in urban road network planning can affect adversely the budget policy of a city authority. Therefore, the development of intelligent systems for

The work was jointly supported by a grant from the Russian Science Foundation (No. 19–71-10012 Multi-agent systems development for automatic remote control of traffic flows in congested urban road networks).

P. Pardalos et al. (Eds.): MOTOR 2021, LNCS 12755, pp. 3–16, 2021.
https://doi.org/10.1007/978-3-030-77876-7_1

decision-making support in the field of large urban road network design seems to be of crucial interest [1, 19].

Comprehensive review on approaches and techniques for transit network planning, operation and control is given in [4]. Transit network planning is commonly divided into subproblems that span tactical, strategical, and operational decisions [3], any of which is an NP-hard computational problem [14]. Thus, since a transit network design problem includes all these items simultaneously, it is can not be solved precisely. Hence, network planners are often equipped only by general recommendations and methodological tools for decision-making support. However, inspired by the recent findings on a composited complex network of multiple subsets [11, 16], and traffic assignment in a network with a transit subnetwork [5, 6, 18], this paper investigates the traffic assignment problem in a multi-subnet composited urban road network, which can be solved precisely.

In the paper we consider a multi-subnet urban road network under arc-additive travel time functions. Section 2 presents a multi-subnet urban road network as a directed graph, while Sect. 3 is devoted to equilibrium traffic assignment in such kind of network. We formulate the equilibrium assignment problem in a multi-subnet urban road network as a non-linear optimization program and prove that its solution corresponds to the equilibrium traffic assignment pattern in a multi-subnet road network. Section 4 gives important analytical results for a road network with disjoint routes, which are directly applied in Sect. 5 for toll road design. Actually, the simple case of a road network topology allows us to consider different scenarios, concerning subnetwork design, and analyze the decision-making process. Section 6 contains the conclusions.

2   Multi-subnet Urban Road Network

Let us consider a multi-subnet urban road network presented by a directed graph $G = (E, V)$, where V represents a set of intersections, while $E \subseteq V \times V$ represents a set of available roads between the adjacent intersections. If we define S as the ordered set of selected vehicle categories, then $G = G_0 \cup \bigcup_{s \in S} G_s$, where $G_0 = (E_0, V_0)$ is the subgraph of public roads, which are open to public traffic, and $G_s = (E_s, V_s)$ is the subgraph of roads, which are open only for the s-th category of vehicles, $s \in S$. Denote $W \subseteq V \times V$ as the ordered set of pairs of nodes with non-zero travel demand $F_0^w > 0$ and/or $F_s^w > 0$, $s \in S$, for any $w \in W$. W is usually called as the set of origin-destination pairs (OD-pairs), $|W| = m$. Any set of sequentially linked arcs initiating in the origin node of OD-pair w and terminating in the destination node of the OD-pair w we call *route* between the OD-pair w, $w \in W$. The ordered sets of all possible routes between nodes of the OD-pair w, $w \in W$, we denote as R_0^w for the subgraph G_0 and R_s^w for the subgraph G_s, $s \in S$. Demand $F_s^w > 0$ for any $s \in S$ and $w \in W$ seeks to be assigned between the available public routes R_0^w and routes for vehicles of s-th category R_s^w. Thus, on the one hand, $\sum_{r \in R_s^w} p_r^w = P_s^w$, where p_r^w is the variable corresponding to the traffic flow of the s-th category vehicles through the route $r \in R_s^w$, while P_s^w is the variable corresponding to the overall traffic flow of the s-th category vehicles through the routes R_s^w. On the other hand, the difference $(F_s^w - P_s^w)$ is the traffic flow of the s-th category vehicles, which can be assigned between the available public routes R_0^w for any $s \in S$ and $w \in W$, since the variable P_s^w satisfies the following condition: $0 \leq P_s^w \leq F_s^w$ for any $s \in S$, $w \in W$. Therefore, demand $F_0^w > 0$ seeks to be assigned between the available public routes R_0^w together with the traffic flow $\sum_{s \in S}(F_s^w - P_s^w)$: $\sum_{r \in R_0^w} f_r^w = F_0^w + \sum_{s \in S}(F_s^w - P_s^w)$, where f_r^w is the variable corresponding to the traffic flow through the public route $r \in R_0^w$ between nodes of OD-pair $w \in W$.

Let us introduce differentiable strictly increasing functions on the set of real numbers $t_e(\cdot)$, $e \in E$. We suppose that $t_e(\cdot)$, $e \in E$, are non-negative and their first derivatives are strictly positive on the set of real numbers. By x_e we denote traffic flow on the edge e, while x is an appropriate vector of arc-flows, $x = (\ldots, x_e, \ldots)^{\mathrm{T}}$, $e \in E$. Defined functions $t_e(x_e)$ are used to describe travel time on arcs e, $e \in E$, and they are commonly called arc *delay*, *cost* or *performance* functions. In this paper we assume that the travel time function of the route $r \in R_0^w \cup \bigcup_{s \in S} R_s^w$ between OD-pair $w \in W$ is the sum of travel delays on all edges belonging to this route. Thus, the travel time through the route $r \in R_0^w \cup \bigcup_{s \in S} R_s^w$ between OD-pair $w \in W$ can be defined as the following sum:

$$\sum_{e \in E} t_e(x_e)\delta_{e,r}^w \quad \forall r \in R_0^w \cup \bigcup_{s \in S} R_s^w, w \in W,$$

where, by definition,

$$\delta_{e,r}^w = \begin{cases} 1, \text{ if edge } e \text{ belongs to the route } r \in R_0^w \cup \bigcup_{s \in S} R_s^w, \\ 0, \text{ otherwise.} \end{cases} \quad \forall e \in E, w \in W,$$

while, naturally,

$$x_e = \sum_{w \in W} \sum_{s \in S} \sum_{r \in R_s^w} p_r^w \delta_{e,r}^w + \sum_{w \in W} \sum_{r \in R_0^w} f_r^w \delta_{e,r}^w, \quad \forall e \in E,$$

i.e., traffic flow on the arc is the sum of traffic flows through those routes, which include this arc.

3   Equilibrium Assignment in a Multi-subnet Road Network

The traffic assignment problem (TAP) is an optimization problem with non-linear objective function and linear constraints, which allows one to find traffic assignment in a road network by given travel demand values. The solution of TAP is proved to satisfy so called user equilibrium (UE) behavioural principle, formulated by J. G. Wardrop as follows: "*The journey times in all routes actually used are equal and less than those that would be experienced by a single vehicle on any unused route*" [13]. Therefore, the *equilibrium traffic assignment problem* is a well-known problem for the urban road network without subnets, i.e., the network consisting of public roads only that serve all the traffic. The first mathematical formulation of such a problem was given by Beckmann *et al.* [2]. Thus, under separable travel time functions, TAP has a form of the following optimization program [9,10]:

$$\min_x \sum_{e \in E} \int_0^{x_e} t_e(u)du, \tag{1}$$

subject to

$$\sum_{r \in R_0^w} f_r^w = F_0^w + \sum_{s \in S} F_s^w, \quad \forall w \in W, \tag{2}$$

$$f_r^w \geq 0 \quad \forall r \in R_0^w, w \in W, \tag{3}$$

where, by definition,

$$x_e = \sum_{w \in W} \sum_{r \in R_0^w} f_r^w \delta_{e,r}^w \quad \forall e \in E. \tag{4}$$

An arc-flow assignment pattern and corresponding route-flow assignment pattern, satisfying (1)–(4), are proved to reflect *user equilibrium* traffic assignment or such an assignment that

$$\sum_{e \in E} t_e(x_e)\delta_{e,r}^w \begin{cases} = \mathfrak{t}^w, \text{ if } f_r^w > 0, \\ \geq \mathfrak{t}^w, \text{ if } f_r^w = 0, \end{cases} \quad \forall r \in R_0^w, w \in W, \tag{5}$$

where $\mathfrak{t}^w$ is called an *equilibrium travel time* or travel time on actually used routes between OD-pair w, $w \in W$ [7].

Let us develop the *equilibrium traffic assignment problem* for a multi-subnet urban road network. For this purpose, we specify the principle, like the user-equilibrium one (5), which should be satisfied by the equilibrium traffic assignment pattern in a multi-subnet urban road network: *"The journey times in all routes actually used are equal and less than those that would be experienced by a single vehicle on any unused route, as well as the journey times in all routes actually used in any subnet less or equal than the journey times in all routes actually used in a public road network"*. The following theorem gives a formulation of the *multi-subnet equilibrium traffic assignment problem* in the form of an optimization program.

Theorem 1. *Equilibrium traffic flow assignment in a multi-subnet urban road network is obtained as a solution of the following optimization program:*

$$\min_x \sum_{e \in E} \int_0^{x_e} t_e(u)du, \tag{6}$$

with constraints $\forall w \in W$

$$\sum_{r \in R_s^w} p_r^w = P_s^w, \quad \forall s \in S, \tag{7}$$

$$\sum_{r \in R_0^w} f_r^w = F^w + \sum_{s \in S} \left(F_s^w - P_s^w \right), \tag{8}$$

$$p_r^w \geq 0 \quad \forall r \in R_s^w, \forall s \in S, \tag{9}$$

$$f_r^w \geq 0 \quad \forall r \in R_0^w, \tag{10}$$

$$0 \leq P_s^w \leq F_s^w \quad \forall s \in S, \tag{11}$$

with definitional constraint

$$x_e = \sum_{w \in W} \sum_{s \in S} \sum_{r \in R_s^w} p_r^w \delta_{e,r}^w + \sum_{w \in W} \sum_{r \in R_0^w} f_r^w \delta_{e,r}^w, \quad \forall e \in E. \tag{12}$$

Proof. Since functions $t_e(x_e)$, $e \in E$, are strictly increasing, then the goal function (6) is convex. Hence, the optimization problem (6)–(11) has the unique solution, which has to satisfy the Karush–Kuhn–Tucker conditions. Let us consider the Lagrangian of the problem:

$$L = \sum_{e \in E} \int_0^{x_e} t_e(u) du + \sum_{w \in W} \left[\sum_{s \in S} t_s^w \left(P_s^w - \sum_{r \in R_s^w} p_r^w \right) \right.$$

$$+ t_0^w \left(F^w + \sum_{s \in S} \left(F_s^w - P_s^w \right) - \sum_{r \in R_0^w} f_r^w \right)$$

$$+ \sum_{s \in S} \sum_{r \in R_s^w} (-p_r^w) \eta_r^w + \sum_{r \in R_0^w} (-f_r^w) \xi_r^w + \sum_{s \in S} \left((-P_s^w) \gamma_s^w + (P_s^w \quad F_s^w) \zeta_s^w \right) \right],$$

where t_0^w, t_s^w, $s \in S$, $\eta_r^w \geq 0$, $r \in R_s^w$ and $s \in S$, $\xi_r^w \geq 0$, $r \in R_0^w$, $\gamma_s^w \geq 0$, $\zeta_s^w \geq 0$, $s \in S$, for any $w \in W$ are Lagrangian multipliers. According to Karush–Kuhn–Tucker conditions, the following equalities hold for any $w \in W$:

$$\frac{\partial L}{\partial p_r^w} = \frac{\partial}{\partial p_r^w} \sum_{e \in E} \int_0^{x_e} t_e(u) du - t_s^w - \eta_r^w = 0, \quad \forall r \in R_s^w, s \in S, \tag{13}$$

$$\frac{\partial L}{\partial f_r^w} = \frac{\partial}{\partial f_r^w} \sum_{e \in E} \int_0^{x_e} t_e(u) du - t_0^w - \xi_r^w = 0, \quad \forall r \in R_0^w, \tag{14}$$

$$\frac{\partial L}{\partial P_s^w} = t_s^w - t_0^w - \gamma_s^w + \zeta_s^w = 0, \quad \forall s \in S. \tag{15}$$

Note that, according to (12), for any $r \in R_s^w$, $s \in S$ and $w \in W$:

$$\frac{\partial}{\partial p_r^w} \sum_{e \in E} \int_0^{x_e} t_e(u) du = \sum_{e \in E} \frac{\partial}{\partial x_e} \left(\int_0^{x_e} t_e(u) du \right) \frac{\partial x_e}{\partial p_r^w} = \sum_{e \in E} t_e(x_e) \delta_{e,r}^w,$$

and for any $r \in R_0^w$ and $w \in W$:

$$\frac{\partial}{\partial f_r^w} \sum_{e \in E} \int_0^{x_e} t_e(u)du = \sum_{e \in E} \frac{\partial}{\partial x_e} \left(\int_0^{x_e} t_e(u)du \right) \frac{\partial x_e}{\partial f_r^w} = \sum_{e \in E} t_e(x_e)\delta_{e,r}^w.$$

Therefore, equations (13)–(15) can be re-written as follows:

$$\sum_{e \in E} t_e(x_e)\delta_{e,r}^w = t_s^w + \eta_r^w, \quad \forall r \in R_s^w, s \in S, w \in W, \tag{16}$$

$$\sum_{e \in E} t_e(x_e)\delta_{e,r}^w = t_0^w + \xi_r^w, \quad \forall r \in R_0^w, w \in W, \tag{17}$$

$$t_s^w = t_0^w + \gamma_s^w - \zeta_s^w, \quad \forall s \in S, w \in W. \tag{18}$$

Moreover, a restriction in the inequality form has to satisfy the complementary slackness condition, i.e., for restrictions (9)–(11) the following equalities hold for any $w \in W$:

$$(-p_r^w)\eta_r^w = 0 \quad \forall r \in R_s^w, s \in S, \quad (-f_r^w)\xi_r^w = 0 \quad \forall r \in R_0^w,$$
$$(-P_s^w)\gamma_s^w = 0 \quad \forall s \in S, \quad (P_s^w - F_s^w)\zeta_s^w = 0 \quad \forall s \in S. \tag{19}$$

Once (19) holds for any $w \in W$, then:

– the equality (16) can be re-written as follows:

$$\sum_{e \in E} t_e(x_e)\delta_{e,r}^w \begin{cases} = t_s^w \text{ for } p_r^w > 0, \\ \geq t_s^w \text{ for } p_r^w = 0, \end{cases} \quad \forall r \in R_s^w, s \in S, w \in W, \tag{20}$$

since $\eta_r^w \geq 0$ for any $r \in R_s^w$, $w \in W$;
– the equality (17) can be re-written as follows:

$$\sum_{e \in E} t_e(x_e)\delta_{e,r}^w \begin{cases} = t_0^w \text{ for } f_r^w > 0, \\ \geq t_0^w \text{ for } f_r^w = 0, \end{cases} \quad \forall r \in R_0^w, w \in W, \tag{21}$$

since $\xi_r^w \geq 0$ for any $r \in R_0^w$, $w \in W$;
– the equality (18) can be re-written for all $w \in W$ as follows:

$$t_s^w \begin{cases} \leq t_0^w \text{ for } P_s^w = F_s^w, \\ = t_0^w \text{ for } 0 < P_s^w < F_s^w, \\ \geq t_0^w \text{ for } P_s^w = 0, \end{cases} \quad \forall s \in S, w \in W, \tag{22}$$

since $\gamma_s^w \geq 0$ and $\zeta_s^w \geq 0$ for any $s \in S$, $w \in W$.

Therefore, the unique solution of the optimization problem (6)–(11) satisfies conditions (20)–(22). In other words, according to (20) and (21), the journey times in all routes actually used are equal and less than those that would be experienced by a single vehicle on any unused route, while, according to (22), the journey times in all routes actually used in any subnet less or equal than the journey times in all routes actually used in a public road network. Thus, the solution of the optimization problem (6)–(11) is indeed the equilibrium traffic assignment in a multi-subnet urban road network. □

The theorem on equilibrium traffic assignment in a multi-subset urban road network is proved. Let us note that the proof of the theorem on equilibrium traffic assignment in the urban road network with only one subnet was earlier given by the author as the contribution to the paper [17].

Remark. We suggest calling Lagrange multiplier t_0^w, $w \in W$, as the *equilibrium travel time* for the public road network, while Lagrange multiplier t_s^w, $s \in S$, $w \in W$, as the *equilibrium travel time* for the subnet s, $s \in S$.

4 Multi-subnet Road Network with Disjoint Routes

Let us consider the particular case of a multi-subnet urban road network presented by the directed graph $G = (V, E)$. The set S is still the set of selected vehicle categories and $G = G_0 \cup \bigcup_{s \in S} G_s$, where $G_0 = (V_0, E_0)$ is the subgraph of public roads, which are open to public traffic, and $G_s = (V_s, E_s)$ is the subgraph of roads, which are open only for the s-th category of vehicles, $s \in S$. We also believe that there is only one OD-pair with non-zero travel demands, i.e., $|W| = 1$, $F_0 > 0$ and $F_s > 0$, $s \in S$. We assume that the topology of the graph G is such that any route initiating in the origin node of the OD-pair and terminating in its destination node has no common arcs with all other available routes between this OD-pair. The ordered sets of all possible routes between nodes of the single OD-pair we denote as R_0, $|R_0| = n_0$, for the subgraph G_0 and R_s, $|R_s| = n_s$, for the subgraph G_s, $s \in S$. Demand $F_s > 0$ for any $s \in S$ seeks to be assigned between the available public routes R_0 and routes for vehicles of s-th category R_s. Thus, on the one hand, $\sum_{r \in R_s} p_r = P_s$, where p_r is the variable corresponding to the traffic flow of the s-th category vehicles through the route $r \in R_s$, while P_s is the variable corresponding to the overall traffic flow of the s-th category vehicles through the routes R_s. On the other hand, the difference $(F_s - P_s)$ is the traffic flow of the s-th category vehicles, which can be assigned between the available public routes R_0 for any $s \in S$, since the variable P_s satisfies the following condition: $0 \leq P_s \leq F_s$ for any $s \in S$. Therefore, demand $F_0 > 0$ seeks to be assigned between the available public routes R_0 together with the traffic flow $\sum_{s \in S}(F_s - P_s)$: $\sum_{r \in R_0} f_r = F_0 + \sum_{s \in S}(F_s - P_s)$, where f_r is the variable corresponding to the traffic flow through the public route $r \in R_0$ between nodes of OD-pair $w \in W$.

Let us also introduce linear strictly increasing functions on the set of real numbers $t_r(\cdot)$, $r \in R_0 \cup \bigcup_{s \in S} R_s$, which are travel cost functions for the defined graph. We assume that $t_r(p_r) = a_r^s + b_r^s p_r$, $a_r^s \geq 0$, $b_r^s > 0$, for any $r \in R_s$, $s \in S$, and $t_r(f_r) = a_r^0 + b_r^0 f_r$, $a_r^0 \geq 0$, $b_r^0 > 0$, for any $r \in R_0$. Moreover, without loss of generality we believe that

$$a_1^0 \leq \ldots \leq a_{n_0}^0 \quad \text{and} \quad a_1^s \leq \ldots \leq a_{n_s}^s \quad \forall s \in S. \tag{23}$$

Fortunately, the equilibrium traffic assignment pattern can be found explicitly for this particular case of multi-subnet road network.

Theorem 2. *Equilibrium in the single-commodity multi-subnet urban road network with disjoint routes and linear cost functions is obtained by the following traffic assignment pattern:*

$$
p_r = \begin{cases} \dfrac{1}{b_r^s} \dfrac{P_s + \sum\limits_{i=1}^{k_s} \frac{a_i^s}{b_i^s}}{\sum\limits_{i=1}^{k_s} \frac{1}{b_i^s}} - \dfrac{a_r^s}{b_r^s} & \text{for } r \le k_s, \\ 0 & \text{for } r > k_s, \end{cases} \qquad \forall r \in R_s, s \in S, \tag{24}
$$

where $0 \le k_s \le n_s$, $s \in S$, is such that

$$
\left. \begin{array}{l} \text{for } r \le k_s, a_r^s < \\ \text{for } r > k_s, a_r^s \ge \end{array} \right\} \dfrac{P_s + \sum\limits_{i=1}^{k_s} \frac{a_i^s}{b_i^s}}{\sum\limits_{i=1}^{k_s} \frac{1}{b_i^s}} \qquad \forall s \in S, \tag{25}
$$

and

$$
f_r = \begin{cases} \dfrac{1}{b_r^0} \dfrac{F_0 + \sum\limits_{s \in S} (F_s - P_s) + \sum\limits_{i=1}^{k_0} \frac{a_i^0}{b_i^0}}{\sum\limits_{i=1}^{k_0} \frac{1}{b_i^0}} - \dfrac{a_r^0}{b_r^0} & \text{for } r \le k_0, \\ 0 & \text{for } r > k_0, \end{cases} \qquad \forall r \in R_0, \tag{26}
$$

where $0 \le k_0 \le n_0$ is such that

$$
\left. \begin{array}{l} \text{for } r \le k_0, a_r^0 < \\ \text{for } r > k_0, a_r^0 \ge \end{array} \right\} \dfrac{F_0 + \sum\limits_{s \in S} (F_s - P_s) + \sum\limits_{i=1}^{k_0} \frac{a_i^0}{b_i^0}}{\sum\limits_{i=1}^{k_0} \frac{1}{b_i^0}}, \tag{27}
$$

while for any $s \in S$:

$$
\dfrac{P_s + \sum\limits_{i=1}^{k_s} \frac{a_i^s}{b_i^s}}{\sum\limits_{i=1}^{k_s} \frac{1}{b_i^s}} \begin{cases} \le \\ = \\ \ge \end{cases} \dfrac{F_0 + \sum\limits_{s \in S} (F_s - P_s) + \sum\limits_{i=1}^{k_0} \frac{a_i^0}{b_i^0}}{\sum\limits_{i=1}^{k_0} \frac{1}{b_i^0}} \quad \begin{array}{l} \text{if } P_s = F_s, \\ \text{if } 0 < P_s < F_s, \\ \text{if } P_s = 0. \end{array} \tag{28}
$$

Proof. According to the proof of Theorem 1, equilibrium traffic flow assignment pattern in a multi-subnet urban road network has to satisfy conditions (20)–(22). For the single-commodity multi-subnet road network with disjoint routes conditions (20)–(22) have the following form:

$$
t_r(p_r) \begin{cases} = t_s \text{ for } p_r > 0, \\ \ge t_s \text{ for } p_r = 0, \end{cases} \qquad \forall r \in R_s, s \in S, \tag{29}
$$

$$t_r(f_r) \begin{cases} = t_0 \text{ for } f_r > 0, \\ \geq t_0 \text{ for } f_r = 0, \end{cases} \quad \forall r \in R_0, \tag{30}$$

$$t_s \begin{cases} \leq t_0 \text{ for } P_s = F_s, \\ = t_0 \text{ for } 0 < P_s < F_s, \quad \forall s \in S. \\ \geq t_0 \text{ for } P_s = 0, \end{cases} \tag{31}$$

Firstly, due to linear travel cost functions, expression (29) can be re-written as follows:

$$a_r^s + b_r^s p_r \begin{cases} = t_s \text{ for } p_r > 0, \\ \geq t_s \text{ for } p_r = 0, \end{cases} \quad \forall r \in R_s, s \in S,$$

or

$$p_r = \begin{cases} \frac{t_s - a_r^s}{b_r^s} \text{ if } a_r^s \leq t_s, \\ 0 \quad \text{ for } a_r^s > t_s, \end{cases} \quad \forall i \in R_s, s \in S.$$

Once condition (23) holds, then there exists k_s, $0 \leq k_s \leq n_s$, such that

$$\left. \begin{array}{l} \text{for } r \leq k_s, a_r^s < \\ \text{for } r > k_s, a_r^s \geq \end{array} \right\} t_s \quad \forall s \in S.$$

Hence, the following equalities hold:

$$\sum_{i=1}^{n_s} p_i = \sum_{i=1}^{k_s} p_i = t_s \sum_{i=1}^{k_s} \frac{1}{b_i^s} - \sum_{i=1}^{k_s} \frac{a_i^s}{b_i^s} = P_s \quad \forall s \in S,$$

thus

$$t_s = \frac{P_s + \sum_{i-1}^{k_s} \frac{a_i^s}{b_i^s}}{\sum_{i=1}^{k_s} \frac{1}{b_i^s}} \quad \forall s \in S. \tag{32}$$

Therefore, conditions (24) and (25) do hold.

Secondly, due to linear travel cost functions, expression (30) can be re-written as follows:

$$a_r^0 + b_r^0 f_r \begin{cases} = t_0 \text{ for } f_r > 0, \\ \geq t_0 \text{ for } f_r = 0, \end{cases} \quad \forall r \in R_0,$$

or

$$f_r = \begin{cases} \frac{t_0 - a_r^0}{b_r^0} \text{ if } a_r^0 \leq t_0, \\ 0 \quad \text{ for } a_r^0 > t_0, \end{cases} \quad \forall r \in R_s, s \in S. \tag{33}$$

Once condition (23) holds, then there exists k_0, $0 \leq k_0 \leq n_0$, such that

$$\left. \begin{array}{l} \text{for } r \leq k_0, a_r^0 < \\ \text{for } r > k_0, a_r^0 \geq \end{array} \right\} t_0.$$

Hence, the following equalities hold:

$$\sum_{i=1}^{n_0} f_i = \sum_{i=1}^{k_0} f_i = t_0 \sum_{i=1}^{k_0} \frac{1}{b_i^0} - \sum_{i=1}^{k_0} \frac{a_i^0}{b_i^0} = F_0 + \sum_{s \in S} (F_s - P_s) \quad \forall s \in S,$$

thus

$$t_0 = \frac{F_0 + \sum_{s \in S} (F_s - P_s) + \sum_{i=1}^{k_0} \frac{a_i^0}{b_i^0}}{\sum_{i=1}^{k_0} \frac{1}{b_i^0}}.\tag{34}$$

Therefore, conditions (26) and (27) do hold.

Eventually, if we substitute (32) and (34) into (31), then we obtain (28). □

The algorithm for the equilibrium traffic assignment search in the multi-subnet road network with only disjoint routes and linear travel cost functions follows directly from the proved theorem. Let us consider its application to toll road design in a simple topology network.

5   Toll Road Design in a Simple Topology Road Network

In Fig. 1, we consider a simple topology network, which consists of 4 nodes and 4 arcs, and single OD-pair (1,3). We assume that the travel demand from origin 1 to destination 3 in the presented network includes drivers who are ready to pay fees for better passage conditions (less travel time) and drivers who are not ready to pay fees for passage. In other words, the overall travel demand from origin 1 to destination 3 is $F_0 + F_1$, where F_1 is drivers who are ready to pay fees for better travel conditions, while F_0 is drivers who are not ready to pay fees. The overall travel demand seeks to be assigned between the available disjoint public routes R_0, where R_0 consists of two routes: $1 \rightarrow 2 \rightarrow 3$ and $1 \rightarrow 4 \rightarrow 3$. We believe that travel time through both alternative routes is modeled by linear functions: $t_r(f_r) = a_r^0 + b_r^0 f_r$, $a_r^0 \geq 0$, $b_r^0 > 0$ for any $r \in R_0$, where f_r is the traffic flow through route r, $r \in R_0$.

Fig. 1. Public road network

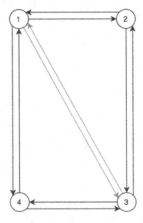

Fig. 2. Toll road subnetwork within the public road network

Suppose that an investor ready to build a toll road with less travel time from origin 1 to destination 3 and wants to evaluate the toll road project from the perspectives of its value for drivers (Fig. 2). In other words, the investor needs to know if drivers will use this toll road in order to decrease their travel time. Therefore, the investor faces the multi-subnet urban road network with disjoint routes and one toll road subnet. Indeed, demand F_1 of drivers who are ready to pay fees for better passage conditions seeks to be assigned between the available disjoint public routes R_0 and routes for toll-paying drivers R_1, where R_1 in our example consists of a single route $1 \to 3$. Thus, on the one hand, $\sum_{r \in R_1} p_r = P_1$, where p_r is the variable corresponding to the traffic flow of the toll-paying drivers through the route $r \in R_1$, while P_1 is the variable corresponding to the overall traffic flow of the toll-paying drivers through the routes R_1. On the other hand, the difference $(F_1 - P_1)$ is the traffic flow of the drivers who are ready to pay fees, but assigned between the available public routes R_0, since the variable P_1 satisfies the following condition: $0 \le P_1 \le F_1$. Therefore, demand $F_0 > 0$ seeks to be assigned between the available public routes R_0 together with the traffic flow $(F_1 - P_1)$: $\sum_{r \in R_0} f_r = F_0 + (F_1 - P_1)$, where f_r is the variable corresponding to the traffic flow through the public route $r \in R_0$ from origin 1 to destination 3. We believe that travel time through subnet routes is modeled by linear functions: $t_r(p_r) = a_r^1 + b_r^1 p_r$, $a_r^1 \ge 0$, $b_r^1 > 0$ for any $r \in R_1$, where p_r is the traffic flow through route r, $r \in R_1$.

According to Theorem 2, the equilibrium traffic assignment pattern in the one-subnet urban road network with disjoint routes and linear cost functions satisfies (24), (26), while actually used routes in toll road subnetwork within the public road network can be found due to (25), (27), and the overall traffic flow of the toll-paying drivers through the routes R_1 can be found due to (28). Let us mention that for one-subnet urban road network with disjoint routes and linear travel time functions, the condition (28) can be relaxed. Indeed, if there exists k_0, $1 \le k_0 \le n_0$, such that

$$a_1^1 > \frac{F_0 + F_1 + \sum_{i=1}^{k_0} \frac{a_i^0}{b_i^0}}{\sum_{i=1}^{k_0} \frac{1}{b_i^0}} \tag{35}$$

then $P_1 = 0$. In other words, condition (35) means that free travel time through toll road subnetwork exceeds the equilibrium travel time in public road network, i.e. no one driver can experience less travel time in toll road subnetwork. However, if there exist k_0, $1 \le k_0 \le n_0$, and k_1, $1 \le k_1 \le n_1$, such that

$$\frac{F_1 + \sum_{i=1}^{k_1} \frac{a_i^1}{b_i^1}}{\sum_{i=1}^{k_1} \frac{1}{b_i^1}} \le \frac{F_0 + \sum_{i=1}^{k_0} \frac{a_i^0}{b_i^0}}{\sum_{i=1}^{k_0} \frac{1}{b_i^0}} \tag{36}$$

then $P_1 = F_1$. In other words, condition (36) means that the equilibrium travel time in toll road subnetwork is less than equilibrium travel time in public road network, i.e. no one toll-paying driver can experience less travel time in public road network. Eventually, if neither condition (35) nor condition (36) holds, then $0 < P_1 < F_1$ and there exist k_0, $1 \leq k_0 \leq n_0$, and k_1, $1 \leq k_1 \leq n_1$, such that

$$
\frac{P_1 + \sum_{i=1}^{k_1} \frac{a_i^1}{b_i^1}}{\sum_{i=1}^{k_1} \frac{1}{b_i^1}} = \frac{F_0 + F_1 - P_1 + \sum_{i=1}^{k_0} \frac{a_i^0}{b_i^0}}{\sum_{i=1}^{k_0} \frac{1}{b_i^0}}
$$

and, hence,

$$
P_1 = \frac{\left(F_0 + F_1 + \sum_{i=1}^{k_0} \frac{a_i^0}{b_i^0} \right) \sum_{i=1}^{k_1} \frac{1}{b_i^1} - \sum_{i=1}^{k_1} \frac{a_i^1}{b_i^1} \sum_{i=1}^{k_0} \frac{1}{b_i^0}}{\left(\sum_{i=1}^{k_1} \frac{1}{b_i^1} + \sum_{i=1}^{k_0} \frac{1}{b_i^0} \right)}. \tag{37}
$$

Table 1. Scenarios for decision-making support.

Evaluation	Scenario	Decision
Inequality (35) holds	Free travel time through toll road subnetwork exceeds the equilibrium travel time in public road network, i.e. no one driver can experience less travel time in toll road subnetwork	Reject the project
Inequality (36) holds	The equilibrium travel time in toll road sub-network is less than equilibrium travel time in public road network, i.e. no one toll-paying driver can experience less travel time in public road network	Accept the project
Inequalities (35) and (36) do not hold, equality (37) holds	The equilibrium travel time in toll road sub-network is equal to the equilibrium travel time in public road network, i.e. the demand of drivers who are ready to pay toll for better passage conditions is not fully satisfied	Improve the project

In other words, condition (37) means that the equilibrium travel time in toll road subnetwork is equal to the equilibrium travel time in public road network, i.e. the demand of drivers who are ready to pay toll for better passage conditions is not fully satisfied.

Therefore, obtained conditions (35)–(37) allow the investor to evaluate the toll road project. Table 1 reflects available scenarios that can support decision-making.

6 Conclusion

The present paper is devoted to the equilibrium traffic flow assignment problem in a multi-subnet urban road network. We formulated this problem as a non-linear optimization program and proved that its solution corresponded to the equilibrium traffic assignment pattern in a multi-subnet road network. Moreover, we proved that obtained equilibrium traffic assignment pattern guaranteed less or equal travel time for selected categories of vehicles in any subnet than experienced by public traffic. The findings of the paper contribute to the traffic theory and give fresh managerial insights for traffic engineers.

References

1. Bagloee, S., Ceder, A.: Transit-network design methodology for actual-size road networks. Transp. Res. Part B **45**, 1787–1804 (2011)
2. Beckmann, M., McGuire, C., Winsten, C.: Studies in the Economics of Transportation. Yale University Press, New Haven (1956)
3. Dijkstra, E.: A note on two problems in connexion with graphs. Num. Math. **1**, 269–271 (1959)
4. Ibarra-Rojas, O.J., Delgado, F., Giesen, R., Munoz, J.: Planning, operation, and control of bus transport systems: a literature review. Transp. Res. Part B **77**(1), 38–75 (2015)
5. Krylatov, A., Zakharov, V.: Competitive traffic assignment in a green transit network. Int. Game Theor. Revi.**18**(2), 3-31 (2016)
6. Krylatov, A., Zakharov, V., Tuovinen, T.: Optimal transit network design. Springer Tracts Transp. Traffic **15**, 141–176 (2020)
7. Krylatov, A., Zakharov, V., Tuovinen, T.: Principles of wardrop for traffic assignment in a road network. Springer Tracts Transp. Traffic **15**, 17–43 (2020)
8. Lampkin, W., Saalmans, P.: The design of routes, service frequencies and schedules for a municipal bus undertaking: a case study. Oper. Res. Q. **18**, 375–397 (1967)
9. Patriksson, M.: The Traffic Assignment Problem: Models and Methods. VSP, Utrecht (1994)
10. Sheffi, Y.: Urban Transportation Networks: Equilibrium Analysis with Mathematical Programming Methods. Prentice-Hall Inc., Englewood Cliffs (1985)
11. Sun, G., Bin, S.: Router-level internet topology evolution model based on multi-subnet composited complex network model. J. Internet Technol. **18**(6), 1275–1283 (2017)
12. Verbas, I., Mahmassani, H.: Integrated frequency allocation and user assignment in multi-modal transit networks: methodology and application to large-scale urban systems. Transp. Res. Record. **2498**, 37–45 (2015)
13. Wardrop, J.G.: Some theoretical aspects of road traffic research. Proc. Inst. Civil Eng. **2**, 325–378 (1952)
14. Wegener, I.: Complexity Theory: Exploring the Limits of Efficient Algorithms, 1st edn. Springer, Heidelberg (2005). https://doi.org/10.1007/3-540-27477-4
15. Xie, F., Levinson, D.: Modeling the growth of transportation networks: a comprehensive review. Netw. Spatial Econ. **9**, 291–307 (2009)
16. Yang, H., An, S.: Robustness evaluation for multi-subnet composited complex network of urban public transport. Alexand. Eng. J. **60**, 2065–2074 (2021)

17. Zakharov, V., Krylatov, A., Volf, D.: Green route allocation in a transportation network. Comput. Methods Appl. Sci. **45**, 71–86 (2018)
18. Zakharov, V.V., Krylatov, A.Y.: Transit network design for green vehicles routing. Adv. Intell. Syst. Comput. **360**, 449–458 (2015)
19. Zakharov, V.V., Krylatov, A.Y., Ivanov, D.A.: Equilibrium traffic flow assignment in case of two navigation providers. IFIP Adv. Inf. Commun. Technol. **408**, 156–163 (2013)
20. Zhao, F., Zeng, X.: Optimization of transit route network, vehicle headways, and timetables for large-scale transit networks. Eur. J. Oper. Res. **186**, 841–855 (2008)

A Local Search Scheme
for the Inequality-Constrained
Optimal Control Problem

A. S. Strekalovsky[(✉)] [iD]

Matrosov Institute for System Dynamics and Control Theory of SB RAS,
Irkutsk, Russia
strekal@icc.ru

Abstract. This paper addresses the nonconvex optimal control (OC) problem with the cost functional and inequality constraint given by the functionals of Bolza. All the functions in the statement of the problem are state-DC, i.e. presented by a difference of the state-convex functions. Meanwhile, the control system is state-linear. Further, with the help of the Exact Penalization Theory we propose the state-DC form of the penalized cost functional and, using the linearization with respect to the basic nonconvexity of the penalized problem, we study the linearized OC problem.

On this basis, we develop a general scheme of the special Local Search Method with a varying penalty parameter. Finally, we address the convergence of the proposed scheme.

Keywords: Nonconvex optimal control · State-convex functions · State-DC functions · Exact penalty · Linearized problem · Local search scheme

1 Introduction

In the last decades, specialists in the optimal control pay more attention to the problems from various applications areas, which are (implicitly or explicitly) nonconvex in the sense that there exists a huge number of local pitfalls from which one can not escape using the standard optimal control (OC) tools [1–11].

Moreover, such objectives as an equilibrium search (say, of Nash), multilevel dynamical optimization, the inverse problem from various applied fields etc., produce generic nonconvexities that are difficult to overcome when it comes to finding a global solution [1–11].

This situation makes change in the field of producing new approaches and generates, for instance, the direct approach, B&B and bioinitiated families of methods etc., which are now so popular and world-spread.

It is worth noting that the demands from the real-world applications [2], which usually have the form of nonlinear control systems or/and nonconvex constraints and cost functionals, include not only a quick solution but also immediate consultations with practical suggestions for a management team.

ⓒ Springer Nature Switzerland AG 2021
P. Pardalos et al. (Eds.): MOTOR 2021, LNCS 12755, pp. 17–31, 2021.
https://doi.org/10.1007/978-3-030-77876-7_2

This leads to the above family of "simple" methods suffering the well-known "course of dimensionality", when the volume of computational efforts increases exponentially with the problem dimension.

On the other hand, the OC theory seems to be in a satisfactory position due to Pontryaguin's Principle (the PMP) and the Dynamical Programming based on the HJB-equation [1–11]. However, from the point of view of numerical methods of the OC it is rather difficult to find a global solution even in the case of nonlinear control systems or, even to escape a stationary control (say, in the sense of the PMP) while improving the cost functionals (see, for instance, examples in [3–11]).

Of course, the new mathematical tools often allow one to do it in some particular OC problems (see [3,5,6]), even in applied problems. However, there exists no general methodology, for instance, for constructing numerical procedures capable of attacking nonlinear, nonconvex OC problem, as it was done, for example, for some cases in [3,5–8,10,11].

It seems that now experts in the OC theory have to separate the set of nonlinear, nonconvex OC problems into several classes to advance different approaches more effectively by finding globally optimal controls as in [3–8,10,11].

It is worth mentioning that for the finite-dimensional optimization the new mathematical tools developed in [23,26–29,32–34] allowed one to successfully solve a specter of different applied problems provided by equilibrium problems, hierarchical optimization problems and even some OC problems [23–26,30,31, 33].

Since it was done for the case of DC optimization problems of various kinds and, taking into account that any continuous function on a compact can be approximated (at any accuracy!) by a DC function, it would be natural to try to apply the advanced methodology for the suitable cases of OC problems.

Here, in this paper, we intend to propose a special local search method for the case of the linear control system and a system of inequality constraints given with the help of DC functions in the terminal and integrand terms.

In addition, our goal is to study convergence of this new numerical tool, which is not a simple problem as was demonstrated by the results obtained.

It is worth noting that special local search procedures play a very important and useful role in the global search in nonconvex optimization problems, providing not only the KKT points or the Pontryaguin's extremal, but often more strong control processes.

Taking this into account, we reduce the original OC problem to the problem without inequality constraints via the Exact Penalization Theory.

In turn, the penalized problem is linearized with respect to the basic nonconvexity, which delivers us the (partially) linearized OC problem. Using the linearization at every current iteration, we obtain a scheme of local search, some convergency points of which allow us to say a few words on some future researches.

2   Statement of the Problem and Exact Penalization

Let us address the state-linear control system (CS) as follows:

$$\dot{x}(t) = A(t)x(t) + B(u(t),t) \ \overset{\circ}{\forall} t \in T :=]t_0,t_1[, \\ x(t_0) = x_0;$$
(1)

$$u(\cdot) \in \mathcal{U} := \{u(\cdot) \in L_\infty^r(T) \mid u(t) \in U \ \overset{\circ}{\forall} t \in T\}$$
(2)

(where $\overset{\circ}{\forall}$ denotes "almost everywhere" in the sense of the Lebesque measure) under standard assumptions [3–11] when the matrix $A(t) = [a_{ij}(t)]_{i,j=1}^n$ and the vector $B(u,t)$ are continuous with respect to the variables $t \in \overline{T} = [t_0,t_1]$ and $u(t) \in U$, $t \in T$, where U is a compact from $\mathbb{R}^r$. Then, as well-known [3–10], for any feasible control $u(\cdot) \in \mathcal{U}$ and $\forall x_0 \in \mathbb{R}^n$ the system of ODEs (1) has a unique absolutely continuous solution

$$x(\cdot,u) \in AC_n(T) =: X, \ \ x(t) = x(t,u), t \in \overline{T}.$$

Furthermore, let us consider the functionals $J_i(x(\cdot),u(\cdot)) := J_i(x,u), i \in \{0\} \cup I, I = \{1,...,m\}$, of the form

$$J_i(x,u) = \varphi_{1i}(x(t_1)) + \int_T \varphi_i(x(t),u(t),t)dt,$$
(3)

where the functions $\varphi_{1i}(x)$, $\varphi_{1i} : \mathbb{R}^n \to \mathbb{R}, i \in \{0\} \cup I$, can be represented as follows

$$\varphi_{1i}(x) := g_{1i}(x) - h_{1i}(x) \ \ \forall x \in \Omega_1 \subset \mathbb{R}^n,$$
(4)

where Ω_1 is an open convex subset of $\mathbb{R}^n$ containing the reachable set $\mathcal{R}(t_1)$ of the control system (1)–(2) at the final moment $t_1 : \mathcal{R}(t_1) \subset \Omega_1$, meanwhile the functions $g_{1i}(\cdot)$ and $h_{1i}(\cdot)$ are convex on Ω_1, so that $\varphi_{1i}(\cdot)$ turns out to be DC functions, $i \in \{0\} \cup I$ [12–15].

In addition, the functions $\varphi_i(x,u,t)$, $\varphi_i : \Omega(t) \times U \times T \to \mathbb{R}$ have the following decompositions

$$\varphi_i(x,u,t) := g_i(x,u,t) - h_i(x,t), \ \ i \in \{0\} \cup I,$$
(5)

$\forall x \in \Omega(t)$, $\forall (u,t) \in U \times T$, where $\Omega(t) \subset \mathbb{R}^n$ is a rather large open convex subset of $\mathbb{R}^n$, such that $\mathcal{R}(t) \subset \Omega(t), t \in T$. Besides, the functions $g_i(x,u,t)$ are continuous in the variables $(x,u,t) \in \mathbb{R}^{n+r+1}$, and the mappings $x \to g_i(x,u,t) : \Omega(t) \to \mathbb{R}$ are convex $\forall (u,t) \in U \times T$ [12–15].

Similarly, the functions $h_i(x,t)$ are continuous on $(x,t) \in \Omega(t) \times T$, and the mappings $x \to h_i(x,t)$ are convex on $\Omega(t) \ \forall t \in T$. forth, we will call the convexity property of the functions $g_{1i}(x)$, $g_i(x,u,t)$, $h_{1i}(x)$, $h_i(x,t)$ with respect to the variable $x \in \mathbb{R}$ as state-convexity, meanwhile, the properties of the functions $\varphi_{1i}(x)$, $\varphi_i(x,u,t)$ to be represented as in (4) and (5), will be said to be state-DC.

On the other hand, taking into account that $x(\cdot) = x(\cdot, u))$, $u \in \mathcal{U}$, is the unique solution to the system (1) of ODEs corresponding to a control $u(\cdot) \in \mathcal{U}$, the following notations look rather natural and comprehensible: $J_i(u) := J_i(x(\cdot, u), u)$, $i \in I \cup \{0\}$. In addition, assume that the data from the above is smooth with respect to the state. Then, due to the state-convexity of the functions above, in particular, the following inequalities hold true ($\nabla := \nabla_x$) [7, 12–15]:

$$
\left.
\begin{aligned}
(a): \quad & \langle \nabla h_{1i}(y), x - y \rangle \leq h_{1i}(x) - h_{1i}(y) \ \forall x, y \in \Omega_1, \\
(b): \quad & \langle \nabla h_i(y(t)), x(t) - y(t) \rangle \leq h_i(x(t), t) - h_i(y(t), t) \\
& \forall x(t), y(t) \in \Omega(t), \ i \in \{0\} \cup I.
\end{aligned}
\right\}
\tag{6}
$$

Let us now address the following optimal control (OC) problem:

$$
(\mathcal{P}): \qquad
\left.
\begin{aligned}
J_0(u) &\overset{\triangle}{=} J_0(x(\cdot, u), u(\cdot)) \downarrow \min_u, \quad u(\cdot) \in \mathcal{U}, \\
J_i(u) &\overset{\triangle}{=} J_i(x(\cdot, u), u(\cdot)) \leq 0, \quad i \in I = \{1, ...m\}.
\end{aligned}
\right\}
\tag{7}
$$

It is clear, that, in virtue of nonconvexity (with respect to the state $x(\cdot, u)$, $(x(t, u) = x(t)$, $t \in T$, $u \in \mathcal{U})$ of the terminal parts $\varphi_{1i}(\cdot)$ and the integrands $\varphi_i(x, u, t)$, every functional $J_i(x, u)$, $i \in \{0\} \cup I$, the feasible region of Problem $(\mathcal{P})$, and Problem $(\mathcal{P})$ itself, as a whole, turn out to be nonconvex. It means that Problem $(\mathcal{P})$ might possess a big number of locally optimal and stationary (say, in the sense of the PMP) processes, which may be rather far from a set $\mathrm{Sol}(\mathcal{P})$ of global solutions (globally optimal controls, processes, if one exists), even with respect to the value of the cost function.

To solve Problem $(\mathcal{P})$, let us now apply a very popular approach of the Exact Penalization [16–22]. To this end, introduce the penalty function $\pi(x, u)$ for Problem $(\mathcal{P})$ in the following way

$$
\pi(x, u) := \pi(u) = \max\{0, J_1(u), ..., J_m(u)\}
\tag{8}
$$

and address the auxiliary (penalized) problem

$$
(\mathcal{P}_\sigma): \qquad J_\sigma(u) := J_\sigma(x(\cdot, u), u(\cdot)) \downarrow \min_u, \quad u(\cdot) \in \mathcal{U},
\tag{9}
$$

with the cost function defined as follows

$$
J_\sigma(u) := J_0(x(\cdot, u), u(\cdot)) + \sigma \pi(x(\cdot, u), u(\cdot)),
\tag{10}
$$

where $\sigma \geq 0$ is a penalty parameter.

Recall that the key feature of the Exact Penalization Theory [16–22] consists in the existence of the threshold value $\sigma_* > 0$ of the penalty parameter for which Problems $(\mathcal{P})$ and $(\mathcal{P}_\sigma)$ are equivalent in the sense that

$$
\mathcal{V}(\mathcal{P}) = \mathcal{V}(\mathcal{P}_\sigma) \text{ and } \mathrm{Sol}(\mathcal{P}) = \mathrm{Sol}(\mathcal{P}_\sigma) \ \forall \sigma > \sigma_*,
\tag{11}
$$

(see [12], Chapter VII, Lemma 1.2.1 and [17–21]).

Hence, the existence of the exact (threshold) value σ_* of the penalty parameter implies that instead of solving a sequence $\{(\mathcal{P}_{\sigma_k})\}$ of unconstrained problems with $\sigma_k \uparrow \infty$ we need to consider a single problem $(\mathcal{P}_\sigma)$ with the penalty parameter $\sigma \geq \sigma_*$.

On the other hand, it is well-known that if a process $(z(\cdot), w(\cdot))$ (a control $w(\cdot)$) is a global solution to Problem $(\mathcal{P}_\sigma)$: $(z, w) \in \mathrm{Sol}(\mathcal{P}_\sigma)$, $z(t) = x(t, w)$, $t \in T$, $w(\cdot) \in \mathcal{U}$, and, besides, $(z(\cdot), w(\cdot))$ is feasible in Problem $(\mathcal{P})$, i.e. $J_i(z, w) \leq 0$, $i \in I$, then $(z(\cdot), w(\cdot))$ is a global solution to Problem $(\mathcal{P})$. It is worth noting that the inverse assertion, in general, does not hold.

Moreover, under various Constraint Qualification (CQ) conditions (e.g. MFCQ, Slater, etc.), the error bound properties, the calmness of the constraint system etc., one can prove the existence of the exact penalty (threshold) value $\sigma_* > 0$ for local and global solutions. In what follows, let us assume that some regularity conditions, ensuring the existence of the threshold value $\sigma_* > 0$ of the penalty parameter in Problem $(\mathcal{P})$, are satisfied (see [16–22]).

3 DC Decomposition of Problem $(\mathcal{P}_\sigma)$

First of all, let us show that every functional $J_i(u) = J_i(x, u)$ defined in (3) can be represented in the form

$$J_i(u) = G_i(x, u) - F_i(x), \quad i \in \{0\} \cup I, \tag{12}$$

where $G_i(\cdot)$ and $F_i(\cdot)$ are state-convex.

Indeed, employing the formulae (3) (5), we have

$$\left.\begin{array}{l} (a): G_i(x, u) :- g_{1i}(x(t_1)) + \int_T g_i(x(t), u(t), t)\, dt, \\[2mm] (b): F_i(x) := h_{1i}(x(t_1)) + \int_T h_i(x(t), t)\, dt, \quad i \in \{0\} \cup I; \end{array}\right\} \tag{12'}$$

which yields the desirable state-convexity property. In particular, for the functions $F_i(x)$ we obtain the feature similar to the convexity inequalities (6).

Actually, under the above assumptions, a differential of the functional $F_i(\cdot)$ can be defined as follows

$$\langle\langle \nabla F_i(y(\cdot)), x(\cdot) \rangle\rangle := \langle \nabla h_{1i}(y(t_1)), x(t_1) \rangle + \int_T \langle \nabla h_i(y(t), t), x(t) \rangle\, dt, \tag{13}$$

where $\langle \cdot, \cdot \rangle_n$ is the inner product in $\mathbb{R}^n$, $x(\cdot)$, $y(\cdot) \in X \overset{\triangle}{=} AC_n(T)$. Therefore, we can consider the pair $(\nabla h_{1i}(y(t_1)), \nabla h_i(y(\cdot), \cdot))$ as a gradient of $F_i(\cdot)$ at a function $y(\cdot) \in X \overset{\triangle}{=} AC_n(T)$: $\nabla F_i(y(\cdot)) := (\nabla h_{1i}(y(t_1)), \nabla h_i(y(t), t), t \in T)$.

As a consequence, due to (6) and (13), the following inequality holds

$$\begin{aligned} \langle\langle \nabla F_i(y(\cdot)), x(\cdot) - y(\cdot) \rangle\rangle &\leq F_i(x(\cdot)) - F_i(y(\cdot)) \\ \forall (x(\cdot), y(\cdot)) &\in X \overset{\triangle}{=} AC_n(T), (i \in \{0\} \cup I). \end{aligned} \tag{6'}$$

Furthermore, it can be readily seen that, thanks to the presentations (5), (8)–(10), (12)–(12'), the cost function $J_\sigma(x, u)$ of the penalized Problem $(\mathcal{P}_\sigma)$–(9), (10) can be represented as follows

$$J_\sigma(x, u) := G_0(x, u) - F_0(x) + \sigma\Pi(x, u) \overset{\triangle}{=}$$
$$= G_0(x, u) - F_0(x) + \sigma\max\{0; [G_i(x, u) - F_i(x)], \ i \in I\}. \tag{10'}$$

Moreover, let us show now that the penalty function $\Pi(x, u)$ defined in (8) can also be represented as a state-DC functional, i.e. $\Pi(x, u) = G_\pi(x, u) - F_\pi(x)$, where $G_\pi(\cdot)$ and $H_\pi(\cdot)$ also have the state-convexity property. Then, obviously, $J_\sigma(x, u)$ will be state-DC (i.e., $x(\cdot) \to J_\sigma(x, u)$ is a DC functional $\forall u(\cdot) \in \mathcal{U}$). Indeed, from (8) it is clear that

$$\Pi(x, u) := \max\{0; [G_i(x, u) - F_i(x)], \ i \in I\} \pm \sum_{j \in I} F_j(x)$$
$$= \max\left\{ \sum_{j \in I} F_j(x); \left[G_i(x, u) + \sum_{p \in I}^{p \neq i} F_p(x)\right], \ i \in I \right\} - \sum_{j \in I} F_j(x). \tag{14}$$

Therefore, using the notations

$$(a): G_\pi(x, u) := \max\left\{ \sum_{i \in I} F_i(x); \left[G_i(x, u) + \sum_{p \in I}^{p \neq i} F_p(x)\right], \ i \in I \right\}$$
$$(b): F_\pi(x) := \sum_{i \in I} F_i(x), \tag{15}$$

one gets the following DC decomposition of the penalty function

$$\Pi(x, u) = G_\pi(x, u) - F_\pi(x), \tag{16}$$

where the functions $G_\pi(\cdot)$ and $F_\pi(\cdot)$ clearly preserve the state-convexity property due to (12), (12') and (15) [7,12–15].

Moreover, as claimed above, the cost function $J_\sigma(x, u)$ defined in (10), because of (10'), (14)–(16) has the following DC-state decomposition

$$J_\sigma(x, u) \overset{\triangle}{=} G_0(x, u) - F_0(x) + \sigma[G_\pi(x, u) - F_\pi(x)]$$
$$= [G_0(x, u) + \sigma G_\pi(x, u)] - [F_0(x) + \sigma F_\pi(x)] = G_\sigma(x, u) - F_\sigma(x), \tag{17}$$

where, thanks to (12'), we have (see (15))

$$G_\sigma(x, u) := G_0(x, u) + \sigma G_\pi(x, u) = g_{10}(x(t_1))$$
$$+ \int_T g_0(x(t), u(t), t)dt + \sigma\max\left\{ \sum_{j \in I} \left[h_{1j}(x(t_1)) + \int_T h_j(x(t), t)dt\right] ; \right.$$
$$\left[g_{1i}(x(t_1)) + \int_T g_i(x(t), u(t), t)dt \right. \tag{18}$$
$$\left. + \sum_{p \in I}^{p \neq i} \left(h_{1p}(x(t_1)) + \int_T h_p(x(t), t)dt \right) \right], \ i \in I \right\};$$

$$F_\sigma(x) := F_0(x) + \sigma F_\pi(x) \triangleq$$

$$= h_{10}(x(t_1)) + \int_T h_0(x(t),t)\,dt + \sigma \sum_{i \in I} \left[h_{1i}(x(t_1)) + \int_T h_i(x(t),t)\,dt \right] \quad (19)$$

$$= h_{10}(x(t_1)) + \sigma \sum_{i \in I} h_{1i}(x(t_1)) + \int_T \left[h_0(x(t),t)dt + \sigma \sum_{i \in I} h_i(x(t),t) \right] dt.$$

It is not difficult to see from (17)–(19) that the functionals $G_\sigma(x,u)$ and $F_\sigma(x)$ are also endowed with the state-convexity property [7,12–15].

On the other hand, from (6) and (13), (19) it can be readily seen that the functional $F_\sigma(x(\cdot))$ is differentiable in the sense that $\forall y(\cdot) \in X$ we have

$$\langle\langle \nabla F_\sigma(y(\cdot)), x(\cdot) \rangle\rangle \triangleq \langle x(t_1), \nabla h_{10}(y(t_1)) + \sigma \sum_{i \in I} \nabla h_{1i}(y(t_1)) \rangle$$

$$+ \int_T \langle x(t), \nabla h_0(y(t),t) + \sigma \sum_{i \in I} \nabla h_i(y(t),t) \rangle dt. \quad (20)$$

Hence, due to the state-convexity of $F_\sigma(\cdot)$, one has the following inequality $(\forall u(\cdot) \in \mathcal{U})$:

$$\langle\langle \nabla F_\sigma(y(\cdot)), x(\cdot,u) - y(\cdot) \rangle\rangle < F_\sigma(x(\cdot,u)) - F_\sigma(y(\cdot)). \quad (21)$$

4   Linearized Problem

Let us return now to the original Problem $(\mathcal{P})$–(7), assuming that the feasible set of $(\mathcal{P})$ is not empty, i.e.

$$\mathcal{F} := \{(x(\cdot),u(\cdot)) \mid x(t) = x(t,u),\ t \in T,\ u(\cdot) \in \mathcal{U};\ J_i(u) \le 0,\ i \in I\} \ne \emptyset, \quad (22)$$

and the optimal value $\mathcal{V}(\mathcal{P})$ of Problem $(\mathcal{P})$ is finite, i.e.

$$(\mathcal{A}_0): \qquad \mathcal{V}(\mathcal{P}) := \inf_{u(\cdot)} \{J_0(u) \mid u(\cdot) \in \mathcal{U},\ (x(\cdot,u),u(\cdot)) \in \mathcal{F}\} > -\infty. \quad (23)$$

Furthermore, let us address the (partially) linearized at $y(\cdot) \in X = AC_n(T)$ optimal control (OC) problem (caused by $(\mathcal{P}_\sigma)$) as follows

$$(\mathcal{P}_\sigma L(y)): \quad \begin{matrix} \Phi_{\sigma y}(x(\cdot),u(\cdot)) := \Phi_{\sigma y}(u) := \\ = G_\sigma(x(\cdot),u(\cdot)) - \langle\langle \nabla F_\sigma(y(\cdot)), x(\cdot) \rangle\rangle \downarrow \min_u,\quad u(\cdot) \in \mathcal{U}, \end{matrix} \quad (24)$$

along the control system (1) of ODEs, i.e. $x(\cdot) = x(\cdot,u)$.

It can be readily seen, that the functional $\Phi_{\sigma y}(u)$ preserve the state-convexity property, because the "anticonvex" term $(-F_\sigma(x(\cdot)))$ in the DC-state-decomposition (17) is linearized at $y(\cdot) \in X$ [12–15].

Moreover, the linearization is performed only with respect to the functional $F_\sigma(x)$ which accumulates all nonconvexities of the original problem $(\mathcal{P}_\sigma)$–(9)–(10). On the other hand, it is worth noting that Problem $(\mathcal{P}_\sigma L(y))$ remains non-differentiable because $G_\sigma(x, u)$ is nonsmooth (see (15) and (18)), so that

$$G_\sigma(x, u) \triangleq G_0(x, u) + \sigma G_\pi(x, u) := G_0(x, u)$$
$$+\sigma \max \left\{ \sum_{j \in I} F_j(x); \left[G_i(x, u) + \sum_{p \in I}^{p \neq i} F_p(x) \right], \ i \in I \right\}. \tag{15'}$$

In order to avoid the obstacle, one can apply Lemma 4.1 from [28] (see also [22]), which allows us to address, instead of Problem $(\mathcal{P}_\sigma L(y))$–(24), the auxiliary OC problem (with state-convex inequality constraints and the supplementary parameter $\gamma \in \mathbb{R}$) of the form

$$\left. \begin{array}{rl} (\mathcal{AP}_\sigma L(y)): & G_0(x(\cdot, u), u) + \sigma\gamma - \langle\langle \nabla F_\sigma(y(\cdot)), x(\cdot, u)\rangle\rangle \downarrow \min_{u, \gamma}, \\[2mm] (a): & \sum_{j \in I} F_j(x(\cdot, u)) \leq \gamma, \ \ \gamma \in \mathbb{R}, u(\cdot) \in \mathcal{U}, \\[2mm] (b): & G_i(x(\cdot, u), u) + \sum_{p \in I}^{p \neq i} F_p(x(\cdot, u)) \leq \gamma, \ \ i \in I; \end{array} \right\} \tag{25}$$

(along the control system (1) of ODEs).

At first, it can be readily seen that Problem $(\mathcal{AP}_\sigma L(y))$–(25) is state-convex due to (4)–(5), the presentations (12'), (15) and the linearization of the functional $F_\sigma(\cdot)$. Further, the data of Problem $(\mathcal{P})$–(7) was assumed to be state-differentiable. Therefore, the data of Problem $(\mathcal{AP}_\sigma L(y))$–(25) remain state-smooth. Henceforth, let us assume that we are able to solve Problem $(\mathcal{AP}_\sigma L(y))$–(25) globally employing its state-convexity. It can be readily seen that then one can obviously calculate the value of $G_\sigma(x_*(\cdot), u_*(\cdot))$ at the solution $(x_*(\cdot), u_*(\cdot))$ to the problem (25) and, moreover, to compute the value of the objective functional $\Phi_{\sigma y}(x_*(\cdot), u_*(\cdot))$ of Problem $(\mathcal{P}_\sigma L(y))$–(24).

5   Local Search Scheme

In this section, we are going to develop a theoretical scheme of Local search for the original Problem $(\mathcal{P})$–(7) via the penalized Problem $(\mathcal{P}_\sigma)$–(9)–(10). The idea of the procedure consists in a consecutive approximate solution to Problem $(\mathcal{P}_\sigma L(y))$–(24) and the usage of the auxiliary Problem $(\mathcal{AP}_\sigma L(y))$–(25) (see [23]).

Let us be given the number sequences

$$\begin{array}{ll} \{\delta_k\}: & \delta_k > 0, \ k = 0, 1, 2, \ldots, \quad \sum_{k=0}^\infty \delta_k < +\infty; \\[2mm] \{\sigma_k\}: & \sigma_k > 0, \ k = 0, 1, 2, \ldots. \end{array} \tag{26}$$

Furthermore, let us be given a starting control $u^0(\cdot) \in \mathcal{U}$ and a current control iterate $u^k(\cdot) \in \mathcal{U}$. Then the corresponding states $x^0(\cdot), x^k(\cdot) \in X$,

$x^0(t) = x(t, u^o)$, and $x^k(t) = x(t, u^k)$, $t \in T$, are the solutions of the control system (CS) (1) of ODEs with $u(\cdot) = u^0(\cdot)$ and $u(\cdot) = u^k(\cdot)$, respectively.

Everywhere below we will use the notation

$$\Phi_k(\cdot) := \Phi_{\sigma_k}(\cdot), \quad G_k(\cdot) := G_{\sigma_k}(\cdot), \quad F_k(\cdot) := F_{\sigma_k}(\cdot), \quad (\mathcal{P}_k L_k) := (\mathcal{P}_{\sigma_k} L(x^k)).$$

Let us now address the following OC problem

$$(\mathcal{P}_k L_k): \quad \Phi_k(x(\cdot), u(\cdot)) := \Phi_{\sigma_k}(u) := G_k(x(\cdot), u(\cdot)) \\ \left. -\langle\langle \nabla F_k(x^k(\cdot)), x(\cdot)\rangle\rangle \downarrow \min_u, \quad u(\cdot) \in \mathcal{U}, \right\} \quad (27)$$

along the control system (1) of ODEs.

Recall that Problem $(\mathcal{P}_k L_k)$–(27) can be solved with the help of the corresponding Problem of type $(\mathcal{AP}_\sigma L(y))$–(25).

Therefore, given an iterate $u^k(\cdot) \in \mathcal{U}$, one can define the next control iterate $u^{k+1}(\cdot) \in \mathcal{U}$ as an approximate solution to Problem $(\mathcal{P}_k L_k)$, i.e. according to the following inequality

$$\Phi_k(x^{k+1}(\cdot), u^{k+1}(\cdot)) \le \mathcal{V}(\mathcal{P}_k L_k) + \delta_k. \quad (28)$$

It means that $u^{k+1}(\cdot)$ solves Problem $(\mathcal{P}_k L_k)$–(27) with the accuracy $\delta_k > 0$, and $x^{k+1}(\cdot) \in X$ is the solution to the CS (1) of ODEs corresponding to $u^{k+1}(\cdot) \in \mathcal{U}$, i.e. $x^{k+1}(t) = x(t, u^{k+1})$, $t \in T$.

Hence, on account (27) and (28), the principal rule (28) of the method can be rewritten as follows

$$G_k(x^{k+1}(\cdot), u^{k+1}(\cdot)) - \langle\langle \nabla F_k(x^k(\cdot)), x^{k+1}(\cdot)\rangle\rangle \\ \le G_k(x(\cdot, u), u(\cdot)) - \langle\langle \nabla F_k(x^k(\cdot)), x(\cdot, u)\rangle\rangle + \delta_k \quad \forall u(\cdot) \in \mathcal{U}. \quad (28')$$

It is possible now to develop the first general variant (a scheme) of a Local Search Method (LSM) for Problem $(\mathcal{P}_\sigma)$–(9), (10) which has some relations to the original Problem $(\mathcal{P})$–(7), due to the properties of the exact Penalization Theory [16–22].

Recall that the principal (and simple) idea of the first version of the LSM for Problem $(\mathcal{P})$ consists in a consecutive solution of the linearized Problem $(\mathcal{P}_k L_k)$–(27) (or $(\mathcal{AP}_\sigma L(x^k))$–(25)) with a variation of the penalty parameter $\sigma > 0$. In addition to the assumptions made above, let there be given an initial value $\sigma_{in} > 0$ (say, $\sigma_{in} = 1$) of the penalty parameter along with two parameters $\eta_1 \in]0, 1[$ and $\eta_2 \in [2, 10]$. Consider now the following procedure.

Local Search Scheme 1(LSSQ1)

Step 0. Set $k := 0$, $u^k(\cdot) := u^0(\cdot)$, $x^k(\cdot) = x(\cdot, u^k)$, $\sigma_k := \sigma_{in}$.

Step 1. Solve Problem $(\mathcal{P}_k L_k)$–(27) to get the control $\bar{u}(\cdot) \in \mathcal{U}$ providing an approximate solution to Problem $(\mathcal{P}_k L_k)$: $(\bar{x}(\cdot), \bar{u}(\cdot)) \in \delta_k - Sol(\mathcal{P}_k L_k)$, where $\bar{x}(t) = \bar{x}(t, \bar{u}), t \in T$, is the solution of the CS (1) with $u(\cdot) = \bar{u}(\cdot)$.

Step 2. IF $\pi(\bar{x}(\cdot), \bar{u}(\cdot)) = 0$, i.e. the process $(\bar{x}(\cdot), \bar{u}(\cdot))$ is feasible in Problem $(\mathcal{P})$–(7), THEN set $\sigma_+ := \sigma_k$, $u_+(\cdot) := \bar{u}(\cdot)$, $x_+(\cdot) := \bar{x}(\cdot)$ and go to Step 6.

Step 3. $(\pi(\bar{x}(\cdot), \bar{u}(\cdot)) > 0)$ IF the inequality

$$\Phi_k(x^k(\cdot), u^k(\cdot)) - \Phi_k(\bar{x}(\cdot), \bar{u}(\cdot)) \geq \eta_1 \sigma_k \left[\pi(x^k(\cdot), u^k(\cdot)) - \pi(\bar{x}(\cdot), \bar{u}(\cdot))\right], \quad (29)$$

is true, THEN set $\sigma_+ := \sigma_k$, $u_+(\cdot) := \bar{u}(\cdot)$, $x_+(\cdot) := \bar{x}(\cdot)$, and go to Step 6.
Step 4. (Else) Increase $\sigma_k > 0$, so that $\sigma_+ := \eta_2 \sigma_k$ with $\eta_2 \in [2, 10]$, and solve the next linearized problem

$$(\mathcal{P}_+L_+): \quad \begin{aligned} \Phi_+(x(\cdot), u), u(\cdot)) &:= G_+(x(\cdot, u), u(\cdot)) \\ &- \langle\langle \nabla F_+(\bar{x}(\cdot)), x(\cdot, u)\rangle\rangle \downarrow \min_{u(\cdot)}, \quad u(\cdot) \in \mathcal{U}_+, \end{aligned} \right\} \quad (30)$$

(along the control system (1)) with $G_+(\cdot) := G_{\sigma_+}(\cdot)$, $F_+(\cdot) := F_{\sigma_+}(\cdot)$.
Let the process $(x_+(\cdot), u_+(\cdot))$ be an approximate (δ_k) solution to Problem $(\mathcal{P}_+L_+)$–(30), so that $(x_+(\cdot), u_+(\cdot)) \in \delta_k - Sol(\mathcal{P}_+L_+)$.
Step 5. Set $\bar{x}(\cdot) := x_+(\cdot)$, $\bar{u}(\cdot) := u_+(\cdot)$, $\sigma_k := \sigma_+$, and go to Step 2.
Step 6. Set $\sigma_{k+1} := \sigma_+$, $x^{k+1}(\cdot) := \bar{x}(\cdot)$, $u^{k+1}(\cdot) := \bar{u}(\cdot)$, $k := k+1$, and loop to Step 1.

It is not difficult to point out that the above scheme is not yet to become a proper algorithm, because, for instance, there is still no stopping criteria for the LSS1 (see [32–34]).

6 Convergence of the LSS1

Let us begin the study of the convergence properties of the LSS1 from above by the next assumption

$$(\mathcal{A}_\pi): \quad \begin{aligned} &(a) \; \xi_k := (\sigma_{k+1} - \sigma_k)\pi(x^k(\cdot), u^k(\cdot)) \geq 0, \; k = 0, 1, 2, \ldots \\ &(b) \; \sum_{k=0}^{\infty} \xi_k \overset{\triangle}{=} \sum_{k=0}^{\infty} (\sigma_{k+1} - \sigma_k)\pi(x^k(\cdot), u^k(\cdot)) < +\infty. \end{aligned} \right\} \quad (31)$$

Whence we immediately derive that

$$\sigma_{k+1} \geq \sigma_k > 0. \quad (31')$$

It is worth noting that in the LSS1 one does not use the cost functional

$$J_\sigma(u) \overset{\triangle}{=} G_\sigma(x(\cdot, u), u) \; - \; F_\sigma(x(\cdot, u))$$

of the penalized Problem $(\mathcal{P}_\sigma)$–(9), (10), but only the linearized one (with respect to the "anticonvex" part $F_\sigma(x(\cdot))$) functional $\Phi_{\sigma_k}(x(\cdot), u(\cdot))$ defined in (27), or, more precisely, $\Phi_k(\cdot) := \Phi_{\sigma_k}(\cdot)$.

Then, it can be readily seen that, employing the principal inequality of the LSS1 (see (28), (28'), with $F_k(\cdot) := F_{\sigma_k}(\cdot)$, $G_k(\cdot) := G_{\sigma_k}(\cdot)$), we have

$$\begin{aligned} \Phi_k(x^{k+1}(\cdot), u^{k+1}(\cdot)) &\overset{\triangle}{=} (G_k(x^{k+1}(\cdot), u^{k+1}(\cdot)) - \langle\langle \nabla F_k(x^k(\cdot)), x^{k+1}(\cdot)\rangle\rangle \\ &\leq \Phi_k(x^k(\cdot), u^k(\cdot)) + \delta_k \overset{\triangle}{=} G_k(x^k(\cdot), u^k(\cdot)) - \langle\langle \nabla F_k(x^k(\cdot)), x^k(\cdot)\rangle\rangle + \delta_k. \end{aligned} \quad (32)$$

The chain (32) implies the solution of Problem $(\mathcal{P}_k L_k)$–(27), and then one can show that, like it was done in [32–34], under the assumption $(\mathcal{A}_\pi)$–(31) and with the help of the convexity inequality (21) we obtain

$$J_{k+1}(u^{k+1}) \leq J_k(u^k) + \xi_k, \ k = 0, 1, 2, \tag{33}$$

(where $J_k(u) := J_{\sigma_k}(u), \ k = 0, 1, 2, \ldots$). The inequality (33) provides the following result (see [32–34]).

Proposition 1. *Let the assumptions* $(\mathcal{A}_0)$–(23) *and* $(\mathcal{A}_\pi)$–(31) *be fulfilled.*
Then the sequence $\{x^k(\cdot), u^k(\cdot)\}$, *produced by the LSS1, satisfies the following conditions.*
The number sequences $\{J_k(u^k)\}$ *and* $\{\Delta\Phi_{k+1}\}$, *where*

$$\Delta\Phi_{k+1} := \Phi_{k+1}(x^k(\cdot), u^k(\cdot)) - \Phi_{k+1}(x^{k+1}(\cdot), u^{k+1}(\cdot)) \overset{\triangle}{=} (G_{k+1}(x^k(\cdot), u^k(\cdot))$$
$$- (G_{k+1}(x^{k+1}(\cdot), u^{k+1}(\cdot)) + \langle\langle \nabla F_{k+1}(x^k(\cdot)), x^{k+1}(\cdot) - x^k(\cdot)\rangle\rangle,$$

converge so that

$$\left.\begin{array}{l} (a): \lim\limits_{k\uparrow\infty} J_k(x^k(\cdot), u^k(\cdot)) =: J_* > -\infty \\[2mm] (b): \lim\limits_{k\uparrow\infty} \Delta\Phi_{k+1} = 0. \end{array}\right\} \tag{34}$$

Furthermore, let the following assumptions hold:

$$(\mathcal{A}_{str}): \left.\begin{array}{l} (a) \text{ At least one of the functions } h_{1i}(\cdot), \ i \in \{0\} \cup \mathcal{I}, \\ \quad \text{ is strongly convex on the convex set } \Omega_1 \subset I\!\!R^n; \\ (b) \text{ at least one of the functions } x \to h_i(x, t), \ i \in \{0\} \cup \mathcal{I}, \\ \quad \text{ is strongly convex on the convex open set } \Omega(t) \in I\!\!R^n \\ \quad \text{ containing a reachable set } \mathcal{R}(t) \subset \Omega(t) \ \forall t \in T. \end{array}\right\} \tag{35}$$

It is not difficult to verify that under the assumptions $(\mathcal{A}_{str})$–(35), the functional $F_\sigma(x(\cdot))$ defined in (19) satisfies the following chain $(x(\cdot), y(\cdot) \in X = AC_n(T))$ (see (20))

$$F_\sigma(x(\cdot)) - F_\sigma(y(\cdot)) \overset{\triangle}{=} h_{10}(x(t_1)) - h_{10}(y(t_1)) + \sigma \sum_{i\in\mathcal{I}} [h_{1i}(x(t_1))$$
$$- h_{1i}(y(t_1))] + \int\limits_T \{h_0(x(t), t) - h_0(y(t), t) + \sigma \sum_{i\in\mathcal{I}} [h_i(x(t), t) - h_i(y(t), t)]\}dt$$
$$\geq \langle\langle \nabla F_\sigma(y(\cdot)), x(\cdot) - y(\cdot)\rangle\rangle + \frac{\rho_1}{2}||x(t_1) - y(t_1)||^2 + \int\limits_T \frac{\rho(t)}{2}||x(t) - y(t)||^2 dt \tag{36}$$

where $\rho_1 > 0$, and $\rho(t) > 0 \ \overset{\circ}{\forall} t \in T$, $\rho(\cdot) \in L_2(T)$. Therefore, with the help of (32), we derive

$$-\delta_s \leq \Phi_k(x^k, u^k) - \Phi_k(x^{k+1}, u^{k+1})$$
$$\overset{\triangle}{=} G_k(x^k, u^k) - G_k(x^{k+1}, u^{k+1}) + \langle\langle \nabla F_k(x^k(\cdot)), x^{k+1} - x^k\rangle\rangle$$

and, on account of (36), we obtain

$$\frac{\rho_1}{2}||x^{k+1}(t_1) - x^k(t_1)||^2 + \frac{1}{2}\int_T \rho(t)||x^{k+1}(t) - x^k(t)||^2 dt$$

$$\leq G_k(x^k, u^k) - G_k(x^{k+1}, u^{k+1}) + F_k(x^{k+1}) - F_k(x^k) + \delta_k$$
$$= J_k(x^k(\cdot), u^k(\cdot)) - J_k(x^{k+1}(\cdot), u^{k+1}(\cdot)) + \delta_k.$$

Hence, in virtue of the relations (26),(33) and (34)(a), we finally obtain

$$\left.\begin{array}{l} (a) : \lim_{k\uparrow\infty} ||x^{k+1}(t_1) - x^k(t_1)|| = 0; \\ (b) : \lim_{k\uparrow\infty} \int_T \rho(t)||x^{k+1}(t) - x^k(t)||^2 dt = 0. \end{array}\right\} \tag{37}$$

It means that the sequence of the states $\{x^k(\cdot)\}$ $x^k(t) = x(t, u^k)$, $t \in T$, $u^k(\cdot) \in \mathcal{U}$, $k = 0, 1, 2, \ldots$ produced by the LSS1, is such that the terminal states $\{x^k(t_1)\}$ form the sequence of Cauchy in $\mathbb{R}^n$ and therefore there exists $x_* \in \mathbb{R}^n$ such that

$$\lim_{k\uparrow\infty} x^k(t_1) = x_*. \quad \text{(in } \mathbb{R}^n). \tag{38}$$

Suppose now that one can find a number $c > 0$ such that

$$\rho(t) \geq c > 0 \quad \overset{\circ}{\forall} t \in T.$$

Then from the condition (37)(b) it follows that

$$\lim_{k\uparrow\infty} \int_T ||x^{k+1}(t) - x^k(t)||^2 dt = 0. \tag{37'}$$

In other words, the sequence $\{x^k(\cdot)\}$ of the states, $x^k(t) = x(t, u^k)$, $u^k(\cdot) \in \mathcal{U}$, $k = 0, 1, 2, \ldots$, produced by the LSS1 turns out to be fundamental (of Cauchy) in $L_2^n(T)$. Therefore, there exists a function $x_*(\cdot) \in L_2^n(T)$ which is a limit function for $\{x^k(\cdot)\}$:

$$\lim_{k\uparrow\infty} x^k(\cdot) = x_*(\cdot) \text{ in } L_2^n(T) \tag{38'}$$

Suppose now that the next assumption holds

$$(\mathcal{A}_\sigma): \quad \exists \sigma_{up} \in \mathbb{R} : \sigma_{up} \geq \sigma_k, \ k = 0, 1, 2, \ldots \tag{39}$$

Then, with the help of (31') and (39), we derive that

$$\exists \sigma_* > 0 : \sigma_* = \lim_{k\uparrow\infty} \sigma_k. \tag{40}$$

One can see, in addition, that the corresponding sequence $\{u^k(\cdot)\} \subset \mathcal{U}$ (i.e. $u^k(t) \in U \overset{\circ}{\forall} t \in T$, where U is a compact set in $\mathbb{R}^r$) is bounded in $L_2^r(T)$, for example. Therefore, it is clear that at the accuracy of a subsequence $\{u^{k_p}, \; p = 1, 2, \ldots\}$, the sequence $\{u^k(\cdot)\}$ turns out to be weekly converging in $L_2^r(T)$ [7].

Hence, we get the number sequences $\{\delta_k\}$, $\{\sigma_k\}$, $\{J_k(u^k)\}$, $\{\Phi_k(x^{k+1}(\cdot), u^{k+1}(\cdot) - V_k\}$, $\{\Phi_k(x^{k+1}(\cdot), u^{k+1}(\cdot))\}$, $\{V_k\}$ which are proved to be converging (or can be proved as was done in [32–34]).

In addition, we have the sequences of functions $\{x^k(\cdot)\}$ and $\{u^k(\cdot)\}$, $x^k(t) = x(t, u^k)$, $t \in T$, $u^k(\cdot) \in \mathcal{U}$, which also converge, but each in its own sense.

Nevertheless, it can be readily seen that all depend on possibility of solving the linearized problem $(\mathcal{P}_k L_k)$–(27) or, more precisely, on the possibility of producing the next control iterate $u^{k+1}(\cdot) \in \mathcal{U}$, using the already known process $(x^k(\cdot), u^k(\cdot))$, $x^k(t) = x(t, u^k)$, $t \in T$, $u^k(\cdot) \in \mathcal{U}$. In order to realize what we need for this, let us recall the precise form of the cost functional

$$\Phi_k(x(\cdot), u(\cdot)) = G_k(x(\cdot), u(\cdot)) - \langle\!\langle \nabla F_k(x^k(\cdot)), x(\cdot) \rangle\!\rangle$$

of the linearized problem $(\mathcal{P}_k L_k)$–(27) .

Using (18)–(20), we derive the next presentation

$$\Phi_k(x(\cdot), u(\cdot)) = [G_0(x(\cdot), u(\cdot)) + \sigma_k G_\pi(x(\cdot), u(\cdot))]$$
$$- \langle\!\langle \nabla F_0(x^k(\cdot)) + \sigma_k \nabla F_\pi(x^k(\cdot)), x(\cdot) \rangle\!\rangle$$

$$= g_{10}(x(t_1)) + \int_T g_0(x(t), u(t), t)dt + \sigma_k \max\left\{ \sum_{j \in I}\left[h_{1j}(x(t_1)) + \int_T h_j(x(t), t)dt \right]; \right.$$

$$\left.\left[g_{1i}(x(t_1)) + \int_T g_i(x(t), u(t), t)dt + \sum_{\substack{p \in I \\ p \neq i}}\left(h_{1p}(x(t_1)) + \int_T h_p(x(t), t)dt \right) \right], i \in I\right\} \quad (41)$$

$$- \langle \nabla h_{10}(x^k(t_1)) + \sigma_k \sum_{i \in I} \nabla h_{1i}(x^k(t_1)), x(t_1) \rangle$$

$$- \int_T \langle \nabla h_0(x^k(t), t) + \sigma_k \sum_{i \in I} \nabla h_i(x^k(t), t), x(t) \rangle dt.$$

Thus, we have to minimize this state-convex functional $\Phi_k(x(\cdot), u(\cdot))$ along the state-linear control system (1) of ODEs with $u(\cdot) \in \mathcal{U}$ defined in (2). To this end, we are going to use the famous Pontryaguin's Principle which for this OC problem seems to be rather relevant, promising and effective. This problem is the object of future investigations.

7   Conclusion

In the present paper, a difficult nonconvex optimal control (OC) Problem $(\mathcal{P})$–(7) with the goal functional and the inequality constraints given by the functionals of Bolza was considered. More precisely, the terminal parts and the integrands of the functionals are state-DC functions, while the control system (1) is state-linear.

Along with Problem $(\mathcal{P})$ we addressed the penalized Problem $(\mathcal{P}_\sigma)$–(9)–(10) which was proven to be also state-DC.

Furthermore, with the help of linearization of the nonconvex part of the cost functional $J_\sigma(\cdot)$ of Problem $(\mathcal{P}_\sigma)$, we obtained the linearized Problem $(\mathcal{P}_\sigma L(y))$. Then the idea of Local Search Scheme consists in a consecutive solution of the linearized Problem $(\mathcal{P}_k L(x^k))$.

After we studied the first convergence properties of the sequence $\{(x^k(\cdot), u^k(\cdot))\}$ produced by the Local Search Scheme, we precised the principal OC problem which one has to solve for constructing the sequence $\{u^k(\cdot)\}$.

Hence, the direction for future research is now defined.

Acknowledgement. The research was funded by the Ministry of Education and Science of the Russian Federation within the framework of the project "Theoretical foundations, methods and high-performance algorithms for continuous and discrete optimization to support interdisciplinary research" (No. of state registration: 121041300065-9).

References

1. Pontryagin, L.S., Boltyanskii, V.G., Gamkrelidze, R.V., Mishchenko, E.F.: The Mathematical Theory of Optimal Processes. Interscience, New York (1976)
2. Marchuk, G.I.: Mathematical Modeling in the Environmental Problem. Nauka, Moscow (1982). (in Russian)
3. Chernousko, F.L., Ananievski, I.M., Reshmin, S.A.: Control of Nonlinear Dynamical Systems: Methods and Applications. Springer, Heidelberg (2008). https://doi.org/10.1007/978-3-540-70784-4
4. Chernousko, F.L., Banichuk, N.V.: Variational Problems of Mechanics and Control. Numerical Methods. Nauka, Moscow (1973). (in Russian)
5. Kurzhanski, A.B., Varaiya, P.: Dynamics and Control of Trajectory Tubes: Theory and Computation. Birkhauser, Boston (2014)
6. Kurzhanski, A.B.: Control and Observation Under Conditions of Uncertainty. Nauka, Moscow (1977). (in Russian)
7. Vasil'ev, F.P.: Optimization Methods. Factorial Press, Moscow (2002). (in Russian)
8. Fedorenko, R.P.: Approximate Solution of Optimal Control Problems. Nauka, Moscow (1978). (in Russian)
9. Gabasov, R., Kirillova, F.M.: Maximum's Principle in the Optimal Control Theory. Nauka i Technika, Minsk (1974). (in Russian)
10. Vasiliev, O.V.: Optimization Methods. Word Federation Publishing Company, Atlanta (1996)
11. Srochko, V.A.: Iterative Solution of Optimal Control Problems. Fizmatlit, Moscow (2000). (in Russian)
12. Hiriart-Urruty, J.-B., Lemarechal, C.: Convex Analysis and Minimization Algorithms. Springer, Heidelberg (1993). https://doi.org/10.1007/978-3-662-02796-7
13. Hiriart-Urruty, J.-B.: Generalized differentiability, duality and optimization for problem dealing with difference of convex functions. In: Ponstein, J. (ed.) Convexity and Duality in Optimization. Lecture Notes in Economics and Mathematical Systems, vol. 256, pp. 37–69. Springer, Heidelberg (1985). https://doi.org/10.1007/978-3-642-45610-7_3
14. Horst, R., Tuy, H.: Global Optimization: Deterministic Approaches. Springer, Heidelberg (1996). https://doi.org/10.1007/978-3-662-02598-7

15. Rockafellar, R.: Convex Analysis. Princeton University Press, Princeton (1970)
16. Eremin, I.: The penalty method in convex programming. Soviet Math. Dokl. **8**, 459–462 (1966)
17. Zangwill, W.: Non-linear programming via penalty functions. Manage. Sci. **13**(5), 344–358 (1967)
18. Zaslavski, A.J.: Exact penalty property in optimization with mixed constraints via variational analysis. SIAM J. Optim. **23**(1), 170–187 (2013)
19. Burke, J.: An exact penalization viewpoint of constrained optimization. SIAM J. Control Optim. **29**(4), 968–998 (1991)
20. Dolgopolik, M.V.: A unifying theory of exactness of linear penalty functions. Optimization **65**(6), 1167–1202 (2016)
21. Dolgopolik, M.V., Fominyh, A.V.: Exact penalty functions for optimal control problems I: Main theorem and free-endpoint problems. Optim. Control Appl. Meth. **40**, 1018–1044 (2019)
22. Izmailov, A.F., Solodov, M.V.: Newton-Type Methods for Optimization and Variational Problems. SSORFE. Springer, Cham (2014). https://doi.org/10.1007/978-3-319-04247-3
23. Strekalovsky, A.S.: On solving optimization problems with hidden nonconvex structures. In: Themistocles, M., Floudas, C.A., Butenko, S. (eds.) Optimization in Science and Engineering, pp. 465–502. Springer, New York (2014). https://doi.org/10.1007/978-3-642-45610-7_3
24. Strekalovsky, A.S.: Global optimality conditions for optimal control problems with functions of A.D. Alexandrov. J. Optim. Theory Appl. **159**, 297–321 (2013)
25. Strekalovsky, A.S.: Maximizing a state convex Lagrange functional in optimal control. Autom. Remote Control **73**(6), 949–961 (2012)
26. Strekalovsky, A.S.: Elements of Nonconvex Optimization. Nauka, Novosibirsk (2003). (in Russian)
27. Strekalovsky, A.S.: Global optimality conditions and exact penalization. Optim. Lett. **13**(2), 597–615 (2019). https://doi.org/10.1007/s11590-017-1214-x
28. Strekalovsky, A.S.: Global optimality conditions in nonconvex optimization. J. Optim. Theory Appl. **173**(3), 770–792 (2017)
29. Strekalovsky, A.S.: New global optimality conditions in a problem with DC constraints. Trudy Inst. Mat. Mekh. UrO RAN **25**, 245–261 (2019). (in Russian)
30. Strekalovsky, A.S., Yanulevich, M.V.: On global search in nonconvex optimal control problems. J. Global Optim. **65**(1), 119–135 (2016). https://doi.org/10.1007/s10898-015-0321-4
31. Strekalovsky, A.S., Yanulevich, M.V.: Global search in the optimal control problem with a terminal objective functional represented as a difference of two convex functions. Comput. Math. Math. Phys. **48**(7), 1119–1132 (2008)
32. Strekalovsky, A.S.: Local search for nonsmooth DC optimization with DC equality and inequality constraints. In: Bagirov, A.M., Gaudioso, M., Karmitsa, N., Mäkelä, M.M., Taheri, S. (eds.) Numerical Nonsmooth Optimization. SSORFE, pp. 229–261. Springer, Cham (2020). https://doi.org/10.1007/978-3-030-34910-3_7
33. Strekalovsky, A.S., Minarchenko, I.M.: A local search method for optimization problem with DC inequality constraints. Appl. Math. Model. **58**, 229–244 (2018)
34. Strekalovsky, A.S.: On local search in DC optimization problems. Appl. Math. Comput. **255**, 73–83 (2015)

Combinatorial Optimization

Serving Rides of Equal Importance for Time-Limited Dial-a-Ride

Barbara M. Anthony[1] , Ananya D. Christman[2] , Christine Chung[3] ,
and David Yuen[4]

[1] Southwestern University, Georgetown, TX, USA
anthonyb@southwestern.edu
[2] Middlebury College, Middlebury, VT, USA
achristman@middlebury.edu
[3] Connecticut College, New London, CT, USA
cchung@conncoll.edu
[4] 1507 Punawainui Street, Kapolei, HI, USA

Abstract. We consider a variant of the offline Dial-a-Ride problem with
a single server where each request has a source, destination, and a prize
earned for serving it. The goal for the server is to serve requests within
a given time limit so as to maximize the total prize money. We consider the variant where prize amounts are uniform which is equivalent
to maximizing the number of requests served. This setting is applicable
when all rides may have equal importance such as paratransit services.
We first prove that no polynomial-time algorithm can be guaranteed to
serve the optimal number of requests, even when the time limit for the
algorithm is augmented by any constant factor $c \geq 1$. We also show that
if $\lambda = t_{max}/t_{min}$, where t_{max} and t_{min} denote the largest and smallest
edge weights in the graph, the approximation ratio for a reasonable class
of algorithms for this problem is unbounded, unless λ is bounded. We
then show that the SEGMENTED BEST PATH (SBP) algorithm from [8] is
a 4-approximation. We then present our main result, an algorithm, k-
Sequence, that repeatedly serves the fastest set of k remaining requests,
and provide upper and lower bounds on its performance. We show k-
Sequence has approximation ratio at most $2 + \lceil \lambda \rceil / k$ and at least $1 + \lambda/k$
and that $1 + \lambda/k$ is tight when $1 + \lambda/k \geq k$. Thus, for the case of $k = 1$,
i.e., when the algorithm repeatedly serves the quickest request, it has
approximation ratio $1 + \lambda$, which is tight for all λ. We also show that
even as k grows beyond the size of λ, the ratio never improves below $9/7$.

1 Introduction

In the Dial-a-Ride Problem (DARP) one or more servers must schedule a collection of pickup and delivery requests, or rides. Each request specifies the pickup
location (or *source*) and the delivery location (or *destination*). In some DARP
variants the requests may be restricted so that they must be served within a
specified time window, they may have weights associated with them, or details

© Springer Nature Switzerland AG 2021
P. Pardalos et al. (Eds.): MOTOR 2021, LNCS 12755, pp. 35–50, 2021.
https://doi.org/10.1007/978-3-030-77876-7_3

about them may be known only when they become available. For most variations the goal is to find a schedule that will allow the server(s) to serve requests within the constraints, while meeting a specified objective. Much of the motivation for DARP arises from the numerous practical applications of the transport of both people and goods, including delivery services, ambulances, ride-sharing services, and paratransit services. For a comprehensive overview of DARP please refer to the surveys *The dial-a-ride problem: models and algorithms* [10] and *Typology and literature review for dial-a-ride problems* [14].

In this work we study offline DARP on weighted graphs with a single server where each request has a source, destination, and prize amount. The prize amount may represent the importance of serving the request in settings such as courier services. In more time-sensitive settings such as ambulance routing, the prize may represent the urgency of a request. In profit-based settings, such as taxi and ride-sharing services, a request's prize amount may represent the revenue earned from serving the request. The server has a specified deadline after which no more requests may be served, and the goal is to find a schedule of requests to serve within the deadline that maximizes the total prize money. We study the variant where prizes are uniform so the goal is equivalent to maximizing the number of requests served within the deadline. This variant is useful for settings where all requests have equal importance, such as nonprofit transportation services for elderly and disabled passengers and courier services where deliveries are not prioritized. For the remainder of this paper, we will refer to this time-limited variant with the objective of maximizing the number requests served as TDARP.

One related problem is the Prize Collecting Traveling Salesperson Problem (PCTSP) where the server earns a prize for every location it visits and a penalty for every location it misses, and the goal is to collect a specified amount of prize money while minimizing travel costs and penalties. PCTSP was introduced by Balas [4] but the first approximation algorithm, with ratio 2.5, was given by Bienstock et al. [5]. Later, Goemans and Williamson [12] developed a primal-dual algorithm to obtain a 2-approximation. Building off of the work in [12], Archer et al. [3] improved the ratio to $2 - \epsilon$, a significant result as the barrier of 2 was thought to be unbreakable. More recently, Paul et al. [15,16] studied a special case of our problem; namely, the *budgeted* variant of PCTSP where the goal is to find a tour that maximizes the number of nodes visited given a bound on the cost of the tour. They present a 2-approximation when the graph is not required to be complete and the tour may visit nodes more than once.

Blum et al. [6] presented the first constant-factor approximation algorithm for a special case of the problem we consider; namely, the *Orienteering Problem* where the input is a weighted graph with rewards on nodes and the goal is to find a path that starts at a specified origin and maximizes the total reward collected, subject to a limit on the path length. Our problem is a generalization of this problem – while the Orienteering Problem has as input a set of points/cities to visit, our problem has a set of requests, each with two distinct points to be visited: a source and a destination.

To our knowledge, despite its relevance to modern-day transportation systems, aside from the work in [1] the request-maximizing time-limited version of DARP we investigate in this paper has not been previously studied in the offline setting. In [1] we presented a 3/2-approximation algorithm, TWOCHAIN, for the more restrictive uniform edge weight version of the problem.

Our Results. In Sect. 2 we begin by establishing some impossibility results. In Sect. 2.1 we prove that no polynomial-time algorithm can be guaranteed to serve as many requests as the optimal schedule, even when the time limit T for the algorithm is augmented by c for any constant $c \geq 1$. We also show that if $\lambda = t_{max}/t_{min}$, where t_{max} and t_{min} denote the largest and smallest edge weights in the graph, the approximation ratio for a reasonable class of TDARP algorithms is unbounded, unless λ is bounded. In Sect. 2.2 we revisit the SEGMENTED BEST PATH (SBP) algorithm that was proposed in [8] for TDARP in the online setting. We show that SBP in the offline setting is a 4-approximation, and we also show this is a tight bound.

In Sect. 3 we present k-*Sequence* (k-SEQ), a family of algorithms parameterized by k, for TDARP on weighted graphs. Informally, the k-SEQ algorithm repeatedly serves the fastest set of k remaining requests where a determination of *fastest* is made by considering both the time to serve the requests and any travel time necessary to serve those requests. Naturally, k is a positive integer. Our approximation ratio depends on λ, a property of the graph, similar to the graph-property dependencies in [7,11]. In many real-world settings, λ may be viewed as a constant [9,13,17]. We prove that k-SEQ has approximation ratio $2 + \lceil \lambda \rceil / k$. In Sect. 3.1 we show that when $1 + \lambda/k \geq k$, the approximation ratio for k-SEQ improves to $1 + \lambda/k$. Thus, for the case of $k = 1$, i.e., the polynomial-time algorithm which repeatedly serves the quickest request, the approximation ratio is $1 + \lambda$ and this is tight. Finally, in Sect. 4, we show that k-SEQ has approximation ratio at least $1 + \lambda/k$, which matches the upper bound for when $1 + \lambda/k \geq k$. We also show that the algorithm has a lower bound of 9/7 for $k > \lambda$.

We summarize our results on the approximation ratio for k-SEQ, for particular λ and k, as follows.

1. When $\lambda \geq k(k-1)$, or equivalently $1 + \lambda/k \geq k$, the ratio is $1 + \lambda/k$, and this is tight. So when $k = 1$ (for any λ), the ratio is $1 + \lambda$, and this is tight.
2. When $k \leq \lambda < k(k-1)$, then the ratio is in the interval $[1 + \lambda/k, 2 + \lceil \lambda \rceil / k]$.
3. When $\lambda < k$, the ratio is in the interval $[\max\{9/7, 1 + \lambda/k\}, 2 + \lceil \lambda \rceil / k]$.

2 Preliminaries

We formally define TDARP as follows. The input is an undirected complete graph $G = (V, E)$ where V is the set of vertices (or nodes) and $E = \{(u, v) : u, v \in V, u \neq v\}$ is the set of edges. For every edge $(u, v) \in E$, there is a distance

$dist(u, v) > 0$, which represents the amount of time it takes to traverse (u, v).[1] We also note that the input can be regarded as a metric space if the weights on the edges are expected to satisfy the triangle-inequality. Indeed, all of our results apply to both complete graphs as well as metric spaces.

One node in the graph, o, is designated as the origin and is where the server is initially located (i.e. at time 0). The input also includes a time limit T and a set of requests, S, that is issued to the server. Each request in S can be considered as simply a pair (s, d) where s is the source node or starting point of the request, and d is the destination node. The output is a schedule of requests, i.e. a set of requests and the time at which to serve each. To serve a request, the server must move from its current location x to s, then from s to d, and remain at d until it is ready to move again. The total time for serving the request is $dist(x, s) + dist(s, d)$, where $dist(x, s) = 0$ if $x = s$.

Every movement of the server can be characterized as either an *empty drive* which is simply a repositioning move along an edge but not serving a request, or a *service drive* in which a request is being served while the server moves. We let *driveTime*(C) denote the minimum total time for the server to travel from its current location and serve the collection of requests $C \subseteq S$, where the minimum is taken over all permutations of the requests in C.

We use $|\text{ALG}(I)|$ to denote the number of requests served by an algorithm ALG on an instance I of TDARP and we drop the I when the instance is clear from context. Similarly we use $|\text{OPT}(I)|$ for the number of requests served by the optimal solution OPT on instance I.

2.1   Impossibility Results

In this section we present two impossibility results. The first is an inapproximability result. The second demonstrates that any algorithm of a class of algorithms that k-SEQ belongs to will have unbounded approximation ratio, unless λ is bounded. Accordingly, this provides some justification for the presence of the parameter λ in our main results.

2.1.1   c-Time Inapproximability

We prove that, unless P = NP, no polynomial-time algorithm can be guaranteed to serve as many requests as the optimal schedule, even when the time limit T for the algorithm is augmented by any constant factor. Let $I = (G, S, T)$ denote an instance of TDARP, where G is the input graph, S is the set of requests, and T is the time limit. We define ALG to be a ρ-*time-approximation* if ALG serves at least as many requests as OPT on the instance $(G, S, \rho T)$. The proof idea is to show that a polynomial-time c-*time-approximation* to TDARP yields a polynomial-time decider for the directed Hamiltonian path problem. Please see the full version of the paper (preprint available at [2]) for the proof.

---

[1] We note that any simple, undirected, connected, weighted graph is allowed as input, with the simple pre-processing step of adding an edge wherever one is not present whose distance is the length of the shortest path between its two endpoints.

Theorem 1. *If $P \neq NP$, then there is no polynomial-time c-time-approximation to TDARP for any constant $c \geq 1$.*

2.1.2   Inductive Stateless Greedy Algorithms

Recall that $\lambda = t_{max}/t_{min}$, where t_{max} and t_{min} denote the largest and smallest edge weights in the graph, respectively. We now show that if a deterministic algorithm satisfies certain properties, then it cannot have a bounded approximation ratio, unless λ is bounded. Consider the following three properties of an algorithm. (Note these are abbreviated summaries of each property; please see the full version of the paper for more detailed definitions and the proof of Theorem 2).

1. *Inductive.* The algorithm chooses paths to take in stages.
2. *Stateless.* In each stage the algorithm does not use state information from a previous stage.
3. *Greedy.* The algorithm makes decisions by optimizing an objective function at each stage, where the function takes as input a set of possible paths to choose from and outputs a chosen path.

Theorem 2. *Let M be a constant and let* ALG *be a deterministic inductive stateless greedy algorithm such that the algorithm considers only candidate paths with at most M edges. If λ is not bounded, then* ALG *has an unbounded approximation ratio.*

2.2   The Segmented Best Path (SBP) Algorithm

Before we present our main results, we will now analyze an algorithm that is based on the previously-studied SEGMENTED BEST PATH (SBP) algorithm from [8], which was proposed for the *online* variant of DARP with non-uniform prize amounts. Specifically, we adapt SBP to apply in our offline setting with uniform prize amounts. Since our problem assumes uniform prizes, we unsurprisingly have a tighter upper bound than the bound of [8], but we show that the lower bound carries over. We note that Theorem 2 does not apply to SBP because there is no constant that bounds the number of edges in the paths considered by SBP in each iteration.

Algorithm 1: SEGMENTED BEST PATH (SBP) Algorithm as adapted from [8]. Input: origin o, time limit $T > 0$, a complete graph G (see footnote 1 in Section 2) with $T \geq 2t_{max}$, and a set of requests S given as source-destination pairs.

1: Let $t_1, t_2, \ldots t_f$ denote time segments of length T/f ending at times
   $T/f, 2T/f, \ldots, T$, respectively, where $f = 2\lfloor T/(2t_{max}) \rfloor$.
2: Let $i = 1$.
3: **while** $i < f$ and there are still unserved requests **do**
4:    At the start of t_i, find the *max-cardinality-sequence*, $\mathcal{R}$.
5:    Move to the source location of the first request in $\mathcal{R}$.
6:    At the start of t_{i+1}, serve the requests in $\mathcal{R}$.
7:    Let $i = i + 2$.
8: **end while**

As described in Algorithm 1, the offline version of SBP starts by splitting the total time T into $f \geq 2$ time segments each of length T/f where f is the largest even integer such that $t_{max} \leq T/f$, which ensures any move, including one serving a request, can be completed entirely in a single segment. At the start of a time segment, the server determines the *max-cardinality-sequence*, $\mathcal{R}$, i.e. the maximum length sequence of requests that can be served within one time segment, and moves to the source of the first request in this set. During the next time segment, it serves the requests in this set. It continues this way, alternating between determining and moving to the source of the first request in $\mathcal{R}$ during one time segment, and serving the requests in $\mathcal{R}$ in the next time segment.[2]

Finding the max-cardinality-sequence may require enumeration of all possible sequences of unserved requests which takes time exponential in the number of unserved requests. However, in many real world settings, the number of requests will be small relative to the input size and in settings where T/f is small, the runtime is further minimized. Therefore it should be feasible to execute the algorithm efficiently in many real world settings.

Let $\text{OPT}(S, T, o)$ and $\text{SBP}(S, T, o)$ denote the schedules returned by OPT and SBP, respectively, on the instance (S, T, o).

Theorem 3. SBP *is a 4-approximation i.e.,* $|\text{OPT}(S, T, o)| \leq 4|\text{SBP}(S, T, o)|$ *for any instance* (S, T, o) *of TDARP, and this is tight.*

Proof. We first note that the lower bound instance of [8], in which SBP earns total prize money of no more than $\text{OPT}/4$ in the online setting, also applies to this offline setting with uniform prizes, since in that instance prize amounts are

---

[2] Note that the algorithm need not take a full time segment to move from one set of requests to another, but it is specified this way for convenience of analysis. Excluding this buffer time in the algorithm specification does not improve its approximation ratio since one can construct an instance where each move requires the full time segment.

uniform and no requests are released after time 0. (For the full proof of the lower bound, please see the full version of the paper.)

For the upper bound, consider a schedule OPT2, which is identical to OPT except it is allowed one extra empty drive at the start that does not add to the overall time taken by the algorithm. More formally, if the first move in OPT is from o to some node n_1, then OPT2 may have an additional (non-time-consuming) move at the start such that its first move is from o to some other node n_1' and its second move is from n_1' to n_1. Since OPT2 is allowed one additional empty drive, we know $|\text{OPT}_2(S,T,o)| \geq |\text{OPT}(S,T,o)|$. We claim that $|\text{SBP}(S,T,o)| \geq |\text{OPT}_2(S,T,o)|/4$, which implies that $|\text{SBP}(S,T,o)| \geq |\text{OPT}(S,T,o)|/4$.

We proceed by strong induction on the number of time windows $w = f/2$ where a time window is two consecutive time segments. For the base case let Q and R denote the set of requests served by OPT2 and SBP, respectively, in the first time window and let q and r denote their respective cardinalities. Recall the greedy nature of SBP which serves requests during every other time segment. If $q = 1$, since $f \geq 2$, SBP can serve the one request in Q within the two time segments so $r = q$. If $q > 1$, then if q is even, $r \geq q/2$ since splitting the window in half leaves at least half of the requests in one of the two time segments, and if q is odd then $r \geq (q-1)/2$.

So if $w = 1$, in all three cases, we have $r \geq q/4$, completing the base case.

For the inductive step, let P denote the path traversed by OPT2, let $p = |\text{OPT}_2| \geq |\text{OPT}|$ denote the number of requests served in P, and let u denote the first node OPT2 visits after the end of the first time window. Consider the subpath, P', of P that starts at u. Since P may contain a request that straddles the first two windows, P' contains at least $p - (q+1)$ requests. Let s_1 denote the last node SBP visits before the start of the second time window. After the first window, SBP is left with a smaller instance of the problem $(S_{new}, T_{new}, o_{new})$ where $S_{new} = S - R$, $T_{new} = T - T/f$, and $o_{new} = s_1$. So P' contains at least $p - (q+1) - r$ requests from this smaller instance and OPT2 on $(S_{new}, T_{new}, o_{new})$ can move from o_{new} to u and serve these requests. By induction, on the smaller instance SBP will serve at least $(p - q - 1 - r)/4$. Thus

$$|\text{SBP}(S,T,o)| = r + |\text{SBP}(S_{new}, T_{new}, o_{new})| \geq r + (p - q - 1 - r)/4$$
$$\geq p/4 + (-q - 1 + 3r)/4. \quad (1)$$

There are three cases for q and r.

1. Case: $q \geq 5$. Then since $r \geq (q-1)/2$, from (1) we have: $|\text{SBP}(S,T,o)| \geq p/4 + (-q - 1 + 3(q-1)/2)/4 \geq p/4 + (q/2 - 5/2)/4 \geq p/4$.
2. Case: $q \leq 4$ and $r \geq 2$. From (1) we have: $|\text{SBP}(S,T,o)| \geq p/4 + (-4 - 1 + 6)/4 \geq p/4$.
3. Case: $q \leq 4$ and $r \leq 1$. If $r = 0$, then $|\text{SBP}(S,T,o)| = |\text{OPT}_2(S,T,o)| = 0$, so the theorem is trivially true, therefore, we assume $r = 1$. We first show by contradiction that every time window in OPT2's schedule has fewer than 4 requests that end in that window. Suppose there is a window i in OPT2's schedule that has 4 or more requests that end in window i. Then there are at

least 3 requests that start and end in window i. This implies that at least one time segment of window i contains at least 2 requests which, by the greediness of SBP, implies $r \geq 2$, which is a contradiction since we are in the case where $r = 1$. Let w' denote the number of windows in which OPT_2 serves at least 1 request. We have $|\mathrm{OPT}_2(S, T, o)| < 4w'$ and $|\mathrm{SBP}(S, T, o)| \geq \min(w, |S|) \geq w'$, so $|\mathrm{SBP}(S, T, o)| \geq |\mathrm{OPT}_2(S, T, o)|/4$. ∎

3 k-Sequence Algorithm and Upper Bound

We now present k-Sequence (k-SEQ), our family of algorithms parameterized by k, for TDARP (see Algorithm 2). For any fixed k, the algorithm repeatedly serves the fastest set of k remaining requests where a determination of *fastest* is made by considering both the time to serve the requests and any travel time necessary to serve those requests. If there are fewer than k requests remaining, the algorithm exhaustively determines how to serve all remaining requests optimally. If the remaining time is insufficient to serve any collection of k requests, the algorithm likewise serves the largest set of requests that can be served within the remaining time. We suggest that when using the algorithm in practice, k can be set as a small constant. The algorithm will run in time $O(|S|^{k+1})$ where S is the set of requests, as each of the at most $|S|$ iterations may require time $O(|S|^k)$.

Algorithm 2: Algorithm k-Sequence (k-SEQ). Input: origin o, time limit $T > 0$, a complete graph G (see footnote 1 in Section 2), and a set of requests S given as source-destination pairs.

1: Set $t := T$.
2: **while** there are at least k unserved requests remaining **do**
3:     Let C be the collection of k requests with fastest $driveTime(C)$, where $driveTime(C)$ denotes the minimum total time to serve C.
4:     **if** $t \geq driveTime(C)$ **then**
5:         Serve C, update $t := t - driveTime(C)$, and update $S = S - C$.
6:     **else**
7:         Exit **while** loop.
8:     **end if**
9: **end while**
10: Find the largest $x \leq k - 1$ s.t. $driveTime(C') \leq t$ for some C' with $|C'| = x$.
11: If $|C'| \neq 0$, serve C'.

Theorem 4. k-SEQ *is a* $(2 + \lceil \lambda \rceil / k)$-*approximation for TDARP.*

Proof. First, note that without loss of generality, we may assume that it is possible to serve k requests during the allotted time T. If there was insufficient time to serve any collection of k requests, then k-SEQ will serve the largest set of requests that can be served within time T, which is thus optimal. If there are fewer than k requests available, k-SEQ will serve all available requests, again achieving an optimal solution.

We now proceed with a proof by induction on an instance in which at least k requests can be served in time T. For the base case, we have $\lfloor T/t_{min} \rfloor = 0$, so $T < t_{min}$ and thus k-SEQ and OPT both serve 0 requests, so we are done. For the inductive case, let $\lfloor T/t_{min} \rfloor = d \geq 1$. Suppose by induction that the theorem is true whenever $\lfloor T/t_{min} \rfloor < d$.

Let $s = o$ be the start location. k-SEQ starts by serving exactly k requests in time T_1, ending at a location we refer to as s_1. Let OPT serve m requests in total. Note that since T_1 is, by construction of k-SEQ, the time required to serve the fastest k requests, OPT serves at most k requests during the initial T_1 time.

Let y' be the location on the OPT path at time T_1, noting that y' need not be at a node. Then define y to be y' if y' is a node, or the next node on OPT's path after y' otherwise.

To develop our inductive argument, we will now create a new instance with new start location s_1, time $T_{new} = T - T_1$, and the k requests that were served by k-SEQ removed from S, leaving us with S_{new}.

We consider P, a feasible path for this new instance (see Fig. 1). This path P starts at s_1, proceeds to y, and then traverses as much as it can of the remainder of the original OPT path from y in the remaining time, that is $T - T_1 - dist(s_1, y) \geq T - T_1 - t_{max}$.[3] Since such a path P is feasible, OPT's path must contain at least as many requests as P. Observe that the segment of the OPT path that P uses from y onward has a distance of at most $T - T_1$.

Since P has time at least $T - T_1 - t_{max}$ left when at y, then P misses at most $(T - T_1) - (T - T_1 - t_{max}) = t_{max}$ time of the tail of the original OPT path in addition to missing the initial T_1 time of the head of the original OPT path. Thus, P misses at most k requests from the head of the original OPT path and at most $\lceil \lambda \rceil$ from the tail of the original OPT path, ensuring that P serves at least $m - k - \lceil \lambda \rceil$ requests from the original instance. Since the new instance had k requests from the original instance removed, we can now say that P serves at least $m - 2k - \lceil \lambda \rceil$ requests from the new instance. Naturally, OPT must also serve at least $m - 2k - \lceil \lambda \rceil$ requests on the new instance.

Note t_{min} and t_{max}, and therefore λ, remain the same in the new instance. The allotted time for the new instance is $T_{new} = T - T_1 \leq T - t_{min}$, giving $\lfloor (T - T_1)/t_{min} \rfloor \leq \lfloor (T - t_{min})/t_{min} \rfloor = \lfloor T/t_{min} \rfloor - 1$. Hence by induction, the theorem is true for this new instance. In other words, the number of requests served by k-SEQ on the new instance is at least $k/(2k + \lceil \lambda \rceil)$ times the number of requests served by OPT on the new instance. Thus, $|k\text{-SEQ}(S, T, o)| = k + |k\text{-SEQ}(S_{new}, T_{new}, s_1)| \geq k + k/(2k + \lceil \lambda \rceil)(m - 2k - \lceil \lambda \rceil) \geq k + km/(2k + \lceil \lambda \rceil) + k(-2k - \lceil \lambda \rceil)/(2k + \lceil \lambda \rceil) = km/(2k + \lceil \lambda \rceil)$, completing the induction.   ∎

---

[3] Note that when the graph is complete, t_{max} (t_{min}) is the maximum (minimum) distance over all pairs of nodes. Otherwise, using the pre-processing described in the footnote 1 in Sect. 2, we have that the distance between any two non-adjacent nodes is the shortest distance between those nodes, and t_{max} (t_{min}) is the maximum (minimum) distance over all of these distances.

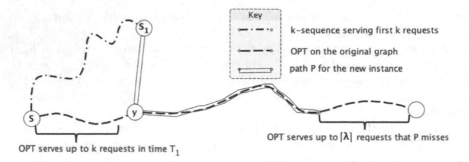

Fig. 1. An illustration of the paths taken by OPT and k-SEQ in Theorem 4. T_1 is the time needed for k-SEQ to serve its initial group of k requests, ending at s_1. The first node on the path of OPT after time T_1 is y. A feasible path P starting at time T_1 is from s_1 to y and then proceeds to the right. (It is possible for s_1 and y to be collocated.)

3.1   k-Sequence Upper Bound for Large λ

We will now show that for sufficiently large constant λ, the k-SEQ algorithm is a $(1 + \lambda/k)$-approximation, a better ratio than the result obtained in Theorem 4 for sufficiently large λ. However, Theorem 4 remains better when $2 + \lambda/k < k$. We note that when $1 + \lambda/k \geq k$, this upper bound of $(1 + \lambda/k)$ matches the lower bound which will be discussed in Sect. 4.

Theorem 5. *For any instance I of TDARP,*

$$|\text{OPT}(I)| \leq \max\{(1 + \lambda/k)|k\text{-SEQ}(I)| + \lambda(k - 1)/k + 1, k|k\text{-SEQ}(I)| + k\}.$$

I.e., when $1 + \lambda/k \geq k$, we have $|\text{OPT}(I)| \leq (1 + \lambda/k)|k\text{-SEQ}(I)| + \max\{\lambda(k - 1)/k + 1, k\}$, so k-SEQ is a $(1 + \lambda/k)$-approximation in this case.[4]

Proof. Let $m = |\text{OPT}|$, and $n = |k\text{-SEQ}|$. Suppose that $m < kn + k$. Then $|\text{OPT}| < kn + k = k|k\text{-SEQ}| + k$, giving our desired result. Thus, for the remainder of the proof, we assume that $m \geq kn + k$, and proceed to show that $|\text{OPT}| < (1 + \lambda/k)|k\text{-SEQ}| + \lambda(k - 1)/k + 1$.

Let the OPT path serve, in order, requests $r_1, r_2, \ldots, r_m$, whose respective service times are $y_1, y_2, \ldots, y_m$. Let x_j be the time taken by an empty drive required between request r_{j-1} and request r_j for $2 \leq j \leq m$, and let x_1 be the time taken to get from the origin to request r_1. Note that any x_j may be 0. Thus, the driveTime taken by the OPT path is:

$$x_1 + y_1 + x_2 + y_2 + \cdots + x_m + y_m \leq T. \tag{2}$$

Let $r_{m+1}, r_{m+2}, \ldots$ be some fixed arbitrary labeling of the requests not served by OPT. Now we consider the k-SEQ algorithm and denote the requests served by k-SEQ as $r_{\alpha_1}, \ldots, r_{\alpha_n}$. Denote by q the number of times that k-SEQ searches

---

[4] Note that if $1 + \lambda/k \geq k$, then $\lambda(k - 1)/k + 1 \geq k$.

for the fastest sequence of k requests to serve; $q = \lceil n/k \rceil$. Then $n = k(q-1) + \rho$ for some $1 \leq \rho \leq k$.

We can find $q - 1$ disjoint subsequences of k consecutive integers from $\{1, \ldots, m\}$, and a qth disjoint subsequence of ρ consecutive integers from $\{1, \ldots, m\}$, i.e. for $i_1 = 1$ we have:

$$i_1, \ldots, i_1 + k - 1, \quad \ldots, \quad i_{q-1}, \ldots, i_{q-1} + k - 1, \quad i_q, \ldots, i_q + \rho - 1 \quad \text{where}$$

$$\{i_j, \ldots, i_j + k - 1\} \cap \{\alpha_1, \ldots, \alpha_{K(j-1)}\} = \emptyset, \text{ for } 2 \leq j \leq q - 1,$$
$$\{i_j, \ldots, i_j + \rho - 1\} \cap \{\alpha_1, \ldots, \alpha_{K(j-1)}\} = \emptyset, \text{ for } j = q.$$

The proof of this existence claim can be found in Lemma 2 of the full version of the paper (preprint available at [2]).

Since both k-SEQ and OPT start at the same origin, the greedy nature of k-SEQ ensures the time k-SEQ spends on its first set of k requests, including any empty drives to those requests, is at most $x_1 + y_1 + \cdots + x_k + y_k$. By the aforementioned Lemma 2 in [2], we know $\{i_j, \ldots, i_j + k - 1\}$ is disjoint from $\{\alpha_1, \ldots, \alpha_{K(j-1)}\}$, so for the jth set of k requests with $2 \leq j \leq q - 1$, the path resulting from going to requests $\{r_{i_j}, \ldots, r_{i_j+k-1}\}$ is available and so by the greedy nature of k-SEQ, the time spent by k-SEQ is at most $t_{max} + y_{i_j} + x_{i_j+1} + \cdots + y_{i_j+k-1}$, since t_{max} is the maximum time needed to get to request r_{i_j}. And finally by the same reasoning the time spent by k-SEQ on the last set of ρ requests, still including any drives to those requests, is at most $t_{max} + y_{i_q} + x_{i_q+1} + \cdots + y_{i_q+\rho-1}$. Thus, the total time spent by k-SEQ is at most

$$T_0 := (q-1)t_{max} + x_1 + y_1 + \cdots + x_k + y_k$$
$$+ \sum_{j=2}^{q-1}(y_{i_j} + x_{i_j+1} + \cdots + y_{i_j+k-1}) + y_{i_q} + x_{i_q+1} + \cdots + y_{i_q+\rho-1}. \quad (3)$$

Now, let r_J be any request served by OPT where J is not any of the indices appearing in the right hand side of (3). If $T_0 + t_{max} + y_J \leq T$, then k-SEQ could have served another request, a contradiction. Therefore, we must have $T_0 + t_{max} + y_J > T$. Combining this observation with (2), we have:

$$(q-1)t_{max} + x_1 + y_1 + \cdots + x_k + y_k + \sum_{j=2}^{q-1}(y_{i_j} + x_{i_j+1} + \cdots + y_{i_j+k-1}) + y_{i_q}$$
$$+ x_{i_q+1} + \cdots + y_{i_q+\rho-1} + t_{max} + y_J > x_1 + y_1 + \cdots + x_m + y_m. \quad (4)$$

By construction, in the left hand side of (4), the x terms all have distinct indices, the y terms all have distinct indices, and these terms also appear on the right hand side.

Let $\mathcal{I}$ be the set of these indices on the left hand side. So $\mathcal{I} \subseteq \{1, \ldots, m\}$. Then subtracting these terms from both sides of the equation yields $qt_{max} > x_J + \sum\{x_j : j \in \{1, \ldots, m\} \backslash \mathcal{I}\} + \sum\{y_j : j \in \{1, \ldots, m\} \backslash \mathcal{I}\}$, so: $qt_{max} > \sum\{y_j : j \in \{1, \ldots, m\} \backslash \mathcal{I}\}$. Since $|\mathcal{I}| = n + 1$, there are $m - n - 1$ of the y_j terms on the right hand side. Since each $y_j \geq t_{min}$, we have $qt_{max} > (m - n - 1)t_{min}$. Thus $q\lambda > m - n - 1$. Because $q = \lceil n/k \rceil$, then $q \leq (n + k - 1)/k$. Then $m \leq (n + k - 1)\lambda/k + n + 1 \leq (1 + \lambda/k)n + \lambda(k - 1)/k + 1$ as desired. ∎

Note that for $k = 1$, k-SEQ is the polynomial time algorithm that repeatedly finds and serves the quickest request. Theorem 5 (in this section) and Theorem 6 (in the next section) yield the following corollary regarding this algorithm.

Corollary 1. 1-SEQ *(i.e., k-SEQ with $k = 1$) has approximation ratio $1 + \lambda$, which is tight for all λ (see Fig. 2 for an illustration of the lower bound).*

4   k-Sequence Lower Bound

We now present lower bounds on k-SEQ; specifically, the lower bound is $1 + \lambda$ for $k = 1$, shrinking to 2 for $k = \lambda$, and shrinking further towards $9/7$ for $k > \lambda$. Note that Theorem 6 matches the upper bound of Theorem 5 when $1 + \lambda/k \geq k$.

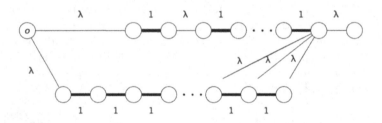

Fig. 2. The instance described in Theorem 6 when $k = 1$. Note that the graph is complete, and any edge (u, v) that is not shown has distance equal to the minimum of λ and the shortest-path distance along the edges shown between u and v. The bold edges represent requests. k-SEQ serves requests along the top path while OPT serves along the bottom.

Theorem 6. *The approximation ratio of k-SEQ for TDARP has lower bound $1 + \lambda/k$.*

Proof. Consider an instance (see Fig. 2 for the case of $k = 1$) where there are two "paths" of interest, both a distance of λ away from the origin. Any edge (u, v) that is not shown has distance equal to the minimum of λ and the shortest-path distance along the edges shown between u and v. There is one long chain of T requests, which is the path chosen by the optimal solution (the bottom path in Fig. 2), and another "broken" chain (the top path in Fig. 2) that consists of k sequential requests at a time with a distance of λ from the end of each chain to any other request in the instance. (Generalizing Fig. 2, for $k > 1$, these requests occur in chains of length k instead of single requests.) Note this graph satisfies the properties of a metric space.

The algorithm may choose to follow the path of the broken chain, serving k requests at a time, but being forced to move a distance of λ between each k-chain. In this manner, for every k requests served, the algorithm requires $k + \lambda$ units of time, while the optimal solution can serve k requests every k time units (after the first λ time units). Thus the approximation ratio of k-SEQ is at least $1 + \lambda/k$. ∎

We now show, however, that as k grows relative to λ the ratio of k-SEQ improves but does not reach (or go below) $9/7$.

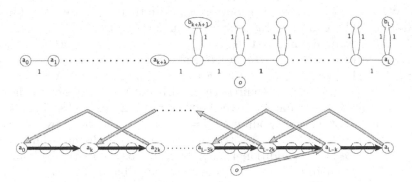

Fig. 3. Top: An instance where OPT serves no fewer than $9/7$ the number of requests served by k-SEQ. Bottom: A depiction of the top instance illustrating the path taken by k-SEQ. Bold edges indicate requests served. k-SEQ starts at o, serves the k requests from a_{L-k} to a_L, and then spends time moving to a_{L-2k} to serve the next collection of k requests, continuing similarly until the time limit. In both figures the graph is complete but only relevant edges are shown.

Theorem 7. *The approximation ratio of k-SEQ has lower bound no better than $9/7$ for any $k > \lambda$.*

Proof. Refer to Fig. 3 with nodes $o, a_0, a_1, \ldots, a_L, b_{k+\lambda+1}, \ldots, b_L$. The distances are: o is λ away from every node, $dist(a_{i-1}, a_i) = 1$ for all $i = 1, \ldots, L$, and $dist(a_i, b_i) = dist(b_i, a_i) = 1$ for all $i = K + \lambda + 1, \ldots, L$.

Consider the edges shown in Fig. 3 (Top) as forming a connected spanning subgraph G'. Define the distance between any two nodes whose distance is not yet defined as the length of the shortest path within G' between the two nodes, capping the distance at λ; that is, for any $i \neq j$, $dist(a_i, a_j) = \min\{\lambda, |i - j|\}$ and $dist(a_i, b_j) = \min\{\lambda, |i - j| + 1\}$. A distance defined this way satisfies the properties of a metric space.

The requests are: (a_{i-1}, i) for $i = 1, \ldots, L$ (the "spine"), (a_i, b_i) and (b_i, a_i) for $i = k + \lambda + 1, k + \lambda + 2, \ldots, L$ (a "loop"). Note there are no loops for $i \leq k + \lambda$.

OPT serves all requests via the path $o, a_0, a_1, a_2, \ldots, a_{k+\lambda}, a_{k+\lambda+1}, b_{k+\lambda+1},$ $a_{k+\lambda+1}, a_{k+\lambda+2}, b_{k+\lambda+2}, \ldots, a_L$. This serves $L + 2(L - k - \lambda) = 3L - 2\lambda - 2k$

requests in time $\lambda + 3L - 2\lambda - 2k = 3L - \lambda - 2k$. Meanwhile, k-SEQ will serve the path that begins with the segment $o, a_{L-k}, a_{L-k+1}, \ldots, a_L$, followed by an empty drive to the segment $a_{L-2k}, a_{L-2k+1}, \ldots, a_{L-k}$, followed by an empty drive, and so on, until the final segment of k requests $a_0, a_1, \ldots, a_k$.

Note that because $2k > \lambda$, then $dist(a_L, a_{L-2k}) = \lambda$. So the entire k-SEQ path then takes total time $(\lambda + k)L/k$ since there are L/k segments, and k-SEQ initially serves L requests during these $(\lambda + k)L/k$ units of time. There is time remaining, namely $T' = 3L - \lambda - 2k - (\lambda + k)L/k = (3 - (\lambda + k)/k)L - \lambda - 2k$. Since $k > \lambda/2$, we have $(\lambda + k)/k < 3$, so T' is positive for large enough L. There are now disconnected two-cycles $(a_i, b_i), (b_i, a_i)$ for $i = \lambda + k + 1, \ldots, L$ left for k-SEQ. With time T' left, k-SEQ is now at a_k. Note that these are all distance λ away from a_k. There are two cases based on the parity of k.

Case 1: k is even. Moving to the group (i.e. a sequence of k requests consisting of $k/2$ consecutive two-cycles) $\{(a_i, b_i), (b_i, a_i)$ for $i = j, \ldots, j + k/2 - 1\}$ for any j with $k + \lambda + 1 \le j \le L - k/2 + 1$ serves k requests in time $k + \lambda + k/2 - 1 = \lambda + 3k/2 - 1$ because of the required empty drive of time λ to get to the first request and the $(k/2 - 1)$ empty drives between the $k/2$ two-cycles. Then k-SEQ from a_k would move to this group with $j = L - k/2 + 1$, followed by this group with $j = L - k + 1$, and so on, subtracting $k/2$ from j each time. Thus k-SEQ serves $T'k/(\lambda + 3k/2 - 1)$ additional requests in the remaining time T'; note for simplicity we can choose L so that $T'k$ is evenly divisible by $(\lambda + 3k/2 - 1)$.

Case 2: k is odd. The behavior of k-SEQ is similar to the even k case except that in each iteration, k-SEQ serves $(k - 1)/2$ two-cycles and one additional request. Specifically, from a_k, k-SEQ would move to the group of k requests

$$\{(a_j, b_j), (b_j, a_j), (a_{j+1}, b_{j+1}), (b_{j+1}, a_{j+1}), \ldots, (a_{j+(k-1)/2-1}, b_{j+(k-1)/2-1}),$$
$$(b_{j+(k-1)/2-1}, a_{j+(k-1)/2-1}), (a_{j+(k-1)/2}, b_{j+(k-1)/2})\}$$

where we here set $j = L - (k - 1)/2$. This serves k requests in time $\lambda + k + (k - 1)/2 = \lambda + 3k/2 - 1/2$. But the next group of k requests would be (still setting $j = L - (k - 1)/2$, and continuing from $b_{j+(k-1)/2}$):

$$\{(b_{j-(k-1)/2}, a_{j-(k-1)/2}), (a_{j-(k-1)/2+1}, b_{j-(k-1)/2+1}), (b_{j-(k-1)/2+1}, a_{j-(k-1)/2+1}),$$
$$(a_{j-(k-1)/2+2}, b_{j-(k-1)/2+2}), \ldots, (b_{j-1}, a_{j-1})\}.$$

This group serves k requests in time $k + \lambda + (k - 1)/2 - 1 = \lambda + 3k/2 - 3/2$. Together these two sequences serves $2k$ requests in time $2\lambda + 3k - 2$. Then k-SEQ repeats these two sequences with j decreasing by k each time until time runs out. Thus k-SEQ serves $T'2k/(2\lambda + 3k - 2)$ additional requests in the remaining time T'. Note this is identical to the case of even k.

In both cases, k-SEQ serves a total of $\frac{L + T' \cdot k}{(\lambda + 3k/2 - 1)} = \frac{(7kL - 2L - 2k\lambda - 4k^2)}{(3k + 2\lambda - 2)}$ requests. Then $\frac{|OPT|}{|k\text{-SEQ}|}$ is $\frac{(3L - 2\lambda - 2k)(3k + 2\lambda - 2)}{7kL - 2L - 2k\lambda - 4k^2}$. As L grows, this approaches $\frac{3(3k + 2\lambda - 2)}{(7k - 2)}$. Note that because $\lambda \ge 1$ and $k \ge 1$, this ratio is $\ge 9/7$; thus $9/7$ is a lower bound. ∎

Note that when $k \leq \lambda$, Theorem 6 gives a lower bound of $1 + \lambda/k \geq 2$; so we have a lower bound of $9/7$ for any k, λ.

5   Final Remarks

Observe that if we let N denote the maximum number of requests that can be served within time t_{max}, then it is possible to show that our upper bound theorems above hold with λ replaced by $N + 1$ and the lower bound theorems hold with λ replaced by N. Note that $N \leq \lambda$; the hypothetically modified upper bound theorems would be improvements in the case where $N + 1 < \lambda$. Additionally, we could have defined t_{min} as the minimum request service time when there is at least one request, leaving t_{max} as the maximum edge weight, and the theorems above would still hold.

It remains open whether our lower bound of $9/7$ from Theorem 7 is tight when $k > \lambda$. Another open problem is to close the gap between the upper and lower bounds in the approximation ratio when $k \leq \lambda \leq k(k - 1)$.

References

1. Anthony, B.M., et al.: Maximizing the number of rides served for dial-a-ride. In: Cacchiani, V., Marchetti-Spaccamela, A. (eds.) 19th Symposium on Algorithmic Approaches for Transportation Modelling, Optimization, and Systems, (ATMOS 2019), vol. 75, pp. 11:1–11:15 (2019)
2. Anthony, B.M., Christman, A.D., Chung, C., Yuen, D.: Serving rides of equal importance for time-limited Dial-a-Ride (2021). https://www.cs.middlebury.edu/~achristman/papers/kscq.pdf. (Online preprint)
3. Archer, A., Bateni, M., Hajiaghayi, M., Karloff, H.: Improved approximation algorithms for prize-collecting Steiner tree and TSP. SIAM J. Comput. **40**(2), 309–332 (2011)
4. Balas, E.: The prize collecting traveling salesman problem. Networks **19**(6), 621–636 (1989)
5. Bienstock, D., Goemans, M.X., Simchi-Levi, D., Williamson, D.: A note on the prize collecting traveling salesman problem. Math. Program. **59**(1–3), 413–420 (1993)
6. Blum, A., Chawla, S., Karger, D.R., Lane, T., Meyerson, A., Minkoff, M.: Approximation algorithms for orienteering and discounted-reward TSP. SIAM J. Comput. **37**(2), 653–670 (2007)
7. Charikar, M., Motwani, R., Raghavan, P., Silverstein, C.: Constrained TSP and low-power computing. In: Dehne, F., Rau-Chaplin, A., Sack, J.-R., Tamassia, R. (eds.) WADS 1997. LNCS, vol. 1272, pp. 104–115. Springer, Heidelberg (1997). https://doi.org/10.1007/3-540-63307-3_51
8. Christman, A., Chung, C., Jaczko, N., Milan, M., Vasilchenko, A., Westvold, S.: Revenue maximization in online dial-a-ride. In: 17th Workshop on Algorithmic Approaches for Transportation Modelling, Optimization, and Systems (ATMOS 2017), Dagstuhl, Germany, vol. 59, pp. 1:1–1:15 (2017)
9. City of Plymouth, Minnesota: Plymouth metrolink dial-a-ride. Personal communication. http://www.plymouthmn.gov/departments/administrative-services-/transit/plymouth-metrolink-dial-a-ride

10. Cordeau, J.F., Laporte, G.: The dial-a-ride problem: models and algorithms. Ann. Oper. Res. **153**(1), 29–46 (2007)
11. Elbassioni, K., Fishkin, A.V., Mustafa, N.H., Sitters, R.: Approximation algorithms for Euclidean group TSP. In: Caires, L., Italiano, G.F., Monteiro, L., Palamidessi, C., Yung, M. (eds.) ICALP 2005. LNCS, vol. 3580, pp. 1115–1126. Springer, Heidelberg (2005). https://doi.org/10.1007/11523468_90
12. Goemans, M.X., Williamson, D.P.: A general approximation technique for constrained forest problems. SIAM J. Comput. **24**(2), 296–317 (1995)
13. Metropolitan Council: Transit link: Dial-a-ride small bus service. Personal communication. https://metrocouncil.org/Transportation/Services/Transit-Link.aspx
14. Molenbruch, Y., Braekers, K., Caris, A.: Typology and literature review for dial-a-ride problems. Ann. Oper. Res. 295–325 (2017). https://doi.org/10.1007/s10479-017-2525-0
15. Paul, A., Freund, D., Ferber, A., Shmoys, D.B., Williamson, D.P.: Prize-collecting TSP with a budget constraint. In: 25th Annual European Symposium on Algorithms (ESA 2017). Schloss Dagstuhl-Leibniz-Zentrum fuer Informatik (2017)
16. Paul, A., Freund, D., Ferber, A., Shmoys, D.B., Williamson, D.P.: Budgeted prize-collecting traveling salesman and minimum spanning tree problems. Math. Oper. Res. **45**(2), 576–590 (2020)
17. Stagecoach Corporation: Dial-a-ride. Personal Communication. http://stagecoach-rides.org/dial-a-ride/

Rig Routing with Possible Returns and Stochastic Drilling Times

Pavel Borisovsky(ID), Anton Eremeev(ID), Yulia Kovalenko$^{(\boxtimes)}$(ID), and Lidia Zaozerskaya(ID)

Sobolev Institute of Mathematics, Novosibirsk, Russia

Abstract. We consider a real-world vehicle routing problem with time windows, arising in drilling rigs routing and well servicing on a set of sites with different geographical locations. Each site includes a predetermined number of wells which must be processed within a given time window. The same rig can visit a site several times, but the overall number of site visits by rigs is bounded from above. Each well is drilled by one rig without preemptions. It is required to find the routes of the rigs, minimizing the total traveling distance. We also consider a stochastic generalization of the problem, where the drilling times are supposed to be random variables with known discrete distributions. New mixed-integer linear programming models are formulated and tested experimentally. A randomized greedy algorithm is proposed for approximate solving the problem in stochastic formulation, if the number of possible realizations of drilling times is so high that existing MIP solvers are not suitable.

Keywords: Vehicle routing problem · Multiple visits · Stochastic duration

1 Introduction

The area of exploration or production of gas and oil raises a number of optimization problems for managing the drilling rigs activities that include drilling and traveling between wells. The widely used approach to modeling and solution of such problems is based on the Mixed Integer Linear Programming (MILP). One of the earliest studies in this direction [4], considers a rather complex and detailed problem of scheduling the drilling and other tasks for several offshore oil production platforms to maximize the total profit. For the rigs, only the number of moves are counted, but not the travel distances. In [14], a simpler model optimizing the drilling durations and travel times is proposed. It also considers the possibility of rigs outsourcing and compatibilities of rigs and wells. No individual time windows for each well are given, but rather a common deadline for the whole project.

© Springer Nature Switzerland AG 2021
P. Pardalos et al. (Eds.): MOTOR 2021, LNCS 12755, pp. 51–66, 2021.
https://doi.org/10.1007/978-3-030-77876-7_4

The problem of our interest was introduced in [7]. In this problem, several rigs travel between a set of sites, and each site has a certain number of wells to drill. This problem was classified as the Split Delivery Vehicle Routing Problem with Time Windows (SDVRPTW). A MIP model based on the classical VRP model with time windows (see e.g. [15]) was proposed, and solutions found by a commercial MIP solver were compared to those from Variable Neighborhood Search metaheuristic [7,8]. In our paper, we consider a generalization of this problem, in which it is allowed to re-visit the same site by the same rig. This feature can be beneficial from the real-life production perspective, but makes the problem more complicated.

In the mentioned above papers, all the necessary data are supposed to be deterministic. Here we extend the study to the case of uncertain drilling durations, assuming that they are random variables with known discrete distributions. Note that a similar assumption is considered in [1] for the scheduling of a set of offshore oil rigs given that the drilling time is a random variable with a known distribution. The authors propose a Monte-Carlo approach, in which the samples of drilling times are simulated and then are used as input data in the GRASP heuristic. As a result, a set of approximate solutions is built and its properties are investigated statistically.

Among the classic problems, the closest one to our formulation seems to be the Split Delivery Vehicle Routing Problem with Time Windows and Uncertain Service Times. There are many papers devoted to some particular aspects of this problems, but we are not aware of any research on the case combining all the indicated problem settings. The problem with random travel and service times originated from [9]. In [3], this case is extended with the time windows constraints. Many papers deal with the robust approach, in which the probability distribution of uncertain parameters are not given, and the solution to be found must be suitable for all possible realizations of uncertain data. Among these papers we can mention [10,13], in which the service times vary within some convex set. A comprehensive survey of stochastic and robust solution of different VRP type problems can be found in [11].

The SDVRPTW with possibility to re-visit sites has a similarity with production scheduling problems, if one considers rigs as machines, wells as product orders, and sites as orders of the same type. The distances between the sites correspond to setup or changeover times, which should be minimized, while all products should be produced within the given time windows. The production scheduling problems of such kind were successfully solved using time-decomposition techniques and MIP-formulations based on the event points approach (see e.g. [2,5,12]).

In our work, we aim at the following three main goals: (i) to compare two different approaches to defining a MIP formulation of the problem, the one based on the classical VRP model with time windows (as in [7,15] etc.) and the one based on the event points approach [5], (ii) to extend the deterministic problem formulation to a stochastic optimization problem where the drilling time at each site is a random value with a known discrete distribution, testing the MIP-solving techniques for finding exact and approximate solutions to this problem, (iii) to develop a heuristic capable of solving approximately the stochastic optimization problems of higher dimension, compared to MIP-solvers.

In order to reach the second goal, we apply a quantille optimization approach from [6]. This approach is more general than required for our stochastic problem formulation, since in our case the objective function of any fixed solution (which consists of a set of rig routes and the assignment of drilling tasks to rigs) is the total traveling length and does not depend on the random variables. The latter ones only influence on the feasibility of a solution with respect to time window constraints. In the stochastic formulation, we assume that a threshold $\alpha \in (0, 1]$ is given and it is required that the obtained solution should satisfy all time window constraints with a probability not less than α.

2   Deterministic Problem Formulation

We have a set of sites $I = \{i_1, \ldots, i_{|I|}\}$ which must be served by a set of vehicles $U = \{u_1, \ldots, u_{|U|}\}$ (drilling rigs). Each site $i \in I$ is characterized by the total number of planned wells n_i and the time window $(a_i, b_i]$, in which all wells should be drilled. A vehicle can visit site $i \in I$ several times, but the total number of visits of i by all vehicles is bounded by $m_i \le n_i$. Each well is drilled by one rig without preemptions. Drilling a well of site $i \subset I$ by vehicle $u \subset U$ requires d_{ui} time units. A subset I_u of sites, that can be served by vehicle $u \in U$, is given. Each vehicle u is initially located at an individual depot id_u. The traveling time between sites i and j for vehicle u is denoted by s_{uij}. It is required to find rigs routes between sites and assignments of wells to rigs minimizing the total traveling time.

In this section, we propose two models for the considered problem. The first one is based on the event point approach and the second one uses the classic approach from VRP theory. Before that, we provide an example which indicates that there are instances where the same rig visits a site several times in any optimal solution.

2.1   Illustrative Example

Consider an instance with 6 sites (see Fig. 1) and the following input data. The number of wells at the sites with odd indices is 5, the number of wells at the sites with even indices is 8. Time windows:

$a_1 = 20$, $b_1 = 30$, $a_2 = 10$, $b_2 = 40$, $a_3 = 30$, $b_3 = 40$,
$a_4 = 20$, $b_4 = 50$, $a_5 = 40$, $b_5 = 50$, $a_6 = 30$, $b_6 = 60$.

There are 3 drilling rigs. The durations of wells drilling at all sites do not depend on the assignment of vehicles to wells and all equal to 2 (i.e. $d_{ui} = 2$). In this example, we suppose that the rigs are identical, i.e., the traveling time between sites i and j is the same for all vehicles (s_{uij} do not depend on u). The distances between pairs of sites are indicated for each edge, if a direct transportation is possible. Direct transportation is prohibited between all other pairs of sites (i.e. $s_{uij} = \infty$). Vertex id corresponds to the initial location of the rigs here (a common depot).

Odd-numbered sites have narrow time windows. For each pair of time windows $(a_i, b_i]$ and $(a_{i+1}, b_{i+1}]$ for $i = 1, 2, 3$, the following condition holds: $(a_i, b_i] \subset (a_{i+1}, b_{i+1}]$, i.e. the i-th window is contained in the $i + 1$-st, dividing it into three parts with durations equal to 10. Thus, if returns of the rigs to previously visited sites are prohibited, then moving from site $i + 1$ to site i is impossible. The optimal solution with $f = 24$ is uniquely determined up to the assignment of rigs to the routes. It is shown in Fig. 1 on the left. The route of each rig has a unique marking.

If returns to the previously visited sites are allowed, then for each pair of sites i, $i + 1$ for $i = 1, 2, 3$, the drilling rig can perform part of the work (drill 4 wells) on site $i + 1$, then move to site i, do all the work there, return to the site $i + 1$ to process the remaining wells there. The optimal solution with $f = 21$ is shown in Fig. 1 on the right. The value of the objective function is smaller by 3, compared to the case where returns are prohibited.

Based on this example, it is easy to build a family of problems with $6k$ sites and $3k$ machines, with the values of objective function $21k$ (if returns are allowed) or $24k$ (if returns are prohibited) for $k \in N$.

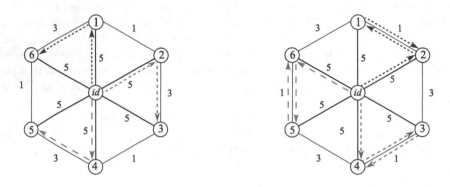

Fig. 1. Optimal solutions in the case of single visits (left) and multiple visits (right).

2.2   MIP Model Based on Event Points

The set of event points for each vehicle u is defined as $K_u = \{1, 2, \ldots, k_u^{\max}\}$, where $k_u^{\max} \leq \sum_{i \in I_u} m_i$. Let U_i denote the subset of rigs suitable for site $i \in I$, i.e. $U_i = \{u \in U : i \in I_u\}$. Introduce the variables:

$x_{uik} \in \{0, 1\}$ such that $x_{uik} = 1$ iff vehicle u visits site i in event point k;
$y_{uik} \in \mathbb{Z}^+$ is the number of wells of site i drilled by vehicle u in event point k;
$t_{uk}^s \geq 0$ is the starting time of works for vehicle u in event point k;
$t_{uk}^f \geq 0$ is the completion time of works for vehicle u in event point k;
$t_{uk}^w \geq 0$ is the traveling time and waiting time between sites in event points $k - 1$ and k.

$t_{uk} \geq 0$ is the traveling time between sites in event points $k - 1$ and k.

Then the set of feasible solutions is defined as follows:

$$1 \leq \sum_{u \in U_i} \sum_{k \in K_u} x_{uik} \leq m_i, \ i \in I, \tag{1}$$

$$\sum_{i \in I_u} x_{uik} \leq 1, \ u \in U, \ k \in K_u, \tag{2}$$

$$\sum_{i \in I_u} x_{u,i,k-1} \geq \sum_{i \in I_u} x_{uik}, \ u \in U, k \in K_u, k > 2, \tag{3}$$

$$x_{u,id_u,1} = 1, \ x_{u,id_u,k} = 0, \ x_{u,i,1} = 0, \ u \in U, \ i \in I_u, \ k \in K_u, k > 1, \tag{4}$$

$$\sum_{u \in U_i} \sum_{k \in K_u} y_{uik} = n_i, \ i \in I, \tag{5}$$

$$y_{uik} \geq x_{uik}, \ i \in I, \ u \in U_i, \ k \in K_u, \tag{6}$$

$$y_{uik} \leq n_i x_{uik}, \ i \in I, \ u \in U_i, \ k \in K_u, \tag{7}$$

$$t_{uk}^w \geq \sum_{i \in I_u \cup \{id_u\}} s_{uij} x_{u,i,k-1} - s_{\max}(1 - x_{ujk}), \tag{8}$$

$$u \in U, \ j \in I_u, \ k \in K_u, \ k > 1,$$

$$\sum_{1 < k' \leq k} \sum_{i \in I_u} d_{ui} y_{u,i,k'} + \sum_{1 < k' \leq k} t_{u,k'}^w \leq \sum_{i \in I_u} b_i x_{uik} + b_{\max}(1 - \sum_{i \in I_u} x_{uik}), \tag{9}$$

$$u \in U, \ k \in K_u, \ k > 1,$$

$$\sum_{1 < k' < k} \sum_{i \in I_u} d_{ui} y_{u,i,k'} + \sum_{1 < k' \leq k} t_{u,k'}^w \geq \sum_{i \in I_u} a_i x_{uik}, \ u \in U, \ k \in K_u, \ k > 1. \tag{10}$$

Here $b_{\max} = \max_{i \in I} b_i$, $s_{\max} = \max_{i,j \in I, \ u \in U_i \cap U_j} s_{uij}$. Inequality (1) provides the upper bound on the number of visits of a site. Constraint (2) implies that in any event point on rig u at most one site may be served. Constraints (3) ensure continuous usage of event points, i.e. if an event point is used for visiting some site, then the previous one is used as well. The initial positions of vehicles are given by constraints (4). Conditions (5) guarantee that all wells of site i will be drilled. If a site i is not served by rig u in the event point k (i.e. $x_{uik} = 0$) then the number of drilled wells should be zero – this is ensured by inequality (7). Constraint (6) indicates that at least one well must be drilled if a rig visits site i. Conditions (9) and (10) ensure rigs routes feasibility with respect to time windows. The traveling time plus waiting time between event points $k - 1$ and k is calculated in (8).

We also can modify constraint (1) for the case of the upper bounds on the number visits m_i' for each rig instead of all rigs:

$$\sum_{k \in K_u} x_{uik} \leq m_i', \ i \in I, \ u \in U_i. \tag{11}$$

The optimization criterion for the presented model is formulated in the following form: minimize

$$f = \sum_{u \in U} \sum_{k \in K_u,\ k>1} t_{uk}, \tag{12}$$

$$t_{uk} \geq \sum_{i \in I_u \cup \{id_u\}} s_{uij} x_{u,i,k-1} - s_{\max}(1 - x_{ujk}), \tag{13}$$

$$u \in U,\ j \in I_u,\ k \in K_u,\ k > 1.$$

The traveling time between event points $k-1$ and k is calculated in (13), and the objective function (12) summarizes the traveling times between all event points.

Using additional variables $t_{uk}^s \geq 0$ and $t_{uk}^f \geq 0$, we can rewrite constraints (9) and (10) in the equivalent form

$$t_{uk}^f \geq t_{uk}^s + \sum_{i \in I_u} d_{ui} y_{uik},\ u \in U,\ k \in K_u, \tag{14}$$

$$t_{uk}^s \geq t_{u,k-1}^f + t_{uk} - b_{\max}\left(1 - \sum_{i \in I_u} x_{uik}\right),\ u \in U,\ k \in K_u, k > 1, \tag{15}$$

$$t_{uk}^f \leq \sum_{i \in I_u} b_i x_{uik},\ u \in U,\ k \in K_u, \tag{16}$$

$$t_{uk}^s \geq \sum_{i \in I_u} a_i x_{uik},\ u \in U,\ k \in K_u. \tag{17}$$

Our preliminary computational experiment shows that model (1)–(7), (12)–(17) is more appropriate for commercial solvers (CPLEX, GUROBI) than model (1)–(10), (12)–(13). The model contains $\sum_{u \in U} \left(\sum_{i \in I_u} m_i \right) \cdot |I_u|$ Boolean variables as well as integer variables.

3   Stochastic Model

In this section we consider a stochastic version of the problem, and construct mixed integer linear programming model similar to model (1)–(7), (12)–(17).

Suppose that drilling times d_{ui} of wells on sites are discrete random variables with values d_{uih} and probabilities p_{ih}, $h = 1, \ldots, v_i$, $\sum_{h=1}^{v_i} p_{ih} = 1$. Here we assume that these probabilities do not depend on u, in other words each outcome h defines the whole vector $(d_{u_1,ih}, ..., d_{u_{|U|},ih})$. Considering all possible combinations of drilling times at sites, we form the total set of possible scenarios SC with cardinality $v = \prod_{i \in I} v_i$. Let $d_{u,i,sc}$ denote the drilling time of a well of site i by rig u in accordance with scenario sc, and p_{sc} be the probability of scenario sc. Now our goal will be to define the rig routes and assign the number of wells drilling to each rig in each visit to a site, minimizing the traveling distance, s.t. the probability of satisfying all time windows constraints is not less than a given threshold level α.

Given some values of all Boolean variables $\mathbf{x} = (x_{uik})$ and integer variables $\mathbf{y} = (y_{uik})$ from MIP problem (1)–(7), (12)–(17) and given a specific realization sc of the random scenario, one can define a function $Q(\mathbf{x}, \mathbf{y}, sc)$ to be 0 if the system of constraints (14)–(17) is consistent for the fixed $\mathbf{x}, \mathbf{y}, sc$, and define $Q(\mathbf{x}, \mathbf{y}, sc) = 1$ otherwise. Using this function, the problem from Sect. 2 for a single fixed scenario sc may be defined as an optimization problem w.r.t. two vectors of variables $\mathbf{x}, \mathbf{y}$ and a vector $\mathbf{t} = (t_{uk})$, asking to minimize the objective function (12), subject to the set of constraints (1)–(7), (13), and $Q(\mathbf{x}, \mathbf{y}, sc) \leq 0$.

Let us denote the system of constraints (1)–(7), (13) on variables $\mathbf{x}, \mathbf{y}, \mathbf{t}$ by $(\mathbf{x}, \mathbf{y}, \mathbf{t})\mathbf{R} \leq \mathbf{r}$, where matrix $\mathbf{R}$ and a row-vector $\mathbf{r}$ are defined appropriately on the basis of the input data. Then the stochastic optimization problem mentioned above may be formulated as

$$\min_{\mathbf{x}, \mathbf{y}, \mathbf{t}} \sum_{u \in U} \sum_{k \in K_u, \ k > 1} t_{uk},$$

$$(\mathbf{x}, \mathbf{y}, \mathbf{t})\mathbf{R} \leq \mathbf{r},$$

$$\Pr\{Q(\mathbf{x}, \mathbf{y}, sc) \leq 0\} \geq \alpha.$$

Let $\mathbf{w} = (w_1, \ldots, w_{|SC|})$ be a vector of scenario indicators. Application of Theorem 1 from [6] shows that the stochastic optimization problem is equivalent to the following deterministic MIP problem.

$$\min_{\mathbf{x}, \mathbf{y}, \mathbf{t}, \mathbf{w}} \sum_{u \in U} \sum_{k \in K_u, \ k > 1} t_{uk},$$

$$(\mathbf{x}, \mathbf{y}, \mathbf{t})\mathbf{R} \leq \mathbf{r},$$

$$Q(\mathbf{x}, \mathbf{y}, sc) \leq 1 - w_{sc}, \quad sc \in SC,$$

$$\sum_{sc \in SC} w_{sc} p_{sc} \geq \alpha.$$

Here the confidence set of level α is formed by the Boolean variables w_{sc} such that $w_{sc} = 1$ if scenario sc belongs to the confidence set, and $w_{sc} = 0$ otherwise.

An equivalent of the constraint $Q(\mathbf{x}, \mathbf{y}, sc) \leq 1 - w_{sc}$ with variables for starting times and completion times may be written as follows.

$$t_{u,k,sc}^{f} \geq t_{u,k,sc}^{s} + \sum_{i \in I_u} d_{u,i,sc} y_{uik}, \ u \in U, \ k \in K_u, \ sc \in SC, \quad (18)$$

$$t_{u,k,sc}^{s} \geq t_{u,k-1,sc}^{f} + t_{uk} - b_{\max}(1 - \sum_{i \in I_u} x_{uik}), \quad (19)$$

$$u \in U, \ k \in K_u, k > 1, \ sc \in SC,$$

$$t_{u,k,sc}^f \le \sum_{i \in I_u} b_i x_{uik} + b_{\max}(1 - w_{sc}), \ u \in U, \ k \in K_u, \ sc \in SC, \qquad (20)$$

$$t_{u,k,sc}^s \ge \sum_{i \in I_u} a_i x_{uik} - b_{\max}(1 - w_{sc}), \ u \in U, \ k \in K_u, \ sc \in SC, \qquad (21)$$

$$\sum_{sc \in SC} w_{sc} p_{sc} \ge \alpha. \qquad (22)$$

3.1   Illustrative Example

Consider an instance with two rigs and one site including two wells. The site has a time window $(0, 4]$, and the drilling time may be 2 or 3 with some probability. The traveling time from rig deports to the site is 1. It is easy to see that if wells are drilled for 2 time units then only one rig visits the site and the objective is 1, but if the wells are drilled for 3 time units then both rigs visit the site and the objective is 2.

4   MIP Model Based on VRP Approach

The proposed VRP-based model is similar to the model from [7], but allows to visit a site by the same rig several times.

For each site $i \in I$ we create m_i copies and introduce a new set of sites I'. All copies of the same original site have identical set of wells. Denote by I_i' all copies of the original site i, $I' = \cup_{i \in I} I_i'$. Traveling times between site copies from I_i' are equal to zero, traveling times between copies of different sites are equal to the traveling times between these sites. Introduce a dummy site f_s corresponding to starting and completion point of the rout of each rig, set $I_f' := I' \cup \{f_s\}$. Put traveling times $s_{u,f_s,i'} := s_{u,id_u,i}$ and $s_{u,i',f_s} := 0$ for $i' \in I_i'$, $i \in I$. All rigs are suitable for the dummy site f_s, i.e. $U_{f_s} = U$. Set $I_u' := \cup_{i \in I_u} I_i' \cup \{f_s\}$ for all $u \in U$ and $U_{i'} = U_i$ for all $i' \in I_i'$, $i \in I$.

Introduce Boolean variables $x_{ui'j'}$ such that $x_{ui'j'} = 1$ if rig u visits site-copy i' and travels to site-copy j', and $x_{ui'j'} = 0$ otherwise. Let Real variables $t_{ui'}^s$ defines the starting time of works for vehicle u on site-copy i', and integer variables $y_{ui'}$ counts the number of wells of site-copy i' drilled by vehicle u. We formulate the following mixed integer linear programming model: minimize

$$f = \sum_{u \in U} \sum_{i' \in I_f'} \sum_{j' \in I_f'} s_{ui'j'} x_{ui'j'}, \qquad (23)$$

$$\sum_{j' \in I_f'} x_{ui'j'} = \sum_{j' \in I_f'} x_{uj'i'}, \ u \in U, \ i' \in I_u' \setminus \{f_s\}, \qquad (24)$$

$$\sum_{u \in U_{j'}} \sum_{i' \in I_u'} x_{ui'j'} \le 1, \ j' \in I'. \qquad (25)$$

$$\sum_{i' \in I_i'} \sum_{u \in U_i} y_{ui'} = n_i, \ i \in I, \tag{26}$$

$$y_{ui'} \geq \sum_{j' \in I_u'} x_{uj'i'}, \ i' \in I', \ u \in U_{i'}, \tag{27}$$

$$y_{ui'} \leq n_i \sum_{j' \in I_u'} x_{uj'i'}, \ i' \in I', \ u \in U_{i'}, \tag{28}$$

$$t_{ui'}^s + y_{ui'} d_{ui} + s_{u,i',j'} \leq t_{u,j'}^s + b_{\max}(1 - x_{ui'j'}), \tag{29}$$
$$i' \neq j' \in I', \ i: \ i' \in I_i', \ u \in U_{i'} \cap U_{j'},$$

$$s_{u,f_s,j'} \leq t_{uj'}^s + b_{\max}(1 - x_{u,f_s,j'}), \ j' \in I', \ u \in U_{j'}, \tag{30}$$

$$t_{ui'}^s \geq \sum_{j' \in I_u'} a_i x_{uj'i'}, \ i \in I, \ i' \in I_i', \ u \in U_i, \tag{31}$$

$$t_{ui'}^s + y_{ui'} d_{ui} \leq \sum_{j' \in I_u'} b_i x_{uj'i'}, \ i \in I, \ i' \in I_i', \ u \in U_i, \tag{32}$$

$$\sum_{i' \in I_u'} x_{u,f_s,i'} - 1, \ u \in U, \tag{33}$$

$$\sum_{i' \in I_u'} x_{u,i',f_s} - 1, \ u \in U. \tag{34}$$

Constraints (24) guarantee that each site-copy has exactly one predecessor and one successor in the route. Inequalities (25) indicate that each rig visits each site-copy at most ones. Conditions (26)–(28) ensure that the required number of wells are drilled at each site, and each well is drilled by one rig. Constraints (29)–(30) set the starting times of the works on sites for rigs. Inequalities (31)–(32) ensure feasibility of rig routes with respect to time windows. Conditions (33)–(34) indicate that each rig starts and completes its rout in depot.

The model contains $\sum_{u \in U} \left(\sum_{i \in I_u} m_i \right)^2$ Boolean variables and $\sum_{u \in U} \sum_{i \in I_u} m_i$ integer variables.

4.1 Stochastic Version

In the stochastic version, as in Sect. 3, we introduce binary variables w_{sc} equipped with the constraint (22), add the scenario index to variables $t_{u,i}^s$ and replace constraints (29)–(32) by the following scenarios-based conditions:

$$t_{u,i',sc}^s + y_{ui'} d_{u,i,sc} + s_{ui'j'} \leq t_{u,j',sc}^s + b_{\max}(2 - x_{ui'j'} - w_{sc}), \tag{35}$$
$$i' \neq j' \in I', \ i: \ i' \in I_i', \ u \in U_{i'} \cap U_{j'}, \ sc \in SC,$$

$$s_{u,f_s,j'} \leq t_{uj'sc}^s + b_{\max}(2 - x_{u,f_s,j'} - w_{sc}), j' \in I', \ u \in U_{j'}, \ sc \in SC, \tag{36}$$

$$t^s_{u,i',sc} \geq \sum_{j' \in I'_u} a_i x_{uj'i'} - b_{\max}(1 - w_{sc}), i \in I, \; i' \in I'_i, \; u \in U_i, \; sc \in SC, \quad (37)$$

$$t^s_{u,i',sc} + y_{ui'} d_{u,i,sc} \leq \sum_{j' \in I'_u} b_i x_{uj'i'} + b_{\max}(1 - w_{sc}), \quad (38)$$

$$i \in I, \; i' \in I'_i, \; u \in U_i, \; sc \in SC.$$

5   Greedy Algorithm for Stochastic Optimization

The number of binary variables w_{sc} grows exponentially in the number of sites with uncertain drilling time. Therefore, a straightforward application of a MIP solver allows to solve only small-sized instances. In order to treat larger instances, we propose a simple randomized greedy heuristic.

Recall that each scenario sc is some realization of random drilling times and it is represented as an $|I|$-dimensional vector $(sc_1, \ldots, sc_{|I|})$. Let us say that scenario sc dominates scenario sc' iff $sc_i \geq sc'_i$ for all i. Clearly, in this case if there is a solution to the considered stochastic problem with $w_{sc} = 1$ one may always set $w_{sc'} = 1$ in this solution without violation of its feasibility or worsening the objective function value. For any subset of scenarios $S \subset SC$ define $D(S) \subset SC$ as the set of all scenarios that are dominated by at least one element of S (note that each scenario dominates itself, so $S \subseteq D(S)$). For a subset S consider a stochastic optimization problem in which the constraint $\sum_{sc} p_{sc} \cdot w_{sc} \geq \alpha$ is excluded, all variables $w_{sc}, sc \in S$ are fixed to one, and all other w_{sc} are fixed to zero. Denote this problem as $P(S)$.

With these notations, we may reformulate our stochastic problem as follows: Find a subset of scenarios S such that the total probability of $D(S)$ is not less than α and an optimal value of the objective function of problem $P(S)$ is minimal. The proposed greedy algorithm is aimed at finding such a subset S and its outline is given below.

Algorithm 1. Randomized Greedy Algorithm

1: Set $S := \emptyset, p := 0$.
2: Repeat until $p \geq \alpha$ or the running time exceeds the given limit.

    2.1 Choose scenarios $sc^1, ..., sc^r \in SC$ uniformly at random.
    2.2 Solve r stochastic optimization problems $P(S \cup \{sc^1\}), ... P(S \cup \{sc^r\})$ and let $f^1, ..., f^r$ be the objective function values for the obtained solutions.
    2.3 Choose sc^j with the minimal value f^j, add it to S, and remove all dominated scenarios: $S := S \cup \{sc^j\}, SC := SC \setminus D(\{sc^j\})$.
    2.4 Update the current value of p as the total probability of $D(S)$.

The solution of stochastic optimization problems at Step 2.2 can be done in parallel. Due to the random nature of the algorithm, it is reasonable to run it several times and choose the best result. One positive feature of the algorithm is that it produces a sequence of solutions with increasing probability values p and the corresponding values of f, which gives a better understanding of the problem structure to a decision maker (this will be illustrated in Sect. 6.3).

6 Implementation and the Computer Experiments

In the experiments, we used a server with two AMD EPYC 7502 processors (each one has 32 cores, hyper-threading mode on), OS Ubuntu 20.04. MILP solver Gurobi (version 9.0.3) was applied to solve MIP problems[1] coded in GAMS.

6.1 Testing Deterministic Models

First, we tested the event-point-based and the VRP-based models on instances with 50 sites from [8], where the results of Gurobi were presented for the VRP-based model version with no returns. Two versions of the models are investigated, when two and when three visits of sites are allowed for rigs. Computational experiment with the same parameters as in [8] did not improve the objective values. We believe that this is due to the structure of the instances from [8], where time windows for objects are uniformly distributed during the planing horizon and have lengths less than or equal to the total drilling time of wells on the objects. The event-point-based model demonstrated slightly worse results than VRP-based model.

Second, we compared the two deterministic models on a family D of problems D_k, $k \in N$, constructed on the basis of the example from Subsect. 2.1. Problem D_k consists of k subproblems with the structure as shown in Fig. 1 and the same initial data. It has $6k$ sites and $3k$ drilling rigs. Each subproblem has the set of sites $G_v = \{6v - 5, 6v - 4, ..., 6v\}$, the set of rigs $U_v = \{3v - 2, \ 3v - 1, 3v\}$ and the point of the initial location of rigs id_v for $v = 1, .., k$. Rigs of U_v can serve all sites from G_v, as well as the first site from the set with the next index (i.e. G_{v+1}). In the distance matrices, we put $s_{u,i,6v+1} = s_{u,6v+1,i} = i + 10$ for $v = 1, ..., k-1$, $i \in G_v$, $u \in U_v$. For all v and $i \in G_v$, $u \in U_v$, put $s_{u,id_v,i} = 5$. Direct transportation is prohibited between all other pairs of sites. For the forbidden movements of drilling rigs, we will assign sufficiently large values as distances. In the experiment, we set this value to 265.

For the VRP-based model, the Gurobi solver was used with the parameters Presolve = 2, GomoryPasses = 0, Method = 0, MinRelNodes = 10627, ImproveStartTime = 8640. For MIP model based on event points we did not find any parameters settings better than the default ones, so the default settings were used. The results for $k = 1, 2, .., 6$ are shown in Table 1. In the case of one visit, these instances required little solving time in both models, although

---

[1] The choice of this solver was based on a preliminary experiment, which indicated that on the MIP instances considered here Gurobi has an advantage to other solvers available to us (e.g. it was approximately twice as fast in comparison with CPLEX 12.10.0.0).

the VRP-based model required less time for most instances. In the case of two visits, as k grows, the problems become more difficult for both models. For all instances, solutions with the optimal value of the objective function were found, but using the VRP-based model, the solver failed to prove the obtained solutions optimality in 10 h of CPU time for $k \geq 4$. The EP-based model yields the best results for this series.

Table 1. Comparison of models on series D.

k	$|I|$	$|U|$	At most 1 visit			At most 2 visits		
			Obj	Time		Obj	Time	
				EP-based	VRP-based		EP-based	VRP-based
1	6	3	24	0,89	**0,21**	21	2,50	**1,82**
2	12	6	48	7,14	**1,55**	42	149,36	**114,93**
3	18	9	72	77,94	**7,41**	63	**576,40**	10502,38
4	24	12	96	328,89	**11,71**	84	**703,38**	>36000
5	30	15	120	97,46	**7,95**	105	**5888,08**	>36000
6	36	18	144	**40,11**	88,36	126	**10695,31**	>36000

6.2   Testing Stochastic Models

The experiments were done on subinstances of the instances from [8]. In two series $S1$ and $S1'$ (10 instances in each series) we take the first seven or the first ten sites, and all given six rigs. Sites have from 5 to 30 wells. Note that instances from $S1'$ are characterized by shorter time windows than instances from $S1$.

For five random sites of each instance we suppose that drilling time can take two values 2 or 3, and the probability of value 2 is generated randomly from the interval $[0.75, 0.85]$. Drilling times are equal to 2 for the rest of the sites. For the sake of simplicity, we assume that $d_{u,i,sc}$ do not depend on u, i.e. they are the same for any fixed pair i and sc.

We test event-point-based and VRP-based models for threshold levels $\alpha = 0.5; 0.6; 0.7; 0.8; 0.9; 0.99$ in two versions, when one and when two visits of sites are allowed for rigs. The results for two instances with seven sites of series $S1$ are presented in Table 2 (the full results for all instances are available at https://gitlab.com/YuliaKovalenko-gl/stochastic-vrp-problem.git). The running time of Gurobi is greater on series $S1'$ than on series $S1$ due to the structure of the instances in this series.

In most of the instances the optimal stochastic solution at level $\alpha = 0.8$ has a lower traveling distance, compared to the worst-case scenario, where all drilling times are equal to 3. In the instances presented in Table 2, the version allowing up to two visits yields a solution with a lower objective traveling distance, compared to the version with no returns. As we can see the running time of Gurobi has no specific tendency as a function of threshold α. In the two-visit-version, none of the considered models clearly dominates the other one in terms of running time of Gurobi. The VRP-based model demonstrates better results in a majority of the cases with no returns.

Table 2. Comparison of models on series with 7 sites.

α	At most 1 visit			At most 2 visits		
	Obj	Time		Obj	Time	
		EP-based	VRP-based		EP-based	VRP-based
Instance 7_2						
0.5	10	34,86	0,63	9	22,09	3,12
0,6	10	24,54	6,70	10	64,34	41,14
0,7	10	29,38	6,07	10	60,62	29,09
0,8	11	74,47	9,65	10	62,67	63,48
0,9	13	410,16	30,88	11	211,69	727,08
0,99	13	133,62	23,92	13	315,74	381,81
Instance 7_3						
0,5	16	20,65	4,35	16	29,59	51,6
0,6	16	4,33	3,07	16	13,42	27,08
0,7	17	8,64	4,64	17	22,56	113,11
0,8	19	62,14	9,95	18	24,28	143,03
0,9	21	21,88	17,35	20	56,21	159,91
0,99	22	13,47	4,06	21	7,03	51,46

6.3 Evaluation of the Greedy Algorithm

For testing the greedy algorithms, two instances from previous section, namely 7_2 and 7_3, were taken and four larger problems were generated on the basis of the instances S1.1, S1.2, S1.3, S1.4 from [8]. The original instances contain 50 sites and 6 rigs. Here only the first 12 sites are extracted. All the travel times are kept unchanged, but the drilling times $d_{u,i}$ now take value 1 with probability 0.8 and value 3 with probability 0.2 for each site. The total number of scenarios is then $2^{12} = 4096$. The MIP model for the first instance has 547980 columns, 7216 discrete-columns, and 3858896 rows. The straightforward application of Gurobi with $\alpha = 0.7$ could not find a feasible solution in five hours.

In the implementation of the greedy algorithm, at Step 2.1, the number of considered scenarios is $r = 3$, and in case of large instances they are chosen at random among the scenarios, in which $d_{u,i,sc}$ have value 3 for more than five sites, otherwise the probability of $D(\{sc\})$ is negligibly small. At Step 2.2, three problems are solved in parallel by Gurobi, each process is allowed to use up to four CPU cores, and the solving time of one problem is limited by 180 s. The algorithm stops when it reaches the level $\alpha = 0.95$, or when the running time exceeds the overall time limit, which was set to one hour. Although the algorithm may work with both EP and VRP based models, the VRP case was chosen, because it showed better performance in earlier tests with no returns.

For each problem instance, five independent runs of the greedy algorithm were made and the best results were collected and summarized in Table 3. As before, the smaller problems were solved in two variants: with at most one or at most two visits of each site (this is marked with "1v" or "2v" in the table). Column "optimistic" shows the objective function values of the solution with the best realization of the drilling times, i.e. the scenario sc with all $d_{u,i,sc} = 1$; similarly, column "pessimistic" corresponds to the worst-case scenario with all $d_{u,i,sc} = 3$. The other columns show the best results provided by the algorithm after reaching the given probability threshold. The cells, for which no feasible solutions were obtained are marked with "–". For example, let us fix $\alpha = 0.8$, then for instance S2.1 there exists a solution with the cost $f = 35$ that is valid with probability at least α. For the smaller problems, the obtained solutions are quite close to the optimal ones (results known to be optimal are marked by "*", compare to Table 2). For problems S2.1...S2.4, which can not be straightforwardly solved by the MIP solver in practically acceptable time, the greedy algorithm still yields reasonable solutions.

Table 3. Results of the greedy algorithm

Instance	Optimistic	p = 0.6	p = 0.7	p = 0.8	p = 0.9	p = 0.95	Pessimistic
7_2(1v)	10	10*	10*	11*	13*	13*	13
7_2(2v)	9	10*	10*	11	11*	13	13
7_3(1v)	16	16*	18	19*	21*	22	22
7_3(2v)	15	16*	18	19	20*	21	21
S2_1(1v)	13	29	33	35	44	–	–
S2_2(1v)	10	21	22	29	31	31	31
S2_3(1v)	16	39	39	42	43	45	45
S2_4(1v)	12	35	40	42	–	–	–

7   Conclusions

In this paper, we have studied a generalization of the drilling rig routing problem suggested by I. Kulachenko and P. Kononova, allowing to re-visit the same site by the same rig and assuming that at some sites the drilling durations are random variables with known discrete distributions. We have compared two different approaches to defining a MIP formulation of the problem, a one based on the classical VRP model with time windows and a one based on the event points approach. Also, we have carried out a computational experiment, comparing the performance of Gurobi solver on these MIP models and found out that in different cases either one of the models has an advantage. To solve approximately the stochastic optimization problems of higher dimension, we have developed a randomized greedy heuristic, which demonstrated promising results.

Acknowledgement. The research is supported by Russian Science Foundation grant 21-41-09017. A server of Sobolev Institute of Mathematics, Omsk Branch is used for computing.

References

1. Bassi, H.V., Ferreira Filho, V.J.M., Bahiense, L.: Planning and scheduling a fleet of rigs using simulation-optimization. Comput. Ind. Eng. **63**(4), 1074–1088 (2012)
2. Borisovsky, P.A., Eremeev, A.V., Kallrath, J.: Multi-product continuous plant scheduling: combination of decomposition, genetic algorithm, and constructive heuristic. Int. J. Prod. Res. **58**(9), 2677–2695 (2020)
3. Errico, F., Desaulniers, G., Gendreau, M., Rei, W., Rousseau, L.-M.: The vehicle routing problem with hard time windows and stochastic service times. EURO J. Transp. Logist. **7**(3), 223–251 (2016). https://doi.org/10.1007/s13676-016-0101-4
4. Iyer, R.R., Grossmann, I.E., Vasantharajan, S., Cullick, A.S.: Optimal planning and scheduling of offshore oil field infrastructure investment and operations. Ind. Eng. Chem. Res. **37**(4), 1380–1397 (1998)
5. Icrapetritou, M.G., Floudas, C.A.: Effective continuous-time formulation for short-term scheduling: I. Multipurpose batch process. Ind. Eng. Chem. Res. **37**(11), 4341–4359 (1998)
6. Kibzun, A.I., Naumov, A.V., Norkin, V.I.: On reducing a quantile optimization problem with discrete distribution to a mixed integer programming problem. Autom. Remote Control **74**, 951–967 (2013)
7. Kulachenko, I., Kononova, P.: A matheuristic for the drilling rig routing problem. In: Kononov, A., Khachay, M., Kalyagin, V.A., Pardalos, P. (eds.) MOTOR 2020. LNCS, vol. 12095, pp. 343–358. Springer, Cham (2020). https://doi.org/10.1007/978-3-030-49988-4_24
8. Kulachenko, I., Kononova, P.: A hybrid algorithm for drilling rig routing problem. Diskretny Analys i Issledovanie Operacii (2021, to appear). (in Russian)
9. Laporte, G., Louveaux, F., Mercure, H.: An integer L-shaped algorithm for the capacitated vehicle routing problem with stochastic demands. Transp. Sci. **26**(3), 161–170 (1992)
10. Nasri, M., Metrane, A., Hafidi, I., Jamali, A.: A robust approach for solving a vehicle routing problem with time windows with uncertain service and travel times. Int. J. Ind. Eng. Comput. **11**, 1–16 (2020)
11. Oyola, J., Arntzen, H., Woodruff, D.L.: The stochastic vehicle routing problem, a literature review, part I: models. EURO J. Transp. Logist. **7**(3), 193–221 (2016). https://doi.org/10.1007/s13676-016-0100-5
12. Shaik, M.A., Floudas, C.A., Kallrath, J., Pitz, H.J.: Production scheduling of a large-scale industrial continuous plant: short-term and medium-term scheduling. Comput. Chem. Eng. **33**, 670–686 (2009)

13. Souyris, S., Cortés, C.E., Ordóñez, F., Weintraub, A.: A robust optimization approach to dispatching technicians under stochastic service times. Optim. Lett. **7**(7), 1549–1568 (2012). https://doi.org/10.1007/s11590-012-0557-6
14. Tavallali, M.S., Zare, M.: Planning the drilling rig activities - routing and allocation. Comput. Aided Chem. Eng. **43**, 1219–1224 (2018)
15. Cordeau, J.-F., Desaulniers, G., Desrosiers, J., Solomon, M.M., Soumis, F.: VRP with time windows. In: Toth, P., Vigo, D. (eds.) The Vehicle Routing Problem, SIAM Monographs on Discrete Mathematics and Applications, pp. 157–193. SIAM (2002)

On Asymptotically Optimal Approach for the Problem of Finding Several Edge-Disjoint Spanning Trees of Given Diameter in an Undirected Graph with Random Edge Weights

Edward Kh. Gimadi[1,2] , Aleksandr S. Shevyakov[2],
and Alexandr A. Shtepa[2(✉)]

[1] Sobolev Institute of Mathematics, Prosp. Akad. Koptyuga, 4, Novosibirsk, Russia
gimadi@math.nsc.ru
[2] Novosibirsk State University, Novosibirsk, Russia
http://www.nsu.ru/
http://www.math.nsc.ru/

Abstract. We consider the intractable problem of finding several edge-disjoint spanning trees of a given diameter in an graph with random edge weights. Earlier, we have implemented an asymptotically optimal approach for this problem in the case of directed graphs. The direct use of this result for the case of undirected graphs turned out to be impossible due to the issues associated with the summation of dependent random variables. In this work we give an $O(n^2)$-time algorithm with conditions of asymptotic optimality for the case of undirected graphs.

Keywords: Given-diameter Minimum Spanning Tree · Approximation algorithm · Probabilistic analysis · Asymptotic optimality.

1 Introduction

The Minimum Spanning Tree (MST) problem is one of the well-known discrete optimization problems. It consists of finding a spanning tree (connected acyclic subgraph) of a minimal weight in a given edge-weighted undirected graph $G = (V, E)$. The polynomial solvability of the problem was shown in the classic algorithms by Boruvka (1926), Kruskal (1956) and Prim (1957). These algorithms have complexities $O(u \log n)$, $O(u \log u)$ and $O(n^2)$, where $u = |E|$ and $n = |V|$. It must be noted that the mathematical expectation for an MST's weight in a graph with edge weights from the class of random variables with a uniform distribution on the interval $(0; 1)$ is close to 2.02 w.h.p. (with a high probability) [3]. Also interested reader may refer to [1,2].

Supported by the program of fundamental scientific researches of the SB RAS (project 0314-2019-0014), and by the Ministry of Science and Higher Education

© Springer Nature Switzerland AG 2021
P. Pardalos et al. (Eds.): MOTOR 2021, LNCS 12755, pp. 67–78, 2021.
https://doi.org/10.1007/978-3-030-77876-7_5

One of the possible generalizations of the above problem may be the problem of finding a bounded diameter MST problem. The diameter of a tree is the number of edges in the longest simple path within the tree connecting a pair of vertices. This problem is as follows. Given an edge-weighted undirected graph and a number $d = d_n$, the goal is to find in the graph a spanning tree T of minimal total weight having its diameter bounded above to given number d, or from below to given number d. Both problems are NP-hard in general.

Earlier in the papers [6,7,9–12], it was studied an asymptotically optimal approach to a bounded MST problem with a graph diameter bounded either from below or above.

Recently, we began to study another modification of the problem, when the diameter of the desired spanning tree is a given number. The work [13] gives a probabilistic analysis of an effective algorithm for solving a given-diameter MST problem in the case of complete directed graph. Unfortunately, the algorithm analysis, presented in this work becomes unacceptable for a problem on undirected graphs. The appearance of the difficulty of probabilistic analysis in the case of the undirected graph arises from the need to take into account the possible dependence between different random variables in the course of the algorithm.

In current paper we consider problem of finding m edge-disjoint MSTs with a given diameter d in the complete undirected graph (m-d-UMST). We introduce a polynomial-time approximation algorithm to solve this problem and provide conditions for this algorithm to be asymptotically optimal. A probabilistic analysis is carried out under conditions that edge weights of given graph are identically independent distributed random variables. Our algorithm can be transformed to solve the problem of finding m edge-disjoint MSTs with bounded diameter from below or above. So all the applications for these problems are valid for m-d-UMST (see, for example, [14]).

2 Finding Several Edge-Disjoined MSTs with a Given Diameter

Given a edge-weighted complete undirected n-vertex graph $G = (V, E)$ and positive integers $m \geq 2$, d such that $m(d + 1) \leq n$, the problem is to find m edge-disjoint spanning trees $T_1, \ldots, T_m$ with a given diameter $d = d_n < \frac{n}{m}$ of minimum total weight. We assume that the weights of the edges are independent and identically distributed random reals, with probability distribution function $f(x)$ defined on $(a_n; b_n)$.

Description of the Algorithm $\mathcal{A}$

Preliminary Step 0. In graph G, choose an arbitrary $(n - m(d + 1))$-vertex subset V', and arbitrary split the remaining $m(d+1)$ vertices into $(d+1)$-vertex subsets $V_1, \ldots, V_m$.

Step 1. For each $l = 1, \ldots, m$, starting at arbitrary vertex in the subgraph $G(V_l)$, construct in it a Hamiltonian path P_l of length $d = d_n$, using the approach "Go to the nearest unvisited vertex". After the construction of entire path P_l we put $T_l = P_l$.

Step 2. We assume without loss of generality that d is odd. For each pair of paths P_i and P_j, $1 \le i < j \le m$, we connect them in a special way by the set E_{ij} of $2(d+1)$ edges, so that the constructed subgraph was composed of two $2(d+1)$-vertex edge-disjoint subtrees with a diameter equals d. We represent each path as two halves (first and second) P_l^1 and P_l^2, $1 \le l \le m$. Each half contains one *end* vertex and $\frac{d-1}{2}$ *inner* vertices, $\frac{d+1}{2}$ vertices totally. We construct the set of connecting edges as follows.

2.1. Connect each inner vertex of P_i^1 by the shortest edge to the inner vertex of P_j^1. So we add this edge to T_j.

2.2. Connect each inner vertex of P_i^2 by the shortest edge to the inner vertex of P_j^2. We add this edge to T_j.

2.3. Connect each inner vertex of P_j^1 by the shortest edge to the inner vertex of P_i^2. Thus, we add this edge to T_i.

2.4. Connect each inner vertex of P_j^2 by the shortest edge to the inner vertex of P_i^1. We add this edge to T_i.

2.5. Connect each end vertex of the path P_i by the shortest edge to the inner vertex of the path P_j. We add this edge to T_j.

2.6. Connect each end vertex of the path P_j by the shortest edge to the inner vertex of the path P_i. So we add this edge to T_i.

Step 3. For $l = 1, \ldots, m$ each vertex of the subgraph $G(V')$ connect by the shortest edge to the inner vertex of the path P_l. Thus, we add this edge to corresponding T_l.

The construction of all m edge-disjoint spanning trees $T_1, \ldots, T_m$ is completed.

Denote by $W_{\mathcal{A}}$ the total weight of all trees $T_1, \ldots, T_m$ constructed by Algorithm $\mathcal{A}$. Denoting summary weights of edges, obtained on Steps 1, 2 and 3 by W_1, W_2 and W_3, we have $W_{\mathcal{A}} = W_1 + W_2 + W_3$.

Let us formulate two statements concerning Algorithm $\mathcal{A}$.

Statement 1. *Algorithm $\mathcal{A}$ constructs a feasible solution for the m-d-UMST.*

Proof. Each of the edge-disjoint constructions consists of n vertices and $n - 1$ edges since we firstly create the tree as the path on $d + 1$ vertices during Step 1 and then we add edges to the tree by connecting all other vertices to the vertices in path on Steps 2–3, totally we obtain m such constructions, and we indeed get feasible solution for the m-d-UMST.

Statement 2. *Running time of Algorithm $\mathcal{A}$ is $\mathcal{O}(n^2)$.*

Proof. Preliminary Step 0 takes $\mathcal{O}(n)$ time.

At Step 1 each path is built in $\mathcal{O}(d^2)$ time, thus, it takes $\mathcal{O}(md^2)$ or $\mathcal{O}(nd)$ time to construct all paths.

At Steps 2.1–2.4 each pair (P_i, P_j), $1 \leq i < j \leq m$, of paths is connected with the edge set E_{ij} in $\mathcal{O}(d^2)$ time, and for all $\frac{m(m-1)}{2}$ pairs of paths it is required $\mathcal{O}(m^2 d^2)$, or (since $m(d+1) \leq n$) $\mathcal{O}(n^2)$-running time.

Steps 2.5–2.6 are carried out in $\mathcal{O}(md)$ time.

Step 3 takes $\mathcal{O}(mn)$ operations since we connect $|G(V')| \leq n$ vertices by the shortest edge to the inner vertex of the path P_l for each spanning tree T_l, $1 \leq l \leq m$.

So, the total time complexity of the Algorithm $\mathcal{A}$ is $\mathcal{O}(n^2)$ (Figs. 1, 2 and 3).

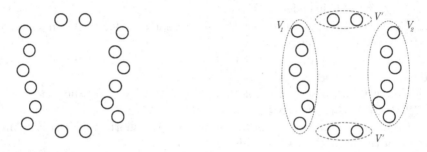

Fig. 1. Initial vertices of the graph and Step 0 of the work of the Algorithm $\mathcal{A}$ on 16-vertex complete graph, $m = 2$, $d = 5$.

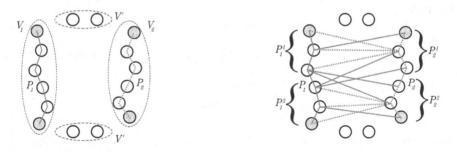

Fig. 2. Step 1 and Step 2 of the work of the Algorithm $\mathcal{A}$ on 16-vertex complete graph, $m = 2$, $d = 5$. The hatched vertices are end vertices. The solid edges are edges of T_1. The dotted edges are edges of T_2.

3   A Probabilistic Analysis of Algorithm $\mathcal{A}$

We perform the probabilistic analysis of Algorithm $\mathcal{A}$ under conditions that weights of graph edges are random variables η from the class $\mathrm{UNI}(a_n; b_n)$,

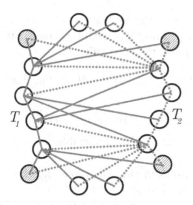

Fig. 3. Step 3 of the work of the Algorithm $\mathcal{A}$ on 16-vertex complete graph, $m = 2$, $d = 5$. The hatched vertices are end vertices. The solid edges are edges of T_1. The dotted edges are edges of T_2.

namely, are independent identically distributed random variables with uniform distribution on a set $(a_n; b_n)$, $0 < a_n \leq b_n < \infty$. Obviously, normalized variables $\xi = \frac{\eta - a_n}{b_n - a_n} \in (0; 1)$ belong to the class UNI$(0; 1)$.

By $F_A(I)$ and $OPT(I)$ we denote respectively the approximate (obtained by some approximation algorithm A) and the optimum value of the objective function of the problem on the input I. An algorithm A is said to have *estimates (performance guarantees)* $(\varepsilon_n, \delta_n)$ on the set of random inputs of the n-sized problem (it is the problem with parameter n, where n is amount of input data required to describe the problem, see [4]), if

$$\mathbf{P}\left\{F_A(I) > \left(1 + \varepsilon_n\right)OPT(I)\right\} \leq \delta_n, \tag{1}$$

where $\varepsilon_n = \varepsilon_A(n)$ is an estimation of *the relative error* of the solution obtained by algorithm A, $\delta_n = \delta_A(n)$ is an estimation of *the failure probability* of the algorithm, which is equal to the proportion of cases when the algorithm does not hold the relative error ε_n or does not produce any answer at all.

Following [5] we say that an approximation algorithm A is called *asymptotically optimal* on the class of input data of the problem, if there exist such performance guarantees that for all input I

$$\varepsilon_n \to 0 \text{ and } \delta_n \to 0 \text{ as } n \to \infty.$$

Further we suppose that the parameter d is defined on the set of values d in two ranges

Case 1: $\ln n \leq d < n/\ln n$ and Case 2: $n/\ln n \leq d < n$.

We denote random variable equal to minimum over k variables from the class UNI$(a_n; b_n)$ (from UNI$(0; 1)$) by η_k (ξ_k, correspondingly).

Further, for simplicity, we assume that the parameter d is odd.

According to the description of Algorithm $\mathcal{A}$ the weight $W_{\mathcal{A}}$ of all constructed spanning trees $T_1, \ldots, T_m$ is a random value equal to

$$W_{\mathcal{A}} = (n-1)a_n + (b_n - a_n)W'_{\mathcal{A}},$$

where

$$W'_{\mathcal{A}} = W'_1 + W'_2 + W'_3.$$

W'_1, W'_2, W'_3 are normalized random variables for values W_1, W_2, W_3, respectively.

Let ξ_k be normalized random variable modeling the construction of the edge during the process of Algorithm $\mathcal{A}$.

$W'_1 = \sum_{i=1}^{m} \sum_{k=1}^{d} \xi_k$ since we construct the path P_i consists of d edges and repeat this construction m times during Step 1.

$W'_2 = C_m^2 \left(4\frac{d-1}{2}\xi_{(d-1)/2} + 4\xi_{(d-1)}\right)$, because for each pair of paths (totally, $C_m^2 = \frac{m(m-1)}{2}$ such pairs) we connect each inner vertex of a one half of a path ($\frac{d-1}{2}$ such vertices) by the shortest edge to the inner vertex of a half of another path on the Step 2.5. Since we connect first and second halves of a path to first and second halves of every another path, multiplication of 4 arises in a resulting estimation for $\xi_{(d-1)/2}$. And finally, we connect each end vertices in each pair of paths by the shortest edge to the inner vertex of path on the Step 2.6. (we can compute this shortest edge by looking over all $(d-1)$ inner vertices of corresponding path). Since every pair of paths has 4 end vertices we multiply the estimation for $\xi_{(d-1)}$ by 4.

$W'_3 = m(n - m(d+1))\xi_{(d-1)}$ since we connect each vertex from $G(V')$ ($|G(V')| = n - m(d+1)$) by the shortest edge to the inner vertex of the path P_l ($(d-1)$ such vertices), $1 \le l \le m$. And we repeat this construction m times.

Lemma 1. *For* $\mathbf{E}(W'_{\mathcal{A}})$ *such inequality is true*

$$\mathbf{E}(W'_{\mathcal{A}}) \le m \ln d + \frac{2mn}{d}.$$

Proof. Consider separately expectations of random variables for values W_1, W_2 and W_3.

$$\mathbf{E}W'_1 = \sum_{i=1}^{m} \sum_{k=1}^{d} \mathbf{E}\xi_k = m \sum_{k=1}^{d} \frac{1}{k+1} \le m \ln d;$$

$$\mathbf{E}W'_2 = C_m^2 \left(4\frac{d-1}{2}\mathbf{E}\xi_{(d-1)/2} + 4\mathbf{E}\xi_{(d-1)}\right) = \frac{m(m-1)}{2}\left(\frac{4(d-1)/2}{(d-1)/2+1} + \frac{4}{d}\right) \le 2m^2.$$

$$\mathbf{E}W'_3 = m(n - m(d+1))\mathbf{E}\xi_{(d-1)} = m\frac{n - m(d+1)}{d} \le \frac{mn}{d} - m^2.$$

From the previous equations we get

$$\mathbf{E}(W'_{\mathcal{A}}) = \mathbf{E}\left(W'_1 + W'_2 + W'_3\right) \le m \ln d + 2m^2 + \frac{mn}{d} - m^2 \le m \ln d + \frac{2mn}{d}.$$

Lemma 2. *Algorithm $\mathcal{A}$ for solving the m-d-UMST on n-vertex complete graph with weights of edges from UNI(a_n; b_n) has the following estimates of the relative error ε_n and the failure probability δ_n:*

$$\varepsilon_n = (1 + \lambda_n)\frac{(b_n - a_n)}{m(n-1)a_n}\widetilde{EW}'_{\mathcal{A}}, \tag{2}$$

$$\delta_n = \mathbf{P}\left\{\widetilde{W}'_{\mathcal{A}} > \lambda_n\widetilde{EW}'_{\mathcal{A}}\right\}, \tag{3}$$

where $\lambda_n > 0$, $\widetilde{EW}'_{\mathcal{A}}$ is some upper bound for expectation $\mathbf{E}W'_{\mathcal{A}}$.

Proof.

$$\mathbf{P}\left\{W_{\mathcal{A}} > (1 + \varepsilon_n)OPT\right\} \leq \mathbf{P}\left\{W_{\mathcal{A}} > (1 + \varepsilon_n)m(n-1)a_n\right\}$$

$$= \mathbf{P}\left\{m(n-1)a_n + (b_n - a_n)W'_{\mathcal{A}} > (1 + \varepsilon_n)m(n-1)a_n\right\}$$

$$= \mathbf{P}\left\{W'_{\mathcal{A}} - \mathbf{E}W'_{\mathcal{A}} > \frac{\varepsilon_n m(n-1)a_n}{(b_n - a_n)} - \mathbf{E}W'_{\mathcal{A}}\right\}$$

$$= \mathbf{P}\left\{\widetilde{W}'_{\mathcal{A}} > \frac{\varepsilon_n m(n-1)a_n}{(b_n - a_n)} - \mathbf{E}W'_{\mathcal{A}}\right\}$$

$$\leq \mathbf{P}\left\{\widetilde{W}'_{\mathcal{A}} > \frac{\varepsilon_n m(n-1)a_n}{(b_n - a_n)} - \widetilde{EW}'_{\mathcal{A}}\right\} = \mathbf{P}\left\{\widetilde{W}'_{\mathcal{A}} > \lambda_n\widetilde{EW}'_{\mathcal{A}}\right\} = \delta_n.$$

Further for the probabilistic analysis of Algorithm $\mathcal{A}$ we use the following probabilistic statement

Petrov's Theorem [15]. *Consider independent random variables $X_1, \ldots, X_n$. Let there be positive constants T and $h_1, \ldots, h_n$ such that for all $k = 1, \ldots, n$ and $0 \leq t \leq T$ the following inequalities hold:*

$$\mathbf{E}e^{tX_k} \leq e^{\frac{h_k t^2}{2}}. \tag{4}$$

Set $S = \sum_{k=1}^n X_k$ and $H = \sum_{k=1}^n h_k$. Then

$$\mathbf{P}\{S > x\} \leq \begin{cases} \exp\{-\frac{x^2}{2H}\}, & \text{if } 0 \leq x \leq HT, \\ \exp\{-\frac{Tx}{2}\}, & \text{if } x \geq HT. \end{cases}$$

Lemma 3. *Let ξ_k be random variable equal to minimum over k independent random variables from the class UNI(0; 1). Given constants $T = 1$ and $h_k = \frac{1}{(k+1)^2}$. Then for variables $\widetilde{\xi}_k = \xi_k - \mathbf{E}\xi_k$ the condition (4) of Petrov's Theorem holds for each $t \leq T$ and $1 \leq k < n$.*

Proof. Evidently, $\mathbf{E}\xi_k = \frac{1}{k+1}$. Denote $\alpha = \frac{t}{k+1}$. Using the formula

$$\mathbf{E}e^{t\xi_k} = \sum_{i=0}^{\infty} \frac{t^i}{(k+1)\cdots(k+i)}$$

(see in the book [8], p. 120), we estimate the value $\mathbf{E}e^{t\xi_k}$ from above:

$$\mathbf{E}e^{t\xi_k} \le 1+\alpha+\alpha^2\frac{(k+1)}{(k+2)}\sum_{i=0}^{\infty}\left(\frac{t}{k+3}\right)^i = 1+\alpha+\alpha^2\cdot Q_{k,t} \le e^{\alpha+\frac{\alpha^2}{2}} = e^{t\mathbf{E}\xi_k}\cdot e^{\frac{h_k t^2}{2}}.$$

where $Q_{k,t} = \frac{(k+1)}{(k+2)(1-\frac{t}{k+3})} \le Q_{k,T} = \frac{(k+1)(k+3)}{(k+2)^2} < 1$ if $k \ge 1$. From this we have

$$\mathbf{E}e^{t(\xi_k - \mathbf{E}\xi_k)} = \mathbf{E}e^{t\widetilde{\xi}_k} \le e^{\frac{h_k t^2}{2}}.$$

Lemma 3 is proved.

Lemma 4. *In the case* $\ln n \le d < n$ *the following upper bound for the sum of constants* $h_k = \frac{1}{(k+1)^2}$ *that correspond to edges of the spanning tree* T_l, $l \in \{1, \ldots, m\}$

$$H \le \frac{mn}{d}.$$

Proof. In the case $\ln n \le d < n$ the parameter H equal to the sum of H_1, H_2 and H_3 according to the steps of Algorithm $\mathcal{A}$ number 1, 2, 3, respectively. Knowing that notation and estimates from above, we obtain

$$H_1 = m\sum_{k=1}^{d} h_k = m\sum_{k=1}^{d} \frac{1}{(k+1)^2} < \psi m,$$

where $\psi = 0.645$. We have used the well-known Euler's estimate for the inverse square equation $1 + \frac{1}{2^2} + \frac{1}{3^2} + \frac{1}{4^2} + \ldots = \frac{\pi^2}{6} < 1.645$ in the calculation of H_1.

$$H_2 = 4C_m^2(d'h_{d'} + h_{d-1}) \le 2m^2\left(\frac{d'}{(d'+1)^2} + \frac{1}{d^2}\right) = 2m^2\left(\frac{2(d-1)}{(d+1)^2} + \frac{1}{d^2}\right) \le 4m^2\frac{d}{(d+1)^2}.$$

$$H_3 = m\big(n - m(d+1)\big)h_{(d-1)} \le \frac{mn}{d^2} - m^2\frac{d}{(d+1)^2}.$$

Since $n \ge m(d+1)$ and $m \ge 2$ we get

$$H = H_1 + H_2 + H_3 < \psi m + 4m^2\frac{d}{(d+1)^2} + \left(\frac{mn}{d^2} - m^2\frac{d}{(d+1)^2}\right) \le$$

$$= \frac{mn}{d}\left(\frac{d\psi}{n} + \frac{1}{d}\right) + 3m^2\frac{d}{(d+1)^2} \le \left(\frac{\psi d}{2(d+1)} + \frac{1}{d} + \frac{3d^2}{(d+1)^3}\right)\frac{mn}{d} \le \frac{mn}{d}.$$

Lemma 4 is proved.

Theorem 1. *Let the parameter $d = d_n$ be defined so that*

$$\ln n \leq d < n, \tag{5}$$

Then Algorithm $\mathcal{A}$ gives asymptotically optimal solutions for the problem m-d-UMST on n-vertex complete undirected graph with weights of edges from $UNI(a_n; b_n)$ with the failure probability

$$\delta_n = n^{-m} \to 0, \ as \ n \to \infty, \tag{6}$$

and the following conditions of asymptotical optimality

$$\frac{b_n}{a_n} = \begin{cases} o(d), & if \ \ln n \leq d < n/\ln n, \\ o(\frac{n}{\ln n}), & if \ n/\ln n \leq d < n. \end{cases} \tag{7}$$

Proof. First of all, we note that in the course of the Algorithm $\mathcal{A}$ we are dealing with random variables of the type ξ_k, $1 \leq k \leq d$. In the case of graphs with weights of edges from $UNI(a_n; b_n)$ these variables satisfy the conditions of the Petrov's theorem for constants $T = 1$ and $h_k = \frac{1}{(k+1)^2}$ (see Lemma 3).

We will carry out a proof for two cases of possible values of the parameter d: $\ln n \leq d < n/\ln n$ and $n/\ln n \leq d < n$.

Case 1: $\ln n \leq d < n/\ln n$.

Lemma 5. *In the case $\ln n \leq d < n/\ln n$ the following upper bound for $\mathbf{E}W'_{\mathcal{A}}$ holds:*

$$\mathbf{E}\widetilde{W}'_{\mathcal{A}} = \frac{3mn}{d}.$$

Proof. Given the fact that $\ln d \leq \ln n$ and $d < \frac{n}{\ln n}$ we have:

$$\mathbf{E}W'_{\mathcal{A}} \leq m \ln d + \frac{2mn}{d} \leq m \ln n + \frac{2mn}{d} < m\frac{n}{d} + \frac{2mn}{d} = \frac{3mn}{d} = \widetilde{\mathbf{E}W}'_{\mathcal{A}}.$$

Lemma 5 is proved.

According to Lemma 5 and the formula (2) for the relative error we have

$$\varepsilon_n = (1+\lambda_n)\frac{(b_n - a_n)}{m(n-1)a_n}\widetilde{\mathbf{E}W}'_{\mathcal{A}} = (1+\lambda_n)\frac{(b_n - a_n)}{m(n-1)a_n}\frac{3mn}{d} \leq (1+\lambda_n)\frac{3n}{(n-1)}\frac{(b_n/a_n)}{d}.$$

Setting $\lambda_n = 1$ we see that $\varepsilon_n \to 0$ under the condition

$$\frac{b_n}{a_n} = o(d_n).$$

Now using Petrov's Theorem and Lemma 5, estimate the failure probability:

$$\delta_n = \mathbf{P}\{\widetilde{W}'_{\mathcal{A}} > \lambda_n \widetilde{\mathbf{E}W}'_{\mathcal{A}}\} = \mathbf{P}\left\{\widetilde{W}'_{\mathcal{A}} > \lambda_n \frac{3mn}{d}\right\} = \mathbf{P}\left\{\widetilde{W}'_{\mathcal{A}} > \frac{3mn}{d}\right\}.$$

Define constants $T = 1$ and $h_k = \frac{1}{(k+1)^2}$ for each variable, whose weight corresponds to a random variable ξ_k, and which are included to the constructed spanning tree.

From Lemma 4 and the inequality: $d < \frac{n}{\ln n}$ we have

$$TH \le \frac{mn}{d} < \frac{3mn}{d} = x$$

According to Petrov's Theorem, we have an estimate for the failure probability of Algorithm $\mathcal{A}$:

$$\delta_n = \mathbf{P}\{\widetilde{W}'_{\mathcal{A}} > x\} \le \exp\left\{-\frac{Tx}{2}\right\}.$$

Since $\ln n < \frac{n}{d}$ it holds that

$$\frac{Tx}{2} = \frac{3mn}{2d} > m\ln n.$$

From this it follows that

$$\delta_n = \mathbf{P}\{W'_{\mathcal{A}} > x\} \le \exp\left\{-\frac{Tx}{2}\right\} < \exp(-m\ln n) = \frac{1}{n^m} \to 0.$$

So in the Case 1 Algorithm $\mathcal{A}$ gives asymptotically optimal solution for the problem m-d-UMST on n-vertex complete graph with weights of edges from UNI$(a_n; b_n)$.

Case 2: $n/\ln n \le d < n$.

Lemma 6. *In the Case 2 $(n/\ln n \le d < n)$ the following inequality holds:*

$$\mathbf{E}W'_{\mathcal{A}} \le 3m\ln n = \widetilde{\mathbf{E}W}'_{\mathcal{A}}.$$

Proof. For all d, $n/\ln n \le d < n$ the following inequality holds:

$$\frac{n}{d} \le \ln n. \qquad (8)$$

According to the Lemma 1, and taking into account the inequality (8), we have

$$\mathbf{E}W'_{\mathcal{A}} \le \left(m\ln d + \frac{2mn}{d}\right) \le 3\,m\ln n = \widetilde{\mathbf{E}W}'_{\mathcal{A}}.$$

Lemma 6 is proved.

According to the Lemma 6 and formula (2), for the relative error we have

$$\varepsilon_n = (1 + \lambda_n)\frac{(b_n - a_n)}{(n-1)a_n}\widetilde{\mathbf{E}W}'_{\mathcal{A}} = (1 + \lambda_n)\frac{(b_n - a_n)}{m(n-1)a_n} \cdot 3m\ln n$$

$$\le (1 + \lambda_n)\frac{3(b_n/a_n)\ln n}{(n-1)}.$$

Setting $\lambda_n = 1$ we see, that $\varepsilon_n \to 0$ under condition

$$\frac{b_n}{a_n} = o\left(\frac{n}{\ln n}\right).$$

Now using Petrov's Theorem and Lemma 6, we estimate the failure probability

$$\delta_n = \mathbf{P}\{\widetilde{W}'_{\mathcal{A}} > \lambda_n \widetilde{\mathbf{E}W}'_{\mathcal{A}}\} = \mathbf{P}\{\widetilde{W}'_{\mathcal{A}} > 3m\ln n\},$$

Set the constants h_k as in the Case 1. Define $T = 1$ and $x = 3m\ln n$. Taking into account Lemma 4 and the values x, T, H and $d \geq \frac{n}{\ln n}$, the following inequality is true:

$$TH \leq \frac{mn}{d} < 3m\ln n = x.$$

According to Petrov's Theorem, we have an estimate for the failure probability of Algorithm $\mathcal{A}$. Since $\frac{Tx}{2} > m\ln n$:

$$\delta_n = \mathbf{P}\{W'_{\mathcal{A}} > x\} \leq \exp\left\{-\frac{Tx}{2}\right\} \leq \exp(-m\ln n) = \frac{1}{n^m} \to 0.$$

From this it follows that in the Case 2 Algorithm $\mathcal{A}$ also gives asymptotically optimal solution for the problem m-d-UMST on n-vertex complete graph with weights of edges from $\mathrm{UNI}(a_n; b_n)$.

We conclude, that within the values of the parameter d for both cases, under conditions (7) we have estimates of the relative error $\varepsilon_n \to 0$ and the failure probability $\delta_n \to 0$ as $n \to \infty$. Theorem 1 is completely proved.

4   Conclusion

In this work, we have described an algorithm for solving several edge-disjoint given-diameter Minimal Spanning Tree problem in a complete edge-weighted undirected graph. We also have obtained asymptotic optimality conditions for this algorithm in the case of a uniform distribution for the weights of the graph edges. It would be interesting to investigate this problem on input data with infinite support like exponential or truncated-normal distribution and on discrete distributions. It is interesting to consider asymptotic optimality of the problem of finding several edge-disjoined spanning trees with a given or bounded diameter using an algorithmic scheme without representing the paths in the form of the corresponding halves, as was done at Step 2 of the described algorithm. It would also be desirable to consider the problem of finding m edge-disjoint Maximum Spanning Trees with given or bounded diameter in future works.

References

1. Angel, O., Flaxman, A.D., Wilson, D.B.: A sharp threshold for minimum bounded-depth and bounded-diameter spanning trees and Steiner trees in random networks. Combinatorica **32**, 1–33 (2012). https://doi.org/10.1007/s00493-012-2552-z
2. Cooper, C., Frieze, A., Ince, N., Janson, S., Spencer, J.: On the length of a random minimum spanning tree. Comb. Probab. Comput. **25**(1), 89–107 (2016)
3. Frieze, A.: On the value of a random MST problem. Discret. Appl. Math. **10**, 47–56 (1985)
4. Garey, M.R., Johnson, D.S.: Computers and Intractability. Freeman, San Francisco (1979). 340 p.
5. Gimadi, E.Kh., Glebov, N.I., Perepelitsa, V.A.: Algorithms with estimates for discrete optimization problems. Problemy Kibernetiki **31**, 35–42 (1975). (in Russian)
6. Gimadi, E.Kh., Istomin, A.M., Shin, E.Yu.: On Algorithm for the minimum spanning tree problem bounded below. In: Proceedings of the DOOR 2016, Vladivostok, Russia, 19–23 September 2016, vol. 1623, pp. 11–17. CEUR-WS (2016)
7. Gimadi, E.Kh., Istomin, A.M., Shin, E.Yu.: On given diameter MST problem on random instances. In: CEUR Workshop Proceedings, pp. 159–168 (2019)
8. Gimadi, E.Kh., Khachay, M.Yu.: Extremal Problems on Sets of Permutations, UrFU Publ., Ekaterinburg (2016). 219 p. (in Russian)
9. Gimadi, E.K., Serdyukov, A.I.: A probabilistic analysis of an approximation algorithm for the minimum weight spanning tree problem with bounded from below diameter. In: Inderfurth, K., Schwödiauer, G., Domschke, W., Juhnke, F., Kleinschmidt, P., Wäscher, G. (eds.) Operations Research Proceedings 1999. ORP, vol. 1999, pp. 63–68. Springer, Heidelberg (2000). https://doi.org/10.1007/978-3-642-58300-1_12
10. Gimadi, E.Kh., Shevyakov, A.S., Shin, E.Yu.: Asymptotically Optimal Approach to a Given Diameter Undirected MST Problem on Random Instances, Proceedings of 15-th International Asian School-Seminar OPCS-2019, Publisher: IEEE Xplore, pp. 48–52 (2019)
11. Gimadi, E.K., Shin, E.Y.: Probabilistic analysis of an algorithm for the minimum spanning tree problem with diameter bounded below. J. Appl. Ind. Math. **9**(4), 480–488 (2015). https://doi.org/10.1134/S1990478915040043
12. Gimadi, E.K., Shin, E.Y.: On given diameter MST problem on random input data. In: Bykadorov, I., Strusevich, V., Tchemisova, T. (eds.) MOTOR 2019. CCIS, vol. 1090, pp. 30–38. Springer, Cham (2019). https://doi.org/10.1007/978-3-030-33394-2_3
13. Gimadi, E.K., Shevyakov, A.S., Shtepa, A.A.: A given diameter MST on a random graph. In: Olenev, N., Evtushenko, Y., Khachay, M., Malkova, V. (eds.) OPTIMA 2020. LNCS, vol. 12422, pp. 110–121. Springer, Cham (2020). https://doi.org/10.1007/978-3-030-62867-3_9
14. Gruber M.: Exact and Heuristic Approaches for Solving the Bounded Diameter Minimum Spanning Tree Problem, Vienna University of Technology, PhD Thesis (2008)
15. Petrov, V.V.: Limit Theorems of Probability Theory. Sequences of Independent Random Variables. Clarendon Press, Oxford (1995). 304 p.

An FPTAS for the Δ-Modular Multidimensional Knapsack Problem

D. V. Gribanov[✉][iD]

National Research University Higher School of Economics,
25/12 Bolshaja Pecherskaja Ulitsa, Nizhny Novgorod 603155, Russian Federation

Abstract. It is known that there is no EPTAS for the m-dimensional knapsack problem unless $W[1] = FPT$. It is true already for the case, when $m = 2$. But, an FPTAS still can exist for some other particular cases of the problem.

In this note, we show that the m-dimensional knapsack problem with a Δ-modular constraints matrix admits an FPTAS, whose complexity bound depends on Δ linearly. More precisely, the proposed algorithm arithmetical complexity is $O(n \cdot (1/\varepsilon)^{m+3} \cdot \Delta)$, for m being fixed. Our algorithm is actually a generalisation of the classical FPTAS for the 1-dimensional case.

Strictly speaking, the considered problem can be solved by an exact polynomial-time algorithm, when m is fixed and Δ grows as a polynomial on n. This fact can be observed combining results of the papers [9,12,28]. We give a slightly more accurate analysis to present an exact algorithm with the complexity bound $O(n \cdot \Delta^{m+1})$, for m being fixed. Note that the last bound is non-linear by Δ with respect to the given FPTAS.

The goal of the paper is only to prove the existence of the described FPTAS, and a more accurate analysis can give better constants in exponents. Moreover, we are not worry to much about memory usage.

Keywords: Multidimensional knapsack problem · Δ-modular integer linear programming · FPTAS · Δ-modular matrix · Approximation algorithm

1 Introduction

1.1 Basic Definitions and Notations

Let $A \in \mathbb{Z}^{m \times n}$ be an integer matrix. We denote by A_{ij} the ij-th element of the matrix, by A_{i*} its i-th row, and by A_{*j} its j-th column. The set of integer values from i to j, is denoted by $i:j = \{i, i+1, \ldots, j\}$. Additionally, for subsets $I \subseteq \{1, \ldots, m\}$ and $J \subseteq \{1, \ldots, n\}$, the symbols A_{IJ} and $A[I, J]$ denote the sub-matrix of A, which is generated by all the rows with indices in I and all

The article was prepared under financial support of Russian Science Foundation grant No 21-11-00194.

P. Pardalos et al. (Eds.): MOTOR 2021, LNCS 12755, pp. 79–95, 2021.
https://doi.org/10.1007/978-3-030-77876-7_6

the columns with indices in J. If I or J are replaced by $*$, then all the rows or columns are selected, respectively. Sometimes, we simply write A_I instead of A_{I*} and A_J instead of A_{*J}, if this does not lead to confusion.

The maximum absolute value of entries in a matrix A is denoted by $\|A\|_{\max} = \max_{i,j} |A_{ij}|$. The l_p-norm of a vector x is denoted by $\|x\|_p$.

Definition 1. *For a matrix $A \in \mathbb{Z}^{m \times n}$, by*

$$\Delta_k(A) = \max\{|\det A_{IJ}| : I \subseteq 1{:}m, \ J \subseteq 1{:}n, \ |I| = |J| = k\},$$

we denote the maximum absolute value of determinants of all the $k \times k$ submatrices of A. Clearly, $\Delta_1(A) = \|A\|_{\max}$. Additionally, let $\Delta(A) = \Delta_{\mathrm{rank}(A)}(A)$.

1.2   Description of Results and Related Work

Let $A \in \mathbb{Z}_+^{m \times n}$, $b \in \mathbb{Z}_+^m$, $c \in \mathbb{Z}_+^n$ and $u \in \mathbb{Z}_+^n$. *The bounded m-dimensional knapsack problem (shortly m-BKP) can be formulated as follows:*

$$c^\top x \to \max$$

$$\begin{cases} Ax \leq b \\ 0 \leq x \leq u \\ x \in \mathbb{Z}^n. \end{cases} \qquad\qquad (m\text{-BKP})$$

It is well known that the m-BKP is NP-hard already for $m = 1$. However, it is also well known that the 1-BKP admits an FPTAS. The historically first FPTAS for the 1-BKP was given in the seminal work of O. Ibarra and C. Kim [17]. The results of [17] were improved in many ways, for example in the works [5, 14, 16, 18, 20–22, 27, 29, 30, 34]. But, it was shown in [25] (see [23, p. 252] for a simplified proof) that the 2-BKP does not admit an FPTAS unless $P = NP$. Due to [26], the 2-BKP does not admit an EPTAS unless $W[1] = FPT$. However, the m-BKP still admits a PTAS. To the best of our knowledge, the state of the art PTAS is given in [4]. The complexity bound proposed in [4] is $O(n^{\lceil \frac{m}{\varepsilon} \rceil - m})$. The perfect survey is given in the book [23].

Within the scope of the article, we are interested in studying m-BKP problems with a special restriction on sub-determinants of the constraints matrix A. More precisely, we assume that all rank-order minors of A are bounded in an absolute value by Δ. We will call this class of m-BKPs as Δ-*modular m-BKPs*. The main result of the paper states that the Δ-modular m-BKP admits an FPTAS, whose complexity bound depends on Δ linearly, for any fixed m.

Theorem 1. *The Δ-modular m-BKP admits an FPTAS with the arithmetical complexity bound*

$$O(T_{LP} \cdot (1/\varepsilon)^{m+3} \cdot (2m)^{2m+6} \cdot \Delta),$$

where T_{LP} is the linear programming complexity bound.

Proof of the theorem is given in Sect. 2.

Due to the seminal work of N. Megiddo [31], the linear program can be solved by a linear-time algorithm if m is fixed.

Corollary 1. *For fixed m the complexity bound of Theorem 1 can be restated as*

$$O(n \cdot (1/\varepsilon)^{m+3} \cdot \Delta).$$

We need to note that results of the papers [9,12,28] can be combined to develop an exact polynomial-time algorithm for the considered Δ-modular m-BKP problem, and even more, for any Δ-modular ILP problem in standard form with a fixed number of constraints m. But, the resulting algorithm complexity contains a non-linear dependence on Δ in contrast with the developed FPTAS. The precise formulation will be given in the following Theorem 2 and Corollary 2. First, we need to make some definitions:

Definition 2. *Let $A \in \mathbb{Z}^{m \times n}$, $b \in \mathbb{Z}^m$, $c \in \mathbb{Z}^n$, $u \in \mathbb{Z}_+^n$, $\text{rank}(A) = m$ and $\Delta = \Delta(A)$. The bounded Δ-modular ILP in standard form (shortly m-BILP) can be formulated as follows:*

$$c^\top x \to \max$$
$$\begin{cases} Ax = b \\ 0 \leq x \leq u \\ x \in \mathbb{Z}^n. \end{cases} \qquad (m\text{ BILP})$$

The main difference between the problems m-BILP and m-BKP is that the input of the problem m-BILP can contain negative numbers. The inequalities of the problem m-BKP can be turned to equalities using slack variables.

Definition 3. *Consider the problem m-BILP. Let z^* be an optimal solution of m-BILP and x^* be an optimal vertex-solution of the LP relaxation of m-BILP. The l_1-proximity bound H of the problem m-BILP is defined by the formula*

$$H = \max_{x^*} \min_{z^*} \|x^* - z^*\|_1.$$

It is proven in [9] that

$$H \leq m \cdot (2m \cdot \Delta_1 + 1)^m, \quad \text{where } \Delta_1 = \Delta_1(A) = \|A\|_{\max}. \qquad (1)$$

It was noted in [28, formula (4)] that this proximity bound (1) of the paper [9] can be restated to work with the parameter $\Delta(A)$ instead of $\Delta_1(A)$. More precisely, there exists an optimal solution z^* of the m-BILP problem such that

$$H \leq m \cdot (2m + 1)^m \cdot \Delta, \quad \text{where } \Delta = \Delta(A).$$

Theorem 2. *The m-BILP problem can be solved by an algorithm with the following arithmetical complexity:*

$$n \cdot O(H + m)^{m+1} \cdot \log^2(H) \cdot \Delta + T_{LP}.$$

The previous complexity bound can be slightly improved in terms of H:

$$n \cdot O(\log m)^{m^2} \cdot (H + m)^m \cdot \log^2(H) \cdot \Delta + T_{LP}.$$

Additionally, for problems with non-negative A, we can remove the $\log^2(H)$ term in the complexity bound:

$$n \cdot O(\log m)^{m^2} \cdot (H + m)^m \cdot \Delta + T_{LP}.$$

The proof can be found in Sect. 3.

Remark 1. The algorithms described in the proof of Theorem 2 are using hash tables with linear expected constructions time and constant worst-case lookup time to store information dynamic tables. An example of a such hash table can be found in the book [7]. So, strictly speaking, algorithms of Theorem 2 are randomized.

Randomization can be removed by using any balanced search-tree, for example, RB-tree [7]. It will lead to additional logarithmic term in the complexity bound.

Applying the proximity bounds (3) and (1) to the previous Theorem 2, we can obtain estimates that are independent of H. For example, we obtain the following corollary:

Corollary 2. *The problem m-BILP can be solved by an algorithm with the following arithmetical complexity bound:*

$$n \cdot O(\log m)^{m^2} \cdot O(m)^{m^2+m} \cdot \Delta^{m+1} \cdot \log^2 \Delta + T_{LP} \quad and$$
$$n \cdot \Delta^{m+1} \cdot \log^2 \Delta, \quad for\ m\ being\ fixed.$$

For problems with non-negative matrix A we can improve the complexity bound by removing of $\log^2 \Delta$ term.

Remark 2. Taking $m = 1$ in the previous corollary we obtain the $O(n \cdot \Delta^2)$ complexity bound for the classical bounded knapsack problem, where Δ is the maximal absolute value of item weights. Our bound is better than the previous state of the art bounds $O(n^2 \cdot \Delta^2)$ and $O(n \cdot \Delta^2 \cdot \log^2 \Delta)$ due to [9].

Better complexity bound for searching of an exact solution can be achieved for the unbounded version of the m-BILP problem. More precisely, for this case, the paper [19] gives the complexity bound

$$O(\sqrt{m}\Delta)^{2m} + T_{LP}.$$

We note that the original complexity bound from the work [19] is stated with respect to the parameter $\Delta_1(A) = \|A\|_\infty$ instead of $\Delta(A)$ (see the next Remark 3), but, due to Lemma 1 of [12], we can assume that $\Delta_1(A) \leq \Delta(A)$.

Remark 3. Another interesting parameter of the considered problems m-BKP and m-BILP is $\Delta_1(A) = \|A\|_{\max}$. Let us denote $\Delta_1 = \Delta_1(A)$. The first exact quasipolynomial-time algorithm for m-BILP was constructed in the seminal work of C. H. Papadimitriou [32]. The result of [32] was recently improved in [9], where it was shown that the m-BILP can be solved exactly by an algorithm with the arithmetical complexity

$$n \cdot O(m)^{(m+1)^2} \cdot \Delta_1^{m(m+1)} \cdot \log^2(m\Delta_1) + T_{LP} \quad \text{and}$$

$$n \cdot \Delta_1^{m(m+1)} \cdot \log^2(\Delta_1), \quad \text{for } m \text{ being fixed.} \tag{2}$$

Due to the results of [19], the unbounded version of the problem can be solved by an algorithm with the arithmetical complexity

$$O(\sqrt{m}\Delta_1)^{2m} + T_{LP}. \tag{3}$$

The results of our note can be easily restated to work with the Δ_1 parameter. Using the Hadamard's inequality, the arithmetical complexity bound of Corollary 2 becomes

$$n \cdot O(\log m)^{m^2} \cdot O(m)^{m^2+3/2m} \cdot \Delta^{m(m+1)} \cdot \log^2(\Delta_1) + T_{LP} \quad \text{and}$$

$$n \cdot \Delta_1^{m(m+1)} \cdot \log^2(\Delta_1), \quad \text{for } m \text{ being fixed,}$$

which is slightly better, than the bound (2) of [9]. For non-negative matrices A the $\log^2 \Delta_1$ term can be removed.

Additionally, Corollary 2 gives currently best bound $O(n \cdot \Delta_1^2)$ for the classical 1-dimensional bounded knapsack problem, see Remark 2.

For our FPTAS we give here a better way, than to use the Hadamard's inequality. Definitely, for $\gamma > 0$ and $M = \{y = Ax : x \in \mathbb{R}_+^n, \|x\|_1 \leq \gamma\}$ we trivially have $|M \cap \mathbb{Z}^m| \leq (\gamma\Delta_1)^m$. Applying the algorithm from Section 2 to this analogue of Corollary 3, it gives an algorithm with the arithmetical complexity

$$O(T_{LP} \cdot (1/\varepsilon)^{m+3} \cdot m^{2m+6} \cdot (2\Delta_1)^m) \quad \text{and}$$

$$O(n \cdot (1/\varepsilon)^{m+3} \cdot \Delta_1^m), \quad \text{for } m \text{ being fixed.}$$

For sufficiently large ε the last bounds give a better dependence on m and Δ_1, than bounds (2) from [9].

Remark 4 (Why Δ-modular ILPs could be interesting?). It is well known that the Maximal Independent Set (shortly MAX-IS) problem on a simple graph $G = (V, E)$ can be formulated by the ILP

$$1^\top x \to \max$$

$$\begin{cases} A(G)\, x \leq 1 \\ x \in \{0, 1\}^{|V|}, \end{cases} \tag{MAX-IS}$$

where $A(G) \in \{0,1\}^{|E| \times |V|}$ is the edge-vertex incidence matrix of G. Due to the seminal work [13]

$$\Delta(A(G)) = 2^{\nu(G)},$$

where $\nu(G)$ is the odd-cycle packing number of G. Hence, the existence of a polynomial-time algorithm for Δ-modular ILPs will lead to the existence of a polynomial-time algorithm for the MAX-IS problem for graphs with a fixed $\nu(G)$ value. Recently, it was shown in [2] that 2-modular ILPs admit a strongly polynomial-time algorithm, and consequently, the MAX-IS $\in P$ for graphs with one independent odd-cycle. But, existence of a polynomial-time algorithms even for the 3-modular or 4-modular ILP problems is an interesting open question, as well as existence of a polynomial-time algorithm for the MAX-IS problem on graphs with $\nu(G) = 2$. Finally, due to [1], if $\Delta(\bar{A})$ is fixed, where $\bar{A} = \left(\begin{smallmatrix} 1^\top \\ A(G) \end{smallmatrix} \right)$ is the extended matrix of the ILP MAX-IS, then the problem can be solved by a polynomial time algorithm. The shorter proof could be found in [10,11], as well as analogue results for vertex and edge Maximal Dominating Set problems. For recent progress on the MAX-IS problem with respect to the $\nu(G)$ parameter see the papers [3,6,15].

Additionally, we note that, due to [3], there are no polynomial-time algorithms for the MAX-IS problem on graphs with $\nu(G) = \Omega(\log n)$ unless the ETH (the Exponential Time Hypothesis) is false. Consequently, with the same assumption, there are no algorithms for the Δ-modular ILP problem with the complexity bound $\text{poly}(s) \cdot \Delta^{O(1)}$, where s is an input size. Despite the fact that algorithms with complexities $\text{poly}(s) \cdot \Delta^{f(\Delta)}$ or $s^{f(\Delta)}$ may still exist, it is interesting to consider existence of algorithms with a polynomial dependence on Δ in their complexities for some partial cases of the Δ-modular ILP problem. It is exactly what we do in the paper while fixing the number of constraints in ILP formulations of the problems m-BKP and m-BILP.

Due to the Hadamard's inequality, the existence of an ILP algorithm, whose complexity depends on Δ linearly, can give sufficiently better complexity bounds in terms of Δ_1, than the bounds of Remark 3.

Remark 5 (Some notes about lower bounds for fixed m.). Unfortunately, there are not many results about lower complexity bounds for the problem m-BILP with fixed m. But, we can try to adopt some bounds based on the Δ_1 parameter to our case. For example, the existence of an algorithm with the complexity bound

$$2^{o(m)} \cdot 2^{o(\log_2 \Delta)} \cdot \text{poly}(s)$$

will contradict to the ETH. It is a straightforward adaptation of [8, Theorem 3].

The Theorem 13 of [19] states that for any $\delta > 0$ there is no algorithm with the arithmetical complexity bound

$$f(m) \cdot (n^{2-\delta} + \Delta_1^{2m-\delta}),$$

unless there exists a truly sub-quadratic algorithm for the $(min, +)$-convolution. Using Hadamard's inequality, it adopts to

$$f(m) \cdot (n^{2-\delta} + \Delta^{2-\delta/m}).$$

The best known bound in terms of m is given in [24, Corollary 2]. More precisely, the existence of an algorithm with the complexity bound

$$2^{o(m \log m)} \cdot \Delta_1^{f(m)} \cdot \text{poly}(s)$$

will contradict to the ETH. But, we does not know how to adopt it for Δ-modular case at the moment.

Unfortunately, all mentioned results are originally constructed for the version of m-BILP with unbounded variables and it is the main reason, why their bounds are probably weak with respect to the dependence on the Δ parameter. And it would be very interesting to construct a lower bound of the form

$$f(m) \cdot \Delta^{\Omega(m)} \cdot \text{poly}(s)$$

for the m-BILP problem. Additionally, at the moment we does not know any FPTAS lower bounds for the m-BKP problem. These questions are good directions for future research.

2 Proof of the Theorem 1

2.1 Greedy Algorithm

The $1/(m + 1)$-approximate algorithm for the m-BKP is presented in [4] (see also [23, p. 252]) for the case $u = 1$. This algorithm can be easily modified to work with a generic upper bounds vector u.

Algorithm 1. The greedy algorithm

Require: an instance of the (m-BKP) problem;
Ensure: return $1/(m + 1)$-approximate solution of the (m-BKP);
  1: compute an optimal solution x^{LP} of the LP relaxation of the (m-BKP);
  2: $y := \lfloor x^{LP} \rfloor$ — a rounded integer solution;
  3: $F := \{i : x_i^{LP} \notin \mathbb{Z}\}$ — variables with fractional values;
  4: **return** $\mathbf{C}_{gr} := \max\{c^\top y, \max_{i \in F}\{c_i\}\}$;

Since the vector x^{LP} can have at most m fractional coordinates and

$$c^\top y + \sum_{i \in F} c_i = c^\top \lceil x^{LP} \rceil \geq c^\top x^{LP} \geq \mathbf{C}_{opt}, \tag{4}$$

we have $\mathbf{C}_{gr} \geq \frac{1}{m+1} \mathbf{C}_{opt}$.

2.2 Dynamic Programming by Costs

The dynamic programming by costs is one of the main tools in many FPTASes for the 1-BKP. Unfortunately, it probably can not be generalized to work with

m-BKPs for greater values of m. However, such generalizations can exist for some partial cases such as the Δ-modular m-BKP.

Suppose that we want to solve the m-BKP, and it is additionally known that $\|x\|_1 \leq \gamma$, for any feasible solution x and some $\gamma > 0$. Then, to develop a dynamic program it is natural to consider only integer points x that satisfy to $\|x\|_1 \leq \gamma$. The following simple lemma and corollary help to define such a program.

Lemma 1. *Let $A \in \mathbb{Z}^{m \times n}$ and $B \in \mathbb{Z}^{m \times m}$ be the non-degenerate sub-matrix of A. Let additionally $\gamma \in \mathbb{R}_{>0}$, $\Delta = \Delta(A)$, $\delta = |\det B|$ and*

$$M = \{y = Ax \colon x \in \mathbb{R}^n, \|x\|_1 \leq \gamma\},$$

then $|M \cap \mathbb{Z}^m| \leq 2^m \cdot \lceil 1 + \gamma \cdot \frac{\Delta}{\delta} \rceil^m \cdot \Delta.$

Points of $M \cap \mathbb{Z}^m$ can be enumerated by an algorithm with the arithmetical complexity bound:

$$O(m^2 \cdot 2^m \cdot D),$$

where $D = \Delta \cdot \left(\gamma \cdot \frac{\Delta}{\delta} \right)^m$.

Proof. W.l.o.g. we can assume that first m columns of A form the sub-matrix B. Consider a decomposition $A = B(I\ U)$, where $(I\ U)$ is a block-matrix, I is the $m \times m$ identity matrix and the matrix U is determined uniquely from this equality. Clearly, $\Delta((I\ U)) = \frac{\Delta}{\delta}$, so $\Delta_k(U) \leq \frac{\Delta}{\delta}$ for all $k \in 1\!:\!m$. Consider the set

$$N = \{y = \lceil 1 + \gamma \cdot \frac{\Delta}{\delta} \rceil Bx \colon x \in (-1, 1)^m\}.$$

Let us show that $M \subseteq N$. Definitely, if $y = Ax$ for $\|x\|_1 \leq \gamma$, then $y = B(I\ U)x = Bt$, for some $t \in [-\gamma, \gamma]^m \cdot \frac{\Delta}{\delta}$. Finally, $\frac{1}{\lceil 1 + \gamma \cdot \frac{\Delta}{\delta} \rceil} t \in (-1, 1)^m$ and $y \in N$.

To estimate the value $|N \cap \mathbb{Z}^m|$ we just note that N can be covered by 2^m parallelepipeds of the form $\{y = Qx \colon x \in [0, 1)^m\}$, where $Q \in \mathbb{Z}^{m \times m}$ and $|\det Q| = \lceil 1 + \gamma \cdot \frac{\Delta}{\delta} \rceil^m \cdot \Delta$. It is well known that the number of integer points in such parallelepipeds is equal to $|\det Q|$, see for example [36] or [35, Section 16.4]. Hence, $|M \cap \mathbb{Z}^m| \leq |N \cap \mathbb{Z}^m| \leq 2^m \cdot \lceil 1 + \gamma \cdot \frac{\Delta}{\delta} \rceil^m \cdot \Delta$. Points inside of the parallelipiped can be enumerated by an algorithm with arithmetical complexity

$$O(m \cdot \min\{\log(|\det Q|), m\} \cdot |\det Q|),$$

see for example [12]. Applying the last formula, we obtain the desired complexity bound to enumerate all integer points inside N.

Corollary 3. *Let $A \in \mathbb{Z}^{m \times n}$, $\gamma \in \mathbb{R}_{>0}$, $\Delta = \Delta(A)$ and*

$$M = \{y = Ax \colon x \in \mathbb{R}^n, \|x\|_1 \leq \gamma\},$$

then $|M \cap \mathbb{Z}^m| \leq 2^m \cdot \lceil 1 + \gamma \rceil^m \cdot \Delta.$

Points of $M \cap \mathbb{Z}^m$ can be enumerated by an algorithm with the arithmetical complexity bound:

$$O(\log m)^{m^2} \cdot \Delta \cdot \gamma^m.$$

Proof. W.l.o.g. we can assume that $\operatorname{rank}(A) = m$. Let us choose $B \in \mathbb{Z}^{m \times m}$, such that $|\det B| = \Delta$, then the desired $|M \cap \mathbb{Z}^m|$-bound follows from the previous Lemma 1. Due to [37], we can compute a matrix $\hat{B} \in \mathbb{Z}^{m \times m}$ such that $\Delta = O(\log m)^m \cdot \delta$, where $\delta = |\det \hat{B}|$, by a polynomial time algorithm. Finally, we take a complexity bound of the previous Lemma 1 with $D = \Delta \cdot \gamma^m \cdot O(\log m)^{m^2}$.

We note that in the current section we need only first parts of these Lemma 1 and Corollary 3 that only estimate number of points nor enumerate them.

Assume that the goal function of the m-BKP is bounded by a constant C. Then, for any $c_0 \in 1 : C$ and $k \in 1 : n$ we denote by $DP(k, c_0)$ the set of all possible points $y \in \mathbb{Z}_+^m$ that satisfy to the system

$$\begin{cases} c_{1k}^\top x = c_0 \\ y = A_{1k} x \\ A_{1k} x \leq b \\ 0 \leq x \leq u_{1k} \\ x \in \mathbb{Z}^k. \end{cases}$$

In particular, the optimal value of the m-BKP can be computed by the formula

$$c^\top x^{opt} = \max\{c_0 \in [1, C] \cap \mathbb{Z} : DP(n, c_0) \neq \emptyset\}.$$

The set $DP(k, c_0)$ can be recursively computed using the following algorithm:

Algorithm 2. An algorithm to compute $DP(k, c_0)$

1: **for all** $z \in [0, \gamma] \cap [0, u_k] \cap \mathbb{Z}$ **do**
2:     **for all** $y \in DP(k - 1, c_0 - z c_k)$ **do**
3:         **if** $y + A_k z \leq b$ **then**
4:             **add** $y + A_k z$ **into** $DP(k, c_0)$
5:         **end if**
6:     **end for**
7: **end for**

By Corollary 3, we have $|DP(k, c_0)| \leq 2^m \cdot \lceil 1 + \gamma \rceil^m \cdot \Delta$. Consequently, to compute $DP(k, c_0)$ we need at most $O(m \cdot (2\gamma)^{m+1} \cdot \Delta)$ arithmetic operations. The total complexity bound is given by the following trivial lemma.

Lemma 2. *The sets $DP(k, c_0)$ for $c_0 \in 1 : C$ and $k \in 1 : n$ can be computed by an algorithm with the arithmetical complexity*

$$O(n \cdot C \cdot m \cdot (2\gamma)^{m+1} \cdot \Delta).$$

2.3 Putting Things Together

Our algorithm is based on the scheme proposed in the seminal work [17] of O. Ibarra and C. Kim. Our choice of an algorithmic base is justified by the fact that it is relatively easy to generalize the approach of [17] to the m-dimensional case. On the other hand, more sophisticated schemes described in the papers [21, 22, 27, 30] give constant improvements in the exponent or improvements in the memory usage only.

First of all, let us define two parameters $\alpha, \beta \in \mathbb{Q}_{>0}$, whose purpose will be explained later. Let $\mathbf{C}^{gr}$ be the value of the greedy algorithm applied to the original Δ-modular m-BKP, x^{opt} be its integer optimal point and $\mathbf{C}^{opt} = c^\top x^{opt}$. As it was proposed in [17], we split items into heavy and light: $H = \{i \colon c_i > \alpha\, \mathbf{C}^{gr}\}$ and $L = \{i \colon c_i \le \alpha\, \mathbf{C}^{gr}\}$.

It can be shown that $\|x_H^{opt}\|_1 \le \frac{m+1}{\alpha}$. Definitely, if $\|x_H^{opt}\|_1 > \frac{m+1}{\alpha}$, then $\mathbf{C}^{opt} = c^\top x^{opt} \ge c_H^\top x_H^{opt} > \alpha\, \mathbf{C}^{gr} \frac{m+1}{\alpha} = (m+1)\, \mathbf{C}^{gr} \ge \mathbf{C}^{opt}$.

Let $s = \beta\, \mathbf{C}^{gr}$, we put $w = \lfloor \frac{c}{s} \rfloor$. Consider a new Δ-modular m-BKP that consists only from heavy items of the original problem with the scaled costs w.

$$w_H^\top x \to \max$$
$$\begin{cases} A_H x \le b \\ 0 \le x \le u_H \\ x \in \mathbb{Z}^{|H|}. \end{cases} \tag{HProb}$$

It follows that $\|x\|_1 \le \frac{m+1}{\alpha}$ for any feasible solution of (HProb). Additionally, we have $w_H^\top x \le \frac{1}{s} c_H^\top x \le \frac{m+1}{s}\, \mathbf{C}^{gr} \le \frac{m+1}{\beta}$, for any x being feasible solution of (HProb). Hence, we can apply Lemma 2 to construct the sets $DP_H(k, c_0)$ for $k \in 1 \colon n$ and $c_0 \in 1 \colon \lceil \frac{m+1}{\beta} \rceil$. Due to Lemma 2, the arithmetical complexity of this computation is bounded by

$$O\left(n \cdot \frac{m(m+1)}{\beta} \cdot \left(2\frac{m+1}{\alpha}\right)^{m+1} \cdot \Delta\right). \tag{5}$$

To proceed further, we need to define a new notation $Pr(I, t)$. For a set of indexes $I \subseteq 1 \colon n$ and for a vector $t \in \mathbb{Z}_+^m$, we denote by $Pr(I, t)$ the optimal value of the sub-problem, induced by variables with indexes in I and by the right hand side vector t. Or by other words, $Pr(I, t)$ is the optimal value of the problem

$$c_I^\top x \to \max$$
$$\begin{cases} A_I x \le t \\ 0 \le x \le u_I \\ x \in \mathbb{Z}^{|I|}. \end{cases}$$

After $DP_H(n, c_0)$ being computed we can construct resulting approximate solution, which will be denoted as $\mathbf{C}^{apr}$, by the following algorithm.

Algorithm 3. An FPTAS for (m-BKP)

1: **for all** $c_0 \in 1 : \lceil \frac{m+1}{\beta} \rceil$ **do**
2:     **for all** $y \in DP_H(n, c_0)$ **do**
3:        compute an approximate solution q of the problem $Pr(L, b - y)$

$$c_L^\top x \to \max$$

$$\begin{cases} A_L x \leq b - y \\ 0 \leq x \leq u_L \\ x \in \mathbb{Z}^{|L|} \end{cases}$$

      using the greedy algorithm.
4:     $\mathbf{C}^{apr} := \max\{\mathbf{C}^{apr}, s\,c_0 + q\}$.
5:    **end for**
6: **end for**

Due to Corollary 3, the arithmetical complexity of the algorithm can be estimated as

$$O(T_{LP} \cdot \frac{m}{\beta} \cdot (2\frac{m+1}{\alpha})^m \cdot \Delta). \tag{6}$$

Clearly, $x^{opt} = x_H^{opt} + x_L^{opt}$. We denote $C_H^{opt} = c_H^\top x_H^{opt}$, $C_L^{opt} = c_L^\top x_L^{opt}$ and $c_0^* = w_H^\top x_H^{opt}$. The value of c_0^* will arise in some evaluation of Line 1 of the proposed algorithm. Or by other words, we will have $c_0 = c_0^*$ in some evaluation of Line 1. Let y^* be the value of $y \in DP_H(n, c_0^*)$ such that $s\,c_0^* + Pr(L, b - y^*)$ is maximized and q^* be the approximate value of $Pr(L, b - y^*)$, given by the greedy algorithm in Line 3. Clearly, $\mathbf{C}^{apr} \geq s\,c_0^* + q^*$, so our goal is to chose parameters α, β in such a way that the inequality $s\,c_0^* + q^* \geq (1 - \varepsilon)\,\mathbf{C}^{opt}$ will be satisfied.

Firstly, we estimate the difference $\mathbf{C}_H^{opt} - s\,c_0^*$:

$$\mathbf{C}_H^{opt} - s\,c_0^* \leq (c_H^\top - s\,w_H^\top)x_H^{opt} \leq s\,\{c_H^\top / s\}x_H^{opt} \leq$$

$$\leq s\frac{m+1}{\alpha} = \frac{(m+1)\beta\,\mathbf{C}_{gr}}{\alpha} \leq \frac{(m+1)\beta\,\mathbf{C}_{opt}}{\alpha}$$

To estimate the difference $\mathbf{C}_L^{opt} - q^*$ we need to note that $Pr(L, b - y^*) \geq Pr(L, b - A_H x_H^{opt}) = \mathbf{C}_L^{opt}$. It follows from optimality of y^* with respect to the developed dynamic program. Next, since $c_i \leq \alpha\,\mathbf{C}^{gr}$ for $i \in L$, due to the inequality (4), we have

$$q^* \geq Pr(L, b - y^*) - m\,\alpha\,\mathbf{C}^{gr} \geq Pr(L, b - y^*) - (m+1)\,\alpha\,\mathbf{C}^{opt}.$$

Finally, we have

$$\mathbf{C}_L^{opt} - q^* \leq Pr(L, b - y^*) - q^* \leq (m+1)\,\alpha\,\mathbf{C}^{opt}.$$

Putting all inequalities together, we have

$$\mathbf{C}^{opt} - \mathbf{C}^{apr} \le (\mathbf{C}_H^{opt} - s\, c_0^*) + (\mathbf{C}_L^{opt} - q^*) \le$$

$$\le (m+1)(\alpha + \frac{\beta}{\alpha})\, \mathbf{C}^{opt}$$

and

$$\mathbf{C}^{apr} \ge (1 - (m+1)(\alpha + \frac{\beta}{\alpha}))\, \mathbf{C}^{opt}. \tag{7}$$

The total arithmetical complexity can be estimated as

$$O(T_{LP} \cdot \frac{1}{\beta} \cdot (2m)^{m+3} \cdot \left(\frac{1}{\alpha}\right)^{m+1} \cdot \Delta). \tag{8}$$

Finally, after the substitution $\beta = \alpha^2$ and $\alpha = \frac{\varepsilon}{2(m+1)}$ to (7) and (8), we have

$$\mathbf{C}^{apr} \ge (1 - \varepsilon)\, \mathbf{C}^{opt}$$

and a complexity bound

$$O(T_{LP} \cdot (1/\varepsilon)^{m+3} \cdot (2m)^{2m+6} \cdot \Delta)$$

that finishes the proof.

3   Proof of Theorem 2

Let x^* be an optimal vertex solution of the LP relaxation of the Δ-modular m-BILP problem. After a standard change of variables $x \to x + \lfloor x^* \rfloor$ the original m-BILP transforms to an equivalent ILP with different upper bounds on variables and a different right-hand side vector b. We can think that lower bounds on variables does not change because a new vertex optimal solution $x^* - \lfloor x^* \rfloor$ of the obtained problem has non-zero coordinates.

Any optimal vertex solution of the LP problem has at most m non-zero coordinates, so we have the following bound on the l_1-norm of an optimal ILP solution $z^* - \lfloor x^* \rfloor$ of the new problem:

$$\|z^* - \lfloor x^* \rfloor\|_1 \le \|x^* - z^*\|_1 + \|x^* - \lfloor x^* \rfloor\|_1 \le H + m.$$

3.1   First Complexity Bound

Consider a weighted digraph $G = (V, E)$, whose vertices are triplets (k, h, l), for $k \in 1\!:\!n$, $l \in 0\!:\!(H + m)$ and $h \in \{Ax : \|x\|_1 \le l\} \cap \mathbb{Z}^m$. Using Corollary 3, we bound the number of vertices $|V|$ by $O(n \cdot 2^m \cdot (H + m)^{m+1} \cdot \Delta)$. By definition, any vertex (k, h, l) has an in-degree equal to $\min\{u_k, l\} + 1$. More precisely, for any $j \in 0\!:\!\min\{u_k, l\}$ there is an arc from $(k - 1, h - A_k j, l - j)$ to (k, h, l), this arc is weighted by $c_k j$. Note that vertex $(k - 1, h - A_k j, l - j)$ exists only if $j \le l$. Additionally, we add to G a starting vertex s, which is connected with all vertices of the first level $(1, *, *)$, weights of this arcs correspond to solutions of 1-dimensional sub-problems. Clearly, the number of arcs can be estimated by

$$|E| = O(|V| \cdot (H + m)) = O(n \cdot 2^m \cdot (H + m)^{m+2} \cdot \Delta).$$

The m-BILP problem is equivalent to searching of the longest path starting from the vertex s and ending at the vertex $(n, b, H + m)$ in G. Since the graph G is acyclic, the longest path problem can be solved by an algorithm with the complexity bound $O(|V| + |E|) = O(n \cdot 2^m \cdot (H + m)^{m+2} \cdot \Delta)$.

We note that during the longest path problem solving, the graph G must be evaluated on the fly. In other words, the vertices and arcs of G are not known in advance, and we build them online. To make constant-time access to vertices we can use a hash-table data structure with constant-time insert and search operations (see Remark 1).

Finally, using the binarization trick, described in the work [9], we can significantly decrease the number of arcs in G. The idea of the trick is that any integer $j \in [0, \min\{u_k, l\}]$ can be uniquely represented using at most $O(\log^2(\min\{u_k, l\})) = O(\log^2(H + m))$ bits. More precisely, for any interval $[0, \min\{u_k, l\}]$ there exist at most $O(\log^2(H + m))$ integers $s(k, i)$ such that any integer $j \in [0, \min\{u_k, l\}]$ can be uniquely represented as

$$j = \sum_i s(k, i) x_i, \quad \text{where } x_i \in \{0, 1\}, \text{ and}$$

$$\sum_i s(k, i) x_i \in [0, \min\{u_k, l\}], \quad \text{for any } x_i \in \{0, 1\}.$$

Using this idea, we replace the part of the graph G connecting vertices of the levels $(k-1, *, *)$ and $(k, *, *)$ by an auxiliary graph, whose vertices correspond to the triplets (i, h, l), where $i \in \{0, 1, \ldots, O(\log^2(H+m))\}$, and any triplet (i, h, l) has in-degree two. More precisely, the vertex (i, h, l) is connected with exactly two vertices: $(i-1, h, l)$ and $(i-1, h - s(k, i)A_k, l - s(k, i))$. The resulting graph will have at most $O(\log^2(H + m)|V|)$ vertices and arcs, where $|V|$ corresponds to the original graph. Total arithmetical complexity can be estimated as

$$O(n \cdot 2^{O(m)} \cdot (H + m)^{m+1} \cdot \log^2 H \cdot \Delta).$$

3.2 Second Complexity Bound

Consider a weighted digraph $G = (V, E)$, whose vertices are pairs (k, h), for $k \in 1:n$ and $h \in \{Ax : \|x\|_1 \leq H + m\} \cap \mathbb{Z}^m$. The edges of G have the same structure as in the graph from the previous subsection. More precisely, for any $j \in 0:u_k$ we put an arc from $(k - 1, h - A_k j)$ to (k, h), if such vertices exist in V, the arc is weighted by $c_k j$.

The main difference here is that we compute all vertices of G directly, using Corollary 3. Arithmetical complexity of this step is bounded by $O(\log m)^{m^2} \cdot (H + m)^m \cdot \Delta$. Due to Corollary 3, $|V| = O(n \cdot 2^m \cdot (H+m)^m \cdot \Delta)$ and $|E| = O(n \cdot 2^m \cdot (H+$

$m)^{m+1} \cdot \Delta)$, since an in-degree of any vertex in G is bounded by $H+m+1$. Again, using the binarization trick, described in the previous subsection, we reduce the total number of arcs and vertices to $O(n \cdot 2^{O(m)} \cdot (H+m)^m \cdot \log(H) \cdot \Delta)$. The total complexity can be roughly estimated as

$$n \cdot O(\log m)^{m^2} \cdot (H+m)^m \cdot \log^2(H) \cdot \Delta.$$

3.3 Third Complexity Bound

Let us show how to remove the $\log^2(H)$ term in the previous complexity bound using non-negativity of elements of A. The main idea is taken from the work [33] (see also [23, Section 7.2.2]).

Consider the graph G, constructed in the previous subsection, without using of the binarization trick. Let us fix a some vertex-level $(k, *)$ of G for some $k \in 1:n$, and consider an auxiliary graph F_k, whose vertices are exactly elements $h \in \{Ax : \|x\|_1 \leq H+m\} \cap \mathbb{Z}^m$. For two vertices h_1, h_2 of F, we put an arc from h_1 to h_2 if $h_2 - h_1 = A_k$. Since the matrix A has non-negative elements and since "in" and "out" degrees of any vertex in F_k are at most one, the graph F_k is a disjoint union of paths. This decomposition can be computed by an algorithm with complexity $O(|V(F_k)|) = O(2^m \cdot (H+m)^m \cdot \Delta)$. Let $(h_1, h_2, \ldots, h_s)$ be some path of the decomposition, and $longest(k, h)$ be the value of the longest path in G starting at s and ending at (k, h). Clearly, for any $i \in 1:s$, the value of $longest(k, h_i)$ can be computed by the formula

$$longest(k, h_i) = \max_{j \in \min\{u_k, i-1\}} longest(k-1, h_{i-j}) + c_k j. \tag{9}$$

Consider a queue Q with operations: $Enque(Q, x)$ that puts an element x into the tail of Q, $Decue(Q)$ that removes an element x from the head of Q, $GetMax(Q)$ that returns maximum of elements of Q. It is known fact that queue can be implemented such that all given operations will have amortized complexity $O(1)$. Now, we compute $longest(k, h_i)$, for $h_i \in (h_1, h_2, \ldots, h_s)$ using the following algorithm:

Algorithm 4. Compute longest path with respect to $(h_1, h_2, \ldots, h_s)$

1: Create an empty queue Q;
2: $t := \min\{u_k, s\}$;
3: **for** $j := 0$ **to** t **do**
4:     $Enque(Q, longest(k-1, h_{s-j}) + c_k j)$;
5: **end for**
6: **for** $i := s$ **down to** 1 **do**
7:     $longest(k, h_i) := GetMax(Q) - c_k(s-i)$;
8:     $Decue(Q)$;
9:     **if** $i \geq t+1$ **then**
10:        $Enque(Q, longest(k-1, h_{i-t-1}) + c_k(s-i+1))$;
11:     **end if**
12: **end for**

Correctness of the algorithm follows from the formula (9). The algorithm's complexity is $O(s)$.

Let us estimate the total arithmetical complexity of the whole procedure. It consists from the following parts:

1. Enumerating of points in the set $M = \{Ax\colon \|x\|_1 \leq H + m\} \cap \mathbb{Z}^m$. Due to Corollary 3, the complexity of this part is $O(\log m)^{m^2} \cdot (H + m)^m \cdot \Delta$;
2. Constructing the graphs F_k for each $k \in 1:n$. The number of edges and vertices in F_k can be estimated as $O(|M|)$. Hence, due to Corollary 3, the complexity of this part can be estimated as $O(n \cdot |M|) = O(n \cdot 2^m \cdot (H+m)^m \cdot \Delta)$.
3. For each F_k, compute a path decomposition of F_k. For each path in the decomposition, apply an Algorithm 4. The complexity of this part is clearly the same as in the previous step.

Therefore, the total complexity bound is roughly

$$n \cdot O(\log m)^{m^2} \cdot (H + m)^m \cdot \Delta.$$

Conclusion

The paper considers the m-dimensional bounded knapsack problem (m-BKP) and the bounded ILP in the standard form (m-BILP). For the problem m-BKP it gives an FPTAS with the arithmetical complexity bound

$$O(n \cdot (1/\varepsilon)^{m+3} \cdot \Delta),$$

where n is the number of variables, m is the number of constraints (we assume here that m is fixed) and $\Delta = \Delta(A)$ is the maximal absolute value of rank-order minors of A. For details see Theorem 1 and Corollary 1.

For the problem m-BILP it gives an exact algorithm with the complexity bound

$$O(n \cdot \Delta^{m+1} \cdot \log^2 \Delta).$$

For the problems with non-negative elements of the matrix A the last bound can be slightly improved:

$$O(n \cdot \Delta^{m+1}).$$

Taking $m = 1$ it gives

$$O(n \cdot \Delta^2)$$

arithmetical complexity bound for the classical bounded knapsack problem. For details see Theorem 2 and Corollary 2.

References

1. Alekseev, V.V., Zakharova, D.V.: Independent sets in the graphs with bounded minors of the extended incidence matrix. J. Appl. Ind. Math. **5**, 14–18 (2011). https://doi.org/10.1134/S1990478911010029

2. Artmann, S., Weismantel, R., Zenklusen, R.: A strongly polynomial algorithm for bimodular integer linear programming. In: Proceedings of 49th Annual ACM Symposium on Theory of Computing, pp. 1206–1219 (2017). https://doi.org/10.1145/3055399.3055473

3. Bock, A., Faenza, Y., Moldenhauer, C., Vargas, R., Jacinto, A.: Solving the stable set problem in terms of the odd cycle packing number. In: Proceedings of 34th Annual Conference on Foundations of Software Technology and Theoretical Computer Science, Leibniz International Proceedings in Informatics (LIPIcs), vol. 29, pp. 187–198 (2014). https://doi.org/10.4230/LIPIcs.FSTTCS.2014.187

4. Caprara, A., Kellerer, H., Pferschy, U., Pisinger, D. Approximation algorithms for knapsack problems with cardinality constraints. Eur. J. Oper. Res. **123**, 333–345 (2000). https://doi.org/10.1016/S0377-2217(99)00261-1

5. Chan, T.: Approximation schemes for $0 - 1$ knapsack. In: Proceedings of the 1st Symposium on Simplicity in Algorithms (SOSA), pp. 5:1–5:12 (2018). https://doi.org/10.4230/OASIcs.SOSA.2018.5

6. Conforti, M., Fiorini, S., Huynh, T., Joret, G., Weltge, S. The stable set problem in graphs with bounded genus and bounded odd cycle packing number. In: Proceedings of the 2020 ACM-SIAM Symposium on Discrete Algorithms (SODA), pp. 2896–2915 (2020). https://doi.org/10.1137/1.9781611975994.176

7. Cormen, T.H., Leiserson, C.E., Rivest, R.L., Stein, C.: Introduction to Algorithms, 3rd edn. MIT Press, Cambridge (2009)

8. Fomin, F. V., Panolan, F., Ramanujan, M. S., Saurabh, S.: On the optimality of pseudo-polynomial algorithms for integer programming. In: ESA 2018, pp. 31:1–31:13 (2018). https://doi.org/10.4230/LIPIcs.ESA.2018.31

9. Eisenbrand, F., Weismantel, R.: Proximity results and faster algorithms for integer programming using the Steinitz lemma. ACM Trans. Algorithms **16**(1) (2019). https://doi.org/10.1145/3340322

10. Gribanov, D.V., Malyshev, D.S.: The computational complexity of three graph problems for instances with bounded minors of constraint matrices. Discrete Appl. Math. **227**, 13–20 (2017). https://doi.org/10.1016/j.dam.2017.04.025

11. Gribanov, D. V., Malyshev, D. S. The computational complexity of dominating set problems for instances with bounded minors of constraint matrices. Discrete Optim. **29**, 103–110 (2018). https://doi.org/10.1016/j.disopt.2018.03.002

12. Gribanov, D.V., Malyshev, D.S., Pardalos, P.M., Veselov, S.I.: FPT-algorithms for some problems related to integer programming. J. Comb. Optim. **35**(4), 1128–1146 (2018). https://doi.org/10.1007/s10878-018-0264-z

13. Grossman, J.V., Kulkarni, D.M., Schochetman, I.E.: On the minors of an incidence matrix and its Smith normal form. Linear Algebra Appl. **218**, 213–224 (1995). https://doi.org/10.1016/0024-3795(93)00173-W

14. Halman, N., Holzhauser, M., Krumke, S.: An FPTAS for the knapsack problem with parametric weights. Oper. Res. Lett. **46**(5), 487–491 (2018). https://doi.org/10.1016/j.orl.2018.07.005

15. Har-Peled, S., Rahul, S.: Two (Known) Results About Graphs with No Short Odd Cycles (2018). https://arxiv.org/abs/1810.01832

16. Holzhauser, M., Krumke, S. An FPTAS for the parametric knapsack problem. Inf. Process. Lett. **126**, 43–47 (2017). https://doi.org/10.1016/j.ipl.2017.06.006

17. Ibarra, O.H., Kim, C.E.: Fast approximation algorithms for the knapsack and sum of subset problem. J. ACM **22**, 463–468 (1975). https://doi.org/10.1287/moor.3.3.197

18. Jansen, K., Kraft, S.: A faster FPTAS for the unbounded knapsack problem. Eur. J. Comb. **68**, 148–174 (2018). https://doi.org/10.1016/j.ejc.2017.07.016

19. Jansen, K., Rohwedder, L.: On Integer Programming, Discrepancy, and Convolution (2018). https://arxiv.org/abs/1803.04744
20. Jin, C.: An improved FPTAS for $0-1$ knapsack. In: 46th International Colloquium on Automata, Languages, and Programming (ICALP 2019), pp. 76:1–76:14 (2019). https://doi.org/10.4230/LIPIcs.ICALP.2019.76
21. Kellerer, H., Pferschy, U.: A new fully polynomial time approximation scheme for the knapsack problem. J. Comb. Optim. **3**, 59–71 (1999). https://doi.org/10.1023/A:1009813105532
22. Kellerer, H., Pferschy, U.: Improved dynamic programming in connection with an FPTAS for the knapsack problem. J. Comb. Optim. **8**, 5–11 (2004). https://doi.org/10.1023/B:JOCO.0000021934.29833.6b
23. Kellerer, H., Pferschy, U., Pisinger, D.: Knapsack Problems. Springer, Heidelberg (2004). https://doi.org/10.1007/978-3-540-24777-7
24. Knop, D., Pilipczuk, M., Wrochna, M.: Tight complexity lower bounds for integer linear programming with few constraints. ACM Trans. Comput. Theory (TOCT) **12**(3), 1–19 (2020). https://doi.org/10.4230/LIPIcs.STACS.2019.44
25. Korte, B., Schrader, R.: On the existence of fast approximation schemes. Nonlinear Programm. **4**, 415–437 (1981). https://doi.org/10.1016/B978-0-12-468662-5.50020-3
26. Kulik, A., Shachnai, H.: There is no EPTAS for two-dimensional knapsack. Inf. Process. Lett. **110**(16), 707–710 (2010). https://doi.org/10.1016/j.ipl.2010.05.031
27. Lawler, E.L.: Fast approximation algorithms for knapsack problems. Math. Oper. Res. **4**, 339–356 (1979). https://doi.org/10.1287/moor.4.4.339
28. Lee, J., Paat, J., Stallknecht, I., Xu, L.: Improving proximity bounds using sparsity. In: Baïou, M., Gendron, B., Günlük, O., Mahjoub, A.R. (eds.) ISCO 2020. LNCS, vol. 12176, pp. 115–127. Springer, Cham (2020). https://doi.org/10.1007/978-3-030-53262-8_10
29. Li, W., Lee, J.A.: Faster FPTAS for Knapsack Problem with Cardinality Constraint (2020). https://arxiv.org/abs/1902.00919
30. Magazine, M.J., Oguz, O.: A fully polynomial approximation algorithm for the $0-1$ knapsack problem. Eur. J. Oper. Res. **8**, 270–273 (1981). https://doi.org/10.1016/0377-2217(81)90175-2
31. Megiddo, N., Tamir, A.: Linear time algorithms for some separable quadratic programming problems. Oper. Res. Lett. **13**, 203–211 (1993). https://doi.org/10.1016/0167-6377(93)90041-E
32. Papadimitriou, C.H.: On the complexity of integer programming. J. Assoc. Comput. Mach. **28**, 765–768 (1981). https://doi.org/10.1145/322276.322287
33. Pferschy, U.: Dynamic programming revisited: improving knapsack algorithms. Computing **63**(4), 419–430 (1999). https://doi.org/10.1007/s006070050042
34. Rhee, D.: Faster fully polynomial approximation schemes for knapsack problems. Master's thesis, Massachusetts Institute of Technology (2015)
35. Schrijver, A.: Theory of Linear and Integer Programming. Wiley, Hoboken (1998)
36. Sebő, A.: An introduction to empty lattice simplices. In: Cornuéjols, G., Burkard, R.E., Woeginger, G.J. (eds.) IPCO 1999. LNCS, vol. 1610, pp. 400–414. Springer, Heidelberg (1999). https://doi.org/10.1007/3-540-48777-8_30
37. Marco, Di S., Friedrich, E., Faenza, Y., Moldenhauer, C.: On largest volume simplices and sub-determinants. In: SODA 2015: Proceedings of the Twenty-Sixth Annual ACM-SIAM Symposium on Discrete Algorithms, pp. 315–323 (2015). https://doi.org/10.5555/2722129.2722152

A Column Generation Based Heuristic for a Temporal Bin Packing Problem

Alexey Ratushnyi[1]($\boxtimes$)(iD) and Yury Kochetov[2](iD)

[1] Novosibirsk State University, Novosibirsk, Russia
[2] Sobolev Institute of Mathematics, Novosibirsk, Russia

Abstract. We introduce a new temporal bin packing problem that originated from cloud computing. We have a finite set of items. For each item, we know an arriving time, processing time, and two weights (CPU, RAM). Some items we call large. Each bin (server) has two capacities and is divided into two identical parts (left and right). A regular item can be placed in one of them. A large item is divided into two identical parts and placed in both parts of a bin. Our goal is to pack all items into the minimum number of bins. For this NP-hard problem, we design a heuristic that is based on column generation to get lower and upper bounds. Preliminary computational experiments for real test instances indicate a small gap between the bounds. The average relative error is at most 0.88% for one week planning horizon and about 50000 items. The average running time is 21 s for a personal computer.

Keywords: Knapsack problem · Column generation · Virtual machine · Bin packing

1 Introduction

We consider a new temporal problem of allocating resources for virtual machines with different configurations and time requirements on identical servers. This problem is strongly NP-hard as the generalized case of d-dimensional bin packing problem $(d \geq 1)$ [8], where d is the number of different types of virtual machine resources. As a rule, only one of all resources is selected as the most demanded and the others have proportional values [1]. However, we use data sets for computational experiments where the combinations of virtual machine (VM) resource requirements are quite diverse, and in different cases, different resources show strong demand. All possible information about VMs is assumed to be known in advance (the creation and deletion times, the required amount of RAM and CPU cores, and the size of VM). Thus, we deal with the problem of specifying the location of VMs on servers for performing computations in such a way as to minimize the maximum number of servers involved for the entire planning horizon. We guess that the model can be useful to analyze possible economic benefits, as well as to identify general patterns to help in the detailed configuration of the online algorithms.

© Springer Nature Switzerland AG 2021
P. Pardalos et al. (Eds.): MOTOR 2021, LNCS 12755, pp. 96–110, 2021.
https://doi.org/10.1007/978-3-030-77876-7_7

The relevance of the temporal bin packing problem gives rise to quite a lot of different variations and studies. In [10], authors suggest an approach based on placement templates to build a fast online algorithm. In [3], several heuristics are proposed including one based on column generation method as well as a way to reduce the number of time points in this problem. In addition to the number of servers involved, a large number of resources are consumed when switching the servers. Thus, the study [1] considers a variation of the temporal bin packing problem with two objectives. In [2], the authors provide a temporal model for this problem and several rules to break the symmetries. A fairly thorough classification of the virtual machines placement problems and solution techniques is offered in [9].

In this paper we present the lower and upper bounds for new temporal bin packing problem with Non-Uniform Memory Access (NUMA) architectures of the servers [7]. We propose an algorithm based on column generation method [4], which can provide suitable results in a short amount of time. First of all, a static problem is solved at a time point with the maximum computational load on the cloud using mathematical modeling and optimization solver [5]. Later on, we apply a procedure that propagates the resulting solution back and forth in time horizon using the First Fit heuristic [6] and design a feasible solution for all time points. Both stages are executed independently of each other, which allows us to carefully configure them or improve if necessary.

The rest of the paper is organized as follows. Section 2 gives the details and notations of the problem (server architecture and components, types of virtual machines, etc.). Section 3 presents models for the two-dimensional packing problem and the bounded knapsack problem that are used in obtaining the lower and upper bounds. It also describes the algorithm for column generation. The algorithm developed for constructing a solution for a temporal problem is given in Sect. 4. Section 5 contains a description of the data sets generated using real data and the results of computational experiments. In Sect. 6, we briefly summarize the paper.

2   Model Description

We consider a set S of identical servers which are located on a data center. Each server has the amount of RAM C_1 and the number of CPU cores C_2. All servers have NUMA architecture (Fig. 1). It means that they consist of two NUMA-nodes with the same resources. Each NUMA-node has $C_1/2$ of RAM and $C_2/2$ of CPU cores. In this problem, we do not take into account any restrictions on the relationship between servers and virtual machines. We assume that each server can serve any number of VMs, regardless of their configurations and duration as long as it has enough resources.

Each VM $m \in M$ is characterized by a corresponding type $i \in L$, which defines the required amount of RAM d_{m1} and CPU cores d_{m2}, for example, 1 core and 1 GB of RAM (small VM) or 32 cores and 64 GB of RAM (large VM). The size of every virtual machine is determined by the number of required

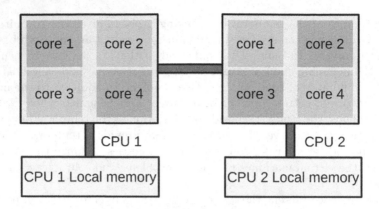

Fig. 1. NUMA architecture

cores. We assume that the set M of VMs consists of two disjoint subsets: large VMs M^l and small VMs M^s. A small VM must be fully placed on one of a server's NUMA-nodes. A large VM is placed on both NUMA-nodes taking half of the VM's requirements from each of the nodes. After placement on the server, the virtual machine cannot be moved. Thus, the VM m takes up the allocated resources at all time points t from the interval $\alpha_m \leq t < \omega_m$. The creation α_m and deletion ω_m times are selected by a user, while the type of a virtual machine is selected from a set of possible options L which is defined by the cloud provider.

Let us introduce the following notations: N is the set of NUMA-nodes ($N = \{1, 2\}$), R is the set of resource types ($R = \{1, 2\}$), $\mathcal{T}$ is the set of time points, M_t^l and M_t^s are the sets of large and small VMs that are used at time point t.

The decision variables are as follows: $x_{msn}, m \in M^s, s \in S, n \in N$ equals 1 if the small virtual machine m is placed on the node n of the server s and 0 otherwise; $y_{ms}, m \in M^l, s \in S$ equals 1 if the large virtual machine m is placed on the server s and 0 otherwise; $z_{st}, s \in S, t \in \mathcal{T}$ equals 1 if the server s is active at the time point t and 0 otherwise; F is the maximum number of simultaneously active servers.

The optimization model takes the following form:

$$\min_{F,(x_{msn}),(y_{ms}),(z_{st})} F \tag{1}$$

$$\text{s.t.} \quad F \geq \sum_{s \in S} z_{st}, \ t \in \mathcal{T}, \tag{2}$$

$$\sum_{s \in S} \sum_{n \in N} x_{msn} = 1, \ m \in M^s, \tag{3}$$

$$\sum_{s \in S} y_{ms} = 1, \ m \in M^l, \tag{4}$$

$$\sum_{n \in N} x_{msn} \leq z_{st}, \ m \in M^s, \alpha_m \leq t \leq \omega_m, s \in S, \tag{5}$$

$$y_{ms} \leq z_{st}, \ m \in M^l, \alpha_m \leq t \leq \omega_m, s \in S, \tag{6}$$

$$\sum_{m \in M_t^s} d_{mr} x_{msn} + \frac{1}{2} \sum_{m \in M_t^l} d_{mr} y_{ms} \leq \frac{C_r}{2} z_{st}, \ t \in T, s \in S, n \in N, r \in R, \tag{7}$$

$$x_{msn}, y_{ms}, z_{st} \in \{0, 1\}. \tag{8}$$

Inequalities (2) define the number of active servers. Constraints (3)–(4) ensure that all VMs are placed on the servers. According to inequalities (5)–(6), the server is active if it serves at least one VM. Constraints (7) limit resource usage on each server.

Despite the small number of NUMA-nodes on each server and the types of resources, the finding optimal solution for the model is time consuming for real world applications. For instances with 2000 VMs, the Gurobi solver (version 9.1 with standard parameters) takes about 5 min. However, the optimal solutions for instances with 50000 VMs (see Sect. 5) could not be obtained in one hour. Therefore, alternative ways to get the lower and upper bounds are proposed below.

3 Lower Bounds

An important part of constructing the algorithm is the way to find lower bounds for evaluating the quality of solutions. To this end, we will use the column generation approach. We apply it to static problem at two time moments: the moment with the highest RAM load t_h^1 and the moment with the highest CPU cores load t_h^2. Since at every time point the current virtual machines with the same type i do not differ from each other, we consider VM types instead of VMs in all the following notations. To describe the algorithm, we need to introduce additional mathematical models.

The first model describes the packing of VMs on the servers at the given time point. Let J denotes the set of all possible packing patterns for a single server and $J' \subset J$ is a subset of this set. We will use this subset to get feasible solutions. Value a_{ij} is the number of virtual machines that are configured according to type $i \in L$ in the pattern $j \in J'$. Thus, the packing pattern j can be associated with the following vector: $(a_{1j}, ..., a_{|L|j})$. The value n_i represents the number of requests to virtual machines with the type i. Variables x_j are equal to the numbers of servers packed according to the pattern $j \in J'$.

The well-known static model for the subset J' takes the following form:

$$\min \sum_{j \in J'} x_j \tag{9}$$

$$\text{s.t.} \ \sum_{j \in J'} a_{ij} x_j \geq n_i, \ i \in L, \tag{10}$$

$$x_j \geq 0, \ j \in J'. \tag{11}$$

The goal function (9) represents the number of active servers. Inequality (10) ensures that all VMs will be packed. Since our goal is to obtain the lower bound, we do not require variables x_j to be integers.

The dual problem is the following:

$$\max \sum_{i \in L} n_i \lambda_i \tag{12}$$

$$\text{s.t.} \ \sum_{i \in L} a_{ij} \lambda_i \leq 1, \ j \in J', \tag{13}$$

$$\lambda_i \geq 0, \ i \in L. \tag{14}$$

Let λ_i^* be the optimal values of dual variables to the problem (12)–(14) and x_j^* be the optimal values to the problem (9)–(11). If the following inequality

$$\sum_{i \in L} a_{ij} \lambda_i^* \leq 1 \tag{15}$$

holds for all possible patterns $j \in J$, then

$$x_j = \begin{cases} x_j^*, & j \in J' \\ 0, & j \in J \backslash J' \end{cases}$$

is the optimal solution to the problem (9)–(11) for the entire set of patterns J and $H = \sum_{j \in J'} x_j^*$ is the desired lower bound.

To check the inequality (15), we consider a new knapsack problem with NUMA-nodes. Let new integer variables y_i define the number VMs with type $i \in L$ in the server and new integer variables z_i^n define the number of small VMs of type $i \in L$ on the NUMA-node $n \in N$. L^s and L^l are the sets of VM types that correspond to the sets of virtual machines M^s and M^l.

Now the knapsack problem with NUMA-nodes takes the following form:

$$\max \alpha = \sum_{i \in L} \lambda_i^* y_i \tag{16}$$

$$\text{s.t.} \ y_i \leq n_i, \ i \in L, \tag{17}$$

$$\sum_{n \in N} z_i^n = y_i, \ i \in L^s, \tag{18}$$

$$\sum_{i \in L^s} d_{ir} z_i^n + \frac{1}{2} \sum_{i \in L^l} d_{ir} y_i \leq \frac{C_r}{2}, \ r \in R, n \in N, \tag{19}$$

$$y_i \geq 0, \ integer, \ i \in L, \tag{20}$$

$$z_i^n \geq 0, \ integer, \ i \in L^s, n \in N. \tag{21}$$

Equalities (18) ensure that each small VM on the server is located on one of the NUMA-nodes and nowhere else. Inequalities (19) check the resources (RAM and CPU cores) on each NUMA-node of the server taking into account the large VMs.

If $\alpha^* \leq 1$ then the inequality (15) holds for all patterns $j \in J$. Otherwise, we have a new pattern $(y_1, ..., y_{|L|})$ and include it into the set J' [8]. The pseudocode of the algorithm for finding the lower bound has the following form:

Algorithm 1: The general scheme for obtaining the lower bound

---

Input: A subset of columns $J' \subset J$, A set of virtual machines VM at moment t;

Output: A lower bound LB, A supplemented set J';

1 **Initialization**: $\alpha^* \leftarrow \infty$;

2 **while** $\alpha^* > 1.0$ **do**

3     $\lambda_i^* \leftarrow solveDualProblem(J', VM)$;   // Solve the problem (12)-(14) and save a vector $\{\lambda_1, ..., \lambda_{|L|}\}$

4     $y_i^*, \alpha^* \leftarrow solveKnapsackModel(\boldsymbol{\lambda}_j^*, VM)$;   // Solve the problem (16)-(21) to get a value of α^* and a column $\{y_1^*, ..., y_{|L|}^*\}$

5     $J'.append(\{y_1^*, ..., y_{|L|}^*\})$;

6 **end**

7 $LB \leftarrow \lceil \sum_{i \in L} n_i \lambda_i^* \rceil$;   // Compute the lower bound

8 **return**: LB, J';

Note that in line 4, we do not need to look for optimal solution for the model (16)–(21). It is sufficient to find a feasible solution such that $\alpha > 1 + \varepsilon$. We include the inequality: $\sum_{i \in L} \lambda_i^* y_i > 1 + \varepsilon$ into the model and also specify an additional parameter for Gurobi to terminate computations when a feasible solution is obtained. Thus, we significantly reduce the running time of the Algorithm 1.

To initialize the set J', we can apply any heuristic for the bin packing problem [6]. Nevertheless, we propose an algorithm based on the model (16)–(21).

Algorithm 2: Generation of an initial set J'

Input: A set of VMs at time moment t;

Output: The set of initial patterns J';

1 **Initialization**: $\lambda_i^* \leftarrow d_{i1}$, $J' \leftarrow empty$, n_i;   // The weights are equal to the number of cores of each VM type, the values of n_i can be calculated using a VM set

2 **while** $\exists\, n_i \neq 0$ **do**

3     $y_i^* \leftarrow solveKnapsackModel(\boldsymbol{\lambda}_j^*, VM)$;   // Solve the problem (16)-(21)

4     $J'.append(\{y_1^*, ..., y_{|L|}^*\})$;   // Add a new pattern

5     $n_i \leftarrow n_i - y_i$, $\forall i \in L$;   // Remove the packed VMs from the set VM and change the corresponding values n_i

6 **end**

7 **return**: The initial set J' of patterns;

We compute the lower bounds at time moments t_h^1 and t_h^2 and choose the largest value: $LB = \max\{H(t_h^1), H(t_h^2)\}$. The values of the weights λ_i^* do not affect the value of the lower bound, but play a role in the construction of the solution with integer variables that will be used later. To solve the models we use the Gurobi solver [5] version 9.1 with default parameters.

4   Upper Bounds

To obtain the lower bound, we create the set of patterns J' by Algorithm 2 and enlarge it with new columns by Algorithm 1. Now we use the final set of patterns J' in the model (22)–(24) which is an integer version of the model (9)–(11).

$$\min \sum_{j \in J'} x_j \tag{22}$$

$$\sum_{j \in J'} a_{ij} x_j \geq n_i, \ i \in L, \tag{23}$$

$$x_j \geq 0, \ integer, \ j \in J'. \tag{24}$$

Our goal is to get a feasible solution to this static problem at each time points t_h^1 or t_h^2. We select point T as one of them (see Fig. 2) and divide the set M into three disjoint sets M_1, M_2 and M_3: $M_1 = \{m \in M | \alpha_m \leq T < \omega_m\}$, $M_2 = \{m \in M | \alpha_m > T\}$, $M_3 = \{m \in M | T \geq \omega_m\}$.

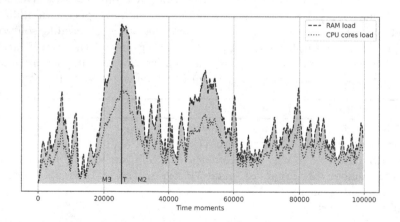

Fig. 2. The total load of servers

For the set M_1 we solve the static problem by Gurobi solver. For the sets M_2 and M_3, we apply the greedy FF heuristic (Algorithm 3). As a result, we get two upper bounds for time points t_h^1 or t_h^2 and select the best of them.

Algorithm 3: First Fit based heuristic

Input: A set of virtual machines VM;
Output: A solution for a temporal bin packing problem;

1 $M_1, M_2, M_3 = $ splitVirtualMachines(VM);
2 $J' \leftarrow Algorithm2(M_1)$;
3 $J' \leftarrow Algorithm1(J', M_1)$;   // Get a supplemented set of columns
4 $servers \leftarrow solveStaticProblem(J', M_1)$;   // Solve (22)-(24)
5 $sort(M_2)$; // Sort the VMs in non-decreasing order of times α_m
6 **for every** $m \in M_2$ **do**
7     $isPlaced \leftarrow false$;
8   **for every server** $s \in servers$ **do**
9   **for every** $m' \in s$ **do**
10   **if** $\omega_{m'} \leq \alpha_m$ **then**
11   remove m' from the server s;
12   **end**
13   **end**
14         $isPlaced \leftarrow Algorithm4(m)$;
15   **if** $isPlaced = true$ **then**
16   break;
17   **end**
18   **end**
19   **if** $isPlaced \neq true$ **then**
20   create $newServer$;
21   place VM m on the first NUMA-node of $newServer$;
22         $isPlaced \leftarrow true$;
23         $servers.append(newServer)$;
24   **end**
25 **end**

Algorithm 3 is described for the set M_2. For the set M_3, the algorithm has minor changes due to the inverted direction of time. First, in line 5, the VMs should be sorted in non-increasing order of times ω_m. Second, in lines 9–13, we remove the virtual machines for which the inequality $\alpha_{m'} \geq \omega_m$ is true. The placement of VMs on the server with NUMA architecture is described by Algorithm 4.

Note that points t_h^1 and t_h^2 may coincide as in Fig. 2. In such a case we can select several local maximum. This approach will significantly increase the running time (about as many times as the number of points) and will give a fairly small improvement in the results, judging by our experiments.

Algorithm 4: Place a virtual machine on a server

Input: A server s and a virtual machine m;

Output: The answer is it possible to put this VM on this server;

1 $isPlaced \leftarrow false$;

2 **if** both NUMA-nodes of the server s have sufficient resources for VM m **then**

3 | select NUMA-node which minimizes the value of max$\{AvailableRAM/C_1, AvailableCores/C_2\}$; // Select the least loaded NUMA-node

4 | place VM m on the selected NUMA-node;

5 | $isPlaced \leftarrow true$;

6 **else if** first NUMA-node of the server s has sufficient resources for VM m **then**

7 | place VM m on the first NUMA-node of the server s;

8 | $isPlaced \leftarrow true$;

9 **else if** second NUMA-node of the server s has sufficient resources for VM m **then**

10 | place VM m on the second NUMA-node of the server s;

11 | $isPlaced \leftarrow true$;

12 **return:** $isPlaced$;

---

5   Computational Experiments

To conduct experiments, we have sampled virtual machines from real data for several data sets, so that each of them satisfies certain rules. Below we present several tables with the results in total for 64 instances. Each of them includes the following columns: $SimpleLB = \max\{\sum_{m \in M_1} d_{m1}/C1; \sum_{m \in M_1} d_{m2}/C2\})$, $ColGenLB$ (lower bounds obtained by Algorithm 1), $CGLB$ Time (the running time for two points t_h^1 and t_h^2), $M_1 Solution$ (the number of active servers at point T), $ColGen$ (the results of Algorithm 3), $ColGen0$ (the results of Algorithm 3 with $T = 0$), $Colgen$ Time (the running time of Algorithm 3 in seconds), $Gap = (ColGen - ColGenLB)/ColGen * 100\%)$. The algorithms are implemented in C++17. All tests are performed on PC with AMD Ryzen 5 3500U, RAM 16 Gb. In the data sets for Tables 1, 2, 3 and 4, we have $5e+4$ virtual machines and 14 different VM types.

Table 1 corresponds to instances with a large number of long-lived VMs, i.e. about 73% of all virtual machines last for almost the entire selected period. Therefore, the final solution strongly depends on the solution of the static problem at point T for the set M_1. The average running time of the algorithm is 30.5 s, the average deviation is 0.347% and the maximum deviation does not exceed 1%.

Table 1. Many VMs with long-time requirements

N	SimpleLB	ColGenLB	CGLB Time, seconds	M_1 Solution	ColGen 0, T=0	ColGen	ColGgen Time, seconds	Gap, %
1	820	820	8	820	872	825	22	0.6
2	847	847	8	848	896	849	32	0.2
3	871	871	8	871	917	876	26	0.57
4	898	898	8	899	927	901	30	0.33
5	919	919	9	919	944	926	28	0.75
6	984	984	9	984	1018	986	36	0.2
7	1047	1047	10	1047	1082	1049	28	0.19
8	1105	1105	9	1106	1173	1112	26	0.62
9	1101	1101	9	1101	1149	1105	34	0.36
10	1227	1227	11	1227	1449	1227	34	0
11	1249	1249	10	1249	1385	1249	40	0

Table 2 shows the results for the data with the ratio of CPU cores require-ments to RAM requirements close to 1 ($\sum_{m \in M} d_{m1} / \sum_{m \in M} d_{m2} \approx 1$). In this way, the effects observed in Fig. 2 can be avoided. For half of the instances, the gap is 0%, although in other cases it is quite high and even reaches 2.71%. The average running time is 23.6 s. In three cases (2,8,9), ColGenLB shows a better lower bound than SimpleLB.

Table 2. Balanced RAM and CPU cores

N	SimpleLB	ColGenLB	CGLB Time, seconds	M_1 Solution	ColGen 0, T=0	ColGen	ColGgen Time, seconds	Gap, %
1	747	747	7	747	842	748	26	0.13
2	660	664	7	665	780	669	22	0.74
3	727	727	8	727	859	727	22	0
4	691	691	8	691	813	696	22	0.72
5	878	878	9	878	1092	878	22	0
6	828	828	8	828	1034	828	36	0
7	823	823	7	823	1053	831	20	0.96
8	751	752	7	752	954	752	18	0
9	713	718	8	719	903	738	16	2.71
10	884	884	10	884	1121	884	22	0
11	1178	1178	10	1178	1382	1178	34	0

For the data set in Table 3, virtual machines were selected from real data in such a way that the ratio of large VMs to all was quite high - about 20%. The average running time is 17.8 s, the average gap is 0.492%, although the maximum is almost four times larger - 1.8%.

The data for Table 4 has the opposite feature—only about 1% of large VMs. The average gap is 0.63%, and the average running time is 20 s. For these

Table 3. Many large VMs

N	SimpleLB	ColGenLB	CGLB Time, seconds	M_1 Solution	ColGen 0, T=0	ColGen	ColGgen Time, seconds	Gap, %
1	676	676	7	677	903	677	20	0.15
2	689	689	6	689	933	691	14	0.2
3	720	720	6	720	986	727	14	0.29
4	695	695	6	695	954	696	14	0.14
5	638	638	8	638	869	638	18	0
6	648	648	15	648	864	660	42	1.8
7	621	621	8	621	804	629	12	1.27
8	643	643	7	644	841	650	18	1.07
9	747	747	9	747	994	747	14	0
10	745	745	7	745	971	745	12	0

Table 4. Few large VMs

N	SimpleLB	ColGenLB	CGLB Time, seconds	M_1 Solution	ColGen 0, T=0	ColGen	ColGgen Time, seconds	Gap, %
1	770	770	6	770	802	772	18	0.26
2	777	777	5	777	811	793	16	2.01
3	833	833	6	833	872	837	18	0.48
4	756	756	6	757	786	757	16	0.13
5	758	758	6	758	781	761	24	0.39
6	826	826	6	826	851	826	22	0
7	731	731	7	732	753	738	16	0.95
8	848	848	7	848	881	854	34	0.7
9	848	848	6	848	885	848	22	0
10	782	782	6	782	817	793	14	1.38

instances, the *ColGen*0 algorithm showed much better results compared to the instances from Table 3. Noted that for both data sets 3 and 4, the lower bound *SimpleLB* and *ColGenLB* coincide.

Table 5. Large virtual machines

N	SimpleLB	ColGenLB	CGLB Time, seconds	M_1Solution	ColGen 0, $T=0$	ColGen	ColGgen Time, seconds	Gap, %
1	533	562	2	562	662	562	2	0
2	541	571	2	571	682	572	2	0.17
3	549	579	3	579	686	579	3	0
4	550	578	2	578	678	578	2	0
5	762	773	4	774	1024	773	4	0
6	734	750	4	750	985	750	4	0
7	784	796	4	796	1056	796	4	0
8	796	808	5	808	1069	808	5	0
9	775	797	4	797	1038	797	4	0
10	819	836	4	836	1108	836	4	0
11	659	705	4	706	869	706	4	0.14
12	616	646	4	646	766	646	4	0

For Table 5, virtual machines of only large sizes were selected from real data. The load on memory and cores also turned out to be quite balanced. The number of virtual machines varies between 7 and 22 thousand for different instances. The number of VM types is only 5. The running time and the gap of the solutions are very small. The $SimpleLB$ shows significantly worse results in all cases compared to $ColGenLB$. In this case, the NUMA architecture does not affect the solution, since there is no choice of NUMA-node for large machines.

For Table 6, virtual machines were randomly selected from real data without taking into account any properties. Thus, these instances are quite similar to the real applications. The number of virtual machines and their types are the same as for Tables 1, 2, 3 and 4. Here we can see a very significant improvement in the $ColGenLB$ compared to the $SimpleLB$. Also, the average gap has become slightly higher than in previous cases and equals 2.55%.

The comparison of the columns $ColGenLB$ and $M_1Solution$ shows that the proposed method for solving the static problem at time moment T reaches the lower bound in most cases. What justifies its rather long running time and allows us to build solutions for the entire temporal problem with small gap.

The $ColGen0$ algorithm in all cases works worse than the $ColGen$ algorithm, although the running time of $ColGen0$ is significantly lower due to the lack of need to solve the model (15)–(17) for a large set of virtual machines. A graphical comparison of the algorithms is shown in Fig. 3 and Fig. 4. It demonstrates the presence of oscillation when using the algorithm $ColGen0$, which leads to a large loss of energy resources when switching on and off servers, and as a result, large costs.

Table 6. Random sampling from real data

N	$SimpleLB$	$ColGenLB$	$CGLB$ Time, seconds	$M_1 Solution$	$ColGen\ 0,$ $T=0$	$ColGen$	$ColGgen$ Time, seconds	Gap, %
1	796	1103	5	1104	1137	1122	10	1.69
2	1253	1739	7	1739	1783	1752	14	0.74
3	1016	1218	4	1219	1231	1219	9	0.08
4	873	1076	3	1077	1094	1092	7	1.46
5	818	948	3	948	1006	998	6	5.27
6	313	314	5	317	358	317	16	0.94
7	1102	1507	5	1507	1595	1569	12	3.95
8	594	797	3	797	892	845	13	5.68
9	167	167	4	167	184	175	11	4.57
10	335	335	6	335	353	339	19	1.17

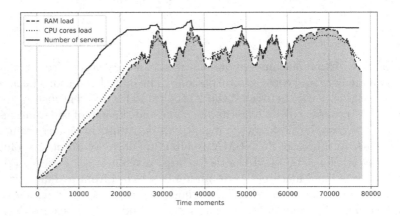

Fig. 3. Solution for instance 9 in Table 2 with $ColGen$ algorithm

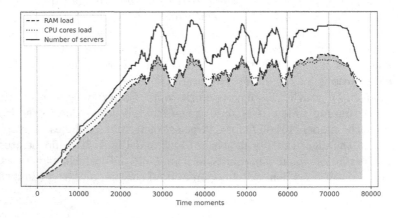

Fig. 4. Solution for instance 9 in Table 2 with $ColGen0$ algorithm

The presence of a significantly larger number of virtual machines, relative to different types, is reflected in the presence of VMs in the data that have the same requirements and follow each other. Thus, we can reduce the number of requests by combining a sequence of such requests into one. However, despite the good results in Table 1 for virtual machines with long-time requirements, this procedure does not provide an advantage for other data sets. A modification of the algorithm that leaves new requests in the same places as the previous requests with the same configurations shows an improvement of no more than 1 server, although in some cases, the opposite situation occurs, and 1 more server is required for the entire problem.

Despite the good results, considering only two points in time to get the lower bound has its drawbacks. One way to generate data in such a way that $ColGenLB$ will show results that are far from optimal is to try to use the dynamic component of the problem. For example, to select virtual machines so that the optimal solutions of a static problems at close time points (t_1 and t_2) will be very different from each other in such a way that it will be impossible to move from the optimal solution at time t_1 to the optimal solution at time t_2. In this case, the dynamic solution will require significantly more servers at one of this two points than the static solution. Thus, the results $ColGenLB$ will not be very accurate.

6   Conclusions

In this paper, we have considered new bin packing problem for virtual machines and servers with NUMA architecture. The heuristic based on column generation and the First Fit algorithm was proposed and tested. Computational experiments were carried out on data sets with different features, and strong results were demonstrated. We also showed the advantage of the considered lower bounds over the primitive lower bounds and confirmed that in most cases it is accurate. The method can also be generalized to other variations of the bin packing problem.

For further research, it may be interesting to modify this algorithm to reduce the number of servers at points where the load reaches a local maximum, see Fig. 3. Such a modification can be a local search for initial solutions at point T or replacing the First Fit algorithm with a more suitable one.

Acknowledgement. The study was carried out within the framework of the state contract of the Sobolev Institute of Mathematics (project no. 0314-2019-0014).

References

1. Aydin, N., Muter, I., Birbil, I.S.: Multi-objective temporal bin packing problem: an application in cloud computing. Comput. Oper. Res. **121** (2020). https://doi.org/10.1016/j.cor.2020.104959

2. de Cauwer, M., Mehta, D., O'Sullivan, B.: The temporal bin packing problem: an application to workload management in data centres. In: 2016 IEEE 28th International Conference on Tools with Artificial Intelligence (ICTAI), pp. 157–164, November 2016. https://doi.org/10.1109/ICTAI.2016.0033

3. Furini, F., Shen, X.: Matheuristics for the temporal bin packing problem. In: Amodeo, L., Talbi, E.-G., Yalaoui, F. (eds.) Recent Developments in Metaheuristics. ORSIS, vol. 62, pp. 333–345. Springer, Cham (2018). https://doi.org/10.1007/978-3-319-58253-5_19

4. Gilmore, R., Gomory, R.: A linear programming approach to the cutting stock problem I. Oper. Res. **9** (1961). https://doi.org/10.1287/opre.9.6.849

5. Gurobi Optimization: Gurobi optimizer reference manual (2021). http://www.gurobi.com

6. Johnson, D.S.: Fast algorithms for bin packing. J. Comput. Syst. Sci. **8**(3), 272–314 (1974). https://doi.org/10.1016/S0022-0000(74)80026-7

7. Manchanda, N., Anand, K.: Non-uniform memory access (NUMA) (2010)

8. Martello, S., Toth, P.: Knapsack Problems: Algorithms and Computer Implementations. Wiley, Hoboken (1990)

9. Pires, F.L., Báran, B.: A virtual machine placement taxonomy. In: Proceedings of the 15th IEEE/ACM International Symposium on Cluster, Cloud, and Grid Computing, CCGRID 2015, pp. 159–168. IEEE Press (2015). https://doi.org/10.1109/CCGrid.2015.15

10. Shi, J., Luo, J., Dong, F., Jin, J., Shen, J.: Fast multi-resource allocation with patterns in large scale cloud data center. J. Comput. Sci. **26**, 389–401 (2018). https://doi.org/10.1016/j.jocs.2017.05.005

On Some Variants of the Merging Variables Based (1+1)-Evolutionary Algorithm with Application to MaxSAT Problem

Alexander Semenov[1,2]([✉]) [iD], Ilya Otpuschennikov[1] [iD], and Kirill Antonov[3] [iD]

[1] ISDCT SB RAS, Irkutsk, Russia
[2] ITMO University, Saint Petersburg, Russia
[3] MEPhI University, Moscow, Russia

Abstract. In this paper we describe a new evolutionary strategy. It is based on the common (1+1) random mutation scheme which was augmented with metaheuristic technique named merging variables principle, that was proposed by us in previous works. We show that the new variant of (1+1)-EA has asymptotically lower worst case estimation than the original (1+1)-EA. In the experimental part we conduct comparison of the proposed strategy with several known variants of (1+1)-EA and demonstrate its practical applicability for a number of hard instances of MaxSAT problem.

Keywords: Boolean satisfiability problem · MaxSAT problem · Evolutionary algorithms · Merging variables principle

1 Introduction

(1+1)-Evolutionary Algorithm ((1+1)-EA) is a fairly popular subject for research in both the theoretical and practical sense. It is impossible to cite every paper that considered various theoretical properties of this algorithm. Let us only mention several key works: [1,3–7,12,19,22,24], etc. The random mutation scheme that lies in the basis of the classical (1+1)-EA is very attractive as a basic operation for many variants of evolutionary and genetic algorithms. In the present paper we propose several new modifications of the random mutation that follow the (1+1)-principle. For the proposed modifications of (1+1)-EA we provide the theoretical estimations of their effectiveness for arbitrary pseudo-Boolean black-box functions.

Let us give a brief outline of our paper. In the next section we cover the background required to evaluate the main results. Based on the ideas from [6] we describe the simple (1+1) Switching Evolutionary Algorithm, that alternates between standard (1+1)-EA and completely random search with some probabilities. In the third section we consider the variant of the (1+1) random mutation

P. Pardalos et al. (Eds.): MOTOR 2021, LNCS 12755, pp. 111–124, 2021.
https://doi.org/10.1007/978-3-030-77876-7_8

operator, which employs the so-called merging variables procedure ((1+1) Merging Variables Evolutionary Algorithm, (1+1)-MVEA), and determine some of its theoretical properties. In the computational part of our work we evaluate the effectiveness of all considered algorithms on a special class of instances of a well known MaxSAT problem. The main feature of the considered class of instances consists in that the problem of finding a Boolean vector, which maximizes the fitness function corresponding to the original problem, can be carried out over the set with significantly smaller cardinality comparing the number of variables in an original CNF. Concretely, as computational tests we used MaxSAT encodings of finding preimages of cryptographic hash functions SHA-1 and SHA-256 with additional conditions for hash value.

2 Preliminaries

Consider the problem of finding a point of the Boolean hypercube $\{0,1\}^n$ at which an arbitrary pseudo-Boolean function [2]

$$f : \{0,1\}^n \to R \tag{1}$$

achieves its maximum. A single iteration of the original (1+1)-EA in application to the problem of finding the maximum of a function (1) looks as follows ([7]):

1. Choose an initial point $\alpha \in \{0,1\}^n$
2. Repeat the following random mutation step: perform n independent Bernoulli trials with success probability $p = \frac{1}{n}$ (this probability is referred to as *mutation rate*). If a trial number $i, i \in \{1, \ldots, n\}$ is successful then change bit α_i in word α to the supplementary bit. Otherwise, leave α_i as is. If as the result of a mutation there was obtained $\alpha' \in \{0,1\}^n : f(\alpha') \geq f(\alpha)$, then $\alpha \leftarrow \alpha'$, else $\alpha \leftarrow \alpha$.

In theory, (1+1)-EA is effective in application to the optimization of some 'model' functions, such as ONEMAX, linear functions [12], [7], but extremely inefficient in the worst case scenario (for any pseudo-Boolean function (1)). According to [7], the corresponding worst-case estimation for (1+1)-EA is defined as the expected value of the number of random mutations until achieving the global extremum of (1), i.e. n^n, which is even worse than the similar estimation for the algorithm, that randomly chooses the vectors from $\{0,1\}^n$ in accordance with a uniform distribution (which can be viewed as the (1+1)-EA variant with mutation rate $p = 1/2$).

Despite its simplicity and theoretical ineffectiveness, (1+1)-EA is sometimes surprisingly good in practice. For some hard pseudo-Boolean optimization problems, (1+1)-EA shows comparable or better results than that demonstrated by local-search-based algorithms or genetic algorithms [18, 25]. As it was mentioned in [22], the practical effectiveness of (1+1)-EA can be partially justified by the fact that on average the algorithm acts just like Hill Climbing that works with Hamming neighborhoods of radius 1 in $\{0,1\}^n$. Indeed, it is quite clear that the expected value of the number of bits flipped during the mutation is 1. Thus, (1+1)-EA is capable of adapting towards the landscape of function (1) graphic. On the other hand, unlike local search methods (1+1)-EA does not suffer once it stumbles upon local extrema.

In the present paper we consider only black box functions of the kind (1), i.e. we assume that the value of this function in an arbitrary point $\alpha \in \{0,1\}^n$ is given by some oracle. Let us consider the question how to improve the effectiveness of the basic (1+1)-EA for any pseudo-Boolean black box function (1). First we would like to clarify some terminological details. Hereinafter, by (1+1)-*mutation operator* we mean the basic mutation scheme: when there is a single predecessor for which its offspring is formed by flipping several bits. By (1+1)-*algorithm* we mean an algorithm in the course of which 1 predecessor transforms into 1 offspring, but the process can be more complex than that in (1+1)-mutation operator. In particular, the algorithm can use (1+1)-mutation operator at some stages.

We are aware of a number of attempts to improve the theoretical effectiveness of the original (1+1)-EA. From our point of view, one of the most interesting approaches was proposed in [6], where the so-called (1+1)-*Fast Evolutionary Algorithm* (or (1+1)-FEA$_\beta$) was proposed. The key feature of this algorithm lies in the use of a variable mutation rate. In particular, the (1+1)-FEA$_\beta$ uses a random variable denoted by λ, that takes value from the set $S_\lambda = \{1, 2, \ldots, \frac{n}{2}\}$ w.r.t. the so-called 'power-law distribution $D^\beta_{n/2}$' [6]. A single mutation in the context of (1+1)-FEA$_\beta$ looks as follows: first, in accordance with the distribution $D^\beta_{n/2}$ (for the fixed β) the value of $\lambda \in \{1, \ldots, \frac{n}{2}\}$ is generated, and then the random mutation with mutation rate $p = \frac{\lambda}{n}$ is applied to the current vector $\alpha \in \{0,1\}^n$.

Among the many properties of (1+1)-FEA$_\beta$ studied in [6], we are interested in the fact that its worst case estimation for any black box function (1) is significantly smaller than that for the original (1+1)-EA. A simple way for constructing such an estimation looks as follows. It is quite easy to see that for the mutation rate $p \in \{\frac{1}{n}, \ldots, \frac{1}{2}\}$ the smallest probability of the transition $\alpha \to \alpha^*$ takes place in the situation when each of the n bits in α must be flipped, i.e. in the situation when $d_H(\alpha, \alpha^*) = n$ (hereinafter by d_H we denote the Hamming distance). In (1+1)-FEA$_\beta$ the probability of transition $\alpha \to \alpha^*$ is the probability of a complex event which occurs together with one of the events from $B_1, \ldots, B_{n/2}$. An arbitrary B_λ, $\lambda \in \{1, \ldots, n/2\}$ is defined as choosing a value of λ in accordance with the distribution $D^\beta_{n/2}$. Thus, $\Pr\{\alpha \to \alpha^*\} = \sum_{\lambda=1}^{n/2} \Pr\{\alpha \to \alpha^* | B_\lambda\} \cdot \Pr\{B_\lambda\}$ from which we have, that

$$\Pr\{\alpha \to \alpha^*\} > \Pr\{\alpha \to \alpha^* \mid B_{n/2}\} \cdot \Pr\{B_{n/2}\} \tag{2}$$

Now let us assume that $d_H(\alpha, \alpha^*) = n$ (i.e. the worst case scenario). Then $\Pr\{\alpha \to \alpha^* | B_{n/2}\} = \frac{1}{2^n}$. Apart from that, $\Pr\{B_{n/2}\} = \left(C^\beta_{n/2}\right)^{-1} \cdot \frac{2^\beta}{n^\beta}$, where $C^\beta_{n/2}$ is a normalization constant [6], which is asymptotically equal to $\zeta(\beta)$, where $\zeta(\cdot)$ is the Riemann zeta function. Therefore, from (2) the probability of transition $\alpha \to \alpha^*$ is at least $\frac{c}{n^\beta \cdot 2^n}$ for some constant c. Thus, we have (w.r.t geometric distribution) the following upper bound for the complexity of $(1+1)$-FEA_β:

$$F_{(1+1)\ FEA_\beta} = O(n^\beta \cdot 2^n) \tag{3}$$

As β is a constant, then (3) is of course not comparable to n^n.

The bound (3) 'is not free': in the case of $(1+1)$-FEA_β the expected value of the number of flipped bits after a single random mutation (let us denote it as $E(d_H(\alpha, \alpha')))$ is greater than 1. For $\beta < 3$ this value increases with the increase of n. For $\beta = 3$ from [6] we have $E[d_H(\alpha, \alpha')] \approx 1.3685$.

Based on the ideas from [6] one can construct different variants of the $(1+1)$-EA algorithm, in which the switch between the different values of the mutation rate happens in accordance to significantly simpler distributions compared to $D_{n/2}^\beta$. Let us describe the simple case, to which we will refer as to $(1+1)$-SEA_δ ($(1+1)$-*Switching Evolutionary Algorithm* with parameter δ). Thus, assume, that similar to $(1+1)$-FEA we first choose the mutation rate p_δ, that, with probability $(1 - \frac{\delta}{n})$, $\delta \in (0, 1]$ is set to $\frac{1}{n}$ and with probability $\frac{\delta}{n}$ is set to $\frac{1}{2}$. Then we perform the random mutation w.r.t common $(1+1)$ scheme using the obtained value of mutation rate. Similar to (2) we have

$$\Pr\{\alpha \to \alpha^*\} = \frac{\delta}{n} \cdot \frac{1}{2^n} + \left(1 - \frac{1}{n}\right) \cdot \frac{1}{n^n} > \frac{\delta}{n \cdot 2^n}$$

Thus, the analogue of (3) for $(1+1)$-SEA_δ looks as follows

$$E_{(1+1)-SEA_\delta} = O(n \cdot 2^n) \tag{4}$$

Now let us estimate $E[d_H(\alpha, \alpha')]$ for $(1+1)$-SEA_δ. Using the conditional expected values we have:

$$E[d_H(\alpha, \alpha')] = E\left[d_H(\alpha, \alpha') \mid p_\delta = \tfrac{1}{2}\right] \cdot \Pr\left\{p_\delta = \tfrac{1}{2}\right\} + E\left[d_H(\alpha, \alpha') \mid p_\delta = \tfrac{1}{n}\right]$$
$$\times \Pr\left\{p_\delta = \tfrac{1}{n}\right\} = \frac{n}{2} \cdot \frac{\delta}{n} + 1 \cdot \left(1 - \frac{\delta}{n}\right) = 1 + \delta \cdot \left(\tfrac{1}{2} - \tfrac{1}{n}\right) \tag{5}$$

Thus, for example, when $\delta = 0.5$, from (5) we have that $E[d_H(\alpha, \alpha')] = 1.25 - o(1)$, i.e. in the context of the above, $(1+1)$-$SEA_{0.5}$ is on average behaves more similarly to Hill Climbing compared to $(1+1)$-FEA_3.

Both $(1+1)$-FEA_β and $(1+1)$-SEA_δ can be referred to as *switching algorithms*. The analysis of the phenomena of significantly lower complexity upper bounds for these algorithms compared to the original $(1+1)$-EA leads to the conclusion that such estimations directly follow from the ability of the algorithm to switch into the mode where it acts as a random search. But, let us suppose that while the algorithm worked as the original $(1+1)$-EA, it was able to approach the global extremum by exploiting the landscape of function (1). In this case switching to the random search mode can negate all the accumulated gains.

Taking into account all said above, we believe that it to be relevant to propose the variant of $(1+1)$-EA with the following properties:

1. Constant mutation rate.
2. The expected value of flipped bits after mutation equal to 1.
3. Worst case estimation asymptotically less than n^n.

We describe such an algorithm in the following section.

3   (1+1)-MVEA and Its Theoretical Properties

For the first time, (1+1)-MVEA was proposed in [20]. In this section we comment on a number of known properties of (1+1)-MVEA, and also establish some new ones.

Consider the problem of finding a global maximum of an arbitrary function (1). Associate with an arbitrary vector $\alpha \in \{0,1\}^n$ the set of Boolean variables $X = \{x_1, \ldots, x_n\}$ and consider an arbitrary subjective mapping

$$\mu : X \to Y \tag{6}$$

where $Y = \{y_1, \ldots, y_r\}$: $1 \le r < n$. Let us refer to any μ of this kind as to *merging mapping*. For an arbitrary $y_j \in Y$, $j = 1, \ldots, r$ denote by X_j the set of preimages of y_j under the mapping (6). Let us refer to such set as to the *j-th basket* and let $l_j = |X_j|$. For a fixed merging mapping μ a single iteration of (1+1)-Merging Variables Evolutionary Algorithm ((1+1)-MVEA) is defined as follows.

1. Choose an initial point $\alpha \in \{0,1\}^n$ and assume that the coordinates of α are the values of variables from X (w.r.t. some fixed order).
2. Define random mutation in the following manner: for each $j \in \{1, \ldots, r\}$ perform a Bernoulli trial with success probability $\frac{1}{r}$. If the trial is successful, then consider the basket X_j, $|X_j| = l_j$ to be *chosen*. Let X_j be an arbitrary chosen basket, $x_1^j, \ldots, x_{l_j}^j$ be the variables that were put in X_j and $\alpha^j = \left(\alpha_1^j, \ldots, \alpha_{l_j}^j\right)$ be the assignment of these variables in α. Let us apply the standard random mutation with rate $\frac{1}{l_j}$ to vector α^j. Perform similar operation for each chosen basket. Let α' be the result of application of the mutation to α. If $f(\alpha') \ge f(\alpha)$ then $\alpha \leftarrow \alpha'$, else $\alpha \leftarrow \alpha$.

Below we present the basic properties of (1+1)-MVEA.

Proposition 1. *Let α' be the result of random mutation of α in the context of (1+1)-MVEA. Then $E[d_H(\alpha, \alpha')] = 1$.*

Proof. Fix an arbitrary merging mapping (6) and consider the following random variables: ζ_j, $j \in \{1, \ldots, r\}$: $\zeta_j \in \{0,1\}$ with distribution $\{1 - \frac{1}{r}, \frac{1}{r}\}$; ξ_j is the number of bits in vector $\left(\alpha_1^j, \ldots, \alpha_{l_j}^j\right)$, which were flipped by standard random mutation with rate $\frac{1}{l_j}$. It is easy to see that $d_H(\alpha, \alpha') = \sum_{j=1}^r \zeta_j \cdot \xi_j$. The variables ζ_j and ξ_j are independent, therefore:

$$E[d_H(\alpha, \alpha')] = \sum_{j=1}^r E[\zeta_j] \cdot E[\xi_j] = \frac{1}{r} \cdot \sum_{j=1}^r 1 = 1$$

The following theorem determines the upper bound for the complexity of $(1+1)$-MVEA (in form of the estimation of the expected value of the number of iterations of the algorithm required to transition into global extremum).

Theorem 1 ([20]). *Let us describe any fixed* $\mu : X \rightarrow Y$, $|X| = n$, $|Y| = r$: $1 \leq r < n$ *for which the following holds:* $l_j = |X_j| \geq 2$ *for all* $j \in \{1, \ldots, r\}$, *and* $l = \max\{l_1, \ldots, l_r\}$. *Then:*

$$E^{\mu}_{(1+1)-MVEA} \leq r^r \cdot l^n \tag{7}$$

By $E^{\mu}_{(1+1)-MVEA}$ in (7) we denote the expected value of the number of iterations of the algorithm (w.r.t. fixed μ) before transitioning into a global extremum of function (1). It is possible to fill the baskets almost uniformly: the maximum number of variables in a basket exceeds their minimum number by at most 1. The corresponding mapping μ is called *uniform merging mapping* [20]. The following corollary holds.

Corollary 1 ([20]). *Let* μ *be an arbitrary fixed uniform merging mapping and* $l_j \geq 2$ *for all* $j \in \{1, \ldots, r\}$. *Then there exists a function* $\epsilon(n) : 1 < \epsilon(n) \leq n$ *that the following estimation holds:*

$$E^{\mu}_{(1+1)-MVEA} \leq n^{n\left(\frac{1}{\epsilon(n)} - \frac{\log_n \epsilon(n)}{\epsilon(n)} + \log_n(\epsilon(n)+1)\right)} \tag{8}$$

In fact, $\epsilon(n) = n/r$, and for the uniform merging mapping we have $\lfloor \frac{n}{r} \rfloor \leq l_j \leq \lceil \frac{n}{r} \rceil$ for all $j \in \{1, \ldots, r\}$. It is always possible to pick such r that the estimation (8) will be asymptotically better than n^n. Therefore, the following question arises: how beneficial (from the point of view of decreasing the upper bound for $E^{\mu}_{(1+1)-MVEA}$) can be such a choice of r? The following theorem gives the answer.

Theorem 2. *For each* $d \in R^+$, $d > 1$ *consider a family of uniform merging mappings for which* $r = \lceil n^{\frac{d-1}{d}} \rceil$, $n \in N$ *and assume that* $l_j \geq 2$ *for all* $j \in \{1, \ldots, r\}$. *Then for any* $d > 1$ *there exists such* $n(d)$ *that for all* $n : n \geq n(d)$ *the following estimation holds:*

$$E^{\mu}_{(1+1)-MVEA} \leq n^{n\left(\frac{1}{d-1} + o(1)\right)} \tag{9}$$

Proof. The estimation (9) can be formed by analyzing the expression from the right part of (8). So, assume that the conditions of the theorem are satisfied and (8) holds. Let $r = \lceil n^{\frac{d-1}{d}} \rceil$ and $\epsilon(n) = \frac{n}{r}$. It is clear that

$$\frac{n}{\lceil n^{\frac{d-1}{d}} \rceil} \leq n^{\frac{1}{d}} \tag{10}$$

Taking (10) into account we have that $\log_n(\epsilon(n) + 1) \leq \log_n\left(n^{\frac{1}{d}} + 1\right)$. It is easy to see, that for any $d > 1$ there must exist such $n(d)$ that for any $n > n(d)$ it will hold that $\log_n\left(n^{\frac{1}{d}}+1\right) < \frac{1}{d-1}$. Now consider the value $\frac{1-\log_n \epsilon(n)}{\epsilon(n)}$. Due to the properties of the function $\lceil\cdot\rceil$ and (10), we have:

$$\frac{1-\log_n \epsilon(n)}{\epsilon(n)} \leq \frac{\left(1-\log_n\left(\frac{n}{\left\lceil n^{\frac{d-1}{d}}\right\rceil}\right)\right)\cdot\left(n^{\frac{d-1}{d}}+1\right)}{n}$$

As $d>1$ we can conclude that the value in the right part of the latter inequality tends to 0 with the increase of n. Thus,

$$\frac{1-\log_n \epsilon(n)}{\epsilon(n)} - o(1)$$

From the above and formula (8) we can conclude that (9) holds.

The next step is outlining the algorithm in which the MVEA technique is combined with switching to the random search mode in the same manner as it is done in (1+1)-SEA. In particular, let $\mu : X \to Y$ be an arbitrary merging mapping and n,r be the parameters of μ. As it follows from the definition of (1+1)-MVEA, in this algorithm there are two types of random mutation: the first one (let us refer to it as a mutation of higher level) corresponds to choosing the baskets in which we perform the mutation of lower level. In the algorithm we consider below, we believe it to be rational to use the standard (1+1) mutation on the lower level, but on the higher level to employ the scheme with switching mutation rate similar to the way it is done in (1+1)-SEA. Let us refer to the corresponding algorithm as (1+1)-SMVEA$_\delta$.

Assume that a merging mapping $\mu : X \to Y$, $|X| = n$, $|Y| = r$, is defined, and it induces the separation of X into baskets X_j, $j \in \{1,\ldots,r\}$.

1. Choose an initial point $\alpha \in \{0,1\}^n$ and suppose that the coordinates of α are the values of variables from X (w.r.t. some fixed order).
2. For an arbitrary fixed $\delta \in (0,1]$ with probability $\left(1-\frac{\delta}{r}\right)$ choose $p_\delta = \frac{1}{r}$, and with probability $\frac{\delta}{r}$ set $p_\delta = \frac{1}{2}$.
3. for each $j \in \{1,\ldots,r\}$ perform Bernoulli trial with success probability p_δ. If the trial is successful, then consider the basket X_j, $|X_j| = l_j$ to be chosen. Let X_j be an arbitrary chosen basket, $x_1^j,\ldots,x_{l_j}^j$ be the variables that were put in X_j and $\alpha^j = \left(\alpha_1^j,\ldots,\alpha_{l_j}^j\right)$ to be the assignment of these variables in α. Apply the standard random mutation with rate $\frac{1}{l_j}$ to vector α^j. Perform the same operation with each chosen basket. Let α' be the result of the application of the mutation to α. If $f(\alpha') \geq f(\alpha)$ then $\alpha \leftarrow \alpha'$, otherwise $\alpha \leftarrow \alpha$.

Let us now state some of the theoretical properties of (1+1)-SMVEA$_\delta$.

Theorem 3. *Let us describe any fixed* $\mu : X \to Y$, $|X| = n$, $|Y| = r$: $1 \leq r < n$ *for which the following holds:* $l_j = |X_j| \geq 2$ *for all* $j \in \{1,\ldots,r\}$, *and* $l = max\{l_1,\ldots,l_r\}$ *then:*

1. $E^{\mu}_{(1+1)-SMVEA_{\delta}} = O(r \cdot 2^r \cdot l^n)$
2. $E[d_H(\alpha, \alpha')] = 1 + \delta \cdot \left(\frac{1}{2} - \frac{1}{r}\right)$

Proof. To determine the validity of the first fact let us modify the proof of Theorem 2 from [20]. By analyzing this proof we can see that for the considered algorithm the situation when $d_H(\alpha, \alpha^*) = n$ yields the smallest probability of transition $\alpha \to \alpha^*$. In this case every basket must be chosen and the values of all variables in each basket must be inverted. If the mutation rate of $p_{\delta} = \frac{1}{2}$ is chosen (remind, that the probability of this is $\frac{\delta}{r}$ then the probability with which all baskets become chosen is $\frac{1}{2^r}$. As the probability of flipping all bits in a basket number j is $l_j^{-l_j}$, it means that the probability of transition $\alpha \to \alpha^*$ for the considered case isn't less than $l^{-\sum_{j=1}^{r} l_j} = l^{-n}$. Taking into account the fact that

$$\Pr\{\alpha \to \alpha^*\} = \Pr\left\{\alpha \to \alpha^* \mid p_{\delta} = \frac{1}{r}\right\} \cdot \left(1 - \frac{\delta}{r}\right) + \Pr\left\{\alpha \to \alpha^* \mid p_{\delta} = \frac{1}{2}\right\} \cdot \frac{\delta}{r}$$

we have that

$$\Pr\{\alpha \to \alpha^*\} \geq \frac{\delta}{r \cdot 2^r \cdot l^n}$$

from which we can conclude that the estimation for $E^{\mu}_{(1+1)-SMVEA_{\delta}}$ from the formulation of the theorem is correct. Now let us construct a similar estimation for the expected value of $d_H(\alpha, \alpha')$. Let us use conditional expected values (similar to (5)):

$$E[d_H(\alpha, \alpha')] = E\left[d_H(\alpha, \alpha') \mid p_{\delta} = \frac{1}{r}\right] \cdot \left(1 - \frac{\delta}{r}\right) + E\left[d_H(\alpha, \alpha') \mid p_{\delta} = \frac{1}{2}\right] \cdot \frac{\delta}{r}$$

Note that the expected value of the number of flipped bits in each chosen basket is 1. Then use the scheme from the proof of Proposition 1. Then

$$E\left[d_H(\alpha, \alpha') \mid p_{\delta} = \frac{1}{r}\right] = 1, \quad E\left[d_H(\alpha, \alpha') \mid p_{\delta} = \frac{1}{2}\right] = \frac{r}{2}$$

Therefore,

$$E[d_H(\alpha, \alpha')] = \left(1 - \frac{\delta}{r}\right) + \frac{r}{2} \cdot \frac{\delta}{r} = 1 + \delta \cdot \left(\frac{1}{2} - \frac{1}{r}\right)$$

Thus, Theorem 3 is proven.

4   On One Special Class of MaxSAT Problem

In this section we consider one special subclass of the well known MaxSAT problem. This problem consists in maximization of a function of the kind (1) $f_C : \{0,1\}^n \to \{1, \ldots, m\}$, the value of which is equal to the number of satisfied clauses in Conjunctive Normal Form (CNF) C over the set of variables X, $|X| = n$; it is assumed that the total number of clauses in C is m [10]. The MaxSAT problem is NP-hard, thus one can effectively reduce to it the combinatorial problems from various classes. In some cases such problems have a number of additional features, and taking them into account often allows one to significantly increase the effectiveness of solving for the corresponding MaxSAT instance. In our case these features were studied in paper [17]. In particular, we will consider a CNF that encodes the problem of finding the preimages of a discrete function of the kind $g : \{0,1\}^p \to \{0,1\}^q$ defined by some algorithm A_g. Based on this algorithm A_g the so-called template CNF C_g is constructed first [21]. Then, once a fixed $\gamma \in Range(g)$ is substituted into C_g, the obtained CNF is denoted as $C_g(\gamma)$. The goal is to find the assignment of variables from X that satisfies $C_g(\gamma)$. If such a satisfying assignment is found, then it is possible to effectively extract from it such an $\alpha \in \{0,1\}^p$, that $g(\alpha) = \gamma$.

If the CNF C_g was constructed in accordance with the rules described in [21], then the problem of finding α can be considered in the form of the problem of maximizing a function

$$\phi : \{0,1\}^p \to \{1,\ldots,m\} \tag{11}$$

where m is the number of clauses in CNF $C_g(\gamma)$.

For this purpose we exploit the fact that the set of variables $X^{in} : |X^{in}| = p$ in $C_g(\gamma)$, that encode the inputs of function g is a Strong Unit Propagation Backdoor Set (SUPBS) [23] for $C_g(\gamma)$. The details describing this fact can be found in [17]. That said, for an arbitrary assignment $\alpha \in \{0,1\}^p$ we can use the simple Unit Propagation rule to effectively (in linear time) derive the assignments of all the other variables from $C_g(\gamma)$. Denote this assignment as λ_α. The value of ϕ on the input α is determined as the number of clauses in $C_g(\gamma)$ that are satisfied over the vector λ_α. In [17] it was shown that function ϕ gets the value of m (i.e. achieves its maximum) only on such an α that $g(\alpha) = \gamma$. Therefore, we can view $\{0,1\}^p$ as a search space for maximizing ϕ on which we can use any pseudo-Boolean optimization algorithms.

In the computational experiments that we show in the next section, we used the benchmarks encoding the problems of inversion of cryptographic functions from SHA family. It should be noted that cryptanalysis of cryptographic hash functions using SAT is a fairly popular research topic [8,9,14,15,21], etc. In our case we considered the problems of finding a 512-bit message, for which the first k bits of the hash value (produced via a corresponding hash function) are zeroes. These problems pose some interest in the context of cryptocurrency mining [13]. They were reduced to SAT using the Transalg system [16].

5   Computational Experiments

In our computational experiments we considered the MaxSAT instances, encoding the problems of finding preimages of the SHA-1 and SHA-256 hash functions for which the hash values were required to have at least k first bits equal to 0. We considered $k \in \{16, 18, 20, 22, 24\}$. In order to study these problems, the following algorithms were applied: (1+1)-EA, (1+1)-FEA$_3$, (1+1)-SEA$_{0.5}$, (1+1)-MVEA and (1+1)-SMVEA$_{0.5}$.

The computational experiments were launched on the computational cluster [11] on the nodes, equipped with 2 Intel Xeon E5-2695 v4 "Broadwell" CPUs (18 cores per CPU).

An interesting fact is that on the considered class of instances all the algorithms stagnated quite fast, by getting into the points, where the value of function (11) differs from its maximum value by just several clauses. In order to traverse a larger part of the search space, all tested algorithms were augmented with a special procedure for exiting such points. In particular, this procedure acts as an extension for any of the listed algorithms.

The resulting algorithm, to which we further refer as $\tilde{A}$ was implemented as a multithreaded application for a computing cluster. In each launch each version of the algorithm used 36 threads of a single cluster node.

We would like to note here, that the theoretical properties of multi-threaded (1+1)-algorithms differ from that of single-threaded, however, showing it is not a goal of the present paper. In fact, we can define q-thread (1+1)-mutation operator as a simultaneous independent application of the standard (1+1)-mutation operator to a considered word. The probability of success of such a multi-threaded mutation is then defined as the probability that the transition $\alpha \to \alpha^*$ occurs in at least one thread. It is easy to see that the lower bound on the probability of success is at least $1 - \prod_{i=1}^{q} (1 - p_i)$, where p_i is the lower bound on the probability of transition $\alpha \to \alpha^*$ for a single-threaded case. The latter expression in case of a standard (1+1)-EA yields the probability of success at least $1 - e^{-\frac{q}{n^n}}$. The same procedure can be applied to evaluate the success probabilities for the remaining algorithms.

Let us give now an informal description of $\tilde{A}$; by A we mean any of the 5 variants of (1+1)-EA mentioned before. We assume that $\tilde{A}$ uses computational threads $T_1, \ldots, T_r$. All threads start from the initial point α_0 (for example, zero vector from $\{0, 1\}^n$), $f(\alpha_0)$ is the current Best Known Value (BKV). The control thread sends the current point to all threads. For an arbitrary T_j, $j \in \{1, \ldots, r\}$ let α be the current point. Assume that $\nu = \alpha$ and view ν as a working point. Initialize the number of mutations with 0. For an arbitrary working point ν mutate ν in accordance with algorithm A. Denote the result of mutation as ν'. Check the following conditions:

1. If $f(\nu') > f(\alpha)$ then consider ν' to be the new current point $\alpha \leftarrow \nu'$, $\nu \leftarrow \alpha$.
2. If $f(\nu') \leq f(\alpha)$ and $d_H(\nu', \alpha) \leq R$ (R is a fixed constant) then $\nu \leftarrow \nu'$.
3. If $f(\nu') \leq f(\alpha)$ and $d_H(\nu', \alpha) > R$ then $\nu \leftarrow \alpha$ (return to the current point).

After each mutation we increase the corresponding counter by 1. Once the value of the counter exceeds some threshold Q, the thread T_j sends the current BKV to the control thread. The control thread chooses among all the obtained values of BKV from different threads the best one and sends the corresponding point from $\{0, 1\}^n$ to all the threads as the new current point. If the new value of BKV did not improve the previous BKV then this situation is viewed as stagnation. In the case of the algorithms that use merging variables, after several stagnations they construct a new merging mapping.

Algorithm 1: Multi-threaded algorithm based on the (1+1)-mutation scheme

Input: Starting point α_0, Hamming distance R, number of mutations Q

```
1  α ← α₀
2  while f(α) < f_max do
3      T₁,...,Tᵣ ← α
4      for Tᵢ ∈ {T₁,...,Tᵣ} do
5          ν ← α
6          for t ∈ {1,...,Q} do
7              ν' ← ApplyMutationAlgorithm(ν)
8              if Distance(ν,ν') = 0 then
9                  continue
10             if f(ν') > f(α) then
11                 α ← ν'
12                 ν ← ν'
13                 continue
14             if Distance(α,ν') ≤ R then
15                 ν ← ν'
16             else
17                 ν ← α
       /* Collect the records from all threads and choose the best.    */
18     α' ← GetCurrentRecordPoint(T₁,...,Tᵣ)
       /* If record updated, then update current point.                */
19     if f(α') > f(α) then
20         α ← α'
21         stagnations ← 0
22     else
           /* Stagnation, remain in the current point                  */
23         stagnations ← stagnations + 1
24     if stagnations = S then
25         α ← MakeRemerging(α)
26         stagnations ← 0
```

For the (1+1)-MVEA algorithm we considered 4 versions with the basket sizes of $2, 4, 8, 32$ (the merge size parameter (ms)). Thus, we tested 8 variants of (1+1)-EA.

In the experiments, for each hash function (SHA-1, SHA-256) and the value of the number k of zeroed first bits in the hash value ($k = 16, 18, 20, 22, 24$) and for each tested algorithm we performed 10 independent launches. The results are shown below on Figs. 1 and 2. For each function we compare all algorithms across all launches. On the plots, the horizontal axis represents the number of solved instances, while the vertical axis shows the time spent. In all cases the runtime of the algorithm was limited by 1800 s.

From the results of experiments it is possible to conclude that none of the considered algorithms dramatically outperforms the others. It is quite interesting that the basic (1+1)-EA showed comparable performance to that of the switching algorithms ((1+1)-FEA$_3$, (1+1)-SEA$_{0.5}$, (1+1)-SMVEA$_{0.5}$) despite having

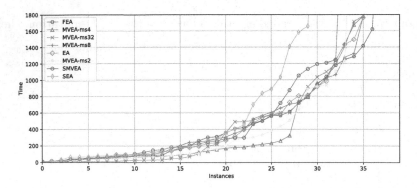

Fig. 1. The results of experiments for SHA-1

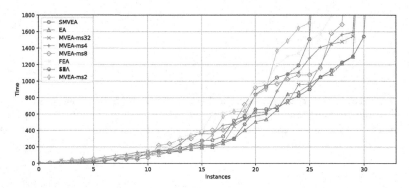

Fig. 2. The results of experiments for SHA-256

significantly larger worst case estimation. At least on the considered class of test problems.

6   Conclusion

In this paper we described a new variant of the (1+1) random mutation strategy that is based on the technique of merging variables proposed in [20]. We determined several theoretical properties of the proposed algorithm and performed computational experiments in which we compare new and existing variants of (1+1)-EA in application to solving special variants of the MaxSAT problem. In particular, we considered the SAT instances encoding the problems of finding preimages of cryptographic hash functions SHA-1 and SHA-256. The proposed algorithms were compared with "classic" (1+1)-EA [12] and (1+1)-FEA_3 [6]. In the computational experiments all algorithms showed comparable performance on this class of test instances.

References

1. Borisovsky, P., Eremeev, A.: Comparing evolutionary algorithms to the (1+1)-EA. Theor. Comput. Sci. **403**(1), 33–41 (2008)
2. Boros, E., Hammer, P.L.: Pseudo-Boolean optimization. Discrete Appl. Math. **123**(1–3), 155–225 (2002)
3. Buzdalov, M., Doerr, B.: Runtime analysis of the $(1 + (\lambda, \lambda))$ genetic algorithm on random satisfiable 3-cnf formulas. In: Bosman, P.A.N. (ed.) Proceedings of the Genetic and Evolutionary Computation Conference, GECCO 2017, Berlin, Germany, 15–19 July, pp. 1343–1350. ACM (2017)
4. Doerr, B.: Analyzing randomized search heuristics via stochastic domination. Theor. Comput. Sci. **773**, 115–137 (2019)
5. Doerr, B., Johannsen, D., Winzen, C.: Multiplicative drift analysis. Algorithmica **64**(4), 673–697 (2012)
6. Doerr, B., Le, H.P., Makhmara, R., Nguyen, T.D.: Fast genetic algorithms. In: Bosman, P.A.N. (ed.) Proceedings of the Genetic and Evolutionary Computation Conference, GECCO 2017, Berlin, Germany, 15–19 July, pp. 777–784. ACM (2017)
7. Droste, S., Jansen, T., Wegener, I.: On the analysis of the (1+1) evolutionary algorithm. Theor. Comput. Sci. **276**(1), 51–81 (2002)
8. Gribanova, I., Semenov, A.: Using automatic generation of relaxation constraints to improve the preimage attack on 39-step MD4. In: 2018 41st International Convention on Information and Communication Technology, Electronics and Microelectronics (MIPRO), pp. 1174–1179 (2018)
9. Gribanova, I., Semenov, A.: Constructing a set of weak values for full-round MD4 hash function. In: 2020 43rd International Convention on Information, Communication and Electronic Technology (MIPRO), pp. 1212–1217 (2020)
10. Li, C.M., Manyà, F.: MaxSAT, hard and soft constraints. In: Biere, A., Heule, M., van Maaren, H., Walsh, T. (eds.) Handbook of Satisfiability, Frontiers in Artificial Intelligence and Applications, vol. 185, pp. 613–631. IOS Press (2009)
11. Irkutsk Supercomputer Center of SB RAS. http://hpc.icc.ru
12. Mühlenbein, H.: How genetic algorithms really work: mutation and hillclimbing. In: Männer, R., Manderick, B. (eds.) Parallel Problem Solving from Nature 2, PPSN-II, Brussels, Belgium, 28–30 September, pp. 15–26. Elsevier (1992)
13. Nakamoto, S.: Bitcoin: a peer-to-peer electronic cash system (2009). http://www.bitcoin.org/bitcoin.pdf
14. Nejati, S., Horáček, J., Gebotys, C., Ganesh, V.: Algebraic fault attack on sha hash functions using programmatic sat solvers. In: Hooker, J. (ed.) Principles and Practice of Constraint Programming, pp. 737–754. Springer International Publishing, Cham (2018)
15. Nejati, S., Newsham, Z., Scott, J., Liang, J.H., Gebotys, C., Poupart, P., Ganesh, V.: A propagation rate based splitting heuristic for divide-and-conquer solvers. In: Gaspers, S., Walsh, T. (eds.) SAT 2017. LNCS, vol. 10491, pp. 251–260. Springer, Cham (2017). https://doi.org/10.1007/978-3-319-66263-3_16
16. Otpuschennikov, I., Semenov, A., Gribanova, I., Zaikin, O., Kochemazov, S.: Encoding cryptographic functions to SAT using TRANSALG system. In: The 22nd European Conference on Artificial Intelligence (ECAI 2016). Frontiers in Artificial Intelligence and Applications, vol. 285, pp. 1594–1595. IOS Press (2016)
17. Otpuschennikov, I.V., Semenov, A.A.: Using merging variables-based local search to solve special variants of maxsat problem. In: Mathematical Optimization Theory and Operations Research. Communications in Computer and Information Science, vol. 1275, pp. 363–378 (2020)

18. Pavlenko, A., Semenov, A., Ulyantsev, V.: Evolutionary computation techniques for constructing SAT-based attacks in algebraic Cryptanalysis. In: Kaufmann, P., Castillo, P.A. (eds.) EvoApplications 2019. LNCS, vol. 11454, pp. 237–253. Springer, Cham (2019). https://doi.org/10.1007/978-3-030-16692-2_16
19. Rudolph, G.: Convergence Properties of Evolutionary Algorithms. Kovač (1997)
20. Semenov, A.A.: Merging variables: one technique of search in pseudo-Boolean optimization. In: Bykadorov, I., Strusevich, V., Tchemisova, T. (eds.) MOTOR 2019. CCIS, vol. 1090, pp. 86–102. Springer, Cham (2019). https://doi.org/10.1007/978-3-030-33394-2_8
21. Semenov, A.A., Otpuschennikov, I.V., Gribanova, I., Zaikin, O., Kochemazov, S.: Translation of algorithmic descriptions of discrete functions to SAT with applications to cryptanalysis problems. Log. Methods Comput. Sci. **16**(1), 29:1–29:42 (2020). https://dblp.org/rec/journals/lmcs/SemenovOG0K19.bib
22. Wegener, I.: Theoretical aspects of evolutionary algorithms. In: Orejas, F., Spirakis, P.G., van Leeuwen, J. (eds.) ICALP 2001. LNCS, vol. 2076, pp. 64–78. Springer, Heidelberg (2001). https://doi.org/10.1007/3-540-48224-5_6
23. Williams, R., Gomes, C.P., Selman, B.: Backdoors to typical case complexity. In: the 18th International Joint Conference on Artificial Intelligence (IJCAI 2003), pp. 1173–1178 (2003)
24. Witt, C.: Runtime analysis of the $(\mu + 1)$-EA on simple pseudo-Boolean functions. Evol. Comput. **14**(1), 65–86 (2006)
25. Zaikin, O.S., Kochemazov, S.E.: On black-box optimization in divide-and-conquer sat solving. Optim. Method Softw. 1–25 (2019)

Mathematical Programming

An Approach for Simultaneous Finding of Multiple Efficient Decisions in Multi-objective Optimization Problems

Konstantin Barkalov⬭, Victor Gergel⬭, Vladimir Grishagin⬭,
and Evgeniy Kozinov(✉)⬭

Lobachevsky State University of Nizhni Novgorod, Nizhni Novgorod, Russia
{konstantin.barkalov,evgeny.kozinov}@itmm.unn.ru, {gergel,vagris}@unn.ru

Abstract. This paper considers computationally intensive multi-objective optimization problems which require computing multiple Pareto-optimal decisions. It is also assumed that efficiency criteria may be multiextremal, and the cost of calculating function values may be quite high. The proposed approach is based on the reduction of multi-objective optimization problems to one-dimensional global optimization problems that can be solved using efficient information-statistical algorithms of global search. One key innovation of the developed approach consists in the possibility of solving several global optimization problems simultaneously, which allows multiple Pareto-optimal decisions to be obtained. Besides, such approach provides for reuse of the computed search information, which considerably reduces computational effort for solving multi-objective optimization problems. Computational experiments confirm the potential of the proposed approach.

Keywords: Multi-objective optimization · Multiple global optimization · Dimensionality reduction · Optimization method · Search information · Computational complexity

1 Introduction

Choosing the optimal decisions (decision making) in situations with many different alternatives is a problem that occurs in almost every domain of human activity. In many cases, decision making problems can be viewed as optimization problems of various kinds such as convex programming, discrete optimization, nonlinear programming, etc. In more complex situations, the objective functions that determine the effectiveness of decisions can be multiextremal, and decision making will require solving global optimization problems. In the most general case, however, the effectiveness of decisions may be determined by several

This work was supported by the Russian Science Foundation, project No. 21-11-00204.

objective criteria, which makes it necessary to solve multi-objective optimization (MOO) problems. Thus, MOO problems are the most common decision making statements that one has to solve in many scientific and technical applications. A large number of approaches and methods have been developed to solve such problems and they have been used to solve many decision making problems in various fields of practical applications – see e.g. [1–9].

One of the approaches most commonly used in the search for efficient decisions is scalarization of the vector efficiency criterion, when a MOO problem is reduced to solving one or several scalar (in the general case, global) optimization problems in which the objective functions are converted to the integrated scalar function. The family of optimization problems generated in such approach is further referred to as multiple global optimization (MGO) problem. Within this approach, we can distinguish lexicographic optimization methods where objective functions are ordered by importance, thus allowing the optimization of the functions to be carried out sequentially as their importance decreases [10]. Possible scalarization methods also include various methods of the efficiency criteria convolution such as the weighted sum method, the compromise programming method, the reference point method, the weighted min-max method, etc. – see, for example, [2, 11, 12].

One common property of methods based on scalarization of the vector efficiency criterion is the existence of some scalarization coefficients that can be varied to obtain different solutions from the Pareto set. Thus, the scalarization coefficients may be interpreted as measures of the importance of the efficiency criteria determined by the decision maker (DM) according to their perception of the required optimality of the decisions to be made. As a result, the general scheme for solving the MOO problem can be represented as a sequence of steps; at each step, DM sets the necessary scalarization coefficients, then the resulting scalar optimization problem is solved, after that DM analyzes the efficient decision found and, if necessary, the above steps are repeated.

The general scheme discussed above can be extended by the possibility of selecting not one but several different scalarization coefficient options at each step. With this possibility, the task of coefficient assignment becomes less complex for DM. Solving several generated scalar optimization problems simultaneously allows one to get efficient decision estimates at the very early stages of computations thus making it possible to change dynamically (in the process of computations) the set of problems being solved: to stop solving obviously unproductive (from DM's point of view) ones or to add new optimization problems.

It is also important to note that by solving simultaneously a large number of scalar optimization problems thus generated it is possible to significantly decrease computational complexity of each separate problem. This effect is achieved due to the fact that all such scalar problems are based on the same MOO problem and, consequently, all computed values of the efficiency criteria of the MOO problem can be reduced to values of any scalar problem being solved simultaneously without any time consuming calculations. In such cases, all the search information obtained in solving any single scalar problem can be used for solving all other scalar problems of the same set.

This paper presents the results of the authors' continuing research aimed to develop efficient methods for solving MOO problems [13–16]. In [13], some methods for solving global optimization problems are considered, in [14], a general decision-making model based on MOO problems is proposed, and in [15], a theoretical analysis is presented regarding the effectiveness of the use of search information when solving MOO problems. In [16], we consider MOO problems in which the efficiency criteria can be ordered by importance (lexicographic multi-objective optimization). In this case, MOO problems are reduced to global optimization problems with non-convex constraints for which constrained optimization methods should be applied. In this paper, we propose new computational schemes for using search information that significantly reduce the amount of calculations required to solve MOO problems.

The structure of the paper is as follows. In Sect. 2, we give the statement of the multi-objective optimization problem. Section 3 presents a scheme for reduction of multi-objective optimization problems to one-dimensional global optimization problems as well as some methods for solving such problems. Section 4 considers the proposed approach for simultaneous solution of several global optimization problems yielding several Pareto-optimal decisions at a time. Section 5 contains the results of numerical experiments confirming the effectiveness of the proposed approach. In conclusion, the results obtained are discussed and possible main directions of further research are outlined.

2   Problem Statement

In the most general form, the MOO problem can be formulated as follows

$$f(y) = (f_1(y), f_2(y), \ldots, f_s(y)) \to \min, y \in D, \tag{1}$$

where $f(y) = (f_1(y), f_2(y), \ldots, f_s(y))$ are objective functions (efficiency criteria), $y = (y_1, y_2, \ldots, y_N)$ is the vector of varied parameters, and N is the dimensionality of the multi-objective optimization problem to be solved. The set of possible parameter values (search domain) D is usually an N-dimensional hyperinterval

$$D = \{y \in R^N : a_i \le y_i \le b_i, 1 \le i \le N\} \tag{2}$$

for given boundary vectors a and b.

Without loss of generality, it is assumed that objective functions should be minimized to improve the decision efficiency $y \in D$. It is also assumed that $f_i(y)$, $1 \le i \le s$, are multiextremal and have the form of time-consuming "black-box" computational procedures. It is also assumed that the objective functions $f_i(y)$, $1 \le i \le s$, satisfy the Lipschitz condition

$$|f_i(y') - f_i(y'')| \le L_i \|y' - y''\|, y', y'' \in D, 1 \le i \le s, \tag{3}$$

where L_i, $1 \le i \le s$, are the Lipschitz constants and $\| * \|$ denotes the Euclidean norm in R^N. Condition (3) means that for variations of the parameter $y \in D$, the corresponding changes in the values of the functions $f_i(y)$, $1 \le i \le s$, are bounded.

3 Reducing the Problem of Multi-objective Optimization to the Problems of One-Dimensional Global Optimization

As mentioned earlier, in the framework of the proposed approach the solution of MOO problems is reduced to solving one or several scalar global optimization problems in which a single efficiency criterion is generated using some scalarization methods of multiple objective functions. Below, we consider a general scheme for such reduction of the MOO problem and present a method for solving the generated global optimization problems.

3.1 Scalarization of Multiple Objective Functions

In the most general form, the global optimization problem generated by the scalarization of multiple objective functions of the MOO problem can be represented in the form

$$\min_{y \in D} \varphi(y) = \min_{y \in D} F(\lambda, y), \tag{4}$$

where F is a scalar multiextremal function generated as a result of scalarization of objective functions f_i, $1 \leq i \leq s$, λ is the vector of parameters of the applied convolution of functions, and D is the search domain from (1). In the proposed approach, we use for scalarization the compromise programming method [2,11], where solution of the MOO problem consists in finding the efficient decision corresponding most closely to the values of the objective functions of the specified reference decision $y^0 \in D$. In this case, a possible statement of scalar optimization problem can have the form

$$\min_{y \in D} F(\lambda, y) = \min_{y \in D} \left\{ \frac{1}{s} \sum_{i=1}^{s} \lambda_i (f_i(y) - f_i(y^0))^2 \right\}, \tag{5}$$

where $F(\lambda, y)$ is the standard deviation of the values of the objective functions f_i, $1 \leq i \leq s$, for the decision $y \in D$ and for the specified reference decision $y^0 \in D$, while the coefficients λ_i, $0 \leq \lambda_i \leq 1$, $1 \leq i \leq s$, are measures of the importance of the approximation accuracy for each variable y_i, $1 \leq i \leq N$, separately. Without loss of generality, we can assume that the domain of possible values of the coefficients λ is a set

$$\lambda = (\lambda_1, \lambda_2, \ldots, \lambda_s) \in \Lambda \subset R^s : \sum_{i=1}^{s} \lambda_i = 1, \lambda_i \geq 0, 1 \leq i \leq s. \tag{6}$$

The reference decision $y^0 \in D$ in (5) can be known *a priori* or determined on the basis of some known prototype. In many cases, an abstract ideal decision $y^0 \in D$ is used as the reference decision $y^* \in D$ in which the objective functions f_i, $1 \leq i \leq s$, have minimal possible values, i.e.

$$f_i^* = f_i(y^*) = \min_{y \in D} f_i(y), 1 \leq i \leq s. \tag{7}$$

Note that by virtue of (3) the scalar function $F(\lambda, y)$ from (5) also satisfies the Lipschitz condition with some constant L, i.e.

$$|F(\lambda, y') - F(\lambda, y'')| \leq L\|y' - y''\|, \; y', y'' \in D. \tag{8}$$

3.2 Dimensionality Reduction

As already mentioned, problem (4) is a multidimensional global optimization problem. Problems of this kind are computationally complex and are known to be subject to the "curse of dimensionality" – computational complexity increases exponentially with increasing dimensionality of the optimization problem being solved [17–24]. Nevertheless, the computational complexity of global optimization algorithms can be significantly reduced by dimensionality reduction based on the use of a Peano space-filling curve (or evolvent) $y(x)$ that uniquely and continuously maps the segment $[0, 1]$ on an N-dimensional domain D – see, for example, [19, 25]. As a result of such reduction, multidimensional global optimization problems (4) are reduced to one-dimensional problems

$$\min_{x \in [0,1]} \varphi(y(x)) = \min_{y \in D} \varphi(y). \tag{9}$$

The resulting one-dimensional functions $\varphi(y(x))$ satisfy the uniform Hölder condition, i.e.

$$|\varphi(y(x')) - \varphi(y(x''))| \leq H|x' - x''|^{1/N}, \; x', x'' \in [0, 1], \tag{10}$$

where the constant H is defined by the relation $H = 2L\sqrt{N + 3}$, L is the Lipschitz constant from (8), and N is the dimensionality of the MOO problem (1).

3.3 Solving the One-Dimensional Reduced Optimization Problem

Using the dimensionality reduction results in one more additional advantage of the proposed approach: many well-known one-dimensional global search algorithms (possibly, after some additional generalisation) can be used to solve the initial multidimensional MOO problem from (1) [26–33]. At the same time, it should be noted that most works where dimensionality reduction is a key feature for solving multiextremal problems like (4) rely on the information-statistical theory of global search [19]. This theory has provided the basis for developing a large number of efficient methods for multiextremal optimization [13–15, 32–38].

Within the framework of information-statistical theory, a general computational scheme of global optimization algorithms was proposed which in brief is as follows [13, 19, 34].

Let k, $k \geq 2$, global search iterations aimed to minimize the function $\varphi(y(x))$ from (9) were completed. Then, to perform adaptive choice of the points of next iterations, the optimization algorithm estimates the possibility that the global minimum is located in the intervals, into which the initial segment $[0, 1]$ is divided by the points of earlier global search iterations

$$x_1 < x_2 < \cdots < x_k. \tag{11}$$

This estimate is determined by means of characteristics $R(i)$ of intervals (x_{i-1}, x_i), $1 < i \le k$, whose values should be proportional to the degree of possibility that the global minimum is located in these intervals. The type of these characteristics depends on the global optimization algorithm used – thus, for example, for an algorithm which constructs a uniform dense grid in the global search domain, the characteristic is simply the interval length

$$R(i) = (x_i - x_{i-1}), 1 < i \le k. \tag{12}$$

For the algorithms proposed in [26,27], when dimensionality reduction is applied, the characteristic is an estimate of the minimum possible value of the function $\varphi(y(x))$ to be minimized on the interval (x_{i-1}, x_i), $1 < i \le k$, i.e.

$$R(i) = 0.5\, H\rho_i - 0.5(z_{i-1} + z_i), \rho_i = (x_i - x_{i-1})^{1N}, 1 < i \le k, \tag{13}$$

where H is the Hölder constant from (10) for the reduced global optimization problem being solved (9), $z_i = \varphi(y(z_i))$, $1 \le i \le k$, and N is the dimensionality of the problem from (1). For the global search algorithm (GSA) [19,28] developed in the framework of the information-statistical approach, the characteristic is

$$R(i) = m\rho_i + \frac{(z_i - z_{i-1})^2}{m\rho_i} - 2(z_{i-1} + z_i), \rho_i = (x_i - x_{i-1})^{1/N}, 1 < i \le k, \tag{14}$$

where m is a numerical estimate of the Hölder constant derived from available search information

$$m = rM, M = \max\{|z_i - z_{i-1}|\rho_i, 1 < i \le k\} \tag{15}$$

($r > 1$ is a parameter of the GSA algorithm).

The presence of interval characteristics makes it possible to describe the procedure of global search iteration as the following sequence of steps [19].

Step 1. Calculate characteristics of the intervals $R(i)$, $1 < i \le k$, and determine the interval with the maximum characteristic

$$R(t) = \max\{R(i), 1 < i \le k\}. \tag{16}$$

Step 2. Select the next iteration point in the interval with the maximum characteristic (the rule X for selecting the point x^{k+1} of the next iteration in the interval (x_{t-1}, x_t) is stated by the global optimization algorithm)

$$x^{k+1} = X(x_{t-1}, x_t), \tag{17}$$

calculate the value z^{k+1} of the function to be minimized at this point (the procedure for calculating the function value will be further referred to as a trial).

Step 3. Check the stopping condition

$$\rho_t \le \varepsilon, \tag{18}$$

where $\rho_t = (x_t - x_{t-1})^{1N}$, t is from (16) and $\varepsilon > 0$ is the specified accuracy of the solution. If the stopping condition (18) is met, then the solving of the optimization problem is stopped, otherwise $k = k + 1$ is assumed and the next global search iteration begins.

After completing the calculations, the lowest computed value of the function being minimized can be taken as the global minimum estimate

$$z_k^* = \min\{z_i, 1 \leq i \leq k\}. \tag{19}$$

It should be noted again that the computational scheme discussed above is quite general. Many global search algorithms can be represented within this scheme, as evidenced, in particular, by examples (12)–(14) and other global optimization algorithms – see, for example, [13–15, 32–38].

Convergence conditions of the algorithms are formulated in the form (16)–(18) depend on the properties of the interval characteristics used. One of sufficient conditions for convergence of algorithms is, for example, the requirement that the characteristic of the interval containing the global minimum point should take on the maximum value at step 1 of the scheme (16)–(18) during the global search iterations. This condition is satisfied, for example, for the multidimensional generalized algorithms proposed in [26, 27] when the Hölder constant from (10) is specified exactly. For GSA, a sufficient condition for convergence is the relation

$$m \geq 2^{3-1N} L\sqrt{N+3}, \tag{20}$$

which must be fulfilled starting from some iteration $k > 1$ of the global search (L is the Lipschitz constant from (8)) [19]. Moreover, if condition (20) is satisfied, only the points of the global minimum of the function $\varphi(y)$ from (4) will be the limit points of the trial sequence $\{y^k = y(x^k)\}$ generated by the GSA algorithm.

4   An Approach for Simultaneous Finding of Multiple Efficient Decisions in Multi-objective Optimization Problems

In this Section, the proposed approach for finding p, $p > 1$ Pareto-optimal decisions of the MOO problem is presented. This problem is formulated as the problem of solving the set $\Phi_p(y)$, $p > 1$ of global optimization problems (see Sect. 4.1). To solve this set of optimization problems, two efficient computational schemes are proposed. The first scheme is based on the traditional sequential procedure for solving multiple problems, however, to solve each subsequent optimization problem, all the search information obtained during previous calculations is taken into account. This accumulated search information can significantly reduce the number of global search iterations required to solve the next optimization problems of the set $\Phi_p(y)$ (see Sect. 4.2 and Sect. 5). In the second scheme, all the problems of the set $\Phi_p(y)$ are solved simultaneously, but the search information obtained when solving each problem is also used for solving all the other problems of the set $\Phi_p(y)$ (see Sect. 4.3). This useful exchange of search information also significantly reduces the amount of required calculations for solving the problems of the set $\Phi_p(y)$.

4.1   The Need to Solve Multiple Global Optimization Problems

As mentioned earlier, in the process of solving the MOO problem, it may be necessary to find several different efficient decisions due to possible changes in the optimality requirements. Obtaining different efficient decisions in the proposed approach is achieved by choosing different convolution coefficients (importance indicators) for the objective functions $f_i(y)$, $1 \leq i \leq s$, which results in obtaining different scalar multiextremal functions $F(\lambda, y)$ from (4). These functions $F(\lambda, y)$ can be optimized sequentially. This determines a multi-stage scheme for solving the MOO problem, when at each stage DM specifies the necessary scalarization coefficients, then the resulting scalar optimization problem is solved, after which DM analyzes the efficient decisions found.

The general scheme considered above can be extended by the possibility of selecting not one but several different scalarization coefficients at each stage. Having such an option makes it easier for DM to specify importance coefficients $\lambda \in \Lambda$ of the objective functions $f_i(y)$, $1 \leq i \leq s$. By solving simultaneously several generated scalar optimization problems, one can obtain the estimates of efficient decisions at the earliest stages of computing, which makes it possible to dynamically (in the process of computations) change the set of problems being solved – to stop solving those that obviously have no prospect of success (from DM's point of view) or to add new optimization problems.

Such generalization of the process of solving a MOO problem means that at each current moment of calculations there is a set of functions being optimized simultaneously having the form

$$\Phi_p(y) = \{F(\lambda_1, y), F(\lambda_2, y), \ldots, F(\lambda_p, y)\}, \lambda_i \in \Lambda, 1 \leq i \leq p, \qquad (21)$$

which can be changed dynamically in the course of the calculations by adding new or removing existing optimization functions $F(\lambda, y)$ from (4).

4.2   Step-by-Step Solution of a Set of Global Optimization Problems

As shown in Sect. 3, the information-statistical multiextremal optimization algorithms used in the proposed approach determine the points of consecutive iterations of the global search taking into account the search information

$$A_k = \{(x_i, z_i, f_i)^T : 1 \leq i \leq k\} \qquad (22)$$

obtained in the calculations (see (14)–(17)). In (22), x_i, $1 \leq i \leq k$, are the reduced points of performed global search iterations ordered in ascending order of coordinates, z_i, f_i, $1 \leq i \leq k$, are the values of the scalar function $F(\lambda, y)$ from (4) and the objective functions $f_i(y)$, $1 \leq i \leq s$, from (1) of the current optimization problem to be solved at the points $x_i = y(x_i)$, $1 \leq i \leq k$. By using search information A_k from (22) when choosing next search iterations it is possible to solve global optimization problems more efficiently and to provide convergence of algorithms only to the global minima of multiextremal functions being minimized.

It is important to note that since the set $\Phi_p(y)$ of functions being optimized simultaneously is generated from the same MOO problem from (1), the existence of the set A_k from (22) allows us to adjust the results of all previously performed calculations of the values of the objective functions $f_i(y)$, $1 \leq i \leq s$, to the values of the next optimized function $F(\lambda, y)$ from (4) without repeating any time consuming calculations of values, i.e.

$$(x_i, f_i) \rightarrow z_i = F(\lambda, y(x_i)), 1 \leq i \leq k. \tag{23}$$

Thus, all the search information A_k from (22), recalculated according to (23), can be reused to continue solving the problems of the set $\Phi_p(y)$. Such a possibility provides a significant reduction in computations up to performing only a limited set of global search iterations (see the results of numerical experiments in Sect. 5).

This type of information connectivity of functions in the set $\Phi_p(y)$ from (21) makes it possible to generalize the computational scheme (16)–(18) for solving a single global optimization problem for the case of optimizing the functions of the set $\Phi_p(y)$ from (21) by adding a preliminary step of the search information transformation.

Step 0. Adjust the state of search information A_k from (22) to the values of the function $F(\lambda, y)$ from the set $\Phi_p(y)$ according to rule (23).

The GSA algorithm applied to optimize the functions of the set $\Phi_p(y)$ from (21) and using the search information A_k will be further referred to as the Multiple Global Search Algorithm (MGSA).

4.3   Simultaneous Solution of a Set of Global Optimization Problems

Information connectivity makes it possible to propose a more general scheme for simultaneous optimization of all the functions of the set $\Phi_p(y)$ from (21). In this case, the search information A_k from (22) will contain the computed values of all simultaneously optimized functions $F(\lambda_i, y)$, $1 \leq i \leq p$, i.e.

$$A_k = \{(x_i, \overrightarrow{z_i}, f_i)^T : 1 \leq i \leq k\}, \tag{24}$$

where the values of $\overrightarrow{z_i}$, $1 \leq i \leq k$, represent vectors

$$\overrightarrow{z_i} = (z_i(1), z_i(2), \ldots, z_i(p)), \ z_i(j) = F(\lambda_j, y(x_i)), 1 \leq i \leq k, 1 \leq j \leq p. \tag{25}$$

Accordingly, for each interval (x_{i-1}, x_i), $1 < i \leq k$, into which the segment [0,1] is divided, the following set of characteristics will be calculated:

$$\overrightarrow{R}(i) = \{R_1(i), R_2(i), \ldots, R_p(i)\}, \tag{26}$$

where

$$R_j(i) = m_j\rho_i + \frac{(z_i(j) - z_{i-1}(j))^2}{m_j\rho_i} - 2(z_{i-1}(j) + z_i(j)), 1 < i \le k, 1 \le j \le p, \quad (27)$$

$$\rho_i = (x_i - x_{i-1})^{1/N}, 1 \le i < k, \quad (28)$$

$$m_j = rM_j, M_j = \left\{ \frac{\max|z_i(j) - z_{i-1}(j)|}{\rho_i}, 1 < i \le k, 1 \le j \le p \right\}. \quad (29)$$

(m_j, $1 \le j \le p$, are the estimates of the Hölder constant in the condition (10) for the functions $F(\lambda_j, y)$, $1 \le j \le p$, of the set $\Phi_p(y)$ from (21)).

The algorithm for simultaneous optimization of all functions of the set $\Phi_p(y)$ from (21) (further denoted as SGSA) can be represented as the following sequence of steps.

Step 1. Compute characteristics of the intervals $R_j(i)$, $1 < i \le k$, $1 \le i \le p$, and determine the function $F(\lambda_j, y)$, $1 \le j \le p$, whose search information contains the interval with the maximum characteristic

$$R_q(t) = \max\{R_j(i), 1 < i \le k, 1 \le i \le p\}. \quad (30)$$

Step 2. Select the point of the next iteration in the interval with the maximum characteristic

$$x^{k+1} = X(x_{t-1}, x_t) \quad (31)$$

and calculate the value $\vec{z}^{k+1}$ of all simultaneously optimized functions $F(\lambda_j, y)$, $1 \le j \le p$, at the point x^{k+1} (when calculating the point x^{k+1} the values $z_i(q)$, $1 \le i \le k$, of the function $F(\lambda_q, y(x_i))$ whose number was determined at Step 1 should be used).

Step 3. Check the stopping condition according to (18)

$$\rho_t \le \varepsilon. \quad (32)$$

When the functions $F(\lambda_j, y)$, $1 \le j \le p$, are optimized simultaneously, it should be kept in mind that the values of these functions at their global minima may differ. To ensure convergence to global minima of all the functions being optimized simultaneously, the SGSA algorithm has to be supplemented by a preliminary step of homogenizing the functions $F(\lambda_j, y)$, $1 \le j \le p$.

Step 0. Convert $F(\lambda_j, y)$, $1 \le j \le p$, according to the rule

$$F'(\lambda_j, y) = \frac{F(\lambda_j, y) - z_{min}(j)}{H_j}, 1 \le j \le p, \quad (33)$$

where $z_{min}(j)$, $1 \le j \le p$, is the minimum value of the function $F(\lambda_j, y)$, $1 \le j \le p$, i.e.

$$z_{min}(j) = \min_{y \in D}\{F(\lambda_j, y)\}, 1 \le j \le p, \quad (34)$$

and H_j, $1 \le j \le p$, are the Hölder constants for the functions $F(\lambda_j, y)$, $1 \le j \le p$.

In the case where the values of $z_{min}(j)$, H_j, $1 \le j \le p$, are not known a priori, these values can be replaced by estimates calculated on the basis of available search information A_k from (22) according to expressions (19) and (29).

5   Results of Numerical Experiments

Numerical experiments were performed on the Lobachevsky supercomputer of the University of Nizhni Novgorod (operating system – CentOS 6.4, management system – SLURM). One supercomputer node has 2 Intel Sandy Bridge E5-2660 2.2 GHz processors, 64 Gb RAM. The CPU is 8-core (i.e. a total of 16 CPU cores are available on the node). The numerical experiments were performed using the Globalizer system [39].

The first series of experiments was performed to compare the MGSA algorithm with a number of well-known multi-objective optimization algorithms by solving a bi-criteria test problem [41]

$$f_1(y) = (y_1 - 1)y_2^2 + 1, f_2(y) = y_2, 0 \leq y_1, y_2 \leq 1. \tag{35}$$

In the course of the experiments, a numerical approximation of the Pareto domain was carried out to solve the problem (35), and the quality of the approximation was evaluated using the hypervolume (HV) and distribution uniformity (DU) indices [15,41]. The first of these indices characterizes the completeness of approximation (a larger value corresponds to a more complete coverage of the Pareto domain), while the second one shows the uniformity of coverage (a smaller value corresponds to a more uniform coverage of the Pareto domain).

Five multi-objective optimization algorithms were compared in this experiment: the Monte-Carlo (MC) method, the genetic algorithm SEMO from the PISA library [43], the Non-Uniform Coverage (NUC) method [41], the Bi-objective Lipschitz Optimization (BLO) method [40] and the MGSA algorithm proposed in this paper. For the first three algorithms, the numerical results were used from [42]. The results of the BLO method were presented in [40].

For MGSA, 100 problems (4) with different values of convolution coefficients λ uniformly distributed in Λ were solved. The results of the experiments performed are presented in Table 1.

Table 1. Comparison of the efficiency of multi-objective optimization algorithms

Solution method	MC	SEMO	NUC	BLO	MGSA
Number of method iterations	500	500	515	498	**338**
Number of points found in the Pareto domain	67	104	29	68	**115**
HV index		0.300	0.312	0.306	0.308	**0.318**
DU index		1.277	1.116	0.210	0.175	**0.107**

As the experimental results show, the MGSA algorithm has a distinct advantage over the other multi-objective optimization methods considered, even for solving relatively simple MOO problems.

In the second series of numerical experiments, we solved bi-criteria two-dimensional MOO problems, i.e. with $N = 2$, $s = 2$. We used multiextremal functions obtained with the help of the GKLS generator [44] as objective functions of the MOO problem. During the experiments, we solved 100 multi-objective problems of this class, for each of them the set $\Phi_p(y)$ from (21) comprised 5, 10 and 25 simultaneously optimized functions respectively (the convolution coefficients $\lambda \in \Lambda$ from (6) for the functions of the set $\Phi_p(y)$ being uniformly distributed in Λ). To check the accuracy of the solution of MOO problems, the computed estimates of the efficient solutions were checked to confirm they belonged to the Pareto domain. For the parameters of the MGSA algorithm, the following values were used: accuracy $\varepsilon = 0.01$, reliability $r = 5.6$. The computational results were averaged over the number of MOO problems solved.

The results of numerical experiments are presented in Table 2 and Fig. 1. The Table shows in the first column the number of functions $\Phi_p(y)$ from (21) being optimized. The columns named Iters contain the average number of iterations performed by the algorithm to solve the MOO problem. The columns named HV and DU contain the values of the HV and DU indicators. The results of the numerical experiments are divided with respect to the three algorithms used: GSA is described in Sect. 3, MGSA is discussed in Subsect. 3.2, SGSA is presented in Subsect. 4.3.

Table 2. Results of numerical experiments on solving bi-criteria two-dimensional MOO problems

Number of functions in $\Phi_p(y)$	Algorithms								
	GSA			MGSA			SGSA		
	Iters	HV	DU	Iters	HV	DU	Iters	HV	DU
5	3771.8	6.455	0.25	1823.4	6.481	0.208	2090.5	6.479	0.217
10	7941.6	6.485	0.183	1955.5	6.486	0.205	2224.4	6.486	0.191
25	20456.2	6.504	0.143	2135.8	6.490	0.209	2456.1	6.493	0.188

The experimental results show that by reusing search information when optimizing 25 functions from the set $\Phi_p(y)$, the total computational iterations can be reduced by more than 9.5 times without resorting to any additional computational resources. At the same time, the MGSA algorithm performs the least number of iterations (the number of calculations of objective function values), while the SGSA algorithm provides a better approximation of the Pareto domain both in terms of approximation completeness and coverage uniformity. It is also noteworthy that the average number of iterations to optimize one function from the set $\Phi_p(y)$ decreases by more than 4 times as the number of functions in the set $\Phi_p(y)$ increases – see Table 3.

In the third series of numerical experiments, 10 bi-criteria five-dimensional MOO problems were solved, i.e. $N = 5$, $s = 2$. The objective functions of the MOO problems were determined, as before, using the GKLS generator [44].

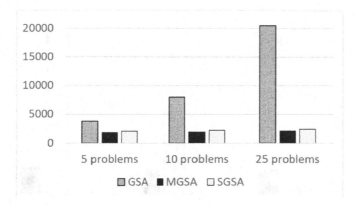

Fig. 1. The average number of iterations for solving bi-criteria two-dimensional MOO problems

Table 3. Average number of iterations required to optimize one function from the set $\Phi_p(y)$

Number of functions in $\Phi_p(y)$	Average number of iterations		
	GSA	MGSA	SGSA
5	754.4	364.7	418.1
10	794.2	195.6	222.4
25	818.2	85.4	98.2

When solving the problems, the accuracy $\varepsilon = 0.05$ and the reliability $r = 5.6$ were used. The results of numerical experiments are presented in Table 4 and Fig. 2.

Table 4. The results of a series of experiments to solve bi-criteria five-dimensional MOO problems

Number of functions in $\Phi_p(y)$	Algorithms								
	GSA			MGSA			SGSA		
	Iters	HV	DU	Iters	HV	DU	Iters	HV	DU
25	1030039.6	2.440	0.543	619755.6	2.441	0.681	669743.8	2.364	0.668
50	2061634.2	2.502	0.572	647508.2	2.533	0.691	710418.6	2.417	0.680
100	4104694.4	2.504	0.686	683410.5	2.555	0.696	728485.3	2.483	0.696

The results of our experiments show that with increasing dimensionality of the MOO problems to be solved, the trend that has been identified continues: the amount of computation (the number of global search iterations) is reduced by more than 5.6 times due to reuse of search information, the MGSA algorithm performs the least number of iterations, and the SGSA algorithm provides the best approximation of the Pareto domain.

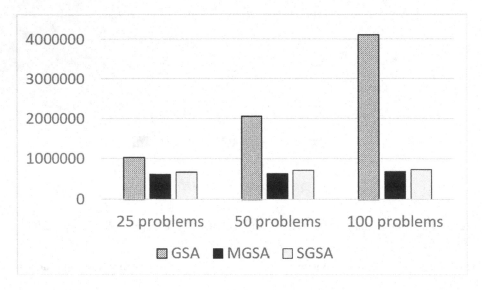

Fig. 2. The average number of iterations for solving bi-criteria five-dimensional MOO problems

Conclusion

This paper proposes an efficient approach for solving computationally intensive multi-objective optimization problems which require computing several Pareto-optimal decisions. It is also assumed that objective functions may be multiextremal, and computation of function values may require a large amount of calculations. The proposed approach is based on the reduction of multi-objective optimization problems to one-dimensional global optimization problems that can be solved using efficient information-statistical algorithms of global search. The novelty of the proposed approach consists in the possibility of solving several global optimization problems simultaneously, which allows multiple Pareto-optimal decisions to be obtained. Besides, such approach provides for reuse of the computed search information, which considerably reduces computational effort for solving multi-objective optimization problems.

The results of numerical experiments show that by using the developed approach it is possible to achieve a significant reduction of computational complexity when solving multi-objective optimization problems.

It should be noted in conclusion that the proposed approach is promising and requires further research in this area. First of all, it is necessary to continue numerical experiments on solving multi-objective optimization problems with a greater number of objective functions and for higher dimensionality of optimization problems to be solved. The possibility of organizing parallel computations for high-performance systems should also be evaluated.

References

1. Miettinen, K.: Nonlinear Multiobjective Optimization. Springer, Boston (1999). https://doi.org/10.1007/978-1-4615-5563-6
2. Ehrgott, M.: Multicriteria Optimization, 2nd edn. Springer, Heidelberg (2010). https://doi.org/10.1007/3-540-27659-9
3. Collette, Y., Siarry, P.: Multiobjective Optimization: Principles and Case Studies. Decision Engineering, Springer, Heidelberg (2004). https://doi.org/10.1007/978-3-662-08883-8
4. Marler, R.T., Arora, J.S.: Multi-Objective Optimization: Concepts and Methods for Engineering. VDM Verlag, Saarbrücken (2009)
5. Pardalos, P.M., Žilinskas, A., Žilinskas, J.: Non-Convex Multi-Objective Optimization. Springer, Cham (2017)
6. Marler, R.T., Arora, J.S.: Survey of multi-objective optimization methods for engineering. Struct. Multidiscip. Optim. **26**, 369–395 (2004)
7. Figueira, J., Greco, S., Ehrgott, M. (eds.): Multiple Criteria Decision Analysis: State of the Art Surveys. Springer, New York (2005). https://doi.org/10.1007/b100605
8. Zavadskas, E.K., Turskis, Z., Kildienė, S.: State of art surveys of overviews on MCDM/MADM methods. Technol. Econ. Dev. Econ. **20**, 165–179 (2014)
9. Hillermeier, C., Jahn, J.: Multiobjective optimization: survey of methods and industrial applications. Surv. Math. Ind. **11**, 1–42 (2005)
10. Collette, Y., Siarry, P.: Multiobjective Optimization: Principles and Case Studies. Decision Engineering. Springer, Heidelberg (2011) https://doi.org/10.1007/978-3-662-08883-8
11. Eichfelder, G.: Scalarizations for adaptively solving multi-objective optimization problems. Comput. Optim. Appl. **44**, 249–273 (2009)
12. Figueira, J., Liefooghe, A., Talbi, E., Wierzbicki, A.: A parallel multiple reference point approach for multi-objective optimization. Eur. J. Oper. Res. **205**(2), 390–400 (2010). https://doi.org/10.1016/j.ejor.2009.12.027
13. Gergel, V.: An unified approach to use of coprocessors of various types for solving global optimization problems. In: 2nd International Conference on Mathematics and Computers in Sciences and in Industry, pp. 13–18 (2015). https://doi.org/10.1109/MCSI.2015.18
14. Gergel, V.P., Kozinov, E.A.: Accelerating multicriterial optimization by the intensive exploitation of accumulated search data. In: AIP Conference Proceedings, vol. 1776, p. 090003 (2016). https://doi.org/10.1063/1.4965367
15. Gergel, V., Kozinov, E.: Efficient multicriterial optimization based on intensive reuse of search information. J. Glob. Optim. **71**(1), 73–90 (2018)
16. Gergel, V., Kozinov, E., Barkalov, K.: Computationally efficient approach for solving lexicographic multicriteria optimization problems. Optim. Lett. (2020). https://doi.org/10.1007/s11590-020-01668-y
17. Zhigljavsky, A.A.: Theory of Global Random Search. Kluwer Academic Publishers, Dordrecht (1991)
18. Pintér, J.D.: Global Optimization in Action (Continuous and Lipschitz Optimization: Algorithms, Implementations and Applications). Kluwer Academic Publishers, Dortrecht (1996)
19. Strongin, R., Sergeyev, Ya.: Global Optimization with Non-Convex Constraints. Sequential and Parallel Algorithms. Kluwer Academic Publishers, Dordrecht (2nd edn. 2013, 3rd edn. 2014)

20. Yang, X.-S.: Nature-inspired Metaheuristic Algorithms. Luniver Press, Frome (2008)
21. Locatelli, M., Schoen, F.: Global Optimization: Theory, Algorithms, and Applications. SIAM (2013)
22. Paulavičius, R., Žilinskas, J.: Simplicial Global Optimization. Springer, New York (2014). https://doi.org/10.1007/978-1-4614-9093-7
23. Floudas, C.A., Pardalos, M.P.: Recent Advances in Global Optimization. Princeton University Press, Princeton (2016)
24. Sergeyev, Y.D., Kvasov, D.E.: Deterministic Global Optimization: An Introduction to the Diagonal Approach. Springer, New York (2017). https://doi.org/10.1007/978-1-4939-7199-2
25. Sergeyev, Y.D., Strongin, R.G., Lera, D.: Introduction to Global Optimization Exploiting Space-Filling Curves. Springer, New York (2013). https://doi.org/10.1007/978-1-4614-8042-6
26. Piyavskij, S.: An algorithm for finding the absolute extremum of a function. Comput. Math. Math. Phys. **12**, 57–67 (1972)
27. Shubert, B.O.: A sequential method seeking the global maximum of a function. SIAM J. Numer. Anal. **9**, 379–388 (1972)
28. Strongin, R.G.: Multiextremal minimization. Autom. Remote. Control. **7**, 1085–1088 (1970)
29. Galperin, E.A.: The cubic algorithm. J. Math. Anal. Appl. **112**, 635–640 (1985)
30. Breiman, L., Cutler, A.: A deterministic algorithm for global optimization. Math. Program. **58**, 179–199 (1993)
31. Baritompa, W.: Accelerations for a variety of global optimization methods. J. Glob. Optim. **4**, 37–45 (1994)
32. Gergel, V.P.: A method of using derivatives in the minimization of multiextremum functions. Comput. Math. Math. Phys. **36**(6), 729–742 (1996)
33. Sergeyev, Y.D.: Global one-dimensional optimization using smooth auxiliary functions. Math. Program. **81**, 127–146 (1998)
34. Sergeyev, Ya.D., Grishagin, V.A.: Sequential and parallel global optimization algorithms. Optim. Methods Softw. **3**, 111–124 (1994)
35. Sergeyev, Y.D.: An information global optimization algorithm with local tuning. SIAM J. Optim. **5**(4), 858–870 (1995)
36. Sergeyev, Y.D., Grishagin, V.A.: Parallel asynchronous global search and the nested optimization scheme. J. Comput. Anal. Appl. **3**(2), 123–145 (2001)
37. Lera, D., Sergeyev, Ya.D.: Deterministic global optimization using space-filling curves and multiple estimates of Lipschitz and Holder constants. Commun. Nonlinear Sci. Numer. Simul. **23**, 328–342 (2015)
38. Grishagin, V., Israfilov, R., Sergeyev, Y.: Comparative efficiency of dimensionality reduction schemes in global optimization. In: AIP Conference Proceedings, vol. 1776 (2016)
39. Gergel, V., Barkalov, K., Sysoyev, A.: Globalizer: a novel supercomputer software system for solving time-consuming global optimization problem. Numer. Algebra Control Opt. **8**(1), 47–62 (2018)
40. Žilinskas, A., Zilinskas, J.: Adaptation of a one-step worst-case optimal univariate algorithm of bi-objective Lipschitz optimization to multidimensional problems. Commun. Nonlinear Sci. Numer. Simul. **21**(1–3), 89–98 (2015). https://doi.org/10.1016/j.cnsns.2014.08.025
41. Evtushenko, Y., Posypkin, M.: A deterministic algorithm for global multiobjective optimization. Optim. Methods Softw. **29**(5), 1005–1019 (2014) https://doi.org/10.1080/10556788.2013.854357

42. Evtushenko, Y., Posypkin, M.: Method of non-uniform coverages to solve the multi-criteria optimization problems with guaranteed accuracy. Autom. Remote Control **75**(6), 1025–1040 (2014)
43. Bleuler, S., Laumanns, M., Thiele, L., Zitzler, E.: PISA—a platform and programming language independent interface for search algorithms. In: Fonseca, C.M., Fleming, P.J., Zitzler, E., Thiele, L., Deb, K. (eds.) EMO 2003. LNCS, vol. 2632, pp. 494–508. Springer, Heidelberg (2003). https://doi.org/10.1007/3-540-36970-8_35
44. Gaviano, M., Kvasov, D.E., Lera, D., Sergeyev, Ya.D.: Software for generation of classes of test functions with known local and global minima for global optimization. ACM Trans. Math. Softw. **29**(4), 469–480 (2003)

One-Point Gradient-Free Methods
for Smooth and Non-smooth Saddle-Point
Problems

Aleksandr Beznosikov[1,2]([⊠]), Vasilii Novitskii[1], and Alexander Gasnikov[1,2,3]

[1] Moscow Institute of Physics and Technology, Moscow, Russia
beznosikov.an@phystech.edu
[2] Higher School of Economics, Moscow, Russia
[3] Caucasus Mathematical Center, Adyghe State University, Maykop, Russia

Abstract. In this paper, we analyze gradient-free methods with one-point feedback for stochastic saddle point problems $\min_x \max_y \varphi(x,y)$. For non-smooth and smooth cases, we present an analysis in a general geometric setup with the arbitrary Bregman divergence. For problems with higher order smoothness, the analysis is carried out only in the Euclidean case. The estimates we have obtained repeat the best currently known estimates of gradient-free methods with one-point feedback for problems of imagining a convex or strongly convex function. The paper uses three main approaches to recovering the gradient through finite differences: standard with a random direction, as well as its modifications with kernels and residual feedback. We also provide experiments to compare these approaches for the matrix game.

Keywords: Saddle-point problem · Zeroth order method · One-point feedback · Stochastic optimization

1 Introduction

This paper is devoted to solving the saddle-point problem:

$$\min_{x \in \mathcal{X}} \max_{y \in \mathcal{Y}} \varphi(x,y). \tag{1}$$

It has many practical applications. These are the already well-known and classic matrix game and Nash equilibrium, as well as modern machine learning problems: Generative Adversarial Networks (GANs) [12] and Reinforcement Learning (RL) [13]. We assume that only zeroth-order information about the function is available, i.e. only its values, not a gradient, hessian, etc. This concept is called a Black-Box and arises in optimization [14], adversarial training [8], RL [10]. To make the problem statement more complex, but close to practice, it is natural

The research of A. Beznosikov and A. Gasnikov was supported by Russian Science Foundation (project No. 21-71-30005).

to assume that we have access inexact values of function $\varphi(x, y, \xi)$, for example, with some random noise ξ. But even with the help of such an oracle, it is possible to recover some estimate of the gradient of a function in terms of finite differences.

Let us highlight two main approaches to such gradient estimates. The first approach is more well researched in the literature and is called a two-point feedback:

$$\frac{n}{2\tau}(\varphi(x + \tau\mathbf{e}_x, y + \tau\mathbf{e}_y, \xi) - \varphi(x - \tau\mathbf{e}_x, y - \tau\mathbf{e}_y, \xi))\begin{pmatrix} \mathbf{e}_x \\ -\mathbf{e}_y \end{pmatrix}.$$

An important feature of this approach is that it is assumed that we were able to obtain the values of the function in points $(x + \tau\mathbf{e}_x, y + \tau\mathbf{e}_y)$ and $(x - \tau\mathbf{e}_x, y - \tau\mathbf{e}_y)$ with the same realization of the noise ξ. From the point of view of theoretical analysis, such an assumption is strong and gives good guarantees of convergence [9,14,17]. But from a practical point of view, this is a very idealistic assumption. Therefore, it is proposed to consider the concept of one-point feedback (which this paper is about):

$$\frac{n}{2\tau}(\varphi(x + \tau\mathbf{e}_x, y + \tau\mathbf{e}_y, \xi^+) - \varphi(x - \tau\mathbf{e}_x, y - \tau\mathbf{e}_y, \xi^-))\begin{pmatrix} \mathbf{e}_x \\ -\mathbf{e}_y \end{pmatrix}.$$

In general $\xi^+ \neq \xi^-$. As far as we know, the use of methods with one-point approximation for saddle-point problems has not been studied at all in the literature. This is the main goal of our work.

1.1 Related Works

Since the use of one-point feedback for saddle-point problems is new in the literature, we present related papers in two categories: two-point gradient-free methods for saddle-point problems, and one-point methods for minimization problems. Partially the results of these works are transferred to Table 1.

Two-Point for Saddle-Point Problems. Here, we first highlight work for non-smooth saddle-point problems [6], as well as work for smooth ones [16]. Note that in these papers an optimal estimate was obtained in the non-smooth case, and in the smooth case only for a special class of "firmly smooth" saddle-point problems. Also note the work devoted to coordinated methods for matrix games [7], which is also close to our topic.

One-Point for Minimization Problems. First of all, we present works that analyze functions with higher order smoothness: [1,2,15]. These works are united by the technique of special random kernels, which allow you to use the smoothness of higher orders. Note that there is an error in work [2], therefore Table 1 shows the corrected result (according to the note from [1]). The special case of higher order smoothness is also interesting – the ordinary smoothness, it is also analyzed in [1,2,15], in addition we note the papers [11,18]. A nonsmooth analysis is presented in [11,18]. Note that in paper [11], not only the Euclidean setup is analyzed, but also the general case with the arbitrary Bregman divergence, which gives additional advantages in the estimates of the convergence (see Table 1).

1.2   Our Contribution

In the nonsmooth case, we consider convex-concave and strongly-convex-strongly-concave problems with bounded $\nabla_x \varphi(x, y)$, $\nabla_y \varphi(x, y)$ on the optimization set. Our algorithm is modification of Mirror Descent with the arbitrary Bregman divergence. The estimates we obtained coincide with the estimates for convex optimization with one-pointed feedback [11,18]. Using the correct geometry helps to reduce the contribution of the problem dimension to the final convergence estimate. In particular, in the entropy setting, convergence depends on the dimension of the problem linearly (see Table 1 for more details in convex-concave case and Table 2 – in strongly-convex-strongly-concave).

In the smooth case we obtained the estimates of the convergence rate with the arbitrary Bregman divergence for convex-concave case and in Euclidean setup for strongly-convex-strongly-concave case. These estimates also coincide with the estimates for convex optimization with one-point feedback [11].

To the best of our knowledge this is the first time when exploiting higher-order smoothness helps to improve performance in saddle-point problems in both strongly-convex-strongly-concave and convex-concave cases. The results also coincide with the estimates for minimization [1,15].

In Tables 1 and 2 one can find a comparison of the oracle complexity of known results with zeroth-order methods for saddle-point problems in related works. Factor q depends on geometric setup of our problem and gives a benefit when we work in the Hölder, but non-Euclidean case (use non-Euclidean prox), i.e. $\|\cdot\| = \|\cdot\|_p$ and $p \in [1; 2]$, then $\|\cdot\|_* = \|\cdot\|_q$, where $1/p + 1/q = 1$. Then q takes values from 2 to ∞, in particular, in the Euclidean case $q = 2$, but when the optimization set is a simplex, $q = \infty$. In higher-order smooth case we consider functions satisfying so called generalized Hölder condition with parameter $\beta > 2$ (see inequality (16) below). Note that it is prefer to use higher-order smooth methods rather than smooth methods only if $\beta > 3$.

2   Preliminaries

To begin with, we introduce some notation and definitions that we use in the work.

2.1   Notation

We use $\langle x, y \rangle \overset{\text{def}}{=} \sum_{i=1}^{n} x_i y_i$ to denote inner product of $x, y \in \mathbb{R}^n$ where x_i is the i-th component of x in the standard basis in $\mathbb{R}^n$. Then it induces ℓ_2-norm

Table 1. Comparison of oracle complexity of one-point/two-point 0th-order methods for non-smooth/smooth **convex** minimization (Min) and **convex-concave** saddle-point (SP) problems under different assumptions. ε means the accuracy of the solution, n – dimension of the problem, $q = 2$ for the Euclidean case and $q = \infty$ for setup of $\|\cdot\|_1$-norm.

Case	Oracle	Prob.	Complexity	Reference
Non-smooth	Two-point	SP	$\mathcal{O}\left(n^{\frac{2}{q}} \cdot \varepsilon^{-2}\right)$	[6]
	One-point	Min	$\mathcal{O}\left(n^{1+\frac{2}{q}} \cdot \varepsilon^{-4}\right)$	[11]
		SP	$\mathcal{O}\left(n^{1+\frac{2}{q}} \cdot \varepsilon^{-4}\right)$	This paper
Smooth	Two-point	SP	$\mathcal{O}\left([n^{\frac{2}{q}} \text{ or } n] \cdot \varepsilon^{-2}\right)$	[16]
	One-point	Min	$\tilde{\mathcal{O}}\left(n^2 \cdot \varepsilon^{-3}\right)$	[11]
		SP	$\tilde{\mathcal{O}}\left(n^2 \cdot \varepsilon^{-3}\right)$	This paper
Higher order smooth	One-point	Min	$\tilde{\mathcal{O}}\left(n^{2+\frac{2}{\beta-1}} \cdot \varepsilon^{-2-\frac{2}{\beta-1}}\right)$	[1,15]
		SP	$\tilde{\mathcal{O}}\left(n^{2+\frac{2}{\beta-1}} \cdot \varepsilon^{-2-\frac{2}{\beta-1}}\right)$	This paper

Table 2. Comparison of oracle complexity of one-point/two-point 0th-order methods for non-smooth/smooth **strongly-convex** minimization (Min) and **strongly-convex-strongly-concave** saddle-point (SP) problems under different assumptions.

Case	Oracle	Prob.	Complexity	Reference
Non-smooth	One-point	Min	$\tilde{\mathcal{O}}\left(n^2 \cdot \varepsilon^{-3}\right)$	[11]
		SP	$\tilde{\mathcal{O}}\left(n^2 \cdot \varepsilon^{-3}\right)$	This paper
Smooth	Two-point	SP	$\mathcal{O}\left(n \cdot \varepsilon^{-1}\right)$	[16]
	One-point	Min	$\tilde{\mathcal{O}}\left(n^2 \cdot \varepsilon^{-2}\right)$	[11]
		SP	$\tilde{\mathcal{O}}\left(n^2 \cdot \varepsilon^{-2}\right)$	This paper
Higher order smooth	One-point	Min	$\tilde{\mathcal{O}}\left(n^{2+\frac{1}{\beta-1}} \cdot \varepsilon^{-\frac{\beta}{\beta-1}}\right)$	[1,15]
		SP	$\tilde{\mathcal{O}}\left(n^{2+\frac{1}{\beta-1}} \cdot \varepsilon^{-\frac{\beta}{\beta-1}}\right)$	This paper

in $\mathbb{R}^n$ in the following way $\|x\|_2 \overset{\text{def}}{=} \sqrt{\langle x, x \rangle}$. We define ℓ_p-norms as $\|x\|_p \overset{\text{def}}{=} \left(\sum_{i=1}^n |x_i|^p\right)^{1/p}$ for $p \in (1, \infty)$ and for $p = \infty$ we use $\|x\|_\infty \overset{\text{def}}{=} \max_{1 \leq i \leq n} |x_i|$. The dual norm $\|\cdot\|_q$ for the norm $\|\cdot\|_p$ is denoted in the following way: $\|y\|_q \overset{\text{def}}{=} \max\{\langle x, y \rangle \mid \|x\|_p \leq 1\}$. Operator $\mathbb{E}[\cdot]$ is full mathematical expectation and operator $\mathbb{E}_\xi[\cdot]$ express conditional mathematical expectation.

Definition 1 (μ-strong convexity). *Function $f(x)$ is μ-strongly convex w.r.t. $\|\cdot\|$-norm on $\mathcal{X} \subseteq \mathbb{R}^n$ when it is continuously differentiable and there is a constant $\mu > 0$ such that the following inequality holds:*

$$f(y) \geq f(x) + \langle \nabla f(x), y - x \rangle + \frac{\mu}{2}\|y - x\|^2, \quad \forall \, x, y \in \mathcal{X}.$$

Definition 2 (Prox-function). *Function* $d(z) : \mathcal{Z} \to \mathbb{R}$ *is called prox-function if* $d(z)$ *is 1-strongly convex w.r.t.* $\| \cdot \|$*-norm and differentiable on* $\mathcal{Z}$.

Definition 3 (Bregman divergence). *Let* $d(z) : \mathcal{Z} \to \mathbb{R}$ *is prox-function. For any two points* $z, w \in \mathcal{Z}$ *we define the Bregman divergence* $V_z(w)$ *associated with* $d(z)$ *as follows:*

$$V_z(w) = d(z) - d(w) - \langle \nabla d(w), z - w \rangle.$$

We denote the Bregman-diameter $\Omega_{\mathcal{Z}}$ of $\mathcal{Z}$ w.r.t. $V_{z_1}(z_2)$ as
$\Omega_{\mathcal{Z}} \overset{\text{def}}{=} \max\{\sqrt{2V_{z_1}(z_2)} \mid z_1, z_2 \in \mathcal{Z}\}$.

Definition 4 (Prox-operator). *Let* $V_z(w)$ *the Bregman divergence. For all* $x \in \mathcal{Z}$ *define prox-operator of* ξ:

$$\operatorname{prox}_x(\xi) = \arg\min_{y \in \mathcal{Z}} \left(V_x(y) + \langle \xi, y \rangle \right).$$

Now we are ready to formally describe the problem statement, as well as the necessary assumptions.

2.2   Settings and Assumptions

As mentioned earlier, we consider the saddle-point problem (1), where $\varphi(\cdot, y)$ is convex function defined on compact convex set $\mathcal{X} \subset \mathbb{R}^{n_x}$, $\varphi(x, \cdot)$ is concave function defined on compact convex set $\mathcal{Y} \subset \mathbb{R}^{n_y}$. For convenience, we denote $\mathcal{Z} = \mathcal{X} \times \mathcal{Y}$ and then $z \in \mathcal{Z}$ means $z \overset{\text{def}}{=} (x, y)$, where $x \in \mathcal{X}$, $y \in \mathcal{Y}$. When we use $\varphi(z)$, we mean $\varphi(z) = \varphi(x, y)$.

Assumption 1 (Diameter of $\mathcal{Z}$). *Let the compact set* $\mathcal{Z}$ *have diameter* Ω.

Assumption 2 (M-Lipschitz continuity). *Function* $\varphi(z)$ *is* M*-Lipschitz continuous in certain neighbourhood of* $\mathcal{Z}$ *with* $M > 0$ *w.r.t. norm* $\| \cdot \|_2$ *when*

$$|\varphi(z) - \varphi(z')| \le M\|z - z'\|_2, \quad \forall\, z, z' \in \mathcal{Z}.$$

One can prove that for all $z \in \mathcal{Z}$ we have

$$\|\tilde{\nabla}\varphi(z)\|_2 \le M. \tag{2}$$

Assumption 3 (μ-strong convexity–strong concavity). *Function* $\varphi(z)$ *is* μ*-strongly-convex-strongly-concave in* $\mathcal{Z}$ *with* $\mu > 0$ *w.r.t. norm* $\| \cdot \|_2$ *when* $\varphi(\cdot, y)$ *is* μ*-strongly-convex for all* y *and* $\varphi(x, \cdot)$ *is* μ*-strongly-concave for all* x *w.r.t.* $\| \cdot \|_2$.

Hereinafter, by $\tilde{\nabla}\varphi(z)$ we mean a block vector consisting of two vectors $\nabla_x\varphi(x,y)$ and $-\nabla_y\varphi(x,y)$. Recall that we do not have access to oracles $\nabla_x\varphi(x,y)$ or $\nabla_y\varphi(x,y)$. We only can use an inexact stochastic zeroth-order oracle $\tilde{\varphi}(x,y,\xi,\delta)$ at each iteration. Our model corresponds to the case when the oracle gives an inexact noisy function value. We have stochastic unbiased noise, depending on the random variable ξ and biased deterministic noise δ. One can write it the following way:

$$\tilde{\varphi}(x,y,\xi) = \varphi(x,y) + \xi + \delta(x,y). \tag{3}$$

Note that δ depends on point (x,y), and ξ is generated randomly regardless of this point.

Assumption 4 (Noise restrictions). *Stochastic noise ξ is unbiased with bounded variance, δ is bounded, i.e. there exists $\Delta, \sigma > 0$ such that*

$$\mathbb{E}\xi = 0, \qquad \mathbb{E}\left[\xi^2\right] \leq \sigma^2, \qquad |\delta| \leq \Delta. \tag{4}$$

3   Theoretical Results

Since we do not have access to $\nabla_x\varphi(x,y)$ or $\nabla_y\varphi(x,y)$, it is proposed to replace them with finite differences. We present two variants: using a random euclidean direction [11,17] in non-smooth case and a kernel approximation [1,15] in smooth. These two concepts will be discussed in more detail later in the respective sections. As mentioned earlier, we work with one-point feedback. We use Mirror Descent as the basic algorithm, but with approximations instead of gradient.

This version of the paper contains no proofs. For the complete version see [5].

3.1   Non-smooth Case

Random Euclidean Direction. For $\mathbf{e} \in \mathcal{RS}_2^n(1)$ (a random vector uniformly distributed on the Euclidean unit sphere) and some constant τ let $\tilde{\varphi}(z+\tau\mathbf{e},\xi) \stackrel{\text{def}}{=} \tilde{\varphi}(x + \tau\mathbf{e}_x, y + \tau\mathbf{e}_y, \xi)$, where $\mathbf{e}_x$ is the first part of $\mathbf{e}$ size of dimension n_x, and $\mathbf{e}_y$ is the second part of dimension n_y. Then define estimation of the gradient through the difference of functions:

$$g(z,\mathbf{e},\tau,\xi^{\pm}) = \frac{n\left(\tilde{\varphi}(z+\tau\mathbf{e},\xi^+) - \tilde{\varphi}(z-\tau\mathbf{e},\xi^-)\right)}{2\tau}\begin{pmatrix}\mathbf{e}_x\\-\mathbf{e}_y\end{pmatrix}, \tag{5}$$

Algorithm 1. zoopMD

Input: z_0, N, γ, τ.
for $k = 0, 1, 2, \ldots, N$ **do**
$\quad z_{k+1} = \text{prox}_{z_k}(\gamma_k \cdot g(z_k, \mathbf{e}_k, \tau, \xi_k^{\pm}))$.
end for
Output: $\bar{z}_N$.

---

where $n = n_x + n_y$. It is important that ξ^+ and ξ^- are different variables – this corresponds to the one-point concept. Next, we present Algorithm 1 – a modification of Mirror Descent with (5). Note that the Bregman divergence can be used in the prox operator. This allows us to take into ac-count the geometric setup of the problem. $\mathbf{e}_k$ and $\xi_k^{\pm}$ are generated independently of the previous iterations and of each other. Here $\bar{z}_N = \frac{1}{N+1}\sum_{i=0}^{N} z_i$. Below we give technical facts about (5).

Lemma 1 (see Lemma 2 from [4] or Lemma 1 from [6]). *For $g(z, \mathbf{e}, \tau, \xi^{\pm})$ defined in (5) under Assumptions 2 and 4 the following inequality holds:*

$$\mathbb{E}\left[\|g(z, \mathbf{e}, \tau, \xi^{\pm})\|_q^2\right] \leq 3a_q^2\left(3nM^2 + \frac{n^2(\sigma^2 + \Delta^2)}{\tau^2}\right), \tag{6}$$

where a_q^2 is determined by $\mathbb{E}[\|e\|_q^2] \leq \sqrt{\mathbb{E}[\|e\|_q^4]} \leq a_q^2$ and the following statement is true

$$a_q^2 = \min\{2q - 1, 32\log n - 8\}n^{\frac{2}{q}-1}, \quad \forall n \geq 3. \tag{7}$$

Next we define an important object for further theoretical discussion – a smoothed version of the function φ (see [14,17]).

Definition 5. *Function $\hat{\varphi}(z)$ defines on set $\mathcal{Z}$ satisfies:*

$$\hat{\varphi}(z) = \mathbb{E}_{\mathbf{e}}\left[\varphi(z + \tau\mathbf{e})\right]. \tag{8}$$

To define smoothed version correctly it is important that the function φ is specified not only on an admissible set $\mathcal{Z}$, but in a certain neighborhood of it. This is due to the fact that for any point z belonging to the set, the point $z + \tau e$ can be outside it.

Lemma 2 (Lemma 8 from [17]). *Let $\varphi(z)$ is μ-strongly-convex-strongly-concave (convex-concave with $\mu = 0$) and $\mathbf{e}$ be from $\mathcal{RS}_2^n(1)$. Then function $\hat{\varphi}(z)$ is μ-strongly-convex-strongly-concave and under Assumption 2 satisfies:*

$$\sup_{z \in \mathcal{Z}} |\hat{\varphi}(z) - \varphi(z)| \leq \tau M. \tag{9}$$

Lemma 3 (Lemma 10 from [17] and Lemma 2 from [4]). *Under Assumption 4 it holds that*

$$\tilde{\nabla}\hat{\varphi}(z) = \mathbb{E}_{\mathbf{e}}\left[\frac{n(\varphi(z + \tau\mathbf{e}) - \varphi(z - \tau\mathbf{e}))}{2\tau}\begin{pmatrix} \mathbf{e}_x \\ -\mathbf{e}_y \end{pmatrix}\right], \tag{10}$$

$$\|\mathbb{E}_{\mathbf{e},\xi}[g(z, \mathbf{e}, \tau, \xi^{\pm})] - \tilde{\nabla}\hat{\varphi}(z)\|_q \leq \frac{\Delta n a_q}{\tau}. \tag{11}$$

Now we are ready to present the main results of this section. Let begin with **convex-concave** case (Assumption 3 with $\mu = 0$)

Theorem 1. *Let problem (1) with function $\varphi(x, y)$ be solved using Algorithm 1 with the oracle (5). Assume, that the set $\mathcal{Z}$, the convex-concave function $\varphi(x, y)$ and its inexact modification $\widetilde{\varphi}(x, y)$ satisfy Assumptions 1, 2, 4. Denote by N the number of iterations and set $\gamma_k = \gamma = \mathrm{const}$. Then the rate of convergence is given by the following expression:*

$$\mathbb{E}\left[\varepsilon_{sad}(\bar{z}_N)\right] \leq \frac{3\Omega^2}{2\gamma(N+1)} + \frac{3\gamma M_{all}^2}{2} + \frac{\Delta\Omega n a_q}{\tau} + 2\tau M.$$

Ω is a diameter of $\mathcal{Z}$, $M_{all}^2 = 3\left(3nM^2 + \frac{n^2(\sigma^2 + \Delta^2)}{\tau^2}\right) a_q^2$ and

$$\varepsilon_{sad}(\bar{z}_N) = \max_{y' \in \mathcal{Y}} \varphi(\bar{x}_N, y') - \min_{x' \in \mathcal{X}} \varphi(x', \bar{y}_N). \tag{12}$$

Let analyze the results:

Corollary 1. *Under the assumptions of Theorem 1 let ε be accuracy of the solution of problem (1) obtained using Algorithm 1. Assume that*

$$\gamma = \Theta\left(\frac{\Omega}{n^{\frac{1}{4}+\frac{1}{2q}}MN^{\frac{3}{4}}}\right), \quad \tau = \Theta\left(\frac{\sigma}{M} \cdot \frac{n^{\frac{1}{4}+\frac{1}{2q}}}{N^{\frac{1}{4}}}\right), \quad \Delta = \mathcal{O}\left(\frac{\varepsilon\tau}{\Omega n a_q}\right), \tag{13}$$

then the number of iterations to find ε-solution

$$N = \mathcal{O}\left(\frac{n^{1+\frac{2}{q}}}{\varepsilon^4}\left[C^4(n, q)M^4\Omega^4 + \sigma^4\right]\right),$$

or with

$$\gamma = \Theta\left(\frac{\Omega}{n^{\frac{1}{q}}MN^{\frac{3}{4}}}\right), \quad \tau = \Theta\left(\frac{\sigma}{M} \cdot \frac{n^{\frac{1}{2}}}{N^{\frac{1}{4}}}\right), \quad \Delta = \mathcal{O}\left(\frac{\varepsilon\tau}{\Omega n a_q}\right),$$

$$N = \mathcal{O}\left(\frac{n^{\frac{4}{q}}C^4(n, q)}{\varepsilon^4}M^4\Omega^4 + \frac{n^2}{\varepsilon^4}\sigma^4\right),$$

where $C(n, q) \overset{def}{=} \min\{2q - 1, 32 \log n - 8\}$.

Analyse separately cases with $p = 1$ and $p = 2$ (Table 3).

Table 3. Summary of convergence estimation for non-smooth case: $p = 2$ and $p = 1$.

p, $(1 \leqslant p \leqslant 2)$	q, $(2 \leqslant q \leqslant \infty)$	N, Number of iterations
$p = 2$	$q = 2$	$\mathcal{O}\left(n^2 \varepsilon^{-4}\right)$
$p = 1$	$q = \infty$	$\mathcal{O}\left(n \log^4 n \cdot \varepsilon^{-4}\right)$

Next we consider μ-**strongly-convex-strongly-concave**. Here we work with $V_z(w) = \frac{1}{2}\|z - w\|_2^2$.

Theorem 2. *Let problem* (1) *with function* $\varphi(x, y)$ *be solved using Algorithm 1 with* $V_z(w) = \frac{1}{2}\|z - w\|_2^2$ *and the oracle* (5). *Assume, that the set* $\mathcal{Z}$, *the function* $\varphi(x, y)$ *and its inexact modification* $\widetilde{\varphi}(x, y)$ *satisfy Assumptions 1, 2, 3, 4. Denote by* N *the number of iterations and* $\gamma_k = \frac{1}{\mu k}$. *Then the rate of convergence is given by the following expression:*

$$\mathbb{E}\left[\varphi(\bar{x}_N, y^*) - \varphi(x^*, \bar{y}_N)\right] \leq \frac{M_{all}^2 \log(N + 1)}{2\mu(N + 1)} + \frac{\Delta n \Omega}{\tau} + 2\tau M$$

Ω *is a diameter of* $\mathcal{Z}$, $M_{all}^2 = 3\left(3nM^2 + \frac{n^2(\sigma^2 + \Delta^2)}{\tau^2}\right)$.

From here one can get

Corollary 2. *Under the assumptions of Theorem 2 let* ε *be accuracy of the solution of problem* (1) *obtained using Algorithm 1. Assume that*

$$\tau = \Theta\left(\sqrt[3]{\frac{\sigma^2}{\mu M}} \cdot \sqrt[3]{\frac{n^2}{N}}\right), \quad \Delta = \mathcal{O}\left(\frac{\varepsilon \tau}{\Omega n}\right),$$

then the number of iterations to find ε-*solution*

$$N = \tilde{\mathcal{O}}\left(\frac{nM^2}{\mu\varepsilon} + \frac{M^2 n^2 \sigma^2}{\mu\varepsilon^3}\right).$$

3.2   Smooth Case

Assumption 5 (Gradient's Lipschitz continuity). *The gradient* $\nabla\varphi(z)$ *of the function* φ *is* L-*Lipschitz continuous in certain neighbourhood of* $\mathcal{Z}$ *with* $L > 0$ *w.r.t. norm* $\|\cdot\|_2$ *when*

$$|\nabla\varphi(z) - \nabla\varphi(z')| \leq L\|z - z'\|_2, \quad \forall z, z' \in \mathcal{Z}.$$

Lemma 4 (Lemma A.3 from [1]**).** *Let* $\varphi(z)$ *be convex-concave (or* μ-*strongly-convex-strongly-concave) and* $\mathbf{e}$ *be from* $\mathcal{RS}_2^n(1)$. *Then function* $\hat{\varphi}(z)$ *is convex-concave (*μ-*strongly-convex-strongly-concave) too and under Assumption 5 satisfies:*

$$\sup_{z \in \mathcal{Z}} |\hat{\varphi}(z) - \varphi(z)| \leq \frac{L\tau^2}{2}. \tag{14}$$

Theorem 3. *Let problem* (1) *with function* $\varphi(x,y)$ *be solved using Algorithm 1 with the oracle* (5). *Assume, that the set* $\mathcal{Z}$, *the convex-concave function* $\varphi(x,y)$ *and its inexact modification* $\widetilde{\varphi}(x,y)$ *satisfy Assumptions 1, 4, 5. Denote by* N *the number of iterations and* $\gamma_k = \gamma = \text{const}$. *Then the rate of convergence is given by the following expression:*

$$\mathbb{E}\left[\varepsilon_{sad}(\bar{z}_N)\right] \leq \frac{3\Omega^2}{2\gamma(N+1)} + \frac{3\gamma M_{all}^2}{2} + \frac{\Delta\Omega na_q}{\tau} + L\tau^2.$$

Ω *is a diameter of* $\mathcal{Z}$, $M_{all}^2 = 3\left(3nM^2 + \frac{n^2(\sigma^2+\Delta^2)}{\tau^2}\right)a_q^2$.

Let's analyze the results:

Corollary 3. *Under the assumptions of Theorem 3 let* ε *be accuracy of the solution of problem* (1) *obtained using Algorithm 1. Assume that*

$$\gamma = \Theta\left(\frac{\Omega}{n^{\frac{1}{3}+\frac{2}{3q}}MN^{\frac{2}{3}}}\right), \quad \tau = \Theta\left(\frac{\sigma}{M}\cdot\frac{n^{\frac{1}{6}+\frac{1}{3q}}}{N^{\frac{1}{6}}}\right), \quad \Delta = \mathcal{O}\left(\frac{\varepsilon\tau}{\Omega na_q}\right), \quad (15)$$

then the number of iterations to find ε-*solution*

$$N - \mathcal{O}\left(\frac{n^{1+\frac{2}{q}}}{\varepsilon^3}\left[M^3\Omega^3 + \frac{L^3\sigma^3}{M^3}\right]\right).$$

Theorem 4. *Let problem* (1) *with function* $\varphi(x,y)$ *be solved using Algorithm 1 with* $V_z(w) = \frac{1}{2}\|z-w\|_2^2$ *and the oracle* (5). *Assume, that the set* $\mathcal{Z}$, *the function* $\varphi(x,y)$ *and its inexact modification* $\widetilde{\varphi}(x,y)$ *satisfy Assumptions 1, 3, 4, 5. Denote by* N *the number of iterations and* $\gamma_k = \frac{1}{\mu k}$. *Then the rate of convergence is given by the following expression:*

$$\mathbb{E}\left[\varphi(\bar{x}_N,y^*) - \varphi(x^*,\bar{y}_N)\right] \leq \frac{M_{all}^2\log(N+1)}{2\mu(N+1)} + \frac{\Delta n\Omega}{\tau} + L\tau^2.$$

Ω *is a diameter of* $\mathcal{Z}$, $M_{all}^2 = 3\left(3nM^2 + \frac{n^2(\sigma^2+\Delta^2)}{\tau^2}\right)$.

Let's analyze the results:

Corollary 4. *Under the assumptions of Theorem 4 let* ε *be accuracy of the solution of problem* (1) *obtained using Algorithm 1. Assume that*

$$\tau = \Theta\left(\sqrt[4]{\frac{\sigma^2}{\mu L}}\cdot\frac{n^{\frac{1}{2}}}{N^{\frac{1}{4}}}\right), \quad \Delta = \mathcal{O}\left(\frac{\varepsilon\tau}{\Omega n}\right),$$

then the number of iterations to find ε-*solution*

$$N = \widetilde{\mathcal{O}}\left(\frac{nM^2}{\mu\varepsilon} + \frac{Ln^2\sigma^2}{\mu\varepsilon^2}\right).$$

3.3 Higher-Order Smooth Case

In this paragraph we study higher-order smooth functions φ functions satisfying so called generalized Hölder condition with parameter $\beta > 2$ (see inequality (16) below).

Higher Order Smoothness. Let l denote maximal integer number strictly less than β. Let $\mathcal{F}_\beta(L_\beta)$ denote the set of all functions $\varphi : \mathbb{R}^n \to \mathbb{R}$ which are differentiable l times and for all $z, z_0 \in U_{\varepsilon_0}(\mathcal{Z})$ satisfy Hölder condition:

$$\left| \varphi(z) - \sum_{0 \le |m| \le l} \frac{1}{m!} D^m \varphi(z_0)(z - z_0)^m \right| \le L_\beta \|z - z_0\|^\beta, \tag{16}$$

where $L_\beta > 0$, the sum is over multi-index $m = (m_1, \ldots, m_n) \in \mathbb{N}^n$, we use the notation $m! = m_1! \cdot \cdots \cdot m_n!$, $|m| = m_1 + \cdots + m_n$ and we defined

$$D^m \varphi(z_0) z^m = \frac{\partial^{|m|} \varphi(z_0)}{\partial^{m_1} z_1 \ldots \partial^{m_n} z_n} z_1^{m_1} \cdot \cdots \cdot z_n^{m_n}, \ \forall z = (z_1, \ldots, z_n) \in \mathbb{R}^n.$$

Let $\mathcal{F}_{\mu,\beta}(L_\beta)$ denote the set of μ-strongly-convex-strongly-concave functions $\varphi \in \mathcal{F}_\beta(L_\beta)$.

To use the higher-order smoothness we propose smoothing kernel though this is not the only way. We propose to use Algorithm 2 which uses the kernel smoothing technique. In fact Algorithm 2 arises from Algorithm 1 in the Euclidean setting ($V_z(w) = \frac{1}{2}\|z - w\|_2^2$).

Algorithm 2. Zero-order Stochastic Projected Gradient

Requires: Kernel $K : [-1, 1] \to \mathbb{R}$, step size $\gamma_k > 0$, parameters τ_k.
Initialization: Generate scalars $r_1, \ldots, r_N$ uniformly on $[-1, 1]$ and vectors $e_1, \ldots, e_N$ uniformly on the Euclidean unit sphere $S_n = \{e \in \mathbb{R}^n : \|e\| = 1\}$.
for $k = 1, \ldots, N$ **do**
  1. $\widetilde{\varphi}_k^+ := \varphi(z_k + \tau_k r_k e_k) + \xi_k^+$, $\widetilde{\varphi}_k^- := \varphi(z_k - \tau_k r_k e_k) + \xi_k^-$
  2. Define $\widetilde{g}_k := \frac{n}{2\tau_k}(\widetilde{\varphi}_k^+ - \widetilde{\varphi}_k^-) \begin{pmatrix} (\mathbf{e}_k)_x \\ -(\mathbf{e}_k)_y \end{pmatrix} K(r_k)$
  3. Update $z_{k+1} := \Pi_Q(z_k - \gamma_k \widetilde{g}_k)$
end for
Output: $\{z_k\}_{k=1}^N$.

To use the higher-order smoothness we propose we need to introduce additional noise assumption:

Assumption 6. *For all $k = 1, 2, \ldots, N$ it holds that*

1. $\mathbb{E}[\xi_k^{+2}] \le \sigma^2$ *and* $\mathbb{E}[\xi_k^{-2}] \le \sigma^2$ *where* $\sigma \ge 0$;
2. *the random variables ξ_k^+ and ξ_k^- are independent from e_k and r_k, the random variables e_k and r_k are independent.*

In other words we assume that $\delta(x, y)$ in (3) is equal to zero. We do not assume here neither zero-mean of ξ_k^+ and ξ_k^- nor i.i.d of $\{\xi_k^+\}_{k=1}^N$ and $\{\xi_k^-\}_{k=1}^N$ as item 2 from Assumption 6 allows to avoid that.

Kernel. For gradient estimator $\widetilde{g}_k$ we use the kernel

$$K : [-1, 1] \to \mathbb{R},$$

satisfying

$$\mathbb{E}[K(r)] = 0, \ \mathbb{E}[rK(r)] = 1, \ \mathbb{E}[r^j K(r)] = 0, \ j = 2, \dots, l, \ \mathbb{E}\left[|r|^\beta |K(r)|\right] \leq \infty, \tag{17}$$

where r is a uniformly distributed on $[-1, 1]$ random variable. This helps us to get better bounds on the gradient bias $\|\widetilde{g}_k - \nabla f(x_k)\|$ (see Theorem 5 for details). The examples of possible kernels are presented in the complete version of this article [5].

For Theorem 5 and Theorem 6 we need to introduce the constants

$$\kappa_\beta = \int |u|^\beta |K(u)| \, du \tag{18}$$

and

$$\kappa = \int K^2(u) \, du. \tag{19}$$

It is proved in [2] that κ_β and κ do not depend on n, they depend only on β:

$$\kappa_\beta \leq 2\sqrt{2}(\beta - 1), \tag{20}$$

$$\kappa \leq \sqrt{3}\beta^{3/2}. \tag{21}$$

Theorem 5. *Let $\varphi \in \mathcal{F}_{\mu,\beta}(L)$ with μ, $L > 0$ and $\beta > 2$. Let Assumption 6 hold and let $\mathcal{Z}$ be a convex compact subset of $\mathbb{R}^n$. Let φ be M-Lipschitz on the Euclidean τ_1-neighborhood of $\mathcal{Z}$ (τ_k when $k = 1$, see parameter τ_k below).*

Then the rate of convergence of Algorithm 2 with parameters

$$\tau_k = \left(\frac{3\kappa\sigma^2 n}{2(\beta - 1)(\kappa_\beta L)^2}\right)^{\frac{1}{2\beta}} k^{-\frac{1}{2\beta}}, \quad \alpha_k = \frac{2}{\mu k}, \quad k = 1, \dots, N$$

is given by the following expression

$$\mathbb{E}\left[\varphi(\overline{x}_N, y^*) - \varphi(x^*, \overline{y}_N)\right] \leq \max_{y \in \mathcal{Y}} \mathbb{E}\left[\varphi(\overline{x}_N, y)\right] - \min_{x \in \mathcal{X}} \mathbb{E}\left[\varphi(x, \overline{y}_N)\right]$$

$$\leq \frac{1}{\mu}\left(n^{2-\frac{1}{\beta}} \frac{A_1}{N^{\frac{\beta-1}{\beta}}} + A_2 \frac{n(1 + \ln N)}{N}\right),$$

where $\overline{x}_N = \frac{1}{N}\sum_{k=1}^{N} x_k$, $\overline{y}_N = \frac{1}{N}\sum_{k=1}^{N} y_k$, $A_1 = 3\beta(\kappa\sigma^2)^{\frac{\beta-1}{\beta}}(\kappa_\beta L)^{\frac{2}{\beta}}$, $A_2 = 9\kappa G^2$, κ_β and κ are constants depending only on β, see (20) and (21).

We emphasize that the usage of kernel smoothing technique, measure concentration inequalities and the assumption that ξ_k is independent from e_k or r_k (Assumption 6) lead to the results better than the state-of-the-art ones for $\beta > 2$. The last assumption also allows us to not assume neither zero-mean of ξ_k^+ and ξ_k^- nor i.i.d of $\{\xi_k^+\}_{k=1}^{N}$ and $\{\xi_k^-\}_{k=1}^{N}$.

Theorem 6. *Let $\varphi \in \mathcal{F}_\beta(L)$ with $L > 0$ and $\beta > 2$. Let Assumption 6 hold and let $\mathcal{Z}$ be a convex compact subset of $\mathbb{R}^n$. Let φ be M-Lipschitz on the Euclidean τ_1-neighborhood of $\mathcal{Z}$.*

Let's define $N(\varepsilon)$:

$$N(\varepsilon) = \max\left\{ \left(R\sqrt{2A_1}\right)^{\frac{2\beta}{\beta-1}} \frac{n^{2+\frac{1}{\beta-1}}}{\varepsilon^{2+\frac{2}{\beta-1}}}, \left(R\sqrt{2c'A_2}\right)^{2(1+\rho)} \frac{n^{1+\rho}}{\varepsilon^{2(1+\rho)}} \right\},$$

where $A_1 = 3\beta(\kappa\sigma^2)^{\frac{\beta-1}{\beta}}(\kappa_\beta L)^{\frac{2}{\beta}}$, $A_2 = 9\kappa G^2$ – constants from Theorem 5, $\rho > 0$ – arbitrarily small positive number, c' – constant which depends on ρ.

Then the rate of convergence is given by the following expression:

$$\mathbb{E}\left[\varphi(\overline{x}_N, y^*) - \varphi(x^*, \overline{y}_N)\right] \leq \max_{y \in \mathcal{Y}} \mathbb{E}\left[\varphi(\overline{x}_N, y)\right] - \min_{x \in \mathcal{X}} \mathbb{E}\left[\varphi(x, \overline{y}_N)\right] \leq \varepsilon \qquad (22)$$

after $N(\varepsilon)$ steps of Algorithm 2 with settings from Theorem 5 for the regularized function: $\varphi_\mu(z) := \varphi(z) + \frac{\mu}{2}\|x - x_0\|^2 - \frac{\mu}{2}\|y - y_0\|^2$, where $\mu \leq \frac{\varepsilon}{R^2}$, $R = \|z_0 - z^*\|$, $z_0 \in \mathcal{Z}$ – arbitrary point. $\overline{x}_N = \frac{1}{N}\sum_{k=1}^{N} x_k$, $\overline{y}_N = \frac{1}{N}\sum_{k=1}^{N} y_k$, τ_1 (τ_k when $k = 1$) is the parameter from Theorem 5 for the regularized function $\varphi_\mu(z)$.

4   Experiments

In our experiments we consider the classical bilinear problem on a probability simplex:

$$\min_{x \in \Delta_n} \max_{y \in \Delta_k} \left[y^T C x\right], \qquad (23)$$

This problem has many different applications and interpretations, one of the main ones is a matrix game (see Part 5 in [3]), i.e. the element c_{ij} of the matrix are interpreted as a winning, provided that player X has chosen the ith strategy and player Y has chosen the jth strategy, the task of one of the players is to maximize the gain, and the opponent's task – to minimize.

The step of our algorithms can be written as follows (see [6]):

$$[x_{k+1}]_i = \frac{[x_k]_i \exp(-\gamma_k[g_x]_i)}{\sum\limits_{j=1}^{n}[x_k]_j \exp(-\gamma_k[g_x]_j)}, \qquad [y_{k+1}]_i = \frac{[y_k]_i \exp(\gamma_k[g_y]_i)}{\sum\limits_{j=1}^{n}[y_k]_j \exp(\gamma_k[g_y]_j)},$$

where under g_x, g_y we mean parts of g which are responsible for x and for y. Note that we do not present a generalization of Algorithm 2 in the arbitrary Bregman setup, but we want to check in practice.

We take matrix 50×50. All elements of the matrix are generated from the uniform distribution from 0 to 1. Next, we select one row of the matrix and generate its elements from the uniform from 5 to 10. Finally, we take one element from this row and generate it uniformly from 1 to 5. Finally, the matrix is normalized. Further, with each call of the function value $y^T C x$ we add stochastic noise with constant variance (which is on average 5% or 10% of the function value).

The main goal of our experiments is to compare three gradient-free approaches: Algorithm 1 with (5) and its modification with residual feedback (see complete paper [5] or original paper [18] for details) approximations, as well as Algorithm 2. We also added a first order method for comparison. Parameters γ and τ are selected with the help of grid-search so that the convergence is the fastest, but stable. See Fig. 1 for results.

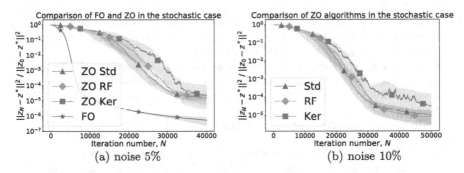

(a) noise 5%   (b) noise 10%

Fig. 1. Algorithm 1 with (5) (ZO Std) approximation and its modification with residual feedback [18] (ZO RF), Algorithm 2 (ZO Ker) and Mirror Descent (FO) applied to solve saddle-problem (23) with noise level: (a) 5%, (b) 10%. The experiment was carried out 10 times for each method: the bold line denotes the mean trajectory, the shaded area denotes the standard deviation from the mean trajectory.

Based on the results of the experiments, we note that the gradient-free methods converge slower than the first-order method – which is predictable. The convergence of zeroth-order methods is approximately the same, the only thing that can be noted is that the method with a kernel is subject to larger fluctuations.

References

1. Akhavan, A., Pontil, M., Tsybakov, A.B.: Exploiting higher order smoothness in derivative-free optimization and continuous bandits. arXiv preprint arXiv:2006.07862 (2020)
2. Bach, F., Perchet, V.: Highly-smooth zero-th order online optimization. In: Conference on Learning Theory, pp. 257–283. PMLR (2016)
3. Ben-Tal, A., Nemirovski, A.: Lectures on modern convex optimization: analysis, algorithms, and engineering applications (2019)
4. Beznosikov, A., Gorbunov, E., Gasnikov, A.: Derivative-free method for composite optimization with applications to decentralized distributed optimization. arXiv preprint arXiv:1911.10645 (2019)

5. Beznosikov, A., Novitskii, V., Gasnikov, A.: One-point gradient-free methods for smooth and non-smooth saddle-point problems. arXiv preprint arXiv:2103.00321 (2021)
6. Beznosikov, A., Sadiev, A., Gasnikov, A.: Gradient-free methods for saddle-point problem. arXiv preprint arXiv:2005.05913 (2020)
7. Carmon, Y., Jin, Y., Sidford, A., Tian, K.: Coordinate methods for matrix games. arXiv preprint arXiv:2009.08447 (2020)
8. Chen, P.Y., Zhang, H., Sharma, Y., Yi, J., Hsieh, C.J.: Zoo. In: Proceedings of the 10th ACM Workshop on Artificial Intelligence and Security - AISec 2017 (2017). https://doi.org/10.1145/3128572.3140448. http://dx.doi.org/10.1145/3128572.3140448
9. Duchi, J.C., Jordan, M.I., Wainwright, M.J., Wibisono, A.: Optimal rates for zero-order convex optimization: the power of two function evaluations. arXiv preprint arXiv:1312.2139 (2013)
10. Fazel, M., Ge, R., Kakade, S., Mesbahi, M.: Global convergence of policy gradient methods for the linear quadratic regulator. In: International Conference on Machine Learning, pp. 1467–1476. PMLR (2018)
11. Gasnikov, A.V., Krymova, E.A., Lagunovskaya, A.A., Usmanova, I.N., Fedorenko, F.A.: Stochastic online optimization. Single-point and multi-point non-linear multi-armed bandits. convex and strongly-convex case. Autom. Remote Control **78**(2), 224–234 (2017)
12. Goodfellow, I.: Nips 2016 tutorial: generative adversarial networks. arXiv preprint arXiv:1701.00160 (2016)
13. Jin, Y., Sidford, A.: Efficiently solving MDPs with stochastic mirror descent. In: Daumé III, H., Singh, A. (eds.) Proceedings of the 37th International Conference on Machine Learning. Proceedings of Machine Learning Research, vol. 119, pp. 4890–4900. PMLR, 13–18 July 2020
14. Nesterov, Y., Spokoiny, V.G.: Random gradient-free minimization of convex functions. Found. Comput. Math. **17**(2), 527–566 (2017)
15. Novitskii, V., Gasnikov, A.: Improved exploiting higher order smoothness in derivative-free optimization and continuous bandit. arXiv preprint arXiv:2101.03821 (2021)
16. Sadiev, A., Beznosikov, A., Dvurechensky, P., Gasnikov, A.: Zeroth-order algorithms for smooth saddle-point problems. arXiv preprint arXiv:2009.09908 (2020)
17. Shamir, O.: An optimal algorithm for bandit and zero-order convex optimization with two-point feedback. J. Mach. Learn. Res. **18**(52), 1–11 (2017)
18. Zhang, Y., Zhou, Y., Ji, K., Zavlanos, M.M.: Improving the convergence rate of one-point zeroth-order optimization using residual feedback. arXiv preprint arXiv:2006.10820 (2020)

Convex Optimization with Inexact Gradients in Hilbert Space and Applications to Elliptic Inverse Problems

Vladislav Matyukhin[1]([✉]) [iD], Sergey Kabanikhin[2,3,4] [iD], Maxim Shishlenin[2,3,4] [iD],
Nikita Novikov[2,3] [iD], Artem Vasin[1] [iD], and Alexander Gasnikov[1,5,6] [iD]

[1] Moscow Institute of Physics and Technology (National Research University),
Dolgoprudny, Russia
matyukhin@phystech.edu
[2] Novosibirsk State University, Novosibirsk, Russia
[3] Institute of Computational Mathematics and Mathematical Geophysics,
Novosibirsk, Russia
[4] Sobolev Institute of Mathematics, Novosibirsk, Russia
[5] Weierstrass Institute for Applied Analysis and Stochastics, Berlin, Germany
[6] Institute for Information Transmission Problems RAS, Moscow, Russia

Abstract. In this paper, we propose the gradient descent type methods to solve convex optimization problems in Hilbert space. We apply it to solve the ill-posed Cauchy problem for the Poisson equation and make a comparative analysis with the Landweber iteration and steepest descent method. The theoretical novelty of the paper consists in the developing of a new stopping rule for accelerated gradient methods with inexact gradient (additive noise). Note that up to the moment of stopping the method "doesn't feel the noise". But after this moment the noise starts to accumulate and the quality of the solution becomes worse for further iterations.

Keywords: Convex optimization · Inexact oracle · Inverse and ill-posed problem · Gradient method

1 Introduction

In this paper, we propose the gradient descent type methods to solve convex optimization problems in Hilbert space. We apply it to solve the ill-posed Cauchy problem for the Poisson equation and make a comparative analysis with the Landweber iteration and steepest descent method. The theoretical novelty of the

The research of V.V. Matyukhin and A.V. Gasnikov in Sects. 1,2,3,4 was supported by Russian Science Foundation (project No. 21-71-30005). The research of S.I. Kabanikhin, M.A. Shishlenin and N.S. Novikov in the last section was supported by RFBR 19-01-00694 and by the comprehensive program of fundamental scientific researches of the SB RAS II.1, project No. 0314-2018-0009. The work of A. Vasin was supported by Andrei M. Raigorodskii Scholarship in Optimization.

P. Pardalos et al. (Eds.): MOTOR 2021, LNCS 12755, pp. 159–175, 2021.
https://doi.org/10.1007/978-3-030-77876-7_11

paper consists in the developing of a new stopping rule for accelerated gradient methods with inexact gradient.

Following the works [6,12,27] we develop new approaches to solve convex optimization problems in Hilbert space [14,22]. The main difference from the existing approaches is that we don't approximate an infinite-dimensional problem by the finite one (see [6,27]). We try to solve the problem in Hilbert space (infinite-dimensional). But we try to do it with the conception of the inexact oracle. That is we use an approximation of the problem only when we calculate the gradient (Frechet derivative) of the functional. This generates inexactness in gradient calculations. We try to combine known results in this area and to understand the best way to solve convex optimization problems in Hilbert space with application to ill-posed and inverse problems [12].

It's important to note, that in the paper we consider only gradient type procedures without 1D-line search. So it means that very popular in practice methods, like steepest descent and conjugate gradient [7,20] and their nonlinear analogues [18], do not take into account. The reason is that we try to develop an approach that justified theoretically. For all of these methods, there exist some troubles with error accumulation [21]. In the worth case, algorithms may diverge. In this paper, we describe how to control this divergence and stop in time for gradient-type methods without a 1D-line search. Fortunately, there exist alternative procedures to 1D-line search (Armijo, Wolf, Nesterov rules [7,17,20]) that perform the same function as 1D-line search. We will use Nesterov's rule [5,7,17]. It allows us to choose an adaptive stepsize policy.

The important part of the paper is an adaptation of the modern results developed for convergence of gradient type methods with inexact gradient for a specific class of inverse ill-posed convex problems in Hilbert space. The basic algorithms are gradient descent [7,20], fast (accelerated) gradient method in variant of Similar Triangles Method (STM) [9] and its combinations [3,13]. For these algorithms, the theory of gradient error accumulation is well developed [3,7,10,13,23,24]. Basically, the theoretical foundation of the facts we use in this paper can be found in the paper [13] and recent arXive preprint [23].

The structure of the paper is as follows. In Sect. 2 we described primal approaches (we solve exactly the problem we have) based on contemporary versions of fast gradient descent methods and their adaptive variants.

In Sect. 3 we described dual approaches (we solve a dual problem) based on the same methods. We try to describe all the methods with the exact estimations of their convergence. But every time we have in mind concrete applications. Since that we include in the description of algorithms such details that allow methods to be more practical.

Section 4 contains a new result about the proper stopping rule for STM when the noise in gradient is additive. This result can be briefly formulated as follows. Having δ-inexact gradient (inexactness is additive), gradient descent and accelerated gradient descent (we consider STM) converge almost like their noise-free analogues up to an accuracy in function $\sim \delta R$, where R corresponds to the size of the solution. After that, we should stop the algorithm, since an error can be further accumulated and caused a divergence of the method [21]. This result seems to be rather unexpected for accelerated algorithms, due to pessimistic results, mentioned in Sect. 3, about the accumulation of the error in another conception of noise.

The rest part of the paper is devoted to applications of the described results to elliptic ill-posed inverse problems. This part experimentally confirms conclusions that have been done in Sects. 2, 3 and 4.

2   Primal Approaches

Assume that $q \in H$, where H is a Hilbert space with a scalar product denoted by $\langle\,,\,\rangle$ (H isn't necessarily finite). In this paper, we investigate the following optimization problem

$$J(q) \to \min_{q \in H}, \tag{1}$$

where J is a convex functional in H.

Let $y^0 \in H$ is a starting point pro an iterative method for solving (1), and

$$R = \left\|y^0 - q_*\right\|_2,$$

where q_* is a solution of (1) that gives R the smallest value. We assume that at least one solution exists [27].

Assume that $J(q)$ has Lipchitz Frechet derivative, i.e. there exist $L > 0$, such that

$$\left\|\nabla J(q_2) - \nabla J(q_1)\right\|_2 \leq L \left\|q_2 - q_1\right\|_2, \tag{2}$$

where $\|q\|_2^2 = \|q\|_H^2 = \langle q, q\rangle_H$. In (2) we also use that due to the Riesz representation theorem [11], one may consider $\nabla J(q)$ to be an element of $H^* = H$.

Example Assume that we have linear operator A from Hilbert space H_1 to another Hilbert space H_2, i.e. $A : H_1 \to H_2$ and $b \in H_2$. Let's consider the following convex optimization problem [12,27]:

$$J(q) = \frac{1}{2}\|Aq - b\|_{H_2}^2 \to \min_{q \in H}.$$

Note that

$$\nabla J(q) = A^*(Aq - b).$$

Formula (2) is equivalent to

$$\langle Aq, Aq\rangle_{H_2} = \|Aq\|_{H_2}^2 \leq L\|q\|_{H_1}^2 = L\langle q, q\rangle_{H_1},$$

i.e. $L = \|A\|_{H_1 \to H_2}^2$. ■

Now following to [9,25,26] (most of the ideas below goes back to the pioneer's works of B.T. Polyak, A.S. Nemirovski and Yu.E. Nesterov) we describe optimal (up to absolute constant factor or logarithmic factor in strongly convex case) numerical methods [16,19,22] (in terms of the number of ideal calculations of $\nabla J(q)$ and $J(q)$) for solving problem (1). The rates of convergence obtained in Theorems 1, 2 can be reached (in the case of Example 2) also by conjugate-gradient methods [12, 27], but we lead these estimates under more general conditions.

Algorithm 1. Similar Triangular Method STM (y^0, L)

Input: $A_0 = \alpha_0 = 1/L, \quad q^0 = u^0 = y^0 - \alpha_0 \nabla J (y^0)$.
1: **Put** k = 0,
2: **Calculate**

$$\alpha_{k+1} = \frac{1}{2L} + \sqrt{\frac{1}{4L^2} + \frac{A_k}{L}}, \quad A_{k+1} = A_k + \alpha_{k+1},$$

$$y^{k+1} = \frac{\alpha_{k+1} u^k + A_k q^k}{A_{k+1}},$$

$$u^{k+1} = u^k - \alpha_{k+1} \nabla J (y^{k+1}),$$

$$q^{k+1} = \frac{\alpha_{k+1} u^{k+1} + A_k q^k}{A_{k+1}}.$$

3: If stopping rule doesn't satisfy, put $k := k + 1$ and **go to** 2.

If $J(q_*) = 0$ (see Example 1) then stopping rule has the form $J(q^k) \leq \varepsilon$ for some $\varepsilon > 0$.

Theorem 1 *(see [9]). Assume that (2) holds. Then after N iterations of STM (y^0, L) we have the following estimate:*

$$J(q^N) - J(q_*) \leq \frac{4LR^2}{N^2}.$$

Sometimes it's hardly possible to estimate the parameter L, used in STM. Moreover, even when we can estimate L we have to use the worth one (the largest one). Is it possible to change the worth case L to the average one (among all the iterations)? The answer is YES [9] (see ASTM below).

Theorem 2 *(see [9]). Assume that (2) holds. Then after N iterations of ASTM (y^0) we have the following estimate:*

$$J(q^N) - J(q_*) \leq \frac{8LR^2}{N^2}.$$

The average number of calculations of $J(q)$ per iteration roughly equals 4 and the average number of calculations Frechet derivative $\nabla J(q)$ per iteration roughly equals 2.

2.1 Gradient Descent

In the literature, one can typically meet non-accelerated simple gradient descent method GD $\left(q^0 = y^0, L\right)$ [3,12,27]

$$q^{k+1} = q^k - \frac{1}{L}\nabla J\left(q^k\right) \tag{3}$$

Algorithm 2. Adaptive Similar Triangular Method ASTM $\left(y^0\right)$

Input: $A_0 = \alpha_0 = 1/L_0^0 = 1, \quad k = 0, \quad j_0 = 0; \quad q^0 := u^0 := y^0 - \alpha_0 \nabla J\left(y^0\right).$

1: **while** $J\left(q^0\right) > J\left(y^0\right) + \left\langle \nabla J\left(y^0\right), q^0 - y^0\right\rangle + \frac{L_0^{j_0}}{2}\left\|q^0 - y^0\right\|_2^2$ **do**

2: $\quad j_0 := j_0 + 1; L_0^{j_0} := 2^{j_0} L_0^0; (A_0 :=) \alpha_0 := \frac{1}{L_0^{j_0}}, \quad q^0 := u^0 := y^0 - \alpha_0 \nabla J\left(y^0\right).$

3: **end while**

4: **Put** $L_{k+1}^0 = L_k^{j_k}\big/2, \ j_{k+1} = 0.$

$$\alpha_{k+1} := \frac{1}{2L_{k+1}^0} + \sqrt{\frac{1}{4\left(L_{k+1}^0\right)^2} + \frac{A_k}{L_{k+1}^0}}, \quad A_{k+1} := A_k + \alpha_{k+1},$$

$$y^{k+1} = \frac{\alpha_{k+1}u^k + A_k q^k}{A_{k+1}},$$

$$u^{k+1} = u^k - \alpha_{k+1}\nabla J\left(y^{k+1}\right),$$

$$q^{k+1} = \frac{\alpha_{k+1}u^{k+1} + A_k q^k}{A_{k+1}}.$$

5: **while** $J\left(y^{k+1}\right) + \left\langle\nabla J\left(y^{k+1}\right), q^{k+1} - y^{k+1}\right\rangle + \frac{L_{k+1}^{j_{k+1}}}{2}\left\|q^{k+1} - y^{k+1}\right\|_2^2 <$ $J\left(q^{k+1}\right)$ **do**

6:

$$j_{k+1} := j_{k+1} + 1; L_{k+1}^{j_{k+1}} = 2^{j_{k+1}}L_{k+1}^0;$$

$$\alpha_{k+1} := \frac{1}{2L_{k+1}^{j_{k+1}}} + \sqrt{\frac{1}{4\left(L_{k+1}^{j_{k+1}}\right)^2} + \frac{A_k}{L_{k+1}^{j_{k+1}}}}, \quad A_{k+1} := A_k + \alpha_{k+1};$$

$$y^{k+1} := \frac{\alpha_{k+1}u^k + A_k q^k}{A_{k+1}}, \quad u^{k+1} := u^k - \alpha_{k+1}\nabla J\left(y^{k+1}\right),$$

$$q^{k+1} := \frac{\alpha_{k+1}u^{k+1} + A_k q^k}{A_{k+1}}.$$

7: **end while**

8: If stopping rule doesn't satisfy, put $k := k + 1$ and **go to** 4.

---

or $\overline{GD}\left(y^0, L\right)$, $q^0 = 0$

$$\begin{cases} y^{k+1} = y^k - \frac{1}{L}\nabla J\left(y^k\right), \\ q^{k+1} = \frac{k}{k+1}q^k + \frac{1}{k+1}y^{k+1}. \end{cases} \tag{4}$$

In case (2) method (4) requires O $\left(LR^2/\varepsilon\right)$ calculations of $\nabla J\left(q\right)$ for $J\left(q^N\right) - J\left(q_*\right) \le \varepsilon$.

One can easily propose an adaptive version of GD and $\overline{GD}$ (AGD and $\overline{AGD}$) see the full version of the paper [15].

3   Dual Approaches

Now we concentrate on Example 1. The described below approaches go back to the Yu.E. Nesterov and A.S. Nemirovski (see historical notes in [1,2]).

Assume that we have to solve the following convex optimization problem

$$g(q) \to \min_{Aq=f}, \tag{5}$$

where $g(q)$ is 1-strongly convex in H_1. We build the dual problem

$$\phi\left(\lambda\right) = \max_q \left\{\langle \lambda, f - Aq\rangle - g\left(q\right)\right\} = \langle \lambda, f - Aq\left(\lambda\right)\rangle - g\left(q\left(\lambda\right)\right) \to \min_{\lambda \ge 0}. \tag{6}$$

Note, that $\nabla\phi\left(\lambda\right) = f - Aq\left(\lambda\right)$.

Let (A)STM with $y^0 = 0$ for problem (6) generates points $\left\{y^k\right\}_{k=0}^N$, $\left\{u^k\right\}_{k=0}^N$ and $\left\{\lambda^k\right\}_{k=0}^N$ (in (A)STM we denote the last ones by $\left\{q^k\right\}_{k=0}^N$). Put

$$q^N = \sum_{k=0}^N \frac{\alpha_k}{A_N}q(y^k).$$

Let q_* be the solution of (5) (this solution is unique due to the strong convexity of $g\left(q\right)$). Then

$$g(q^N) - g(q_*) \le \phi(\lambda^N) + g(q^N).$$

The next theorem [1,2] allows us to calculate the solution of (5) with prescribed precision.

Theorem 3. *Assume that we want to solve problem (5) by passing to the dual problem (6), according to the formulas mentioned above. Let's use (A)STM to solve (6) with the following stopping rule*

$$\phi\left(\lambda^N\right) + g\left(q^N\right) \le \varepsilon, \quad \left\|Aq^N - f\right\|_{H_2} \le \tilde{\varepsilon}.$$

Then (A)STM stops by making no more than

$$6 \cdot \max\left\{\sqrt{\frac{L\tilde{R}^2}{\varepsilon}}, \sqrt{\frac{L\tilde{R}}{\tilde{\varepsilon}}}\right\} \tag{7}$$

iterations, where $L = \|A^*\|^2_{H_2 \to H_1} = \|A\|^2_{H_1 \to H_2}$, $\breve{R} = \|\lambda_*\|_{H_2}$, λ_* *is a solution of problem* (6) *(if the solution is not unique then we can choose such a solution* λ_* *that minimizes* $\breve{R}$).

For ASTM the average number of calculations of $\phi(\lambda)$ per iteration roughly equals 4 and the average number of calculations Frechet derivative $\nabla \phi(\lambda) = f - Aq(\lambda)$ per iteration roughly equals 2.

Example 2 (see [1,8]**).** Let us consider the following optimization problem

$$\frac{1}{2} \|q\|^2_{H_1} \to \min_{Aq=f} .$$

One can build the dual problem

$$\min_{Aq=f} \frac{1}{2} \|q\|^2_{H_1} = \min_q \max_\lambda \left\{ \frac{1}{2} \|q\|^2_{H_1} + \langle f - Aq, \lambda \rangle \right\}$$

$$= \max_\lambda \min_q \left\{ \frac{1}{2} \|q\|^2_{H_1} + \langle f - Aq, \lambda \rangle \right\} = \max_\lambda \left\{ \langle f, \lambda \rangle - \frac{1}{2} \|A^*\lambda\|^2_{H_1} \right\}. \quad (8)$$

We assume that $Aq = f$ is compatible, hence for Fredgolm's theorem it's not possible that there exists such a λ: $A^*\lambda = 0$ and $\langle b, \lambda \rangle > 0.$[1] Hence the dual problem is solvable (but the solution isn't necessarily unique). Let's denote λ_* to be the solution of the dual problem

$$\phi(\lambda) = \frac{1}{2} \|A^*\lambda\|^2_{H_1} - \langle f, \lambda \rangle \to \min_\lambda$$

with minimal H_2-norm. Let's introduce (from the optimality condition in (8) for q): $q(\lambda) = A^*\lambda$. Using (A)STM for the dual problem one can find (Theorem 3)

$$\|Aq^N - f\|_{H_1} = O\left(\frac{L\breve{R}}{N^2} \right), \quad (9)$$

where $L = \|A^*\|^2_{H_2 \to H_1} = \|A\|^2_{H_1 \to H_2}$ (as in Example 2), $\breve{R} = \|\lambda_*\|_{H_2}$.
If one will try to solve the primal problem in Example (1)

$$\frac{1}{2} \|Aq - f\|^2_{H_2} \to \min_q$$

---

[1] Indeed, if there exists q such that $Aq = f$ then for all λ we have $\langle Aq, \lambda \rangle = \langle f, \lambda \rangle$. Hence, $\langle q, A^*\lambda \rangle = \langle f, \lambda \rangle$. Assume that there exists a λ, such that $A^*\lambda = 0$ and $\langle f, \lambda \rangle > 0$. If it is so we observe a contradiction:

$$0 = \langle q, A^*\lambda \rangle = \langle f, \lambda \rangle > 0.$$

by (A)STM, one can obtain the following estimate

$$\left\| Aq^N - f \right\|_{H_2} = O\left(\frac{\sqrt{L}R}{N} \right), \tag{10}$$

where $L = \|A\|^2_{H_1 \to H_2}$, $R = \|q_*\|_{H_1}$. The estimate (10) seems worse than (9). But the estimate (10) cannot be improving up to a constant factor [16]. There is no contradiction here since in general $\tilde{R}$ can be big enough, i.e. this parameter is uncontrollable. But in real applications, we can hope that this (dual) approach leads us to a faster convergence rate (9). ∎

Indeed, all the mentioned above methods (expect (A)GD) are primal-dual ones [1,2,8] (if we use their non strongly convex variants). That is for these methods analogues of Theorem 3 holds true with proper modification of (7) for (A)$\overline{\text{GD}}$

$$3 \cdot \max\left\{ \frac{L\tilde{R}^2}{\varepsilon}, \frac{L\breve{R}}{\tilde{\varepsilon}} \right\}.$$

This means that we can apply the results of Sect. 2 for this approach.

Now let's describe the main motivating example for this paper.

Example 3 (inverse problem for elliptic initial-boundary value problem). Let u be the solution of the following problem (P)

$$u_{xx} + u_{yy} = 0, \quad x, y \in (0, 1),$$

$$u_x(0, y) = 0, \quad y \in (0, 1),$$

$$u(1, y) = q(y), \quad y \in (0, 1),$$

$$u(x, 0) = u(x, 1) = 0, \quad x \in (0, 1).$$

And the corresponding dual problem (D)

$$\psi_{xx} + \psi_{yy} = 0, \quad x, y \in (0, 1),$$

$$\psi_x(0, y) = \lambda(y), \quad y \in (0, 1),$$

$$\psi(1, y) = 0, \quad y \in (0, 1),$$

$$\psi(x, 0) = \psi(x, 1) = 0, \quad x \in (0, 1).$$

Let's introduce the operator

$$A: \quad q(y) := u(1, y) \mapsto u(0, y).$$

Here $u(x, y)$ is a solution of problem (P). It was shown in [12] that

$$A: L_2(0, 1) \to L_2(0, 1).$$

The conjugate is defined as follows operator

$$A^* : \quad \lambda(y) := \psi_x(0, y) \mapsto \psi_x(1, y), \qquad A^* : L_2(0,1) \to L_2(0,1).$$

Here $\psi(x, y)$ is the solution of problem (D) [12]. To obtain these formulas one may use the general approach, described, for example, in § 7, Chap. 8 [27] (see also Chap. 4 in [6]).

Let us formulate the inverse problem [12]: find the function q by known additional information

$$u(0, y) = f(y).$$

The inverse problem is reduced to the optimization problem of the following cost functional

$$J(q) = \|Aq - f\|_{L_2(0,1)} \to \min_q,$$

$$\|q\|_H^2 \to \min_{Aq=f} \quad // \quad \phi(\lambda) = \|A^*\lambda\|_H^2 - \langle f, \lambda \rangle \to \min_\lambda.$$

It is obvious that $\nabla J(q) = A^*(Aq - f)$ and $\nabla J(q)$ can be found by the following formula:

$$\nabla J(q)(y) = \psi_x(1, y).$$

Here $\psi(x, y)$ is the solution of (D) with $\lambda(y) = 2(u(0, y) - f(y))$.

For Example 2 one can obtain that $\nabla \phi(\lambda) = 2(f - \Lambda(\Lambda^*\lambda))$, $\nabla \phi(\lambda) \subset L_2(0,1)$ and

$$\nabla \phi(\lambda)(y) = 2(f(y) - u(0, y)).$$

Note, that for this example $L = 1$ [12], see (2) for definition of L. ∎

4   Stopping Rule for STM

Let us consider $STM(y^0, L)$ in the following conception of inexact oracle [21]: for all $q_1, q_2 \in H$

$$\|\nabla J(q_1) - \tilde{\nabla} J(q_2)\|_2 \leqslant \tilde{\delta}.$$

Using inequality (2), convexity of J and Fenchel inequality we can get, that: for all $q_1, q_2 \in H$

$$J(q_1) \leqslant J(q_2) + \langle J(q_2), q_1 - q_2 \rangle + \frac{2L}{2}\|q_1 - q_2\|_2^2 + \frac{\tilde{\delta}^2}{2L},$$

$$J(q_2) + \langle \tilde{\nabla} J(q_2), q_1 - q_2 \rangle - \tilde{\delta}\|q_1 - q_2\|_2 \leqslant J(q_1).$$

If we introduce:

$$\psi_k(q) = \frac{1}{2}\|q - y^0\|_2^2 + \sum_{j=0}^{k} \left(J(y^j) + \langle \tilde{\nabla} J(y^j), q - y^j \rangle \right).$$

It can be shown, that in general degenerate situation (not strongly convex case) the following estimates hold [4, 23]:[2]

$$A_k J(q_k) \leqslant \psi_k(u_k) + \frac{\tilde{\delta}^2}{2L} \sum_{j=0}^{k} A_j + \tilde{\delta} \sum_{j=1}^{k} \alpha_j \|y^j - u^{j-1}\|_2^2,$$

$$J(q^N) - J(q_*) \leqslant \frac{4LR^2}{N^2} + 3\tilde{R}\tilde{\delta} + N\frac{\tilde{\delta}^2}{2L}, \quad (11)$$

$$\tilde{R} = \max_{k \leqslant N}\{\|q_* - y^k\|_2, \|q_* - u^k\|_2, \|q_* - q^k\|_2\}.$$

Note, that if we know, that $\|u^k - q_*\|_2 \leqslant R$ we can easily show that $\|y^k - q_*\|_2 \leqslant R$ and $\|q^k - q_*\|_2 \leqslant R$:

$$\|y^k - q_*\|_2 \leqslant \frac{A_{k-1}}{A_k}\|q^{k-1} - q_*\|_2 + \frac{\alpha_k}{A_k}\|u^{k-1} - q_*\|_2 \leqslant R.$$

Similarly for the sequence q^k. Therefore, we show how, using the stopping criterion, to obtain this inequality for the sequence u^k. If we know the value of $J(q_*)$ and such bound $R_* > 0$, that $\|q_*\|_2 \leqslant R_*$. Then by choosing $y^0 = 0$ (obviously, that in this case $R \leq R_*$), we can formulate a computable stopping criterion: for all $\zeta > 0$:

$$J(q_k) - J(q_*) \leqslant k\frac{\tilde{\delta}^2}{2L} + 3R_*\tilde{\delta} + \zeta.$$

Then, using the convexity of the function ψ_k we get:

$$A_k J(q^k) + \frac{1}{2}\|u^k - q_*\|_2^2 \leqslant \frac{1}{2}\|u^k - q_*\|_2 + \psi_k(u^k) + \frac{\tilde{\delta}^2}{2L}\sum_{j=0}^{k} A_j$$

$$+ \tilde{\delta}\sum_{j=1}^{k}\alpha_j\|\tilde{q}^j - u^{j-1}\|_2 \leqslant \psi_k(q_*) + \tilde{\delta}\sum_{j=1}^{k}\alpha_j\|\tilde{q}^j - u^{j-1}\|_2 \leqslant \frac{\tilde{\delta}^2}{2L}\sum_{j=0}^{k} A_j$$

$$+ \frac{1}{2}R^2 + A_k J(q_*) + \tilde{\delta}\sum_{j=1}^{k}\alpha_j\|\tilde{q}^j - u^{j-1}\|_2 + \tilde{\delta}\sum_{j=0}^{k}\alpha_j\|u^k - q_*\|_2$$

$$\leqslant \frac{1}{2}R^2 + A_k 3\tilde{\delta} R_* + \frac{\tilde{\delta}^2}{2L}\sum_{j=0}^{k} A_j + A_k J(q_*)$$

$$\Rightarrow \frac{1}{2}(R^2 - \|u^k - q_*\|_2) \geqslant A_k\left(\left(J(q^k) - J(q_*)\right) - \left(k\frac{\tilde{\delta}^2}{2L} + 3R_*\tilde{\delta} + \zeta\right)\right) \geqslant 0.$$

Also this criterion is achievable:

$$J(q^N) - J(q_*) \leqslant \frac{4LR^2}{N^2} + 3R_*\tilde{\delta} + N\frac{\tilde{\delta}^2}{2L},$$

$$\frac{4LR^2}{N^2} \leqslant \zeta, N \geqslant 2\sqrt{\frac{LR^2}{\zeta}}, N = O\left(\sqrt{\frac{LR^2}{\zeta}}\right).$$

---

[2] Recall that $R = \|q_* - y^0\|_2$.

That is N iterations is enough to reach the stopping criterion. Finally we get the following theorem:

Theorem 4. *Assume that we solve problem* (1) *and we know the value* $J(q_*)$ *and the bound* $R_* > 0$ *for* $\|q_*\|_2$. *Using STM(0, L) with stopping rule:*

$$J(q^k) - J(q_*) \leqslant \frac{\tilde{\delta}^2}{2L}k + 3R_*\tilde{\delta} + \zeta.$$

We get the following estimation:

$$\tilde{R} \leqslant R, \tilde{R} = \max_{k \leqslant N}\{\|q_* - y^k\|_2, \|q_* - u^k\|_2, \|q_* - q^k\|_2\}.$$

And it is guaranteed, that the criteria will be reached by:

$$N = O\left(\sqrt{\frac{LR^2}{\zeta}}\right).$$

Using this theorem we can get, that solving the problem:

$$J(q^N) - J(q_*) \leqslant \varepsilon. \tag{12}$$

We can choose $\zeta \sim \varepsilon$ and $\tilde{\delta} \sim \frac{\varepsilon}{R_*}$ and the number of iterations will be estimated as

$$N = O\left(\sqrt{\frac{LR^2}{\varepsilon}}\right). \tag{13}$$

Similar results can be formulated for all other methods from Sect. 2 including adaptive ones, since (11) holds for this generality too [23]. As an example, we remark that for GD (12) takes place with the same requirement for $\tilde{\delta} \sim \frac{\varepsilon}{R_*}$, but with a worse bound on

$$N = O\left(\frac{LR^2}{\varepsilon}\right). \tag{14}$$

From this, we may expect that with the same level of noise in the gradient $\tilde{\delta}$ accelerated algorithms (STM) reach the same quality $J(q^N) - J(q_*) \simeq \tilde{\delta}R$ as non-accelerated ones (GD), but they do it faster (compare (13) and (14)). Therefore we may expect that with a proper stopping rule STM must outperform GD for degenerate (non strongly convex) problems. That was confirmed in the numerical experiments described in Sect. 5.

5 Numerical Results

We consider the following continuation problem in the domain $\Omega = \{(x, y, z) \in [0, 1]^2 \times [0, H]\}$:[3]

$$\Delta u(x, y, z) \equiv u_{xx} + u_{yy} + u_{zz} = h(x, y, z), \quad (x, y, z) \in \Omega$$

$$u|_{x=0} = u|_{x=1} = u|_{y=0} = u|_{y=1} = 0,$$

$$u_z|_{z=0} = 0, u|_{z=H} = q(x, y)$$

---

[3] The mathematical background of the described example see in the full version of the paper [15].

The problem is to determine the unknown function $q(x,y)$ by using the additional information of the function $u(x,y,z)$ on the boundary $z = 0$:

$$u|_{z=0} = f(x,y).$$

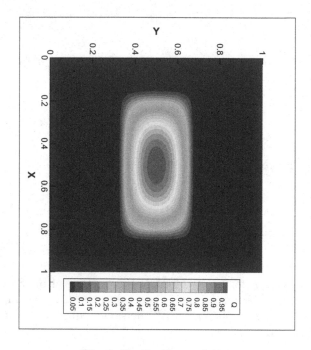

Fig. 1. Test1 - True solution

We solve the formulated problem by using different versions of STM and GD methods. The structure of the gradient of the functional has the like in Example 3. We choose the test solution as follows:

$$q(x,y) = \begin{cases} e^{l_1(x)+l_2(y)}, (x,y) \in [0.1, 0.9] \times [0.3, 0.7] \\ 0, \text{if else} \end{cases}$$

Here $l_1(x) = 1 + \frac{0.16}{(x-0.5)^2-0.16}, l_2(y) = 1 + \frac{0.04}{(y-0.5)^2-0.04}$. The structure of this function is presented on the Fig. 1. We solve the direct problem, using $q(x,y)$ as a given function, to calculate true data $f(x,y)$.

For the first series of tests we use the following parameters: $H = 0.5, N_{iter} = 1000$. We consider the similar triangles method (STM), simple gradient descent method and steepest descent method. We used initial approximation $q(x,y) = 0$ for all methods. Due to the fact, that it is hard to get the accurate estimation for the norm of the operator, we couldn't get the precise values for the parameters L, α of the STM and GD methods correspondingly during the numerical

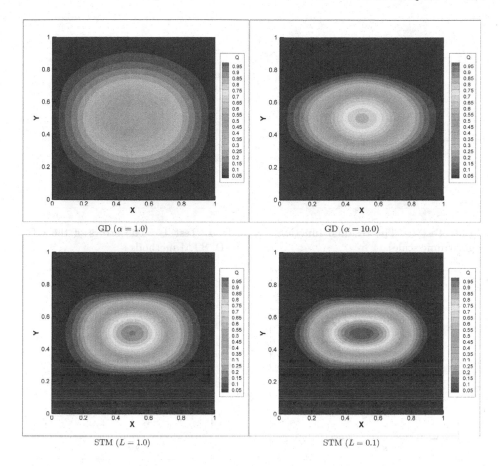

Fig. 2. Test 1 - solution of continuation problem by GD and STM methods

solution. Thus, we choose the parameters of STM and GD methods by trials and errors. However, in the case of the homogeneous right hand side and boundary conditions we use analytic expression for the descent parameters of the steepest descent method: $\alpha_n = \frac{|J'(q_n)|^2}{2|A(J'(q_n))|^2}$. The results of computations are presented on Figs. 2, 3 and 4.

The similar triangles method provides the most efficient results of the considered methods. The steepest descent methods converge faster on the first iterations, but eventually, the STM method provides better results in terms of both the residual and errors. The accuracy of the methods is acceptable (if suitable parameters of the methods were chosen). In order to illustrate the influence of the parameter L on the problem, we considered two different values of the parameter of the STM method during this experiment.

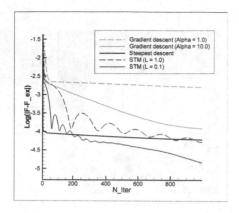

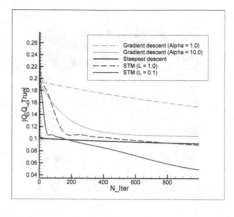

Fig. 3. Test 1 - The residual functional (logarithmic scale)

Fig. 4. Test 1 - The errors of the GD, SGD, STM methods

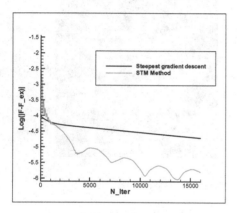

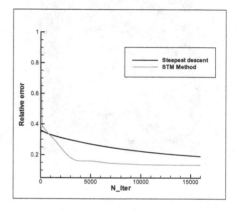

Fig. 5. Test 3 (increased number of iterations) - The residual functional (logarithmic scale)

Fig. 6. Test 3 (increased number of iterations) - The residual functional (logarithmic scale)

For the second series of tests, we added some non-homogeneous boundary conditions, and the right-hand side of the following form:

$$h(x, y, z) = (1 - z)cos(\pi x)cos(\pi y)$$

We increased the depth to $H = 1.0$. The structure of the function $q(x, y)$ remains the same. However, the increased depth significantly decreases the influence of data, that we have during the experiments with synthetic data, that we balance by increasing the number of iterations to $N_{iter} = 16000$. The behavior of the functional is presented on the Figs. 6 and 7. This allows us to provide the solution, almost identical to exact one. During the last series of test we considered the medium number of iterations $N_{iter} = 6000$ and the depth $H = 1.25$ to study the variation of STM method with restarts. The computational results

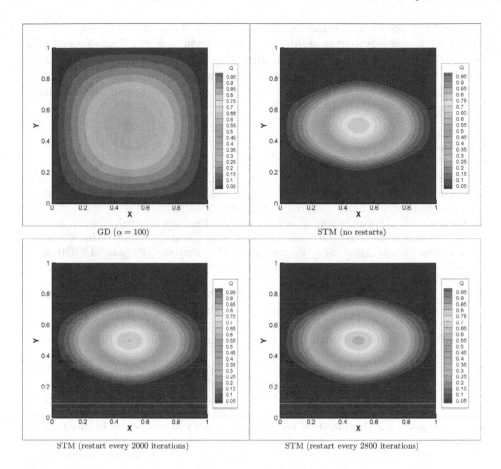

Fig. 7. Test 2 (increased depth) - solution of continuation problem by GD and STM methods.

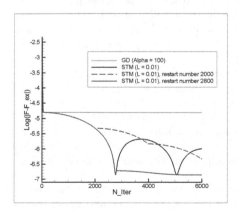

Fig. 8. Test 2 - The residual functional (logarithmic scale)

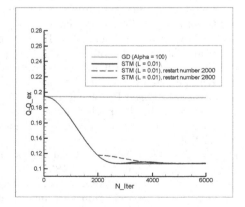

Fig. 9. Test 2 (increased depth) - The errors of the methods

are presented on Figs. 7, 8 and 9. We notice that the similar triangles method provides significantly better results, compared to simple gradient descent. The usage of restart technique allows obtaining better results in terms of residual, but the effects of the restarts are much less significant in terms of errors. The cause of this difference is the ill-posedness of the problem, which becomes more noticeable with the increase of depth.

References

1. Anikin, A., Gasnikov, A., Dvurechensky, P., Turin, A., Chernov, A.: Dual approaches to the strongly convex simple function minimization problem under affine restrictions. arXiv preprint arXiv:1602.01686 (2016)
2. Chernov, A., Dvurechensky, P., Gasnikov, A.: Fast primal-dual gradient method for strongly convex minimization problems with linear Constraints. In: Kochetov, Y., Khachay, M., Beresnev, V., Nurminski, E., Pardalos, P. (eds.) DOOR 2016. LNCS, vol. 9869, pp. 391–403. Springer, Cham (2016). https://doi.org/10.1007/978-3-319-44914-2_31
3. Devolder, O.: Exactness, inexactness and stochasticity in first-order methods for large-scale convex optimization. Ph.D. thesis (2013)
4. Dvinskikh, D., Gasnikov, A.: Decentralized and parallel primal and dual accelerated methods for stochastic convex programming problems. J. Inverse Ill-posed Problems (2021)
5. Dvurechensky, P.E., Gasnikov, A.V., Nurminski, E.A., Stonyakin, F.S.: Advances in low-memory subgradient optimization. In: Bagirov, A.M., Gaudioso, M., Karmitsa, N., Mäkelä, M.M., Taheri, S. (eds.) Numerical Nonsmooth Optimization, pp. 19–59. Springer, Cham (2020). https://doi.org/10.1007/978-3-030-34910-3_2
6. Evtushenko, Y.: Optimization and fast automatic differentiation. Preprint CCAS (2013)
7. Gasnikov, A.: Universal gradient descent. arXiv preprint arXiv:1711.00394 (2017)
8. Gasnikov, A., Dvurechensky, P., Nesterov, Y.: Stochastic gradient methods with inexact oracle. TRUDY MIPT 8(1), 41–91 (2016)
9. Gasnikov, A.V., Nesterov, Y.E.: Universal method for stochastic composite optimization problems. Comput. Math. and Math. Phys. 58(1), 48–64 (2018)
10. Gasnikov, A., Tyurin, A.: Fast gradient descent for convex minimization problems with an oracle producing a (δ, l)-model of function at the requested point. Comput. Math. Math. Phys. 59(7), 1085–1097 (2019)
11. Halmos, P.R.: A Hilbert Space Problem Book, vol. 19. Springer, Heidelberg (2012)
12. Kabanikhin, S.I.: Definitions and examples of inverse and ill-posed problems. J. Inverse Ill-Posed Probl. 16(4), 317–357 (2008)
13. Kamzolov, D., Dvurechensky, P., Gasnikov, A.V.: Universal intermediate gradient method for convex problems with inexact oracle. Optim. Methods Softw. 1–28 (2020)
14. Kantorovich, L.V.: Functional analysis and applied mathematics. Uspekhi Matematicheskikh Nauk 3(6), 89–185 (1948)
15. Matyukhin, V., Kabanikhin, S., Shishlenin, M., Novikov, N., Vasin, A., Gasnikov, A.: Convex optimization with inexact gradients in hilbert space and applications to elliptic inverse problems. WIAS Preprint 2815 (2021)
16. Nemirovskij, A.S., Yudin, D.B.: Problem complexity and method efficiency in optimization. Wiley-Interscience (1983)

17. Nesterov, Y.: Universal gradient methods for convex optimization problems. Math. Program. **152**(1), 381–404 (2015)
18. Nesterov, Y., Gasnikov, A., Guminov, S., Dvurechensky, P.: Primal-dual accelerated gradient methods with small-dimensional relaxation oracle. Optim. Methods Softw. 1–38 (2020)
19. Nesterov, Y., et al.: Lectures on Convex Optimization, vol. 137. Springer, Cham (2018). https://doi.org/10.1007/978-3-319-91578-4
20. Nocedal, J., Wright, S.: Numerical Optimization. Springer, New York (2006)
21. Poljak, B.: Iterative algorithms for singular minimization problems. In: Nonlinear Programming 4, pp. 147–166. Elsevier (1981)
22. Polyak, B.T.: Introduction to Optimization. Optimization software. Inc., Publications Division, New York 1 (1987)
23. Stonyakin, F., et al.: Inexact relative smoothness and strong convexity for optimization and variational inequalities by inexact model. arXiv preprint arXiv:2001.09013 (2020)
24. Stonyakin, F.S., et al.: Gradient methods for problems with inexact model of the objective. In: Khachay, M., Kochetov, Y., Pardalos, P. (eds.) MOTOR 2019. LNCS, vol. 11548, pp. 97–114. Springer, Cham (2019). https://doi.org/10.1007/978-3-030-22629-9_8
25. Tseng, P.: On accelerated proximal gradient methods for convex-concave optimization. SIAM J. Optim. 1 (2008)
26. Tyurin, A.: Mirror version of similar triangles method for constrained optimization problems. arXiv preprint arXiv:1705.09809 (2017)
27. Vasiliev, F.: Optimization methods: In: MCCME, vol. 1053, Moscow (2011). 2 books

On the Computational Efficiency of Catalyst Accelerated Coordinate Descent

Dmitry Pasechnyuk$^{(\boxtimes)}$ and Vladislav Matyukhin

Moscow Institute of Physics and Technology (National Research University),
Dolgoprudny, Russia
{pasechniuk.da,matyukhin}@phystech.edu

Abstract. This article is devoted to one particular case of using universal accelerated proximal envelopes to obtain computationally efficient accelerated versions of methods used to solve various optimization problem setups. We propose a proximally accelerated coordinate descent method that achieves the efficient algorithmic complexity of iteration and allows taking advantage of the data sparseness. It was considered an example of applying the proposed approach to optimizing a SoftMax-like function, for which the described method allowing weaken the dependence of the computational complexity on the dimension n in $\mathcal{O}(\sqrt{n})$ times and, in practice, demonstrates a faster convergence in comparison with standard methods. As an example of applying the proposed approach, it was shown a variant of obtaining on its basis some efficient methods for optimizing Markov Decision Processes (MDP) in a minimax formulation with a Nesterov smoothed target function.

Keywords: Proximal accelerated method · Catalyst · Accelerated coordinate descent method · SoftMax · Markov decision processes

1 Introduction

One of the most important theoretical results in convex optimization was the development of accelerated optimization methods [23]. At the initial stage of implementation of this concept, many accelerated algorithms for different problem setups were proposed. But each such case required special consideration of the possibility of acceleration. Therefore, the proposed designs were significantly different and did not allowing assume a way to generalize them. A significant step towards the development of a universal scheme for accelerating optimization

D.A. Pasechnyuk's research was supported by the A.M. Raigorodsky Scholarship in the field of optimization and RFBR grant 19-31-51001 (Scientific mentoring). The work of V.V. Matyukhin was supported by the Ministry of Science and Higher Education of the Russian Federation (state assignment) No. 075-00337-20-03, project number 0714-2020-0005.

P. Pardalos et al. (Eds.): MOTOR 2021, LNCS 12755, pp. 176–191, 2021.
https://doi.org/10.1007/978-3-030-77876-7_12

methods was the paper in which an algorithm called Catalyst proposed, based on the idea of [26, 27] and allowing to accelerate other optimization methods, using them for the sequential solving of several Moreau–Yosida regularized auxiliary problems [18, 19]. Following these ideas, many variants of the applications of this method and its modifications [14, 17, 25] were proposed. Among the most recent results until the time of writing this paper, the generalizations of the discussed approach to tensor methods [5, 10, 11, 20] were also described. The corresponding form of the accelerated proximal envelope to the authors' knowledge is the most general of those described in the literature, and therefore, this paper is focused primarily on the methods proposed in the papers [10, 11].

main motivation of this paper is to describe the possibilities of the practical application of universal accelerated proximal envelopes for constructing computationally and oracle efficient optimization methods. Let us consider the classical coordinate descent method [4], the iteration of which for the convex function $f : \mathbb{R}^n \to \mathbb{R}$ is of the form:

$$x^i_{k+1} = x^i_k - \eta \nabla_i f(x_k), \quad i \sim \mathcal{U}\{1, ..., n\}, \quad \eta > 0.$$

One of the many applications of this method is the optimization of functions, for which the calculation of the one component of the gradient is significantly more efficient than the calculation of the full gradient vector (in particular, many problems in the case of sparse formulations satisfy this condition). The oracle complexity of this method, provided that the method stops when the ε-small function value residual is reached, is $\mathcal{O}\left(n\frac{\overline{L}R^2}{\varepsilon}\right)$, where $R^2 = \|x_0 - x_*\|_2^2$, $\overline{L} = \frac{1}{n}\sum_{i=1}^n L_i$ is the average of the Lipschitz constants of the gradient components. However, this estimate is not optimal for the class of convex problems. Let us now consider the accelerated coordinate descent method proposed by Yu.E. Nesterov [24]. The oracle complexity of this method corresponds to the optimal bound: $\mathcal{O}\left(n\sqrt{\frac{\widetilde{L}R^2}{\varepsilon}}\right)$, where $\sqrt{\widetilde{L}} = \frac{1}{n}\sum_{i=1}^n \sqrt{L_i}$ is the mean of square roots of the Lipschitz constants of the gradient components. At the same time, the situation changes drastically when the algorithmic complexity of the method considered. Namely, even if the computation of one component of the gradient has the complexity $\mathcal{O}(s)$, $s \ll n$, the complexity of the whole iteration of the accelerated coordinate descent method will be $\mathcal{O}(n)$, unlike the standard method, the iteration complexity of which is $\mathcal{O}(s)$. It means that the sparseness of the problem when using the accelerated coordinate descent method does not significantly affect the complexity of the algorithm, and besides, the complexity in this case quadratically depends on the dimension of the problem. Together, this somewhat devalues the use of the coordinate descent method in this case. Thus, an interesting problem is the construction of an accelerated coordinate descent method, the iteration complexity of which, as in the standard version of the method, is $\mathcal{O}(s)$. This is possible due to the application of the universal accelerated proximal envelope "Accelerated Meta-algorithm" [11].

This article consists of an introduction, conclusion and the main Sect. 2. It describes the theoretical results on the convergence and algorithmic complexity of the coordinate descent method, accelerated by using the "Accelerated Meta-algorithm" envelope (Sect. 2.1). Using the example of the SoftMax-like function optimization problem, it was experimentally tested method's effectiveness with relation to its working time. There were also described the possibilities of its computationally efficient implementation, and carried out a comparison with standard methods (Sect. 2.2). Further, as an example of applying the proposed approach, it was provided a method for optimizing Markov Decision Processes in a minimax formulation, based on applying the method introduced in this paper to the Nesterov smoothed target function. The proposed approach obtains estimates close to that for several efficient and practical methods for optimizing the discounted MDP and matches the best estimates for the averaged MDP problem (Sect. 2.3).

2 Accelerated Meta-algorithm and Coordinate Descent

2.1 Theoretical Guarantees

Let us consider the following optimization problem of the function $f : \mathbb{R}^n \to \mathbb{R}$:

$$\min_{x \in \mathbb{R}^n} f(x),$$

subject to:

1. f is differentiable on $\mathbb{R}^n$;
2. f is convex on $\mathbb{R}^n$;
3. $\nabla_i f$ is component-wise Lipschitz continuous, i.e. $\forall x \in \mathbb{R}^n$ and $u \in \mathbb{R}$, $\exists L_i \in \mathbb{R}$ $(i = 1, \dots, n)$, such that

$$|\nabla_i f (x + u e_i) - \nabla_i f(x)| \le L_i |u|,$$

   where e_i is the i-th unit basis vector, $i \in \{1, \dots, n\}$;
4. ∇f is L-Lipschitz continuous.

Let us turn to the content of the paper [11], where a general version of the "Accelerated Meta-algorithm" for solving convex optimization problems for composite functionals in form of $F(x) = f(x) + g(x)$ was proposed. For the considered formulation of the problem, such generality is not required. It is sufficient to apply a special case of the described scheme for $p = 1$, $f \equiv 0$ (using the designations of the corresponding paper), in which the described envelope takes the form of an accelerated proximal gradient method. The method used is listed as Algorithm 1.

Before formulating any results on the convergence of the proposed accelerated coordinate descent method described below, it is necessary to start with a detailed consideration of the process of solving the auxiliary problems, where its analytical solution is available only in rare cases. Therefore, one should apply numerical methods to find its approximate solution, and that is inaccurate. The

---

Algorithm 1: Accelerated Meta-algorithm for First-order Method $\mathcal{M}$

Input: $H > 0$, $x_0 \in \mathbb{R}^n$;

$\lambda \leftarrow 1/2H$;

$A_0 \leftarrow 0$; $v_0 \leftarrow x_0$;

for $k = 0, ..., \widetilde{N} - 1$ **do**

$\quad a_{k+1} \leftarrow \dfrac{\lambda + \sqrt{\lambda^2 + 4\lambda A_k}}{2}$;

$\quad A_{k+1} \leftarrow A_k + a_{k+1}$;

$\quad \widetilde{x}_k \leftarrow \dfrac{A_k v_k + a_{k+1} x_k}{A_{k+1}}$;

$\quad$ By running the method $\mathcal{M}$,
$\quad$ find the solution of the following auxiliary problem
$\quad$ with an accuracy ε by the argument:

$\quad v_{k+1} \in \text{Arg}^\varepsilon \min\limits_{y \in \mathbb{R}^n} \left\{ f(y) + \dfrac{H}{2} \|y - \widetilde{x}_k\|_2^2 \right\}$;

$\quad x_{k+1} \leftarrow x_k - a_{k+1} \nabla f(v_{k+1})$;

end

return $v_{\widetilde{N}}$;

---

process of solving the auxiliary problem should continue until the following stopping condition is satisfied ([16], Appendix B):

$$\left\| \nabla \left\{ F(y_\star) := f(y_\star) + \frac{H}{2} \|y_\star - \widetilde{x}_k\|_2^2 \right\} \right\|_2 \leq \frac{H}{2} \|y_\star - \widetilde{x}_k\|_2, \tag{1}$$

where $y_\star$ is an approximate solution of the auxiliary problem, returned by internal method $\mathcal{M}$. Due to the $\|\nabla F(y_*)\|_2 = 0$ (where y_* denotes an exact solution of the considered problem), and due to the $(L + H)$-Lipschitz continuity of ∇F, we have got:

$$\|\nabla F(y_\star)\|_2 \leq (L + H)\|y_\star - y_*\|_2. \tag{2}$$

Writing out the triangle inequality: $\|\widetilde{x}_k - y_*\|_2 - \|y_\star - y_*\|_2 \leq \|y_\star - \widetilde{x}_k\|_2$, and using it together the inequalities (1) and (2), we have got the final form of the stopping condition:

$$\|y_\star - y_*\|_2 \leq \frac{H}{3H + 2L} \|\widetilde{x}_k - y_*\|_2. \tag{3}$$

This implies that the required argument accuracy of solving the auxiliary problem does not depend on the accuracy required for the main problem solution. That makes it possible to significantly simplify the inference of further results.

Let us now consider the main method used for solving auxiliary problems. Coordinate Descent Method in version of [22] (in the particular case, when $\gamma = 1$) is listed as Algorithm 2. For this method, in the case of the considered auxiliary problems, the following result holds:

Algorithm 2: Coordinate descent method

Input: $y_0 \in \mathbb{R}^n$;

$Z \leftarrow \sum_{i=1}^{n}(H + L_i)$;
$p_i \leftarrow (H + L_i)/Z, \quad i \in \{1, ..., n\}$;
Discrete probability distribution π
with probabilities p_i;

for $k = 0, ..., N - 1$ **do**
    $i \sim \pi\{1, ..., n\}$;
    $y_{k+1} \leftarrow y_k$;
    $y_{k+1}^i = y_k^i - \dfrac{1}{H + L_i}\nabla_i F(y_k)$;
end
return y_N;

Theorem 1 *([4], theorem 6.8). Let F be H-strongly convex function with respect to $\| \cdot \|_2$. Then for the sequence $\{y_k\}_{k=1}^{N}$ generated by the described coordinate descent algorithm 2, it holds the following inequality:*

$$\mathbb{E}[F(y_N)] - F(y_*) \leq \left(1 - \frac{1}{\kappa}\right)^N (F(y_0) - F(y_*)), \tag{4}$$

$$\text{where} \quad \kappa = \frac{H}{Z}, \quad Z = \sum_{i=1}^{n}(H + L_i), \tag{5}$$

where $\mathbb{E}[\cdot]$ denotes the mathematical expectation of the specified random variable with respect to the randomness of methods trajectory induced by a random choice of components i at each iteration.

Using this result, lets formulate the following statement on the number of iterations of the coordinate descent method sufficient to satisfy the stopping condition (3).

Corollary 1. *The expectation $\mathbb{E}[y_N]$ of the point resulting from the coordinate descent method (Algorithm 2) satisfies the condition (3) if the following inequality on iterations number holds:*

$$N \geq N(\widetilde{\varepsilon}) = \left\lceil \frac{Z}{H} \ln \left\{ \left(1 + \frac{L}{H}\right)\left(3 + \frac{2L}{H}\right)^2 \right\} \right\rceil, \tag{6}$$

$$\text{where} \quad \widetilde{\varepsilon} = \frac{H}{2}\left(\frac{H}{3H + 2L}\right)^2 \|y_0 - y_*\|_2^2. \tag{7}$$

Proof. Is in Appendix A[1].

---

[1] For the detailed proofs see appendices in full paper version on https://arxiv.org/abs/2103.06688.

Now that the question of the required accuracy and oracle complexity of solving the auxiliary problem using the proposed coordinate descent method is clarified, we can proceed to the results on the convergence of the Accelerated Meta-algorithm. For the used stopping condition (3), the following result on the convergence of the Accelerated Meta-algorithm holds.

Theorem 2 *([11], theorem 1). For $H > 0$ and the sequence $\{v_k\}_{k=1}^{\widetilde{N}}$ generated by the Accelerated Meta-algorithm with some non-stochastic internal method, it holds the following inequality:*

$$f(v_{\widetilde{N}}) - f(x_*) \leq \frac{48}{5} \frac{H\|x_0 - x_*\|_2^2}{\widetilde{N}^2}. \tag{8}$$

Based on the last statement, one can formulate a theorem on the convergence of the Accelerated Meta-algorithm in the case of using the stochastic method and, in particular, coordinate gradient descent method.

Theorem 3. *For $H > 0$ and some $0 < \delta < 1$, the point $v_{\widetilde{N}}$ resulting from the Accelerated Meta-algorithm using coordinate descent method to solve the auxiliary problem, solving it within N_δ iterations, satisfies the condition*

$$Pr(f(v_{\widetilde{N}}) - f(x_*) < \varepsilon) \geq 1 - \delta,$$

where $Pr(\cdot)$ denotes the probability of the specified event, if

$$\widetilde{N} \geq \left\lceil \frac{4\sqrt{15}}{5} \sqrt{\frac{H\|x_0 - x_*\|_2^2}{\varepsilon}} \right\rceil, \tag{9}$$

$$N_\delta \geq N\left(\frac{\widetilde{\varepsilon}\delta}{\widetilde{N}}\right) = \left\lceil \frac{Z}{H} \ln\left\{ \frac{\widetilde{N}}{\delta}\left(1 + \frac{L}{H}\right)\left(3 + \frac{2L}{H}\right)^2 \right\} \right\rceil. \tag{10}$$

Proof. The Corollary 1 presents an estimate of the number of iterations sufficient to satisfy the following condition for the expected value of the function at the resulting point of the method:

$$\mathbb{E}[F(y_{N(\widetilde{\varepsilon})})] - F(y_*) \leq \widetilde{\varepsilon}.$$

Lets use the Markov inequality and obtain the formulation of this condition in terms of the bound for the probability of large deviations [1]: deliberately choose the admissible value of the probability of non-fulfilment of the stated condition, so that $0 < \delta/\widetilde{N} < 1$, where $\widetilde{N}$ expressed from (8); then

$$Pr\left(F\left(y_{N(\widetilde{\varepsilon}\delta/\widetilde{N})}\right) - F(y_*) \geq \widetilde{\varepsilon}\right) \leq \frac{\delta}{\widetilde{N}} \cdot \frac{\mathbb{E}\left[F\left(y_{N(\widetilde{\varepsilon}\delta/\widetilde{N})}\right)\right] - F(y_*)}{\widetilde{\varepsilon} \cdot \delta/\widetilde{N}} = \frac{\delta}{\widetilde{N}}.$$

Since the probability that the obtained solution of some separately taken auxiliary problem will not satisfy the stated condition is equal to $\delta/\widetilde{N}$. It means that the probability that for $\widetilde{N}$ iterations of the Accelerated Meta-algorithm the condition will not be satisfied for at least one of the problems is $\widetilde{N} \cdot \delta/\widetilde{N} = \delta$, whence the proved statement follows.

Further, combining the estimates given in Theorem 3, we can obtain an asymptotic estimate for the total number of iterations of the coordinate descent method, sufficient to obtain a solution of the considered optimization problem with a certain specified accuracy, as well as an estimate for the optimal H value:

Corollary 2. *In order to the point $v_{\widetilde{N}}$, which is the result of the Accelerated Meta-algorithm, to satisfy the condition*

$$Pr(f(v_{\widetilde{N}}) - f(x_*) < \varepsilon) \geq 1 - \delta,$$

it is sufficient to perform a total of

$$\hat{N} \geq \widetilde{N} \cdot N_\delta = \mathcal{O}\left(\frac{Z\|x_0 - x_*\|_2}{\sqrt{H}} \cdot \frac{1}{\varepsilon^{1/2}} \log\left\{\frac{1}{\varepsilon^{1/2}\delta}\right\}\right) \tag{11}$$

iterations of coordinate descent method to solve the auxiliary problem. In this case, the optimal value of the regularization parameter H of the auxiliary problem should be chosen as $H \simeq \frac{1}{n}\sum_{i=1}^{n} L_i$ ($\simeq$ denotes equality up to a small factor of the $\log$ order).

Proof. The expression for $\hat{N}$ can be obtained by the direct substitution of one of the estimates given in (9) into another, and their subsequent multiplication. If we exclude from consideration a small factor of order $\log(L/H)$, the constant in the estimate will depend on H as:

$$\sqrt{H} \cdot \frac{Z/n}{H} = \sqrt{H}\left(1 + \frac{\frac{1}{n}\sum_{i=1}^{n} L_i}{H}\right).$$

By minimizing the presented expression by H, we get the specified result.

A similar statement can be formulated for the expectation of the total iterations number without resorting to bound for the probabilities of large deviations, following the reasoning scheme proposed in [9]:

Theorem 4. *The expectation $\mathbb{E}[\hat{N}]$ of the sufficient total number of the iterations of the coordinate descent method, to obtain a point $v_{\widetilde{N}}$ satisfying the following condition:*

$$f(v_{\widetilde{N}}) - f(x_*) < \varepsilon$$

can be bounded as follows:

$$\mathbb{E}[\hat{N}] \leq \widetilde{N} \cdot (N(\widetilde{\varepsilon}) + 1) = \mathcal{O}\left(\sqrt{\frac{\overline{L}\|x_0 - x_*\|_2^2}{\varepsilon}}\right), \quad \text{where} \quad \overline{L} = Z/n = \frac{1}{n}\sum_{i=1}^{n} L_i.$$

Proof. Is in Appendix B.

As we can see, the proposed scheme of reasoning allows us to reduce the logarithmic factor in estimate for the number of iterations of the method. However, this result is less constructive than the presented one in Corollary 2. Indeed, in the above reasoning, we operated with the number of iterations N, after which the stopping condition for the internal method is satisfied. But in the program implementation, stopping immediately after fulfilment of this condition is not possible, if only because it is impossible to verify this criterion due to the natural lack of information about y_*. So the last result is more relevant from the point of view of evaluating the theoretical effectiveness of the method, while when considering specific practical cases, one should rely on the estimate (11).

Let us now consider in more detail the issue of the algorithmic complexity of the proposed accelerated coordinate descent method.

Theorem 5. *Let the complexity of computing one component of the gradient of* f *is* $\mathcal{O}(s)$. *Then the algorithmic complexity of the Accelerated Meta-algorithm with coordinate descent as internal method is*

$$ T = \mathcal{O}\left(s \cdot n \cdot \sqrt{\frac{\overline{L}\|x_0 - x_*\|_2^2}{\varepsilon}} \log\left\{ \frac{1}{\varepsilon^{1/2}\delta} \right\} \right). $$

Proof. Is in Appendix C.

Note also that the memory complexity of the method is $\mathcal{O}(n)$, as well as the complexity of the preliminary calculations (for the coordinate descent method, there is no need to perform them again for every iteration).

Let us compare the estimates obtained for proposed approach (Catalyst CDM) with estimates for other methods that can be used to solve problems in the described setting: Fast Gradient Method (FGM), classical Coordinate Descent Method (CDM) and Accelerated Coordinate Descent Method in the version of Yu.E. Nesterov (ACDM). The estimates are shown in Table 1 below. As can be seen from the above asymptotic estimates of the computational complexity, the proposed method allows to achieve a convergence rate that is not inferior to other methods with respect to the dependence on the dimension n and the required accuracy ε (for a certain price, in the form of additional logarithmic factor). Note, in addition, that despite the significant similarity of estimates, between the two most efficient methods in the table (FGM and Catalyst CDM) there is also a difference in the constants characterising the smoothness of the function. Namely, L in FGM and $\overline{L}$ in Catalyst CDM. Thus the behaviour of the considered method for various problems directly depends on the character of its component-wise smoothness.

2.2 Numerical Experiments

This section describes the character of the practical behaviour of the proposed method by the example of the following SoftMax-like function optimization problem:

$$ \min_{x \in \mathbb{R}^n} \left\{ f(x) = \gamma \ln\left(\sum_{j=1}^{m} \exp\left(\frac{[Ax]_j}{\gamma} \right) \right) - \langle b, x \rangle \right\}, \tag{12} $$

Table 1. Comparison of the effectiveness of methods

Algorithm	Iteration complexity	Comp. complexity	Source
FGM	$\mathcal{O}(s \cdot n)$	$\mathcal{O}\left(s \cdot n \cdot \dfrac{1}{\varepsilon^{1/2}}\right)$	[23]
CDM	$\mathcal{O}(s)$	$\mathcal{O}\left(s \cdot n \cdot \dfrac{1}{\varepsilon}\right)$	[4]
ACDM	$\mathcal{O}(n)$	$\mathcal{O}\left(n^2 \cdot \dfrac{1}{\varepsilon^{1/2}}\right)$	[24]
Catalyst CDM	$\mathcal{O}(s)$	$\widetilde{\mathcal{O}}\left(s \cdot n \cdot \dfrac{1}{\varepsilon^{1/2}}\right)$	This paper

where $b \in \mathbb{R}^n$, $A \in \mathbb{R}^{m \times n}, \gamma \in \mathbb{R}_+$. Similar problems are essential for many applications. In particular, they arise in entropy-linear programming problems as a dual problem [8,12], in particular in optimal transport problem, is also a smoothed approximation of the max function (which gave the function the name SoftMax) and, accordingly, of the norm $\|\cdot\|_\infty$, which may be needed in some formulations of the PageRank problem or for solving systems of linear equations. Moreover, in all the described problems, an important special case is the sparse setting, in which the matrix A is sparse, that is, the average number of nonzero entries in the row A_j does not exceed some $s \ll n$ (it will also be convenient to assume that the one of the rows A_j is non-sparse).

Let us formulate the properties possessed by the function f[13]:

1. f is differentiable;
2. ∇f satisfies the Lipschitz condition with the constant $L = \max_{j=1,\dots,m} \|A_j\|_2^2$;
3. $\nabla_i f$ satisfy the component-wise Lipschitz condition with the constants $L_i = \max_{j=1,\dots,m} |A_{ji}|$.

Let us write the expression for the i-th component of the gradient of the function f:

$$\nabla_i f(x) = \frac{\sum_{j=1}^m A_{ji} \exp\left([Ax]_j\right)}{\sum_{j=1}^m \exp\left([Ax]_j\right)}.$$

As we can see, the naive calculation of this expression can take time comparable to the calculation of the whole gradient and it will significantly affect the computational complexity and the working time of the method. At the same time, many terms in this expression can be recalculated either infrequently or in a component-wise manner, and used as additional sequences when performing a step of method. Using this approach, the complexity of the iteration will remain efficient, and the use of the coordinate descent methods will be justified. For the convenience of describing the computational methods used, we write the step of the coordinate descent algorithm in the form of

$$y_{k+1} = y_k + \eta e_i,$$

where η is the step size, multiplied by the corresponding gradient component, e_i is the i-th unit basis vector.

So, let us describe the additionally introduced computational procedures:

1. We will store a sequence of values $\left\{\exp\left([Ay_k]_j\right)\right\}_{j=1}^m$, used to calculate the sum in the numerator. Updating these values after executing a method step takes $\mathcal{O}(s)$ algorithmic complexity, due to the $Ay_{k+1} = Ay_k + \eta A_i$ and that A_i has at most s nonzero components, which means that it will be necessary to calculate not more than s correcting factors and multiply the corresponding values from the sequence by them.

2. From the first point, it can be understood that the multiplication of sparse vectors should be performed in $\mathcal{O}(s)$, considering only nonzero components. In terms of program implementation, this means the need to use a sparse representation for cached values and for rows of the matrix A, that is, storing only index-value pairs for all nonzero elements. Then, obviously, the complexity of arithmetic operations for such vectors will be proportional to the complexity of a loop with elementary arithmetic operations, the number of iterations of which is equal to the number of nonzero elements (in the python programming language, for example, this storage format is implemented in the method scipy.sparse.csr_matrix [28]).

3. Similarly, we will store the value $\sum_{j-1}^m \exp\left([Ay_k]_j\right)$, which is the denominator of the presented expression. Its updating is carried out with the same complexity as updating a sequence (by calculating the sum of nonzero terms added to each value from the sequence).

4. Since evaluating the specified expression requires evaluating exponent values, the type overflow errors can occur. To solve this problem, it will be used the standard technique of exp-normalize trick [3]. However, to use it, one should also store the value $\max_{j=1,\dots,m} [Ay_k]_j$. At the same time, there is no need to maintain exactly this value. Indeed, it is sufficient to use only an approximation of it to keep the exponent values small, so this value can be recalculated much rarely: for example, once in m iterations (in this case, the amortized complexity will also be equal to $\mathcal{O}(s)$).

So, in the further reasoning, one can assume that the iteration of the coordinate descent algorithm for solving the corresponding auxiliary problem has amortized complexity $\mathcal{O}(s)$.

Further, let us consider in more detail the question of the values of the smoothness constants of this function. We can write down asymptotic formulas for L and $\overline{L} = \frac{1}{n}\sum_{i=1}^n L_i$:

$$L = \max_{j=1,\dots,m} \|A_j\|_2^2 = \mathcal{O}(n), \quad \overline{L} = \frac{1}{n}\sum_{i=1}^n \max_{j=1,\dots,m} |A_{ji}| = \mathcal{O}(1).$$

Using these estimates, let us refine the computational complexity of the FGM and Catalyst CDM (CCDM) methods as applied to this problem:

$$T_{FGM} = \mathcal{O}\left(s \cdot n^{3/2} \cdot \frac{1}{\varepsilon^{1/2}}\right), \quad T_{CCDM} = \tilde{\mathcal{O}}\left(s \cdot n \cdot \frac{1}{\varepsilon^{1/2}}\right).$$

Thus, in theory, the application of the Catalyst CDM method for solving this problem allows, in comparison with FGM, to reduce the factor of order $\mathcal{O}(\sqrt{n})$ in the asymptotic estimate of the computational complexity. In practice, this means that it is very reasonable to apply the proposed method to problems of large dimensions.

Let us now compare the performance of the proposed approach (Catalyst CDM) with a number of alternative approaches: Gradient Method (GM), Fast Gradient Method (FGM), Coordinate Descent Method (CDM) and Accelerated Coordinate Descent Method (ACDM), by the example of the problem (12) with an artificially generated matrix A in two different ways. Figures 1 and 2 present plots of the convergence of the methods under consideration: the x-axis shows the working time of the methods in seconds, and the y-axis shows the function value residual in a logarithmic scale (f_* calculated by searching for the corresponding point x_* using the FGM method, tuned for an accuracy that is obviously much higher than that possible to achieve at the selected time interval).

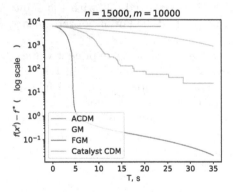
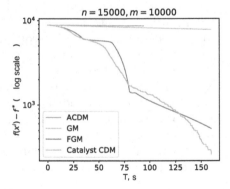

Fig. 1. Convergence of methods for the SoftMax problem (12) with a uniformly sparse random matrix.

Fig. 2. Convergence of methods for the SoftMax problem (12) with heterogeneously sparse matrix.

In Fig. 1, the case is presented for which all elements of the matrix A are i.i.d. random variables from the discrete uniform distribution $A_{ji} \in \mathcal{U}\{0, 1\}$, the number of nonzero elements is $s \approx 0.2m$, and the parameter $\gamma = 0.6$ (as well as in the second case). In this setting, the proposed method demonstrates faster convergence compared to all methods under consideration, except the FGM. At the same time, in the setting shown in Fig. 2, in which the number of nonzero elements, in comparison with the first case, is increased to $s \approx 0.75\,m$, and the matrix is generated heterogeneously in accordance with the rule: $0.9m$ rows with $0.1n$ nonzero elements and $0.1m$ rows with $0.9n$ nonzero elements, and also one row of the matrix is completely nonsparse, the proposed method (Accelerated Meta-algorithm with coordinate descent as internal method) converges faster than FGM. This is explained by the fact that in this case $L = n$, but $\overline{L}$ is still quite small and, as a result, the constant in the proposed method has a noticeably smaller effect on the computational complexity than in the case of FGM. From the results of the experiment, it can also be noted that the character of its componentwise smoothness affects the efficiency of the proposed method much more significantly than the sparseness of the problem.

2.3   Application to Optimization of Markov Decision Processes

We denote an MDP instance by a tuple $\mathcal{MDP} := (\mathcal{S}, \mathcal{A}, \mathcal{P}, r, \gamma)$ with components defined as follows:

1. $\mathcal{S}$ is a finite set of states, $|\mathcal{S}| = n$;
2. $\mathcal{A} = \bigcup_{i \in \mathcal{S}} \mathcal{A}_i$ is a finite set of actions that is a collection of sets of actions $\mathcal{A}_i$ for states $i \in \mathcal{S}$, $|\mathcal{A}| = m$;
3. $\mathcal{P}$ is the collection of state-to-state transition probabilities where $\mathcal{P} := \{p_{ij}(a_i)|i, j \in \mathcal{S}, a_i \in \mathcal{A}_i\}$;
4. r is the vector of state-action transitional rewards where $r \in [0,1]^{\mathcal{A}}$, r_{i,a_i} is the instant reward received when taking action a_i at state $i \in \mathcal{S}$;
5. γ is the discount factor of MDP, by which one down-weights the reward in the next future step. When $\gamma \in (0,1)$, we call the instance a discounted MDP (DMDP) and when $\gamma = 1$ we call the instance an average-reward MDP (AMDP).

Let as denote by $P \in \mathbb{R}^{\mathcal{A} \times \mathcal{S}}$ the state-transition matrix where its (i, a_i)-th row corresponds to the transition probability from state $i \in \mathcal{S}$ where $a_i \in \mathcal{A}_i$ to state j. Correspondingly we use $\hat{I}$ as the matrix with a_i th row corresponding to e_i, for all $i \in \mathcal{S}$, $a_i \in \mathcal{A}_i$. Our goal is to compute a random policy which determines which actions to take at each state. A random policy is a collection of probability distributions $\pi := \{\pi_i\}_{i \in \mathcal{S}}$, where $\pi_i \in \Delta_{|\mathcal{A}_i|}$, $\pi_i(a_j)$ denotes the probability of taking $a_j \in \mathcal{A}_i$ at state i. One can extend π_i to the set of Δ_m by filling in zeros on entries corresponding to other states $j \neq i$. Given an MDP instance $\mathcal{MDP} = (\mathcal{S}, \mathcal{A}, \mathcal{P}, r, \gamma)$ and an initial distribution over states $q \in \Delta_n$, we are interested in finding the optimal $\pi^\star$ among all policies π that maximizes the following cumulative reward $\overline{v}_\pi$ of the MDP:

$$\pi^\star := \arg\max_\pi \overline{v}^\pi,$$

$$\overline{v}^\pi := \begin{cases} \mathbb{E}^\pi \left[\sum_{t=1}^{\infty} \gamma^{t-1} r_{i_t, a_t} \big| i_1 \sim q \right] & \text{in the case of } DMDP, \\ \lim_{T \to \infty} \frac{1}{T} \mathbb{E}^\pi \left[\sum_{t=1}^{T} r_{i_t, a_t} \big| i_1 \sim q \right] & \text{in the case of } AMDP. \end{cases}$$

For AMDP, $\overline{v}^{\star}$ is the optimal average reward if and only if there exists a vector $v^{\star} = (v_i^{\star})_{i \in \mathcal{S}}$ satisfying its corresponding Bellman equation [2]:

$$\overline{v}^{\star} + v_i^{\star} = \max_{a_i \in \mathcal{A}_i} \left\{ \sum_{j \in \mathcal{S}} p_{ij}(a_i) v_j^{\star} + r_{i,a_i} \right\}, \forall i \in \mathcal{S}.$$

For DMDP, one can show that at optimal policy $\pi^{\star}$, each state $i \in \mathcal{S}$ can be assigned an optimal cost-to-go value $v_i^{\star}$ satisfying the following Bellman equation:

$$v_i^{\star} = \max_{a_i \in \mathcal{A}_i} \left\{ \sum_{j \in \mathcal{S}} \gamma p_{ij}(a_i) v_j^{\star} + r_{i,a_i} \right\}, \forall i \in \mathcal{S}.$$

One can further write the above Bellman equations equivalently as the following primal linear programming problems.

$$\min_{\overline{v},v} \quad \overline{v} \qquad \qquad \text{s.t. } \overline{v}\mathbf{1} + (\hat{I} - P)v - r \geq 0 \quad \textbf{(LP AMDP)}$$

$$\min_{v} \quad (1 - \gamma)q^{\top}v \qquad \text{s.t. } (\hat{I} - \gamma P)v + r \geq 0 \quad \textbf{(LP DMDP)}$$

By standard linear duality, we can recast the problem formulation using the method of Lagrangian multipliers, as bi-linear saddle-point (minimax) problem. The equivalent minimax formulations are

$$\min_{v \in \mathbb{R}^n} \left\{ F(v) = \max_{\mu \in \Delta_m} \left(\mu^{\top}((P - \hat{I})v + r) \right) \right\} \qquad \textbf{(AMDP)}$$

$$\min_{v \in \mathbb{R}^n} \left\{ F_{\gamma}(v) = \max_{\mu \in \Delta_m} \left((1 - \gamma)q^{\top}v + \mu^{\top}((\gamma P - \hat{I})v + r) \right) \right\} \qquad \textbf{(DMDP)}$$

Then, one can apply the Nesterov smoothing technique to the presented max-type function, according to [21]. (The calculation is presented for the case of DMDP, as a more general one. A smoothed version of the AMDP function is obtained similarly with the re-designations $A := P - \hat{I}$ and $\gamma = 1$):

$$F_{\gamma}(v) = \max_{\mu \in \Delta_m} \left(\underbrace{(1 - \gamma)q^{\top}}_{b} v + \mu^{\top}(\underbrace{(\gamma P - \hat{I})}_{A}v + r) \right) = \max_{\mu \in \Delta_m} \left(\sum_{j=1}^{m} \mu_j([Av]_j + r_j) \right) + \langle b, v \rangle$$

$$\to \max_{\mu \in \Delta_m} \left(\sum_{j=1}^{m} \mu_j([Av]_j + r_j) - \sigma \sum_{j=1}^{m} \mu_j \ln\left(\frac{\mu_j}{1/m}\right) \right) + \langle b, v \rangle$$

$$= \sigma \ln\left(\sum_{j=1}^{m} \exp\left(\frac{[Av]_j + r_j}{\sigma} \right) \right) - \sigma \ln m - \langle b, v \rangle =: f_{\gamma}(v), \text{ where } \sigma := \varepsilon/(2\ln m).$$

The resulting problem is of the form of SoftMax function, discussed in detail in the previous section. Taking into account the form of the matrix A, we can calculate the average component-wise Lipschitz constant:

$$P_{ji} \in [0,1], \hat{I}_{ji} \in \{0,1\} \implies \overline{L} = \frac{1}{\sigma}\frac{1}{n} \sum_{i=1}^{n} \max_{j=1,\ldots,m} |A_{ji}| \leq \frac{\gamma}{\sigma} = \frac{2\gamma \ln m}{\varepsilon}$$

So, one can get the following estimates, which are also given in the Tables 2 and 3 (transition from the duality gap accuracy ε in matrix game setting to the $\widetilde{\varepsilon}$ accuracy to obtain the $\widetilde{\varepsilon}$-approximate optimal policy satisfying the condition in expectation $\mathbb{E}\bar{v}^{\pi} \geq \bar{v}^{\star} - \widetilde{\varepsilon}$ is carried out according to the rules described in more detail in the paper [15]):

$$\varepsilon \sim \frac{1}{2} \cdot \frac{\widetilde{\varepsilon}}{3} \quad \Longrightarrow \quad T_{CCDM} = \widetilde{\mathcal{O}}\left(\text{nnz}(P)\sqrt{\log m} \cdot \widetilde{\varepsilon}^{-1}\right),$$

$$\varepsilon_{\gamma} \sim \frac{1}{2} \cdot \frac{(1-\gamma)\widetilde{\varepsilon}}{3} \quad \Longrightarrow \quad T_{CCDM,\gamma} = \widetilde{\mathcal{O}}\left(\gamma^{1/2}(1-\gamma)^{-1} \cdot \text{nnz}(P)\sqrt{\log m} \cdot \widetilde{\varepsilon}^{-1}\right),$$

where $\text{nnz}(P) \leq n \cdot m$ denotes the number of nonzero elements in matrix P. In addition to the result corresponding to the considered method, the Tables 2 and 3 contain complexity bounds of other known approaches to solve matrix games and MDP problems (for the sake of compactness, it is used the notation $\text{nnz}'(P) = \text{nnz}(P) + (m+n)\log^3(mn)$). It can be seen that for the case of $\gamma = 1$, the described approach allowing to obtain one of the best among the known estimates, and in the case of $\gamma < 1$ it is close in efficiency to many modern approaches. Moreover, to describe the method used in this article, it was enough to apply only a special case of the universal accelerated proximal envelope for the classical coordinate descent method. This approach is conceptually much simpler than the other methods cited here (which, by the way, are often applicable only to very particular settings), and allows one to obtain complexity bounds for AMDP problem that notedly competitive with the best alternatives.

Table 2. Comparison of the effectiveness of approaches ($\gamma = 1$ case)

Computational complexity	Source
$\widetilde{\mathcal{O}}\left(\text{nnz}(P)\sqrt{\log m} \cdot \widetilde{\varepsilon}^{-1}\right)$	This paper
$\widetilde{\mathcal{O}}\left(\text{nnz}(P)\sqrt{m/n} \cdot \widetilde{\varepsilon}^{-1}\right)$	[6]
$\widetilde{\mathcal{O}}\left(\log^3(mn)\sqrt{\text{nnz}(P) \cdot \text{nnz}'(P)} \cdot \widetilde{\varepsilon}^{-1}\right)$	[7]

Table 3. Comparison of the effectiveness of approaches ($\gamma \in (0,1)$ case)

Computational complexity	Source
$\widetilde{\mathcal{O}}\left(\gamma^{1/2}(1-\gamma)^{-1}\,\text{nnz}(P)\sqrt{\log m} \cdot \widetilde{\varepsilon}^{-1}\right)$	This paper
$\widetilde{\mathcal{O}}\left(\gamma(1-\gamma)^{-1}\,\text{nnz}(P)\sqrt{m/n} \cdot \widetilde{\varepsilon}^{-1}\right)$	[6]
$\widetilde{\mathcal{O}}\left(\gamma(1-\gamma)^{-1}\,\log^3(mn)\sqrt{\text{nnz}(P) \cdot \text{nnz}'(P)} \cdot \widetilde{\varepsilon}^{-1}\right)$	[7]
$\widetilde{\mathcal{O}}\left(nm\left(n + (1-\gamma)^{-3}\right) \cdot \log\left(\widetilde{\varepsilon}^{-1}\right)\right)$	[29]

3   Conclusion

In this paper, we propose a version of the Coordinate Descent Method, accelerated using the universal proximal envelope "Accelerated Meta-algorithm". The performed theoretical analysis allows us to assert that the dependence of its computational complexity on the dimensionality and required solution accuracy is not inferior to other methods used to optimize convex Lipschitz smooth functions. Moreover, its computational complexity is comparable to that of the Fast Gradient Method. At the same time, the proposed scheme retains the properties of the classical Coordinate Descent Method, including the possibility of using the properties of component-wise smoothness of the function. The given numerical experiments confirm the practical efficiency of the method, and also emphasize the particular relevance of the proposed approach for the SoftMax-like function optimization problem that often arises in various applications. As an example of such an application, it was considered the problem of optimizing the MDP, and using the described approach, a method was proposed for solving the averaged version of the MDP problem, which gives a complexity bound that competes with the most efficient known ones.

References

1. Anikin, A., et al.: Modern efficient numerical approaches to regularized regression problems in application to traffic demands matrix calculation from link loads. In: Proceedings of International conference ITAS-2015. Russia, Sochi (2015)
2. Bertsekas, D.P.: Dynamic Programming and Optimal Control. Athena Scientific, Belmont, MA (1995)
3. Blanchard, P., Higham, D.J., Higham, N.J.: Accurately Computing the Log-Sum-Exp and Softmax Functions (2019)
4. Bubeck, S.: Convex optimization: algorithms and complexity. arXiv preprint arXiv:1405.4980 (2014)
5. Bubeck, S., Jiang, Q., Lee, Y.T., Li, Y., Sidford, A.: Near-optimal method for highly smooth convex optimization. In: Conference on Learning Theory, pp. 492–507. PMLR (2019)
6. Carmon, Y., Jin, Y., Sidford, A., Tian, K.: Variance reduction for matrix games. In: Advances in Neural Information Processing Systems, pp. 11381–11392 (2019)
7. Carmon, Y., Jin, Y., Sidford, A., Tian, K.: Coordinate methods for matrix games. arXiv preprint arXiv:2009.08447 (2020)
8. Chernov, A.: Direct-dual method for solving the entropy-linear programming problem. Intell. Syst. Theor Appl. **20**(1), 39–59 (2016)
9. d'Aspremont, A., Scieur, D., Taylor, A.: Acceleration methods (2021)
10. Doikov, N., Nesterov, Y.: Contracting proximal methods for smooth convex optimization. SIAM J. Optim. **30**(4), 3146–3169 (2020)
11. Dvinskikh, D., et al.: Accelerated meta-algorithm for convex optimization. arXiv preprint arXiv:2004.08691 (2020)
12. Gasnikov, A., Gasnikova, E., Nesterov, Y., Chernov, A.: Efficient numerical methods for entropy-linear programming problems. Comput. Math. Math. Phys. **56**(4), 523–534 (2016)

13. Gasnikov, A.: Universal gradient descent. arXiv preprint arXiv:1711.00394 (2017)
14. Ivanova, A., Pasechnyuk, D., Grishchenko, D., Shulgin, E., Gasnikov, A., Matyukhin, V.: Adaptive catalyst for smooth convex optimization. arXiv preprint arXiv:1911.11271 (2019)
15. Jin, Y., Sidford, A.: Efficiently solving MDPs with stochastic mirror descent. In: International Conference on Machine Learning, pp. 4890–4900. PMLR (2020)
16. Kamzolov, D., Gasnikov, A., Dvurechensky, P.: On the optimal combination of tensor optimization methods. arXiv preprint arXiv:2002.01004 (2020)
17. Kulunchakov, A., Mairal, J.: A generic acceleration framework for stochastic composite optimization. In: Advances in Neural Information Processing Systems. pp. 12556–12567 (2019)
18. Lin, H., Mairal, J., Harchaoui, Z.: A universal catalyst for first-order optimization. Adv. Neural Inf. Process. Syst. **28**, 3384–3392 (2015)
19. Lin, H., Mairal, J., Harchaoui, Z.: Catalyst acceleration for first-order convex optimization: from theory to practice. J. Mach. Learn. Res. **18**(1), 7854–7907 (2017)
20. Monteiro, R.D., Svaiter, B.F.: An accelerated hybrid proximal extragradient method for convex optimization and its implications to second-order methods. SIAM J. Optim. **23**(2), 1092–1125 (2013)
21. Nesterov, Y.: Smooth minimization of non-smooth functions. Math. Program. **103**(1), 127–152 (2005)
22. Nesterov, Y.: Efficiency of coordinate descent methods on huge-scale optimization problems. SIAM J. Optim. **22**(2), 341–362 (2012)
23. Nesterov, Y.: Lectures on Convex Optimization. SOIA, vol. 137. Springer, Cham (2018). https://doi.org/10.1007/978-3-319-91578-4
24. Nesterov, Y., Stich, S.U.: Efficiency of the accelerated coordinate descent method on structured optimization problems. SIAM J. Optim. **27**(1), 110–123 (2017)
25. Paquette, C., Lin, H., Drusvyatskiy, D., Mairal, J., Harchaoui, Z.: Catalyst acceleration for gradient-based non-convex optimization. arXiv preprint arXiv:1703.10993 (2017)
26. Parikh, N., Boyd, S.: Proximal algorithms. Found. Trends Optim 1(3), 127–239 (2014)
27. Rockafellar, R.T.: Monotone operators and the proximal point algorithm. SIAM j. Control Optim. **14**(5), 877–898 (1976)
28. SciPy.org: Python Scipy documentation: scipy.sparse.csr_matrix (2020) https://doc s.scipy.org/doc/scipy/reference/generated/scipy.sparse.csr_matrix.html. Accessed 5 Jan 2021
29. Sidford, A., Wang, M., Wu, X., Ye, Y.: Variance reduced value iteration and faster algorithms for solving Markov decision processes. In: Proceedings of the Twenty-Ninth Annual ACM-SIAM Symposium on Discrete Algorithms. pp. 770–787. SIAM (2018)

Duality Gap Estimates for a Class of Greedy Optimization Algorithms in Banach Spaces

Sergei Sidorov[ID] and Kirill Spiridinov[(✉)]

Saratov State University, Saratov 410012, Russian Federation
sidorovsp@sgu.ru
http://www.sgu.ru

Abstract. The paper examines a class of algorithms called Weak Biorthogonal Greedy Algorithms (WBGA) designed for the task of finding the approximate solution to a convex cardinality-constrained optimization problem in a Banach space using linear combinations of some set of "simple" elements of this space (a dictionary), i.e. the problem of finding the infimum of a given convex function over all linear combinations of the dictionary elements with the given cardinality. An important issue when one computationally solves optimization problems is to obtain an estimate of the proximity to an optimal solution, that can be used to effectively check that the approximate solution is with a given accuracy. A similar idea that has already been applied to solving some optimization problems, in which such an estimate of the proximity (certificate) to the optimal solution is called the "duality gap". We introduce the notion of the duality gap for greedy optimization in Banach Spaces and obtain dual convergence estimates for sparse-constrained optimization by means of algorithms from the WBGA class.

Keywords: Convex optimization · Greedy algorithms · Duality gap · Cardinality-constrained optimization problem · Banach space

1 Introduction

Let X be a Banach space with norm $\| \cdot \|$. A set of elements $\mathcal{D}$ from the space X is called a *dictionary* (see, e.g. [25]) if each element $g \in \mathcal{D}$ has norm bounded by one, $\|g\| \leq 1$, and the closure of span $\mathcal{D}$ is X, i.e. $\overline{\text{span}}\,\mathcal{D} = X$. A dictionary $\mathcal{D}$ is called symmetric if $-g \in \mathcal{D}$ for every $g \in \mathcal{D}$. In this paper we assume that the dictionary $\mathcal{D}$ is symmetric.

Let E be a convex function defined on X. In many applied problems it is often necessary to find an approximate representation of the solution to the

This work was supported by the Ministry of Science and Higher Education of the Russian Federation in the framework of the basic part of the scientific research state task, project FSRR-2020-0006.

P. Pardalos et al. (Eds.): MOTOR 2021, LNCS 12755, pp. 192–205, 2021.
https://doi.org/10.1007/978-3-030-77876-7_13

minimization problem

$$E(x) \to \min_{x \in X} \tag{1}$$

as a linear combination of elements from the dictionary $\mathcal{D}$. Moreover, it is highly desirable that the solutions of the optimization problem to be sparse with respect to the dictionary $\mathcal{D}$, i.e. the aim is to solve the following problem:

$$E(X) \to \inf_{x \in \Sigma_m(\mathcal{D})}, \tag{2}$$

where $\Sigma_m(\mathcal{D})$ is the set of all m-term polynomials with respect to $\mathcal{D}$:

$$\Sigma_m(\mathcal{D}) = \left\{ x \in X \ : \ x = \sum_{i=1}^{m} c_i g_i, \ g_i \in \mathcal{D} \right\}. \tag{3}$$

In many real applications the dimension of the search space while is finite, but is too large. Therefore, our interest lies in obtaining estimates on the rate of convergence not depending on the dimension of X. Obviously, results for the infinite Banach spaces provide such estimates on the convergence rate. Following [25], we examine the problem in an infinite dimensional Banach space setting.

One of the apparent choices among constructive methods for finding the best m-term optimizations are greedy algorithms. Applicability of the greedy algorithms on a finite combinatorial structure is connected with Matroid Theory. The design of greedy algorithms allows us to obtain sparse solutions with respect to $\mathcal{D}$. Obtaining sparse solutions is of interest in many real applications including compressed sensing in which one need to decode a sparse signal by means of a small number of linear measurements [2,8], portfolio selection and economic applications [9,23] and many other.

Perhaps, the Frank-Wolfe method [10], which is also known as the 'conditional gradient' method [17], is one of the most prominent algorithms for finding optimal solutions of constrained convex optimization problems. Important contributions to the development of Frank-Wolfe type algorithms can be found in [3,11,14]. The paper [14] provides general primal-dual convergence results for Frank-Wolfe-type algorithms by extending the duality concept presented in the work [3]. Recent convergence results for greedy algorithms one can find in the works [1,4,6,7,12,13,15,16,18,21,22,26,27].

The paper [5] presents a unified way of examining a certain type of greedy algorithms in Banach spaces. The authors of [5] describe the class of Weak Biorthogonal Greedy Algorithms (WBGA) and obtain the convergence results for the corresponding algorithms belonging to the class.

This paper also examines algorithms from the WBGA class which are aimed in finding solutions of optimization problem (3), which are sparse with respect to a dictionary, in Banach spaces. Primal convergence results for the weak relaxed greedy algorithms were obtained in [5].

In this paper, extending the ideas of [3,14] we introduce the notion of the duality gap for algorithms from the WBGA class to obtain dual convergence estimates for sparse-constrained optimization problems of type (3). In contrast to paper [5], in this paper we focus on obtaining dual convergence results based on duality gap analysis. Note that estimates for the duality gap for some algorithms were obtained in papers [18,22,24]. However, in this paper we present estimates for the duality gap for a wider class of algorithms class of algorithms called Weak Biorthogonal Greedy Algorithms (WBGA), rather that for particular algorithms.

2   Greedy Optimization Algorithms in Banach Spaces

A greedy algorithm is to iteratively construct an approximate solution (G_m) to the problem (1) by a linear combination of appropriate elements of a given set $D \subset X$ (the dictionary) in accordance with the following general steps:

- (A) Find a fitting element $\phi_m \in \mathcal{D}$;
- (B) Update a solution G_m to the problem (1) using G_{m-1} and ϕ_m.

Let E be a convex function defined on X. In this paper it is supposed that function E is Fréchet differentiable. Let $E'(x)$ denote Fréchet differential of E at x. Then it follows from convexity of E that

$$E(y) \geq E(x) + \langle E'(x), y - x \rangle,$$

for any x, y.

One of the approaches for choosing ϕ_m is based on the use of the Fréchet differential, so ϕ_m is selecting from the solution of the problem

$$\langle -E'(x), \phi_m - G_{m-1} \rangle = \sup_{s \in \mathcal{D}} \langle -E'(G_{m-1}), s - G_{m-1} \rangle. \tag{4}$$

The algorithms that exploit the Fréchet differential for selecting of ϕ_m at step (A) are belonged to the class of dual greedy algorithms.

Let $\Omega := \{x \in X : E(x) \leq E(0)\}$ and suppose that Ω is bounded. As it turns out, the convergence analysis of greedy algorithms essentially depends on a measure of non-linearity of the objective function E over set Ω, which can be depicted via the modulus of smoothness of function E.

Let us remind that the modulus of smoothness of function E on the bounded set Ω can be defined as

$$\rho(E, u) = \frac{1}{2} \sup_{x \in \Omega, \|y\|=1} |E(x + uy) + E(x - uy) - 2E(x)|, \ u > 0. \tag{5}$$

E is called uniformly smooth function on Ω if $\lim_{u \to 0} \rho(E, u)/u = 0$.

The paper [5] introduces the class of Weak Biorthogonal Greedy Algorithms (WBGA) that is consisted of algorithms in which sequences of approximate solutions $\{G_m\}$, and selected elements ϕ_m satisfy the following conditions at every iteration $m \geq 1$:

- (1) Greedy selection:
  $$\langle -E'(G_{m-1}), \phi_m - G_{m-1} \rangle \geq t_m \sup_{s \in \mathcal{D}} \langle -E'(G_{m-1}), s - G_{m-1} \rangle;$$

- (2) Error reduction: $E(G_m) \leq \inf_{\lambda \geq 0} E(G_{m-1} + \lambda \phi_m)$;
- (3) Bi-orthogonality: $\langle E'(G_m), G_m \rangle = 0$.

The well-known examples of algorithms from the WBGA class are Week Chebyshev Greedy Algorithm (WCGA), Weak Greedy Algorithm with Free Relaxation (WGAFR) and Rescaled Weak Relaxed Greedy Algorithm (RWRGA) (see Algorithms 1, 2 and 3, respectively).

Algorithms from the WBGA for solving (2) have some advantages compared to other greedy algorithms. To solve the sub-problem $\sup_{s \in \mathcal{D}} \langle -E'(G_{m-1}), s - G_{m-1} \rangle$ exactly may be too expensive in many real cases. WBGA algorithms use the weakness sequence in the greedy selection step to finding an approximate minimizer instead, which has approximation quality at least t_m in step m. Another advantage, for example, in comparison with the Frank-Wolfe algorithm, is that algorithms from the WBGA are aimed to solve unconditional optimization problems, whereas the Frank-Wolfe algorithm solves optimization problems with linear constraints.

It was shown in [5] that the WCGA, the WGAFR and the RWRGA belong to the class WBGA. Moreover, they proved that the algorithms converge to the solution of the unconstrained problem (1) as $m \to \infty$ and find the rate of the convergence. Let us present their results in more detail.

Let $\tau = \{t_m\}$ be a weakness sequence, $0 < \theta \leq \frac{1}{2}$. Let $\xi_m := \xi_m(\rho, \tau, \theta)$ be defined as the root of the equation

$$\rho(u) = \theta t_m u. \tag{6}$$

Theorem 1. *Let E be a uniformly smooth function on Ω with the modulus of smoothness $\rho(u)$. Let $\tau = \{t_m\}$ be a sequence that for any $\theta > 0$ the equality $\sum_{m=1}^{\infty} t_m \xi_m(\rho, \tau, \theta) = \infty$ holds. Then an algorithm from the WBGA with the weakness sequence τ converges to the solution of the unconstrained problem (1) as $m \to \infty$.*

Algorithm 1: WEEK CHEBYSHEV GREEDY ALGORITHM (WCGA)

---

begin
  · Let $G_0 = 0$;
  for each $m = 1, 2, \ldots, M$ **do**
    · (*Greedy selection*) Find the element $\phi_m \in \mathcal{D}$ such that
    $\langle -E'(x), \phi_m - G_{m-1} \rangle = \sup_{s \in \mathcal{D}} \langle -E'(G_{m-1}), s - G_{m-1} \rangle$;
    · (*Chebyshev-type search*) Find $S_m \in \Phi_m := \text{span}(\phi_1, \ldots, \phi_m)$, such that
    $S_m = \text{argmin}_{G \in \Phi_m} E(G)$;
    · (*Update step*) $G_m = S_m$;
end

Denote $A_1(\mathcal{D})$ the convex hull of the dictionary $\mathcal{D}$. Denote

$$x^* := \arg \inf_{x \in X} E(x),$$

Theorem 2. *Let E be a uniformly smooth convex function on Ω with the modulus of smoothness $\rho(u)$ such that $\rho(u) \leq \gamma u^q$ with $1 < q \leq 2$. Let $\epsilon \geq 0$ and $x^\epsilon \in X$ be such that*

$$E(x^\epsilon) - E(x^*) \leq \epsilon, \quad x^\epsilon/A(\epsilon) \in \mathcal{A}_1(\mathcal{D})$$

with some $A(\epsilon) > 1$. Then for an algorithm from the WBGA with the weakness sequence τ we have

$$E(G_m) - E^* \leq \max\left\{ 2\epsilon, C(q,\gamma)A(\epsilon)^q \left(C(E,q,\gamma) + \sum_{k=1}^m t_k^p \right)^{1-q} \right\},$$

where $p = \frac{q}{q-1}$.

Corollary 1. *Let E be a uniformly smooth convex function on Ω, with the modulus of smoothness $\rho(u)$ such that $\rho(u) \leq \gamma u^q$, $1 < q \leq 2$. If $\operatorname{argmin}_{x \in X} E(x) \in \mathcal{A}_1(\mathcal{D})$ then for any algorithm from the class WBGA we have*

$$E(G_m) - E^* \leq C(q,\gamma) \left(C(E,q,\gamma) + \sum_{k=1}^m t_k^p \right)^{1-q},$$

where $p = \frac{q}{q-1}$.

Algorithm 2: WEAK GREEDY ALGORITHM WITH FREE RELAXATION (WGAFR)

---

begin
    · Let $G_0 = 0$;
    for each $m = 1, 2, \ldots, M$ **do**
        · (*Greedy selection*) Find the element $\phi_m \in \mathcal{D}$ such that
          $\langle -E'(x), \phi_m - G_{m-1} \rangle = \sup_{s \in \mathcal{D}} \langle -E'(G_{m-1}), s - G_{m-1} \rangle$;
        · (*Two-dimensional search*) Find $\omega_m \in \mathbb{R}$ and $\lambda_m \geq 0$ such that
          $E((1 - \omega_m)G_{m-1} + \lambda_m \phi_m) = \inf_{\lambda \geq 0, \omega \in \mathbb{R}} E((1 - \omega)G_{m-1} + \lambda \phi_m)$;
        · (*Update step*) Let $G_m = (1 - \omega_m)G_{m-1} + \lambda_m \phi_m$;
end

---

3   Duality Gap Estimates for Greedy Approximation in Banach Spaces

3.1   Duality Gap

This section presents dual results on convergence for algorithms from the class of Weak Biorthogonal Greedy Algorithms (WBGA).

Algorithm 3: RESCALED WEAK RELAXED GREEDY ALGORITHM (RWRGA)

---

begin

$\quad$ · Let $f_0 = f$, $G_0 = 0$;

$\quad$ **for each** $m = 1, 2, \ldots, M$ **do**

$\qquad$ · (*Greedy selection*) Find the element $\phi_m \in \mathcal{D}$ such that

$\qquad \langle -E'(x), \phi_m - G_{m-1} \rangle = \sup_{s \in \mathcal{D}} \langle -E'(G_{m-1}), s - G_{m-1} \rangle;$

$\qquad$ · (*Two one-dimensional searches*) (a) Find $\lambda_m \geq 0$ such that

$\qquad E(G_{m-1} - \lambda_m \phi_m) = \inf_{\lambda \geq 0} E(G_{m-1} - \lambda \phi_m);$

$\qquad$ (b) Find $\mu_m \in \mathbb{R}$ such that

$\qquad E(\mu_m(G_{m-1} + \lambda_m \phi_m)) = \inf_{\mu \in \mathbb{R}} E(\mu(G_{m-1} + \lambda_m \phi_m))$;

$\qquad$ · (*Update step*) Let $G_m = \mu_m(G_{m-1} + \lambda_m \phi_m);$

end

An important problem when performing computations when solving optimization problems is to obtain an estimate of the proximity to the optimal solution, which can be used to effectively check that the approximate solution is built with a given accuracy. The idea has already been applied to solving optimization problems, in which such an estimate of the proximity (certificate) to the optimal solution is called the "duality gap". For example, [19] examines the concept of certificates in the context of convex computational problems. Optimization methods, in which it is possible to obtain estimates of the proximity to the optimal solution, belong to the class of primal-dual methods.

It should be noted that the accuracy certificate, on the one hand, allows substantiating the direct duality of the investigated method [19], and on the other hand, the accuracy certificate is computable and therefore can be used as an estimate of the proximity of the current solution to the optimal. Further development of these ideas can be found in [20].

If a convex function f is defined on finite dimensional Euclidean space $X = \mathbb{R}^n$ and $A \subset \mathbb{R}^n$ is a compact convex set, then the duality gap for the constrained optimization problem $f(x) \rightarrow \min_{x \in A}$ is defined as (see e.g. [14])

$$g(G) = \max_{s \in A} \langle -\nabla f(G), s - G \rangle, \tag{7}$$

where $\nabla f(G)$ is the gradient of f at G.

In this paper, we propose to use the "proximity" certificate to optimal solutions of the optimization problem (2) based on the duality gap. We need a definition different from (7), since the problem (2) is unconstrained and its solution does not have to lie in $\mathcal{A}_1(\mathcal{D})$.

Let E be a uniformly smooth convex function. Take $\epsilon \geq 0$ and let x^ϵ from X be such that $E(x^\epsilon) - E(x^*) \leq \epsilon$ and $x^\epsilon / A_\epsilon \in \mathcal{A}_1(\mathcal{D})$.

Definition 1. *Let the duality gap for function $E : X \to \mathbb{R}$, element $G \in \Omega$ and error $\epsilon \geq 0$ be defined as follows:*

$$g(G) = g(G, \epsilon) =: \sup_{s \in \mathcal{D}} \langle -E'(G), A_\epsilon s - G \rangle. \tag{8}$$

The duality gap value is implicitly computed at each iteration of the greedy algorithm. As will be shown below, the value of the duality gap estimates from above the error between the current approximate solution G_m and the optimal optimization. In other words, the duality gap is the upper bound for the difference between the value of goal function E at the current iteration of the algorithm and the value of goal function E at the optimal point. Since we do not know the optimal solution, as well as the corresponding error, the current values of the duality gap can be used, for example, in the criteria for stopping a greedy algorithm. In this paper, we find estimates for the values of the duality gap depending on the number of iterations for the considered class of greedy algorithms.

The notion of the duality gap for algorithms from the WBGA class similar to the notion in [14], but has a key difference. A_ϵ is a scaling parameter, we use it to associate each element of a dictionary to an element of a unit ball that contains the optimal solution.

We need the following well-known Lemma (see, e.g. [25]).

Lemma 1. *For any bounded linear functional F and any dictionary $\mathcal{D}$, we have*

$$\|F\|_{\mathcal{D}} = \sup_{g \in \mathcal{D}} F(g) = \sup_{g \in A_1(\mathcal{D})} F(g).$$

A useful property of the duality gap is described by the following proposition.

Proposition 1. *Let E be a uniformly smooth convex function defined on Banach space X. Let x^* denote the optimal solution to the problem (1). Then for any $x \in \Omega := \{x \in X : E(x) \leq E(0)\}$ we have*

$$E(x) - E(x^*) \leq g(x, \epsilon) + \epsilon.$$

Proof. Let x^ϵ be such that $E(x^\epsilon) - E(x^*) < \epsilon$ and $x^\epsilon / A_\epsilon \in A_1(\mathcal{D})$, i.e. $x^\epsilon \in \mathcal{L}_{A_\epsilon} := A_\epsilon \cdot A_1(\mathcal{D})$. First we will show that

$$E(x) - E(x^\epsilon) \leq g(x).$$

It follows from the convexity of E on X that for any $x \in \Omega$ we get

$$E(x^\epsilon) \geq E(x) + \langle E'(x), x^\epsilon - x \rangle. \tag{9}$$

From inclusion $x^\epsilon \in \mathcal{L}_{A_\epsilon}$ we have

$$E(x) + \langle E'(x), x^\epsilon - x \rangle \geq$$
$$E(x) - \sup_{s \in \mathcal{L}_{A_\epsilon}} \langle E'(x), x - s \rangle = E(x) - A_\epsilon \sup_{s \in \mathcal{L}_{A_\epsilon}} \langle E'(x), x A_\epsilon^{-1} - s A_\epsilon^{-1} \rangle$$
$$= E(x) - A_\epsilon \sup_{s' \in A_1(\mathcal{D})} \langle E'(x), x A_\epsilon^{-1} - s' \rangle$$
$$= E(x) - A_\epsilon \sup_{s' \in \mathcal{D}} \langle E'(x), x A_\epsilon^{-1} - s' \rangle, \tag{10}$$

with use of Lemma 1.

It follows from (9) and (10) that

$$E(x) - E(x^\epsilon) \leq A_\epsilon \sup_{s' \in \mathcal{D}} \langle E'(x), xA_\epsilon^{-1} - s' \rangle =: g(x), \tag{11}$$

and then the proposition follows from the inequality $E(x^\epsilon) - E(x^*) < \epsilon$.

$\square$

Thus, the practical applicability of the duality gap is related to the fact that the duality gap $g(x)$ is an estimate of how close the current solution $E(G)$ is to the best optimization E^*.

We need the following auxiliary result.

Lemma 2. *Let X be a uniformly smooth convex function defined on Banach space X, with the modulus of smoothness $\rho(E, u)$, $\rho(E, u) \leq \gamma u^q$, $1 < q \leq 2$. Let $\tau = \{t_m\}_{m=1}^\infty$ be a weakness sequence, $0 < t_k \leq 1$, $k = 1, 2, \ldots$. Take $\epsilon \geq 0$ and let f, f^ϵ from X be such that $E(f) - E(f^\epsilon) \leq \epsilon$ and $f^\epsilon/A_\epsilon \in \mathcal{A}_1(\mathcal{D})$. Then for any algorithm from WBGA, the inequality*

$$E(G_m) \leq E(G_{m-1}) + \inf_{\lambda \geq 0} \left(-\lambda t_m A_\epsilon^{-1} g(G_{m-1}) + 2\rho(E, \lambda) \right)$$

holds for any $m = 1, 2, \ldots$.

Proof. It follows from the definition of modulus of smoothness (8) that

$$E(G_{m-1} + \lambda\phi_m) \leq 2E(G_{m-1}) - E(G_{m-1} - \lambda\phi_m) + 2\rho(E, \lambda).$$

From the convexity of E on X we have

$$E(G_{m-1} - \lambda\phi_m) \geq E(G_{m-1}) + \lambda\langle -E'(G_{m-1}), \phi_m \rangle.$$

Combining together two above inequalities we get

$$E(G_{m-1} + \lambda\phi_m) \leq E(G_{m-1}) - \lambda\langle -E'(G_{m-1}), \phi_m \rangle + 2\rho(E, \lambda). \tag{12}$$

The first step of iteration m for an algorithm from the WBGA implies

$$\langle -E'(G_{m-1}), \phi_m \rangle \geq t_m \sup_{s \in \mathcal{D}} \langle -E'(G_{m-1}), s \rangle$$

$$= t_m A_\epsilon^{-1} \sup_{s \in \mathcal{D}} \langle -E'(G_{m-1}), A_\epsilon s - G_{m-1} \rangle = t_m A_\epsilon^{-1} g(G_{m-1}),$$

where we use the equality $\langle E'(G_m), G_m \rangle = 0$, i.e. the biorthogonality of E'. Substitute previous result in (12) we have

$$E(G_{m-1} + \lambda\phi_m) \leq E(G_{m-1}) - \lambda t_m A_\epsilon^{-1} g(G_{m-1}) + 2\rho(E, \lambda).$$

Error reduction step for an algorithm from the WBGA:

$$E(G_m) \leq \inf_{\lambda \geq 0} E(G_{m-1} + \lambda\phi_m). \tag{13}$$

Then lemma follows from (12) and (13).

$\square$

Denote

$$\mathcal{L}_M := \{s \in X \; : \; s/M \in A_1(\mathcal{D})\},$$

$$A_\epsilon := A(f, \epsilon) = \inf\{M \; : \; \exists y \in \mathcal{L}_M \text{ s.t. } E(y) - E^* \le \epsilon\},$$

$$A_0 := \inf\{M \; : \; x^* \in \mathcal{L}_M\}.$$

The following dual result is valid for the algorithms from the class of Weak Biorthogonal Greedy Algorithms (WBGA).

Theorem 3. *Let X be a uniformly smooth convex function with the modulus of smoothness $\rho(E, u)$, $\rho(E, u) \le \gamma u^q$, $1 < q \le 2$. Let $\tau = \{t_m\}_{m=1}^\infty$ be a weakness sequence, $0 < \theta < t_k < 1$, $k = 1, 2, \dots$. Take $\epsilon \ge 0$ and let x, x^ϵ from X be such that $E(x) - E(x^\epsilon) \le \epsilon$ and $x^\epsilon/A_\epsilon \in A_1(\mathcal{D})$. Suppose that for an algorithm from WBGA runs for $N > 2$ iterations. Then there exists an iteration $1 \le \tilde{m} \le N$ such that*

$$g(G_{\tilde{m}}) \le \beta C(q, \gamma) A_0^q N^{1-q}, \tag{14}$$

where $\beta > 0$ does not depend on N.

Proof. Suppose that

$$g(G_m) > \beta C(q, \gamma) A_0^q N^{1-q}, \tag{15}$$

for all $m \in [\lfloor \mu N \rfloor, N]$. Parameter $0 < \mu < 1$ is arbitrary fixed and will be defined later.

By Lemma 2 we have

$$E(G_m) - E^* \le E(G_{m-1}) - E^* + \inf_{\lambda \ge 0} \left(-\lambda t_m A_\epsilon^{-1} g(G_{m-1}) + 2\rho(E, \lambda) \right)$$

$$\le E(G_{m-1}) - E^* - \lambda t_m A_\epsilon^{-1} g(G_{m-1}) + 2\rho(E, \lambda).$$

By assumption (15) we have

$$E(G_m) - E^* < E(G_{m-1}) - E^* - \lambda t_m A_\epsilon^{-1} \beta C(q, \gamma) A_0^q N^{1-q} + 2\rho(E, \lambda).$$

Weakness sequence τ satisfies the condition $0 < \theta < t_k < 1$, hence

$$E(G_m) - E^* < E(G_{m-1}) - E^* - \lambda \theta A_\epsilon^{-1} \beta C(q, \gamma) A_0^q N^{1-q} + 2\rho(E, \lambda).$$

Modulus of smoothness has upper bound and therefore

$$E(G_m) - E^* < E(G_{m-1}) - E^* - \lambda \theta A_\epsilon^{-1} \beta C(q, \gamma) A_0^q N^{1-q} + 2\gamma \lambda^q.$$

Denote $m_0 = \lfloor \mu N \rfloor$ and sum up above inequality for all $m \in [m_0, N]$. Note that $N - m_0 + 1 = N - \lfloor \mu N \rfloor + 1 \ge (1 - \mu)N$.

$$E(G_N) - E^* < E(G_{m_0}) - E^* - (N - m_0 + 1) \left(\lambda \theta A_\epsilon^{-1} \beta C(q, \gamma) A_0^q N^{1-q} - 2\gamma \lambda^q \right)$$

$$\le E(G_{m_0}) - E^* - (1 - \mu)N \left(\lambda \theta A_\epsilon^{-1} \beta C(q, \gamma) A_0^q N^{1-q} - 2\gamma \lambda^q \right).$$

By Corollary 1 and take $\lambda = N^{-1}$ we have

$$E(G_N) - E^* < C(q,\gamma) \left(C(E,q,\gamma) + \sum_{k=1}^{m} t_k^p \right)^{1-q}$$
$$-(1-\mu)N^{1-q} \left(\theta A_\epsilon^{-1} \beta C(q,\gamma) A_0^q - 2\gamma \right). \quad (16)$$

It follows from condition of the Theorem that $1 < q \leq 2, p = \frac{q}{q-1}$ and $N > 2$, therefore $-1 \leq 1 - q < 0$ and

$$\left(C(E,q,\gamma) + \sum_{k=1}^{m} t_k^p \right)^{1-q} \leq (C(E,q,\gamma) + m_0 \theta^p)^{1-q}$$

$$\leq (C(E,q,\gamma) + (\mu N - 1)\theta^p)^{1-q} = ((C(E,q,\gamma) - \theta^p) + \mu N \theta^p)^{1-q} \leq (\mu N)^{1-q} \theta^{-q}.$$

Substitute above estimate in (16)

$$E(G_N) - E^* < C(q,\gamma)(\mu N)^{1-q}\theta^{-q} - (1-\mu)N^{1-q}\left(\theta A_\epsilon^{-1}\beta C(q,\gamma)A_0^q - 2\gamma\right)$$
$$= N^{1-q}(C(q,\gamma)\mu^{1-q}\theta^{-q} - (1-\mu)\left(\theta A_\epsilon^{-1}\beta C(q,\gamma)A_0^q - 2\gamma\right)).$$

We can choice β an arbitrary large,

$$\beta > \frac{2\gamma + \frac{C(q,\gamma)\mu^{1-q}\theta^{-q}}{(1-\mu)}}{\theta A_\epsilon^{-1} C(q,\gamma) A_0^q},$$

and therefore $E(G_N) - E^* < 0$. We arrive at the contradiction that the primal error become negative. Parameter μ can be such that:

$$\mu^* - \operatorname*{argmin}_{0<\mu<1} \left\{ 2\gamma + \frac{C(q,\gamma)\mu^{1-q}\theta^{-q}}{(1-\mu)} \right\}.$$

$\square$

4   Empirical Results

In this section let us consider a simple example to illustrate the dynamic behavior of the duality gap and to demonstrate that the empirical behavior of the duality gap is well predicted by Theorem 3 from the previous section.

Let $\mathbb{B}[0,1]$ be the Banach space of bounded functions with norm $\|f\| = \max_{x\in[0,1]} |f(x)|$. A function $z \in \mathbb{B}[0,1]$ is said to be monotone if for every pair $x_2 > x_1$ the inequality $z(x_2) - z(x_1) \geq 0$ holds. Denote Δ_1 the set of all monotone functions from $\mathbb{B}[0,1]$.

The problem of constructing a monotone regression can be formulated as a convex optimization problem as follows: it is necessary to find the function

$z \in \mathbb{B}[0,1]$ with the smallest approximation error of the function $y \in \mathbb{B}[0,1]$ in the norm $L_q[0,1]$, $q \in (1,2]$, provided that z is monotonic:

$$E(z) = \int_0^1 (z(x) - y(x))^q dx \rightarrow \min_{z \in \Delta_1}. \tag{17}$$

Denote

$$\theta_a(x) = \begin{cases} 0, & x < a, \\ 1, & x \geq a. \end{cases}$$

Then the set $\mathcal{D} = \{\theta_a : a \in (0,1)\}$. is a dictionary in $\mathbb{B}[0,1]$, since the norm of each function g from $\mathcal{D}$ is bounded by 1, $\|g\|_1 \leq 1$, and the closure of all linear combinations of elements from $\mathcal{D}$ is $\mathbb{B}[0,1]$, i.e. $\overline{\text{span}\,\mathcal{D}} = \mathbb{B}[0,1]$.

Therefore, the solution of the problem (17) should be a linear combination of elements from the dictionary: $z = \sum_{i=1}^n b_i \theta_{a_i}$, where $a_1 < a_2 < \ldots < a_n$ and $b_i \geq 0$, $i = 1, \ldots, n$, $n \in \mathbb{N}$, and

$$z(1) - z(0) = \sum_{i=1}^n b_i = \sup_{x \in [0,1]} y(x) - \inf_{x \in [0,1]} y(x).$$

Let $S(\mathcal{D})$ be the closure all such z in $\mathbb{B}[0,1]$.

In this section, we consider a greedy algorithm for finding solutions to the convex optimization problem (17), which are sparse with respect to the dictionary, in the space $\mathbb{B}[0,1]$.

Note that the function E is Fréchet differentiable. Let $\langle E'(x), y \rangle$ denote the value of functional $E'(x)$ at point y. To solve the problem (17), we will use Algorithm 4. For each $m \geq 1$, the algorithm finds the next element G_m by induction with the use of the current iteration function G_{m-1} and a dictionary element θ_{a_m}, obtained on the greedy step.

Algorithm 4: Weak Greedy Algorithm for finding monotone regression

> **begin**
> > · Let $G_0 = 0$;
> > **for each** $m = 1, 2, \ldots, M$ **do**
> > > · (*Greedy step*) Find $a_m \in (0,1)$ (i.e. $\theta_{a_m} \in \mathcal{D}$) such that
> > > $\langle -E'(G_{m-1}), \theta_{a_m} \rangle \geq t_m \sup_{s \in \mathcal{D}} \langle -E'(G_{m-1}), s \rangle$;
> > > · (*Linear search*) Find $0 \leq \lambda_m \leq 1$ such that
> > > $E((1 - \lambda_m)G_{m-1} + \lambda_m \theta_{a_m}) = \inf_{0 \leq \lambda \leq 1} E((1 - \lambda)G_{m-1} + \lambda \theta_{a_m})$;
> > > · (*Update step*) $G_m = (1 - \lambda_m)G_{m-1} + \lambda_m \theta_{a_m}$;
> **end**

The algorithm was implemented in the R language. The algorithm was run for synthetic data generated as follows: we took some monotonically increasing function f and added to its values a random normally distributed variable. The obtained in such a way element from space $\mathbb{B}[0,1]$ is not monotonic, but the original function f could be used as an unknown ("ideal") solution. Figure 1 shows the dynamics of

- the primal error (defined as the norm of the difference between the solution obtained at an iteration and the ideal solution f),
- the dynamics of the duality gap.

For this example we constructed a input data as follows: $f = \ln(x + 10\xi)$, $\xi \sim N(0,1)$. The figure shows that the duality gap is always greater than the corresponding primal error. However, no matter what iteration we take, there is a preceding iteration, for which the value of the duality gap is close to the value of the corresponding primal error.

It should be noted that the value of the duality gap depends on $E'(G)$ and $A_\epsilon s - G$ which are changing at each iteration. Hence, the duality gap may oscillates as it can be seen at Fig. 1.

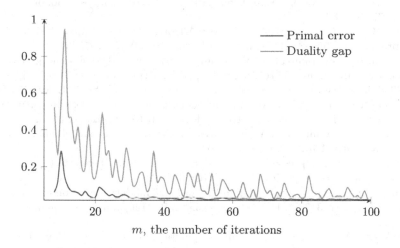

Fig. 1. Dynamics of primal errors and duality gaps

5   Conclusion

All algorithms from the Weak Biorthogonal Greedy Algorithms class (WCGA, WGAFR, RWRGA) have primal errors that depend on geometric properties of the objective function E (Theorem 2 and Corollary 1). In many real applications very often value of goal function E at the optimal point is unknown, and therefore, estimates for the quality of current approximation to optimal solution are considerably in demand. Following ideas of [3,14], we introduced the notion of the duality gap for algorithms from the WBGA class by the equality (8). The values of duality gap are calculating on each iteration of WCGA, WGAFR and RWRGA at the greedy selection step, and therefore, they are upper bounds for primal errors, i.e. differences between values of objective function at current and optimal points on each step. We obtain dual convergence estimates for the above-mentioned algorithms. Our empirical results in Sect. 4 demonstrated that the empirical behavior of the duality gap is well predicted by Theorem 3.

References

1. Barron, A.R., Cohen, A., Dahmen, W., DeVore, R.A.: Approximation and learning by greedy algorithms. Ann. Stat. **36**(1), 64–94 (2008). https://doi.org/10.1214/009053607000000631
2. Chena, Z., Huang, C., Lin, S.: A new sparse representation framework for compressed sensing MRI. Knowl. Based Syst. **188**, 104969 (2020). https://doi.org/10.1016/j.knosys.2019.104969
3. Clarkson, K.L.: Coresets, sparse greedy approximation, and the frank-wolfe algorithm. ACM Trans. Algorithms **6**(4), 1–30 (2010). https://doi.org/10.1145/1824777.1824783
4. Davis, G., Mallat, S., Avellaneda, M.: Adaptive greedy approximation. Constr. Approx. **13**(1), 57–98 (1997). https://doi.org/10.1007/bf02678430
5. Dereventsov, A., Temlyakov, V.: A unified way of analyzing some greedy algorithms. J. Funct. Anal. **277**(12), 108286 (2019). https://doi.org/10.1016/j.jfa.2019.108286, http://www.sciencedirect.com/science/article/pii/S0022123619302496
6. DeVore, R.A., Temlyakov, V.N.: Some remarks on greedy algorithms. Adv. Comput. Math. **5**, 173–187 (1996). https://doi.org/10.1007/bf02124742
7. DeVore, R.A., Temlyakov, V.N.: Convex optimization on banach spaces. Found. Comput. Math. **16**(2), 369–394 (2016)
8. Donoho, D.L.: Compressed sensing. IEEE Trans. Inf. Theory **52**(4), 1289–1306 (2006). https://doi.org/10.1109/TIT.2006.871582
9. Fan, J., Lv, J., Qi, L.: Sparse high-dimensional models in economics. Ann. Rev. Econ. **3**(1), 291–317 (2011). https://doi.org/10.1146/annurev-economics-061109-080451
10. Frank, M., Wolfe, P.: An algorithm for quadratic programming. Naval Res. Logis. Q. **3**(1–2), 95–110 (1956). https://doi.org/10.1002/nav.3800030109
11. Freund, R.M., Grigas, P.: New analysis and results for the frank-wolfe method. Math. Program. **155**(1), 199–230 (2016)
12. Friedman, J.: Greedy function approximation: a gradient boosting machine. Ann. Stat. **29**(5), 1189–1232 (2001)
13. Huber, P.J.: Projection pursuit. Ann. Stat. **13**, 435–525 (1985)
14. Jaggi, M.: Revisiting Frank-Wolfe: projection-free sparse convex optimization. In: Proceedings of the 30th International Conference on Machine Learning (ICML-2013), pp. 427–435 (2013)
15. Jones, L.: On a conjecture of Huber concerning the convergence of projection pursuit regression. Ann. Stat. **15**(2), 880–882 (1987). https://doi.org/10.1214/aos/1176350382
16. Konyagin, S.V., Temlyakov, V.N.: A remark on greedy approximation in banach spaces. East J. Approx. **5**(3), 365–379 (1999)

17. Levitin, E.S., Polyak, B.T.: Constrained minimization methods. USSR Comput. Math. Math. Phys. **6**(5), 1–50 (1966)
18. Mironov, S.V., Sidorov, S.P.: Duality gap estimates for weak Chebyshev greedy algorithms in Banach spaces. Comput. Math. Math. Phys. **59**(6), 904–914 (2019). https://doi.org/10.1134/S0965542519060113
19. Nemirovski, A., Onn, S., Rothblum, U.G.: Accuracy certificates for computational problems with convex structure. Math. Oper. Res. **35**(1), 52–78 (2010). https://doi.org/10.1287/moor.1090.0427
20. Nesterov, Y.: Complexity bounds for primal-dual methods minimizing the model of objective function. Math. Program. **171**(1-2), 311–330 (2017). https://doi.org/10.1007/s10107-017-1188-6
21. Nguyen, H., Petrova, G.: Greedy strategies for convex optimization. Calcolo **54**(1), 207–224 (2016). https://doi.org/10.1007/s10092-016-0183-2
22. Sidorov, S.P., Mironov, S.V., Pleshakov, M.G.: Dual convergence estimates for a family of greedy algorithms in Banach spaces. In: Nicosia, G., Pardalos, P., Giuffrida, G., Umeton, R. (eds.) MOD 2017. LNCS, vol. 10710, pp. 109–120. Springer, Cham (2018). https://doi.org/10.1007/978-3-319-72926-8_10
23. Sidorov, S.P., Faizliev, A.R., Khomchenko, A.A.: Algorithms for l_1-norm minimisation of index tracking error and their performance. Int. J. Math. Oper. Res. **11**(4), 497–519 (2017). https://ideas.repec.org/a/ids/ijmore/v11y2017i4p497-519.html
24. Sidorov, S.P., Mironov, S.V.: Duality gap analysis of weak relaxed greedy algorithms. In: Battiti, R., Kvasov, D.E., Sergeyev, Y.D. (eds.) LION 2017. LNCS, vol. 10556, pp. 251–262. Springer, Cham (2017). https://doi.org/10.1007/978-3-319-69404-7_18
25. Temlyakov, V.N.: Greedy approximation in convex optimization. Constr. Approx. **41**(2), 269–296 (2015). https://doi.org/10.1007/s00365-014-9272-0
26. Temlyakov, V.N.: Dictionary descent in optimization. Anal. Mathematica **42**(1), 69–89 (2016). https://doi.org/10.1007/s10476-016-0106-0
27. Zhang, Z., Shwartz, S., Wagner, L., Miller, W.: A greedy algorithm for aligning DNA sequences. J. Comput. Biol. **7**(1–2), 203–214 (2000). https://doi.org/10.1089/10665270050081478

A Penalty Approach to Linear Programs with Many Two-Sided Constraints

Petro Stetsyuk[1], Andreas Fischer[2], and Oksana Pichugina[3]

[1] V.M. Glushkov Institute of Cybernetics, National Academy of Sciences of Ukraine, Kyiv, Ukraine
[2] Faculty of Mathematics, Technische Universität Dresden, Dresden, Germany
Andreas.Fischer@tu-dresden.de
[3] National Aerospace University – "Kharkiv Aviation Institute", Kharkiv, Ukraine
o.pichugina@khai.edu
http://incyb.kiev.ua/viddil-metodiv-negladkoi-optimizacii
https://tu-dresden.de/mn/mathe/numerik/fischer
https://khai.edu/

Abstract. The paper focuses on linear programming problems with two-sided constraints whose number is much larger than the number of variables. The solution approach is based on a non-smooth convex penalty function. An appropriate penalty parameter ensures the equivalence between the original problem and the problem of minimizing the penalty function. The latter problem is tackled by means of a modification of the r-algorithm, i.e., of an iterative subgradient method with an adaptive step adjustment and a constant coefficient of space dilation in the direction of the difference of two successive subgradients. Based on an implementation of this LPralg algorithm with GNU Octave, computational results on randomly generated instances with 20.000 to 1.500.000 two-sided constraints and up to 300 variables are presented. The results turn out to be very promising compared to well-known linear programming software, such as the GLPK package as well as CPLEX and Gurobi solvers. Among other problems, the new approach can be applied to robust linear programs with a finite uncertainty set.

Keywords: Linear programming · Non-smooth penalty method · r-algorithm · GNU Octave

1 Introduction

Linear programs (LPs) that possess much more constraints than variables play a significant role in several fields. Examples are robust linear optimization problems, Danzig-Wolfe and Benders decomposition schemes, minimax and maximin problems, approximation problems with Chebyshev's minimax criterion, and Boolean estimation problems. Here, we are interested in robust linear optimization (for more general cases, see [1,8], for example), i.e., we consider problems of the form

This work was supported by the Volkswagen Foundation under grant No. 97 775.

P. Pardalos et al. (Eds.): MOTOR 2021, LNCS 12755, pp. 206–217, 2021.
https://doi.org/10.1007/978-3-030-77876-7_14

$$\min_{x \in \mathbb{R}^n} \left\{ c^\top x : \ B\left(\xi\right) x - b\left(\xi\right) \leq 0 \ \forall \xi \in U \right\}. \tag{1}$$

The vector $c \in \mathbb{R}^n$ is given, the parameter vector ξ belongs to an uncertainty set U, and the matrix $B(\xi)$, as well as the right-hand-side $b(\xi)$ may depend on $\xi \in U$. Let us further assume that the set U is finite, but contains up to thousands or millions of given parameters. Then, the robust LP (1) is nothing else than a usual LP, but may have a relatively small number of variables compared to a very large number of constraints. Due to this, using standard software for such LPs is either impossible or impractical because it would require significant computing resources. Therefore, new approaches to the construction of algorithms for solving such LPs become of interest.

In the following, we suggest a solution approach for robust LPs with two-sided constraints. This approach is based on a non-smooth penalty function [5,6,10–13]. A reasonable choice of the penalty parameter ensures the equivalence between the original LP and the problem of minimizing the penalty function. Our main contribution is to tackle the latter problem by a modification of the r-algorithm [12–17], i.e., an iterative subgradient method with an adaptive step adjustment and a constant coefficient of space dilation towards the difference of two successive subgradients. This approach yields the algorithm LPralg and is implemented with GNU Octave [4]. Based on this, computational results on randomly generated LP-instances with up to 1.500.000 two-sided constraints and up to 300 variables are presented. The results are promising if compared to well-known linear programming software, such as the GLPK package [20], Gurobi and CPLEX solvers [18,19].

The paper is organized as follows. In Sect. 2, after stating the LP with two-sided constraints, an equivalent auxiliary problem of unconstrained minimization is constructed based on a non-smooth penalty function. Thereafter, in Sect. 3, the $r(\alpha)$-algorithm is described, which results in the LPralg algorithm and its Octave implementation for solving the auxiliary problem. Finally, results of test experiments on applying LPralg algorithm to problems with a large number of two-sided constraints are presented in Sect. 4.

2 Linear Programs with Two-Sided Constraints and Their Equivalence to a Non-smooth Optimization Problem

Let a matrix $A \in \mathbb{R}^{m \times n}$, vectors $l, u \in \mathbb{R}^m$ of lower and upper bounds, and $c \in \mathbb{R}^n$ be given. Then, we consider the following LP with two-sided constraints

$$\min_{x \in \mathbb{R}^n} \quad c^\top x \qquad \text{subject to} \qquad l \leq Ax \leq u. \tag{2}$$

Throughout, we assume that (2) has a solution x^*, where $c_* := c^\top x^*$ denotes the optimal value, and $X_* \subset \mathbb{R}^n$ is the set of all solutions (minimizers).

For a moment, let us consider an LP with only one-sided constraints, say $Ax \le u$, then such constraints can be reformulated by means of a sufficiently large $M > 0$, e.g., by $l := -Me \le Ax \le u$, where $e = (1, \ldots, 1)^\top \in \mathbb{R}^m$. An appropriate number M can be computed if lower and upper bounds for a solution $x^* \in X^*$ are known. In a simple case, let $\|x^*\|_\infty \le \mu$ be satisfied for some known value $\mu > 0$. Then, $\|Ax^*\|_\infty \le \|A\|_\infty \|x^*\|_\infty \le \|A\|_\infty \mu$ yields $M := \|A\|_\infty \mu$. Note that replacing $Ax \le u$ by $-Me \le Ax \le u$ may yield a smaller solution set that, however, at least contains x^*.

Our aim is to develop a subgradient method for the solution of problem (2). To this end, we first introduce the penalty function $c_P : \mathbb{R}^n \to \mathbb{R}$ by

$$c_P(x) := c^\top x + P \cdot \max \left\{ \|(Ax - u)_+\|_\infty, \|(l - Ax)_+\|_\infty \right\},$$

where $v_+ := (\max\{0, v_1,\}, \ldots, \max\{0, v_m\})^\top$ for any $v \in \mathbb{R}^m$ and $P > 0$ denotes a penalty parameter. Obviously, $c_P(x) = c^\top x$ holds if and only if x belongs to the feasible set of (2). The function c_P is convex and piecewise linear. Moreover let X_P denote the set of minimizers of c_P, i.e.,

$$X_P = \left\{ \hat{x} \in \mathbb{R}^n : \ c_P(\hat{x}) = \min_{x \in \mathbb{R}^n} c_P(x) \right\}.$$

Note that this set might be empty for smaller P, see Theorem 1 below.

Due to the above assumption that the LP (2) has at least one solution x^*, there are Lagrange multiplier vectors $\lambda^* \in \mathbb{R}^m$ (associated to the constraints $l - Ax \le 0$) and $\Lambda^* \in \mathbb{R}^m$ (associated to $Ax - u \le 0$) so that $(x^*, \lambda^*, \Lambda^*)$ satisfy the Karush-Kuhn-Tucker conditions

$$c - A^\top \lambda + A^\top \Lambda = 0,$$
$$(l - Ax)^\top \lambda = 0, \quad (Ax - u)^\top \Lambda = 0, \quad \lambda \ge 0, \quad \Lambda \ge 0, \quad l \le Ax \le u \qquad (3)$$

for problem (2). Now, the following theorem can be obtained by applying Theorem 27 in [13] to problem (2).

Theorem 1. *Let x^* be a solution of problem (2) and λ^*, Λ^* Lagrange multiplier vectors so that $(x^*, \lambda^*, \Lambda^*)$ satisfies the Karush-Kuhn-Tucker conditions (3).*

(a) If $P \ge P_* := \|\lambda^*\|_1 + \|\Lambda^*\|_1$, then $\min_{x \in \mathbb{R}^n} c_P(x) = c_*$ holds.
(b) If $P > P_*$, then the set X_P of minimizers of c_P coincides with X^*.

In general, a number $P > P_*$ is not known in advance. Conceptually, P can be updated as follows. If, for some $P > 0$, the vector x_P is a solution of $\min_{x \in \mathbb{R}^n} c_P(x)$ and if the condition $\max\{0, Ax - u, l - Ax\} \le 0$ is fulfilled, then this P is used. Otherwise, the value of P is multiplied by a predefined factor larger than 1 until the condition is met.

We now provide just one subgradient of c_P at some $x \in \mathbb{R}^n$. The function c_P is convex, but nondifferentiable at certain points so that the subdifferential $\partial c_P(x)$

may consist of infinitely many subgradients for those points. Given $x \in \mathbb{R}^n$, we use the following subgradient $g_{c_P}(x) \in \partial c_P(x)$:

$$g_{c_P}(x) = c + P \cdot \begin{cases} 0 \in \mathbb{R}^n, & \text{if } t_1 \leq 0 \text{ and } t_2 \leq 0, \\ (a_{i_1 1}, \ldots, a_{i_1 n})^\top, & \text{if } t_1 > 0 \text{ and } t_1 \geq t_2, \\ -(a_{i_2 1}, \ldots, a_{i_2 n})^\top, & \text{if } t_2 > 0 \text{ and } t_2 > t_1, \end{cases} \quad (4)$$

where $A = (a_{ij}) \in \mathbb{R}^{m \times n}$ and the numbers t_1, t_2 as well as the indices i_1, i_2 are defined by

$$t_1 := (Ax - u)_{i_1} \geq (Ax - u)_i \quad \forall i = 1, \ldots, m,$$
$$t_2 := (l - Ax)_{i_2} \geq (l - Ax)_i \quad \forall i = 1, \ldots, m.$$

In Table 1, we show the Octave function **fgLP**, which implements formula (4) (with g denoting $g_{c_P}(x)$ as well as the formula for computing the function value $c_P(x)$ denoted by f). Note that the parameters c, A, l, u, n, m, and P are defined as global variables.

Table 1. Octave code **fgLP** for the implementation of formula (4)

```
1 function [f,g] = fgLP(x)
2 global c A l u n m P
3 f = sum(c.*x); g = c;
4 tmp0 = A*x;
5 tmp = [l-tmp0; tmp0-u];
6 [tmpmax imax] = max(tmp);
7 if (tmpmax > 0.d0)
8     if (imax <= m)
9         f = f + P*tmpmax;
10        g = g - P*A(imax,1:n)';
11    endif
12    if (imax > m)
13        f = f + P*tmpmax;
14        g = g + P*A(imax-m,1:n)';
15    endif
16 endif
17 endfunction #fgLP
```

3   The $r(\alpha)$-Algorithm with Adaptive Steps

In this section, the general problem of iteratively approximating a minimizer of a convex function $f : \mathbb{R}^n \to \mathbb{R}$ is considered, see [15,16]. In Sect. 4, the $r(\alpha)$-algorithm we suggest below will be applied to the problem $\min_{x \in \mathbb{R}^n} c_P(x)$ described in Sect. 2. In addition to the mathematical formulation of the algorithm, we also provide the Octave source code as it will be used in Sect. 4.

Definition 1. *Let $f : \mathbb{R}^n \to \mathbb{R}$ be a convex function and $g_f(x)$ an element of the subdifferential $\partial f(x)$ for $x \in \mathbb{R}^n$. Moreover, let $x_0 \in \mathbb{R}^n$, $h_0 > 0$, $B_0 := I \in \mathbb{R}^{n \times n}$, and $\alpha > 1$ be given. Then, the $r(\alpha)$-algorithm is an iterative procedure that constructs sequences $\{x_k\} \subset \mathbb{R}^n$, $\{h_k\} \subset (0, \infty)$, $\{\xi_k\} \subset \mathbb{R}^n$, $\{\eta_k\} \subset \mathbb{R}^n$, and $\{B_k\} \subset \mathbb{R}^{n \times n}$ according to the following scheme:*

$$x_{k+1} := x_k - h_k B_k \xi_k, \quad B_{k+1} := B_k R_\beta(\eta_k), \quad k = 0, 1, 2, \ldots, \tag{5}$$

where

$$\xi_k := \frac{B_k^\top g_f(x_k)}{\|B_k^\top g_f(x_k)\|}, \quad h_k \geq h_k^* := \underset{h \geq 0}{\arg\min}\, f(x_k - h B_k \xi_k), \tag{6}$$

$$\eta_k := \frac{B_k^\top r_k}{\|B_k^\top r_k\|}, \quad r_k := g_f(x_{k+1}) - g_f(x_k), \tag{7}$$

$$R_\beta(\eta_k) := I + (\beta - 1)\eta_k \eta_k^\top, \quad \beta := \tfrac{1}{\alpha},$$

provided that no zero denominator appears.

Obviously, x_0 is the starting point of the $r(\alpha)$-algorithm. For scaling purposes, the matrix $B_0 \in \mathbb{R}^{n \times n}$ can be any diagonal matrix with only positive coefficients on the diagonal. Moreover, h_k plays the role of a step size with the aim of minimizing f in direction $-B_k \xi_k$. The operator $R_\beta(\eta) = I + (\beta - 1)\eta\eta^\top$ is used for updating B_k and yields a compression of the transformed subgradient space (spanned by $B_k^\top(g_f(x_{k+1}) - g_f(x_k))$) in the normalized direction η with the coefficient $\beta = 1/\alpha < 1$, see also Remark 1 below. Moreover, the definition of $R_\beta(\eta)$ guarantees that all matrices B_k generated by the $r(\alpha)$-algorithm are nonsingular. To stop the $r(\alpha)$-algorithm after x_k is obtained, the termination criteria

- $\|x_k - x_{k-1}\|_2 \leq \varepsilon_x$,
- $\|g_f(x_k)\|_2 \leq \varepsilon_g$,
- $k \geq maxitn$

are used. If one of them is satisfied, we set $k_* := k$. The parameters $\varepsilon_x > 0$, $\varepsilon_g > 0$, and a large natural number $maxitn$ are predefined values. If the last criterion is satisfied, the function f might be unbounded from below.

For further details, in particular on the choice of the step lengths h_k, we refer to [13, Section 3.4]. Details related to the implementation can be found in Sect. 4.

Remark 1. For any fixed k, let us define the function $\varphi : \mathbb{R}^n \to \mathbb{R}$ by $\varphi(y) := f(B_k y)$. One can easily check that φ inherits the convexity of f and that a subgradient $g_f(x_k) \in \partial f(x_k)$ provides a subgradient $g_\varphi(y_k) := B_k^\top g_f(x_k)$ in

$\partial\varphi(y_k)$. Since the matrices B_k are nonsingular, the mapping $x \mapsto y := B_k^{-1}x$ is bijective and we can transform the iteration (5) to

$$y_{k+1} = B_k^{-1}x_{k+1} = B_k^{-1}x_k - h_k\xi_k = y_k - h_k\frac{B_k^\top g_f(x_k)}{\|B_k^\top g_f(x_k)\|} = y_k - h_k\frac{g_\varphi(y_k)}{\|g_\varphi(y_k)\|} \quad (8)$$

and back.

The $r(\alpha)$-algorithm as described above and with the step length adjustment in [13, Section 3.4] has been implemented in Octave as **ralgb5** [15]. In the implementation **ralgb5a** [17], the step length adjustment has been simplified. The letter "**b**" refers to the use of the matrices B_k instead of employing $H_k := B_k B_k^\top$, cf. [13, Section 3.4]. Moreover, "**5**" means that one step by **ralgb5** (without the expenses for computing function values and subgradients) requires $5n^2$ multiplications, where other arithmetic operations are neglected. The code of the Octave function **ralgb5a** is given in Table 2. This function calls the function **calcfg** that, for $x \in \mathbb{R}^n$, provides $f(x)$ as well as a subgradient $g_f(x) \in \partial f(x)$. Since we are interested in minimizing the function c_P and in solving the LP (2), the function **calcfg** appearing as an input parameter of **ralgb5a** is substituted by the function **fgLP** in Sect. 4. The Octave code of the latter function is shown in Table 1. The input and output parameters of **ralgb5** are

- **x** (starting point x_0),
- **alpha** (parameter α for space dilation, see above),
  h0 and **q1** (parameters for computing the step lengths h_k),
- **epsx**, **epsg**, **maxitn** (termination parameters ε_x, ε_g, $maxitn$, see above),
- **intp** (after each **intp** steps, some intermediate progress report is printed),
- **xr**, **fr** (record point x_r with best found function value f_r),
- **itn** (number of iterations),
- **nfg** (number of calls of the function **calcfg** or **fgLP**, resp.),
  ist (shows the reason for termination).

Finally, we note that putting the Octave functions **ralgb5a** and **fgLP** (instead of the function **calcfg**) together yields our implementation of algorithm LPralg. This implementation is used for experiments in Sect. 4. For an empirical study on the computational expense of the $r(\alpha)$-algorithm, see [14].

4   Computational Experiments

In this section, results of two computational experiments on test LP-problems of type (2) with a quite large number of constraints and a small number of variables are presented. The first experiment is done, on the one hand, with the Octave implementation of LPralg as described above (GNU Octave version 5.1.0). On the other hand, we apply the GLP_PRIMAL solver from the GNU Linear Programming Kit (GLPK) to directly solve these test problems. Both algorithms were run on the same machine, an Intel Core i5-9400f processor with 2.9 GHz and 16 GB RAM. Moreover, for the second experiment, we run test problems on the NEOS server [2,3,9] with solvers from CPLEX [19] and Gurobi [18].

Table 2. Octave code **ralgb5a** for the algorithm LPralg

```
1  function [xr,fr,itn,nfg,ist] = ralgb5a(calcfg,x,alpha,
2                                   h0,q1,epsg,epsx,maxitn,intp);
3  itn = 0; B = eye(length(x)); hs = h0; lsa = 0; lsm = 0;
4  xr = x; [fr,g0] = calcfg(xr); nfg = 1;
5  if (intp>0)
6    printf("itn %4d f%16.6e fr%16.6e nfg %4d\n",itn,fr,fr,nfg);
7  endif
8  if(norm(g0) < epsg) ist = 2; return; endif
9  for (itn = 1:maxitn)
10    dx = B * (g1 = B' * g0)/norm(g1);
11    d = 1; ls = 0; ddx = 0;
12    while (d > 0)
13      x -= hs * dx; ddx += hs * norm(dx);
14      [f, g1] = calcfg(x); nfg ++;
15      if (f < fr) fr = f; xr = x; endif
16      if(norm(g1) < epsg) ist = 2; return; endif
17      ls ++; (mod(ls,3) == 0) && (hs *= 1.1);
18      if(ls > 500) ist = 5; return; endif
19      d = dx' * g1;
20    endwhile
21    (ls == 1) && (hs *= q1); lsa=lsa+ls; lsm=max(lsm,ls);
22    if(mod(itn,intp)==0)
23      if (intp>0)
24        printf("itn %4d f %14.6e fr %14.6e", itn, f, fr);
25        printf(" nfg %4d lsa %3d lsm %3d\n", nfg, lsa, lsm);
26      endif
27      lsa=0; lsm=0;
28    endif
29    if(ddx < epsx) ist = 3; return; endif
30    xi = (dg = B' * (g1 - g0) )/norm(dg);
31    B += (1 / alpha - 1) * B * xi * xi';
32    g0 = g1;
33  endfor
34  ist = 4;
35  endfunction #ralgb5a
```

For both experiments, the entries of the vector c and the matrix $A = (a_{ij})$ in the test problems were generated randomly by means of the standard uniform distribution $U(0, 1)$, namely

$$c_j \sim U(0, 1), \quad a_{ij} \sim U(0, 1) \qquad \text{for } i, \dots, m \quad \text{and} \quad j = 1, \dots, n.$$

In addition, the entries of the lower and upper bound vectors l and u were defined by

$$l_i := 0.9 \sum_{j=1}^{n} a_{ij}, \quad u_i := 1.1 \sum_{j=1}^{n} a_{ij} \qquad \text{for } i, \dots, m.$$

Individual details for the two experiments will be given below.

For the first experiment, the following parameters were used:

$$\mathtt{x} := 0 \in \mathbb{R}^n, \quad \mathtt{alpha} := 4, \quad \mathtt{q1} := 1.0, \quad \mathtt{h0} := 1. \tag{9}$$

The termination parameters are given by

$$\mathtt{epsx} := 10^{-10}, \quad \mathtt{epsg} := 10^{-8}, \quad \mathtt{maxitn} := 10.000 \quad \mathtt{intp} := 1.000.$$

The meaning of these input parameters for **ralgb5a** can be found in Sect. 3. The penalty parameter P (to be used in the function **fgLP**) was always set to 10. All generated test problems are characterized by a small number $n \in \{100, 200, 300\}$ of variables and a much larger number m of two-sided constraints with $m \in \{20.000, 50.000, 100.000, 200.000\}$. The Octave implementation of **ralgb5a** is shown in Table 2, whereas Table 3 presents the code for setting up and running the first experiment.

The run times for approximately solving the test problems are denoted by t_{GLPK} for the GPLK solver and by t_{LPralg} for the LPralg algorithm. In addition to these run times, Table 4 also shows the Euclidean distance Δ between the approximate solutions obtained by the two methods. The computational results in all tables are rounded. Table 4 demonstrates that the GLPK solver needs 4–12 times longer for test problems than LPralg. The last column of this table shows that the deviation between the two approximate solutions obtained after GLPK and LPralg terminated is very small, which means that both programs yield very similar approximate solutions of the test problems.

In the second experiment, we intend to compare run times spent by LPralg and commercial solvers CPLEX and Gurobi. Currently, we were just able to run the latter two solvers on the NEOS server. To compare the performance of CPLEX and Gurobi on NEOS and on our machine described above, we performed computational tests, which showed that the NEOS run times are close to those produced on our machine. Therefore, we first present the results of applying LPralg to 3 test examples of different size on this machine (Table 5). Then, Table 6 shows the times for the NEOS server for the same test problems.

For LPralg, the same parameters as in (9) are used. The termination parameters are now $\mathtt{epsx} := 10^{-9}$, $\mathtt{epsg} := 10^{-8}$, $\mathtt{maxitn} := 20.000$, $\mathtt{intp} := 1.000$.

Table 3. Octave code for the first experiment

```
1  # Code for First Experiment: LPralg and GLPK
2  global c A l u n m P
3  printf("\n"); # set parameters for r-algorithm
4  alpha = 4.0, h0 = 1.0, q1 = 1.0,
5  epsx = 1.e-10, epsg = 1.e-8, maxitn = 20000, intp = 1000,
6  P = 10, nmtest = [100 20000; 100 50000; 100 100000; 100 200000],
7  #P = 10, nmtest = [200 20000; 200 50000; 200 100000; 200 200000],
8  #P = 10, nmtest = [300 20000; 300 50000; 300 100000; 300 200000],
9  rand("seed", 2020); time12 = fopt = delta = [];
10 for itest = 1:rows(nmtest)
11     printf("\n");
12     n = nmtest(itest,1), m = nmtest(itest,2),
13     c = rand(n,1); A = rand(m,n); x0 = ones(n,1);
14     rhs = A*x0; l = 0.9*rhs; u = 1.1*rhs;
15     # run glpk
16     a=[A; -A]; b=[u; -l];
17     lb = -10*ones(n,1); ub = 10*ones(n,1);
18     ctype = repmat('U',1,2*m); vartype = repmat('C',1,n);
19     sense=1, tstart=time();
20     [xmin, fmin, status] = glpk(c, a, b, lb, ub, ctype, vartype, sense);
21     time1 = time()-tstart;
22     fmin, status, time1,
23     # set start point and run r-algorithm
24     x0 = zeros(n,1); tstart=time();
25     [xr,fr,itn,nfg,ist] = ralgb5a(@fgLP,x0,alpha,h0,q1,
26                                             epsg,epsx,maxitn,intp);
27     time2=time()-tstart,
28     printf("..itn %4d fr %23.15e ist %d nfg %4d\n",itn,fr,ist,nfg);
29     fr, ctxr = c'*xr, time2,
30     time12 = [time12; time1 time2]; dnorm1 = norm(xmin-xr),
31     fopt = [ fopt; fmin c'*xr fr]; delta = [delta dnorm1];
32 endfor
33 nmtest, fopt,
34 printf("\n n n m .t1.. .t2.. t1/t2 delta ");
35 for itest = 1:rows(nmtest)
36     n = nmtest(itest,1); m = nmtest(itest,2);
37     t1 = time12(itest,1); t2 = time12(itest,2);
38     dn = dnorm(1, itest);
39     printf("\n %3d %6d %7.2f %7.2f %7.2f %9.2e",n,m,t1,t2,t1/t2,dn),
40 endfor
41 printf("\n");
```

Table 4. First experiment: results for GLPK and LPralg (run time in seconds)

n	m	t_{GLPK}	t_{LPralg}	$t_{\text{GLPK}}/t_{\text{LPralg}}$	Δ
100	20.000	23.7	3.4	7.0	2.7e−08
	50.000	81.0	6.5	12.5	2.0e−08
	100.000	107	17.6	6.1	8.5e−09
	200.000	271	35.5	7.6	2.8e−08
200	20.000	58.6	13.2	4.4	4.4e−08
	50.000	155	30.5	5.1	2.3e−08
	100.000	325	61.6	5.3	4.2e−08
	200.000	708	139	5.1	5.1e−08
300	20.000	125	27.6	4.5	3.3e−07
	50.000	323	65.5	4.9	4.1e−08
	100.000	672	172	3.9	1.4e−07
	200.000	2 025	388	5.2	4.4e−07

Table 5. Second experiment: run times for LPralg (in seconds)

n	m	q1 = 0.95	q1 = 1.0
100	300.000	164	58.3
50	600.000	59.7	46.4
20	1.500.000	33.5	24.3

Table 5 shows the run times of LPralg not only for q1 := 1.0 as in (9) but also for q1 := 0.95. In addition to the results of the computations shown in Table 5, we remark that the corresponding approximate solutions x_r satisfy $\max \{\|(Ax_r - u)_+\|_\infty, \|(l - Ax_r)_+\|_\infty\} < 10^{-5}$.

To finally apply CPLEX and Gurobi to the same test problems as used in Table 5, Table 6 shows the results obtained with the NEOS server. To reduce time differences, which result from concrete situations at NEOS (like the number of parallel jobs, machine used), we have performed 4 runs for the same test problem for each of the two solvers. Table 6 presents the average run time. Note that the code, we provided to the NEOS server, is in AMPL [7] and uses two constraints for each two-sided constraint. If we compare the results in Table 6 with those in Table 5, we see that LPralg is comparable and often seems to be faster by a factor of 2–4 than CPLEX or Gurobi.

Table 6. Second experiment: averaged run times (in seconds) on the NEOS server

CPLEX		
n	m	Averaged run time
100	300.000	202
50	600.000	151
20	1.500.000	142
Gurobi		
n	m	Averaged run time
100	300.000	108
50	600.000	148
20	1.500.000	84

5   Conclusion

The paper presents the algorithm LPralg (Linear Programming by r-algorithm) and its software implementation in the Octave programming language. LPralg is particularly intended to solve linear programs with many two-sided constraints. The algorithm applies a penalty approach to reformulate a linear program equivalently as an unconstrained problem of minimizing a non-smooth convex penalty function. The later is tackled by a modified version of Shor's r-algorithm with an adaptive step length adjustment. The results for cases with much more two-sided constraints than variables and comparisons with linear programming codes are limited but quite promising. Moreover, there is plenty of things that remain to investigate. For example, how LPralg can deal with more general linear programs, with programs less restricted in the choice of data, and how sensitive LPralg is with respect to parameters like q1 and P.

References

1. Ben-Tal, A., Ghaoui, L.E., Nemirovski, A.: Robust Optimization. Princeton University Press, Princeton (2009). https://doi.org/10.1515/9781400831050
2. Czyzyk, J., Mesnier, M.P., Moré, J.J.: The NEOS server. IEEE J. Comput. Sci. Eng. **5**, 68–75 (1998). https://doi.org/10.1109/99.714603
3. Dolan, E.: The NEOS Server 4.0 Administrative Guide. Argonne National Laboratory, Mathematics and Computer Science Division, Technical Memorandum ANL/MCS-TM-250 (2001)
4. Eaton, J.W., Bateman, D., Hauberg, S.: GNU Octave Manual Version 3. Network Theory Ltd. (2008)
5. Eremin, I.I.: Method of the "penalties" in convex programming. Dokl. Akad. Nauk SSSR **173**(4), 748–751 (1967). (in Russian)
6. Eremin, I.I.: About the penalty method in mathematical programming. Dokl. Akad. Nauk **346**(4), 459–461 (1996). (in Russian)

7. Fourer, R., Gay, D.M., Kernighan, B.W.: AMPL: A Modeling Language for Mathematical Programming. Duxbury Press, Pacific Grove (2002)
8. Goryashko, A., Nemirovski, A.: Robust energy cost optimization of water distribution system with uncertain demand. Autom. Remote Control. **75**(10), 1754–1769 (2014). https://doi.org/10.1134/S000511791410004X
9. Gropp, W., Moré, J.J.: Optimization environments and the NEOS server. In: Buhmann, M.D., Iserles, A. (eds.) Approximation Theory and Optimization, pp. 167–182. Cambridge University Press, Cambridge (1997)
10. Polyakova, L.N.: Nonsmooth penalty functions. IFAC Proc. Vol. **33**(16), 287–291 (2000). https://doi.org/10.1016/S1474-6670(17)39644-1
11. Polyakova, L., Karelin, V.: Exact penalty methods for nonsmooth optimization. In: 20th International Workshop on Beam Dynamics and Optimization (BDO), pp. 1–2 (2014). https://doi.org/10.1109/BDO.2014.6890067
12. Shor, N.Z.: Minimization Methods for Non-Differentiable Functions. Springer, Berlin (1985). https://doi.org/10.1007/978-3-642-82118-9
13. Shor, N.Z.: Nondifferentiable Optimization and Polynomial Problems. Kluwer Academic Publishers, Dordrecht (1998). https://doi.org/10.1007/978-1-4757-6015-6
14. Shor, N.Z., Zhurbenko, N.G., Likhovid, A.P., Stetsyuk, P.I.: Algorithms of nondifferentiable optimization: development and application. Cybern. Syst. Anal. **39**(4), 537–548 (2003). https://doi.org/10.1023/B:CASA.0000003503.25710.84
15. Stetsyuk, P.I.: Theory and software implementations of Shor's r-algorithms. Cybern. Syst. Anal. **53**(5), 692–703 (2017). https://doi.org/10.1007/s10559-017-9971-1
16. Stetsyuk, P.I.: Shor's r-algorithms: theory and practice. In: Butenko, S., Pardalos, P.M., Shylo, V. (eds.) Optimization Methods and Applications. SOIA, vol. 130, pp. 495–520. Springer, Cham (2017). https://doi.org/10.1007/978-3-319-68640-0_24
17. Stetsyuk, P., Fischer, A.: Shor's r-algorithms and octave-function ralgb5a. In: International Conference on "Modern Informatics: Problems. Achievements and Prospects for Development", pp. 143–146. V.M. Glushkov Institute of Cybernetics of the NAS of Ukraine, Kyiv (2017). (in Russian)
18. Gurobi Optimization Inc.: Gurobi Optimizer Reference Manual. http://www.gurobi.com/documentation/
19. IBM: CPLEX User's Manual, Version 12 Release 8. https://www.ibm.com/support/knowledgecenter/SSSA5P_12.8.0/ilog.odms.studio.help/pdf/usrcplex.pdf
20. Free Software Foundation Inc.: GNU Linear Programming Kit (GLPK). http://www.gnu.org/software/glpk/glpk.html

Bilevel Optimization

Sample Approximations of Bilevel Stochastic Programming Problems with Probabilistic and Quantile Criteria

Sergey V. Ivanov$^{(\boxtimes)}$ (ID) and Aleksei N. Ignatov

Moscow Aviation Institute (National Research University),
Volokolamskoe Shosse, 4, Moscow 125993, Russia

Abstract. In this paper, bilevel stochastic programming problems with probabilistic and quantile criteria are considered. The lower level problem is assumed to be linear for fixed leader's (upper level) variables and fixed realizations of the random parameters. The objective function and the constraints of the lower level problem depend on the leader's strategy and random parameters. The objective function of the upper level problem is defined as the value of the probabilistic or quantile functional of the random losses on the upper level. We suggest conditions guaranteeing that the objective function of the upper level is a normal integrand. It is shown that these conditions are satisfied for a class of problems with positive coefficients of the lower level problem. This allows us to suggest sufficient conditions of the existence of a solution to the considered problem. We construct sample approximations of these problems. These approximations reduce to mixed integer nonlinear programming problems. We describe sufficient conditions of the convergence of the sample approximations to the original problems.

Keywords: Bilevel programming · Sample approximation · Stochastic programming · Value-at-Risk · Probabilistic criterion · Quantile criterion

1 Introduction

Bilevel programming problems describe hierarchical interaction between two subjects. The subject making decision first is called a leader. The second subject is called a follower. Their decisions are solutions to upper and lower level problems respectively. The parameters of the lower level problem depend on the leader's strategy. The leader takes into account the optimal follower's solution when the upper level strategy is selected. The theory of bilevel problems is described in monographs [1–3] and in the review [4].

In this paper, we study stochastic bilevel programming problems with probabilistic and quantile criteria. These criteria are used in stochastic models for taking into account reliability requirements [5]. The probabilistic criterion is

© Springer Nature Switzerland AG 2021
P. Pardalos et al. (Eds.): MOTOR 2021, LNCS 12755, pp. 221–234, 2021.
https://doi.org/10.1007/978-3-030-77876-7_15

defined as the probability of successful work of the system modeled. The quantile criterion (also known as Value-at-Risk) is the minimal losses that cannot be exceeded with a given probability. There are a few works on using the probabilistic and quantile criteria in the stochastic bilevel optimization. The probabilistic criterion for a linear problem with fuzzy random data is studied in [6], where an algorithm to solve the problem is suggested. The quantile criterion for a stochastic bilevel problem with quantile criterion and discrete distribution of the random parameters is studied in [7]. A method to reduce this problem to a mixed integer programming problem is suggested in [7]. Another approach in stochastic programming to deal with reliability requirements is using coherent risk measures [8]. Properties of stochastic bilevel problems with criteria involving coherent risk measures and optimality conditions in these problems are studied in [9,10].

When the exact distribution of the random parameters is unknown, the objective function and the constraints of a stochastic problem can be estimated by using a sample. Thus, original problems are replaced by their approximations. The properties of the obtained sample approximations are studied in [11] for the expectation criterion and in [8,12] for problems with probabilistic constraints. In [13], the convergence of this method is studied for problems with probabilistic and quantile criteria.

In this paper, we study sample approximations of bilevel stochastic programming problems with probabilistic and quantile criteria. We reduce the sample approximations to mixed integer programming problems and give sufficient conditions of their convergence. Also, we describe conditions guaranteeing that the loss function of the problem is a normal integrand. These conditions are required to formulate results on the existence of an optimal solution and on the convergence of the sample approximations.

2   Statement

Let X be a random vector defined on a probability space $(\mathcal{X}, \mathcal{F}, \mathbf{P})$, where $\mathcal{X}$ is a closed subset of $\mathbb{R}^m$. The σ-algebra $\mathcal{F}$ is assumed to be complete, i.e., $S' \in \mathcal{F}$ if there exists a set $S \in \mathcal{F}$ such that $S' \subset S$ and $\mathbf{P}(S) = 0$. For simplicity, we assume that $X(x) = x$ for all $x \in \mathcal{X}$. This means that the sample space $\mathcal{X}$ is considered as the space of realizations of the random vector X.

Let $U \subset \mathbb{R}^r$ be a set feasible values of leader's variables. The follower's problem is defined by the linear programming problem

$$\mathcal{Y}^*(u, x) := \operatorname*{Arg\,min}_{y \in \mathbb{R}^s} \left\{ c(u, x)^\top y \mid y \in \mathcal{Y}(u, x) \right\}, \tag{1}$$

$$\mathcal{Y}(u, x) := \left\{ y \in \mathbb{R}^s \mid A(u, x)y \geq b(u, x), \ y \geq 0 \right\}. \tag{2}$$

where $u \in U$ is the leader's variable, y is the follower's variable, $A \colon U \times \mathcal{X} \to \mathbb{R}^{k \times s}$, $b \colon U \times \mathcal{X} \to \mathbb{R}^k$, $c \colon U \times \mathcal{X} \to \mathbb{R}^s$ are a matrix and vectors depending on u and x. Thus, the leader's variable u and the realization x of the random vector X define the constraints and the objective function of the follower's problem.

The set-valued mappings $\mathcal{Y}^*, \mathcal{Y} \colon U \times \mathcal{X} \to 2^{\mathbb{R}^s}$ are introduced by (1) and (2). We note that $\mathcal{Y}^*(u, x) = \emptyset$ if problem (1) is infeasible or unbounded.

Let $\Psi \colon U \times Y \times \mathcal{X} \to \mathbb{R}^*$ be a leader's loss function, where Y is the closure of the set

$$\bigcup_{u \in U, \, x \in \mathcal{X}} \mathcal{Y}(u, x),$$

$\mathbb{R}^* = \mathbb{R} \cup \{-\infty, +\infty\}$ is the extended real line.

To describe the losses of the leader when the follower chooses an optimal variable $y \in \mathcal{Y}^*(u, x)$, we introduce the function $\Phi \colon U \times \mathcal{X} \to \mathbb{R}^*$ by the rule

$$\Phi(u, x) := \inf_{y \in \mathcal{Y}^*(u,x)} \Psi(u, y, x). \tag{3}$$

The infimum in (3) means that the follower chooses the best decision for the leader among the optimal decisions $y \in \mathcal{Y}^*(u, x)$. Thus, the optimistic statement of the bilevel problem formulated below will be studied. We call Φ the optimistic leader's loss function.

Let us consider the probability function

$$P_\varphi(u) := \mathbf{P}\{\Phi(u, X) \leq \varphi\}, \tag{4}$$

where $\varphi \in \mathbb{R}^*$ is a fixed level of the optimistic leader's loss function Φ. The value $P_\varphi(u)$ in (4) is well defined when the function $x \mapsto \Phi(u, x)$ is measurable. Sufficient conditions for this will be suggested below.

The quantile function is defined by the equality

$$\varphi_\alpha(u) := \min\{\varphi \in \mathbb{R}^* \mid P_\varphi(u) \geq \alpha\}, \tag{5}$$

where $\alpha \in (0, 1]$ is a fixed probability level.

In this paper, we study the probability maximization problem

$$U^* := \operatorname{Arg\,max}_{u \in U} P_\varphi(u), \quad \alpha^* := \sup_{u \in U} P_\varphi(u), \tag{6}$$

and the quantile minimization problem

$$V^* := \operatorname{Arg\,min}_{u \in U} \varphi_\alpha(u), \quad \varphi^* := \inf_{u \in U} \varphi_\alpha(u). \tag{7}$$

Problems (6) and (7) are optimistic bilevel stochastic programming problems with probabilistic and quantile criteria respectively.

3 Existence of Optimal Solution

It is known [5, 14] that problems (6) and (7) are well defined and have optimal solutions if the function $(u, x) \mapsto \Phi(u, x)$ is a normal integrand. When the σ-algebra $\mathcal{F}$ is complete, the normal integrand can be defined as a lower semicontinuous in $u \in U$ and $\mathcal{B}(U) \times \mathcal{F}$-measurable function, where $\mathcal{B}(U)$ is the Borel

σ-algebra of subsets U. In this section, we suggest conditions under which the function $(u, x) \mapsto \Phi(u, x)$ is a normal integrand.

Let us consider the follower's problem (1). According to duality theory, the follower's variable $y \in \mathbb{R}^s$ is optimal in (1) if $y \in \mathcal{Y}(u, x)$ and there exists a vector $\lambda \in \mathbb{R}^k$ such that

$$A(u, x)^\top \lambda \leq c(u, x), \ \lambda \geq 0, \tag{8}$$

$$(A(u, x)y - b(u, x))^\top \lambda = 0, \tag{9}$$

$$\left(A(u, x)^\top \lambda - c(u, x)\right)^\top y = 0. \tag{10}$$

Denote by $\Lambda(u, y, x)$ the set of $\lambda \in \mathbb{R}^k$ satisfying (8)–(10). Let us introduce the function

$$\delta_\Lambda(u, y, \lambda, x) = \begin{cases} 0 & \text{if } y \in \mathcal{Y}(u, x) \text{ and } \lambda \in \Lambda(u, y, x), \\ +\infty & \text{otherwise.} \end{cases}$$

Then the optimistic loss function Φ can be represented in the form

$$\Phi(u, x) = \min_{y \in Y} \ \min_{\lambda \in \Lambda(u, y, x)} \ \{\Psi(u, y, x) + \delta_\Lambda(u, y, \lambda, x)\} . \tag{11}$$

We use the convention $-\infty + \infty = +\infty$ in (11) and below.

Let us denote by Y^* the closure of the set $\bigcup_{u \in U x \in \mathcal{X}} \mathcal{Y}^*(u, x)$.

Theorem 1. *Let the following conditions hold:*

(i)   *the function $(u, y, x) \mapsto \Psi(u, y, x)$ is lower semicontinuous in $(u, y) \in U \times Y$ and $\mathcal{B}(U) \times \mathcal{B}(Y) \times \mathcal{F}$-measurable;*

(ii) *the functions $(u, x) \mapsto A(u, x)$, $(u, x) \mapsto b(u, x)$, $(u, x) \mapsto c(u, x)$ are continuous in $u \in U$ and measurable in x;*

(iii) *the set Y^* is bounded;*

(iv) *there exists a compact set Λ^* such that $\Lambda^* \cap \Lambda(u, y, x) \neq \emptyset$ if and only if $\Lambda(u, y, x) \neq \emptyset$.*

Then the function $(u, x) \mapsto \Phi(u, x)$ is a normal integrand.

Proof. Taking into account (iii), (iv) and (11), the optimistic loss function can be rewritten in the form

$$\Phi(u, x) = \min_{y \in Y^*, \lambda \in \Lambda^*} \ \{\Psi(u, y, x) + \delta_\Lambda(u, y, \lambda, x)\} . \tag{12}$$

From (i) it follows that the function $((u, y), x) \mapsto \Psi(u, y, x)$ is a normal integrand. From (ii) it follows that functions $(u, x) \mapsto A(u, x)$, $(u, x) \mapsto b(u, x)$, $(u, x) \mapsto c(u, x)$ are normal integrands [15, Example 14.29]. Therefore, the set

$$\{(u, y, \lambda, x) \in U \times Y^* \times \Lambda^* \times \mathcal{X} \mid \lambda \in \Lambda(u, y, x)\}$$

is $\mathcal{B}(U) \times \mathcal{B}(Y^*) \times \mathcal{B}(\Lambda^*) \times \mathcal{F}$-measurable and its sections $\{(u, y, \lambda) \mid \lambda \in \Lambda(u, y, x)\}$ are closed for all $x \in \mathcal{X}$. Hence, the function

$$((u, y, \lambda), x) \mapsto \Psi(u, y, x) + \delta_\Lambda(u, y, \lambda, x)$$

defined on the set $U \times Y^* \times \Lambda^* \times \mathcal{X}$ is a normal integrand. Since the set $Y^* \times \Lambda^*$ is compact, the minimum in (12) can be written and the function $(u, x) \mapsto \Phi(u, x)$ is a normal integrand [15, Proposition 14.47]. Theorem 1 is proved.

Let us consider the case when the values of functions b and c are positive.

Corollary 1. *Let conditions (i), (ii) of Theorem 1 hold. Suppose that*

(v) there exist $\underline{c}, \underline{b} \in \mathbb{R}$ such that

$$\inf_{u \in U, x \in \mathcal{X}} \min_{i=\overline{1,s}} c_i(u, x) > \underline{c} > 0,$$

$$\inf_{u \in U, x \in \mathcal{X}} \min_{j=\overline{1,k}} b_j(u, x) > \underline{b} > 0;$$

(vi) there exists $\bar{c} \in \mathbb{R}$ such that

$$\sup_{u \in U, x \in \mathcal{X}} \min_{y \in \mathcal{Y}(u,x)} c(u, x)^\top y < \bar{c};$$

Then the function $(u, x) \mapsto \Phi(u, x)$ is a normal integrand.

Proof. Let us notice that from (vi) it follows that $\mathcal{Y}(u, x) \neq \emptyset$ for all $u \subset U$, $x \in \mathcal{X}$. From (v) and (vi) we get

$$\bigcup_{u \in U x \in \mathcal{X}} \mathcal{Y}^*(u, x) \subset \left\{ y \in \mathbb{R}^s \mid \sup_{u \in U, x \in \mathcal{X}} c(u, x)^\top y \leq \bar{c}, \, y \geq 0 \right\}$$

$$\subset \left\{ y \in \mathbb{R}^s \mid \max_{i=\overline{1,s}} y_i \leq \frac{\bar{c}}{\underline{c}}, \, y \geq 0 \right\}.$$

Thus, condition (iii) of Theorem 1 is satisfied.

From duality theory it is known that for $y \in \mathcal{Y}^*(u, x)$ there exists a vector $\lambda \in \Lambda(u, y, x)$ such that $c(u, x)^\top y = b(u, x)^\top \lambda$. Hence, the set Λ^* satisfying condition (iv) of Theorem 1 can be taken in the form

$$\Lambda^* = \left\{ \lambda \in \mathbb{R}^k \mid \max_{j=\overline{1,k}} \lambda_j \leq \frac{\bar{c}}{\underline{b}}, \, \lambda \geq 0 \right\}.$$

By Theorem 1, the function $(u, x) \mapsto \Phi(u, x)$ is a normal integrand. Corollary 1 is proved.

Example 1. Let us consider the follower's problem

$$\mathcal{Y}^*(u, x) = \operatorname*{Arg\,min}_{y \in \mathbb{R}} \{\max\{0, u\} y \mid y \geqslant 0\},$$

where $u \in U = [-1, 1]$. It easily seen that $\mathcal{Y}^*(u, x) = \{0\}$ if $u > 0$ and $\mathcal{Y}^*(u, x) = [0, +\infty)$ if $u \leq 0$. Then for the normal integrand $\Psi(u, y, x) = e^{uy}$ the infimum-function $\Phi(u, x) = \inf_{y \in \mathcal{Y}^*(u,x)} \Psi(u, y, x)$ is not lower semicontinuous, because $\Phi(u, x) = 1$ if $u \geq 0$ and $\Phi(u, x) = 0$ if $u < 0$. This example shows that condition (v) of Corollary 1 cannot be replaced by the conditions

$$\inf_{u \in U, x \in \mathcal{X}} \min_{i=\overline{1,s}} c_i(u, x) \geq 0, \qquad \inf_{u \in U, x \in \mathcal{X}} \min_{j=\overline{1,k}} b_j(u, x) \geq 0.$$

Example 2. Let us consider a production planning model. In this model, the leader and the follower are the head office and the production division of a company. The leader can get a contract for the production of several types of products. The prices of these products manufactured according to the contract are deterministic. The leader gives the follower a task to produce the products. The leader supplies resources to the follower. The follower pays for using resources. The prices for using resources are known when the follower produces products, but the prices are considered to be random when the task for the follower is stated. The leader's variable $u \in \mathbb{R}^r$ ($u \geq 0$) consists of production volumes required by the contract. The follower's variable $y \in \mathbb{R}^s$ ($y \geq 0$) consists of volumes of required resources. Let $c(u, X) = \tilde{c}(X)$ be a random vector of prices for using resources such that $\bar{c} > c_i(u, x) = \tilde{c}_i(x) > \underline{c} > 0$ for all $x \in \mathcal{X}$. The matrix $A(u, x)$ is constant such that $A(u, x)y$ is the vector of manufactured products. Let $b(u, x) = u$. Thus, the constraint $A(u, x)y \geq b(u, x)$ means that the follower must produce the products according to the leader's decision u. The leader's loss function has the form $\Psi(u, y, x) = -\pi^\top u - \tilde{\pi}(x)^\top (A(u, x)y - b(u, x)) + f(y)$, where $f(y)$ is the cost of buying resources y, π is the vector of prices for manufactured products according to the contract, $\tilde{\pi}(x)$ is a random vector of prices for additionally manufactured products. The function f can be linear or convex (if big volumes of resources require additional costs). Notice that the conditions of Corollary 1 are satisfied for this model if $U = \{u \in \mathbb{R}^r \mid \underline{u} \leq u \leq \bar{u}\}$, where $0 < \underline{u} < \bar{u}$.

Let us formulate a corollary from Theorem 1 on the existence of optimal solutions to problems (6) and (7).

Corollary 2. *Let the conditions of Theorem 1 (or the conditions of Corollary 1) hold. Let the set U be compact. Then the set U^* of optimal solutions to problem (6) is nonempty. If there exists a point $u \in U$ such that $\varphi_\alpha(u) < +\infty$, then the set V^* of optimal solutions to problem (7) is nonempty.*

Proof. It is proved in [14, Theorem 6] that, if the function $(u, x) \mapsto \Phi(u, x)$ is a normal integrand, then the probability function $u \mapsto P_\varphi(u)$ is upper semicontinuous for all $\varphi \in \mathbb{R}$ and the quantile function $u \mapsto \varphi_\alpha(u)$ is lower semicontinuous for all $\alpha \in (0, 1]$ for any normal integrand $(u, x) \mapsto \Phi(u, x)$. The conditions of Theorem 1 (or the conditions of Corollary 1) guarantees that the function $(u, x) \mapsto \Phi(u, x)$ is a normal integrand. Thus, the assertion of Corollary 2 follows from the Weierstrass theorem.

4   Sample Approximations

In this section, we construct sample approximations of the probability maximization problem (6) and the quantile minimization problem (7) by using a sample $(X^1, X^2, \ldots, X^N)$ generated by the random vector X. The sequence of random vectors (X^N), $N \in \mathbb{N}$, is defined on a complete probability space $(\Omega, \mathcal{F}', \mathbf{P}')$. The distribution functions of independent random variables X^N coincide with the distribution function of X.

Let us estimate the probability function (4) by the frequency of the event $\{\Phi(u, X) \leq \varphi\}$:

$$P_\varphi^{(N)}(u) := \frac{1}{N} \sum_{\nu=1}^N \chi_{[-\infty, \varphi]}(\Phi(u, X^\nu)), \ N \in \mathbb{N}, \tag{13}$$

where

$$\chi_S(x) := \begin{cases} 1, & x \in S; \\ 0, & x \notin S. \end{cases}$$

Replacing the probability function in (5) by the estimator (13), we obtain the sample estimator of the quantile function:

$$\varphi_\alpha^{(N)}(u) := \min \left\{ \varphi \in \mathbb{R}^* \mid P_\varphi^{(N)}(u) \geq \alpha \right\}.$$

We consider the sample approximation of the probability maximization problem in the form

$$U^{(N)} := \underset{u \in U}{\mathrm{Arg\,max}}\, P_\varphi^{(N)}(u), \quad \alpha_N := \sup_{u \in U} P_\varphi^{(N)}(u), \tag{14}$$

and the sample approximation of the quantile minimization problem in the form

$$V^{(N)} := \underset{u \in U}{\mathrm{Arg\,min}}\, \varphi_\alpha^{(N)}(u), \quad \varphi_N := \inf_{u \in U} \varphi_\alpha(u). \tag{15}$$

When a realization $(x^1, x^2, \ldots, x^N)$ of the sample $(X^1, X^2, \ldots, X^N)$ is fixed, problems (14) and (15) can be considered as stochastic programming problems with discrete distribution of the random parameters. This allows us to use the technique suggested in [16,17] for reducing the problems to deterministic mixed integer programming problems.

Recall that Y^* is the closure of the set $\bigcup_{u \in U, \, x \in \mathcal{X}} Y^*(u, x)$. Suppose that a set $\Lambda^* \subset \mathbb{R}^k$ is chosen in such a way that, for all $u \in U$, $y \in Y^*$, $x \in \mathcal{X}$, the set $\Lambda^* \cap \Lambda(u, y, x) \neq \emptyset$ if and only if $\Lambda(u, y, x) \neq \emptyset$.

Let functions $\gamma_1 \colon U \times Y \times \mathcal{X} \times \mathbb{R} \to \mathbb{R}$, $\gamma_2 \colon U \times Y \times \mathcal{X} \to \mathbb{R}^k$, $\gamma_3 \colon U \times Y \times \mathcal{X} \to \mathbb{R}^k$, $\gamma_4 \colon \mathbb{R}^k \mapsto \mathbb{R}^k$, $\gamma_5 \colon U \times \mathbb{R}^k \times \mathcal{X} \to \mathbb{R}^s$, $\gamma_6 \colon U \times \mathbb{R}^k \times \mathcal{X} \to \mathbb{R}^s$, $\gamma_7 \colon Y \to \mathbb{R}^s$ satisfying the following conditions be known:

1. $\Psi(u, y, x) - \varphi \leq \gamma_1(u, y, x, \varphi)$ for all $u \in U$, $y \in Y^*$, $x \in \mathcal{X}$, $\varphi \in \mathbb{R}$;
2. $-\gamma_2(u, y, x) \leq A(u, x)y - b(u, x) \leq \gamma_3(u, y, x)$ for all $u \in U$, $y \in Y^*$, $x \in \mathcal{X}$;

3. $\lambda \le \gamma_4(\lambda)$ for all $\lambda \in \Lambda^*$.
4. $-\gamma_6(u, \lambda, x) \le A(u, x)^\top \lambda - c(u, x) \le \gamma_5(u, \lambda, x)$ for all $u \in U$, $\lambda \in \Lambda^*$, $x \in \mathcal{X}$;
5. $y \le \gamma_7(y)$ for all $y \in Y^*$.

Let us introduce the variables y^ν and λ^ν corresponding to realizations x^ν, $\nu = \overline{1, N}$, and the vectors of auxilary binary variables $\delta \in \{0,1\}^N$, $\eta^\nu \in \{0,1\}^k$, $\zeta^\nu \in \{0,1\}^s$, $\nu = \overline{1, N}$. The sense of these variables follows from the proof of Theorem 2 given below. The value of δ_ν is equal to 1 if $\Phi(u, x^\nu) \le \varphi$. If $\delta_\nu = 1$, then zero elements of η^ν correspond to nonzero elements of the dual variables for constraints $A(u, x^\nu)y^\nu \ge b(u, x^\nu)$, zero elements of ζ^ν correspond to nonzero elements of the vector y^ν. Problem (14) reduces to the problem

$$\frac{1}{N} \sum_{\nu=1}^N \delta_\nu \to \max_{u \in U,\, y^\nu \in Y,\, \lambda^\nu \in \mathbb{R}^k,\, \delta \in \{0,1\}^N,\, \eta^\nu \in \{0,1\}^k,\, \zeta^\nu \in \{0,1\}^s,\, \nu=\overline{1,N}} \qquad (16)$$

subject to

$$\Psi(u, y^\nu, x^\nu) \le \varphi + (1 - \delta_\nu)\gamma_1(u, y^\nu, x^\nu, \varphi), \qquad (17)$$

$$A(u, x^\nu)y^\nu > b(u, x^\nu) - (1 - \delta_\nu)\gamma_2(u, y^\nu, x^\nu), \qquad (18)$$

$$A(u, x^\nu)y^\nu \le b(u, x^\nu) + (\eta^\nu + (1 - \delta_\nu)e_k) \circ \gamma_3(u, y^\nu, x^\nu), \qquad (19)$$

$$0 \le \lambda^\nu \le ((2 - \delta_\nu)e_k - \eta^\nu) \circ \gamma_4(\lambda^\nu), \qquad (20)$$

$$A(u, x^\nu)^\top \lambda^\nu \le c(u, x^\nu) + (1 - \delta_\nu)\gamma_5(u, \lambda^\nu, x^\nu), \qquad (21)$$

$$A(u, x^\nu)^\top \lambda^\nu \ge c(u, x^\nu) - (\zeta^\nu + (1 - \delta_\nu)e_s) \circ \gamma_6(u, y^\nu, \lambda^\nu, x^\nu), \qquad (22)$$

$$0 \le y^\nu \le ((2 - \delta_\nu)e_s - \zeta^\nu) \circ \gamma_7(y^\nu), \quad \nu = \overline{1, N}, \qquad (23)$$

where e_s, e_k are vectors consisting of ones with dimension s and k respectively, $\circ$ denotes the element-wise product of two vectors.

Denote by $(\bar{u}, (\bar{y}^\nu), (\bar{\lambda}^\nu), \bar{\delta}, (\bar{\eta}^\nu), (\bar{\zeta}^\nu))$ the optimal solution to problem (16), where $(\bar{y}^\nu) := (y^1, y^2, \ldots, y^N)$. Notation $(\bar{\lambda}^\nu)$, $(\bar{\eta}^\nu)$, $(\bar{\zeta}^\nu)$ has the same sense.

Theorem 2. *Let the conditions of Theorem 1 hold. Suppose that Λ^* satisfies condition (iv) of Theorem 1. Then,*

1. *if $\bar{u}$ is an optimal value of the variable u in problem (16), then $\bar{u} \in U^{(N)}$;*
2. *if $\bar{\delta}$ is an optimal value of the variable δ in problem (16), then*

$$\alpha_N = \frac{1}{N} \sum_{\nu=1}^N \bar{\delta}_\nu;$$

3. *for any optimal $\tilde{u} \in U^{(N)}$ there exist values $\tilde{y}^\nu \in Y$, $\tilde{\lambda}^\nu \in \mathbb{R}^k$, $\tilde{\delta} \in \{0,1\}^N$, $\tilde{\eta}^\nu \in \{0,1\}^k$, $\tilde{\zeta}^\nu \in \{0,1\}^s$, $\nu = \overline{1, N}$ such that*

$$\left(\tilde{u}, (\tilde{y}^\nu), (\tilde{\lambda}^\nu), \tilde{\delta}, (\tilde{\eta}^\nu), (\tilde{\zeta}^\nu)\right)$$

is an optimal solution to problem (16).

Proof. Let $\left(\bar{u}, (\bar{y}^\nu), (\bar{\lambda}^\nu), \bar{\delta}, (\bar{\eta}^\nu), (\bar{\zeta}^\nu)\right)$ be an optimal solution to problem (16). Let

$$K := \left\{ \nu \mid \bar{\delta}_\nu = 1, \nu = \overline{1, N} \right\}.$$

It follows from inequalities (18)–(23) that, for all $\nu \in K$, $\bar{y}^\nu$ belongs to $\mathcal{Y}(\bar{u}, x^\nu)$, and inequalities (8)–(10) hold for $\lambda = \bar{\lambda}^\nu$, $y = \bar{y}^\nu$. Therefore, $\bar{y}^\nu \in \mathcal{Y}^*(\bar{u}, x^\nu)$. Hence, for any $\nu \in K$,

$$\Phi(\bar{u}, x^\nu) \leq \Psi(\bar{u}, \bar{y}^\nu, x^\nu) \leq \varphi.$$

Thus,

$$P_\varphi^{(N)}(\bar{u}) = \frac{1}{N} \sum_{\nu=1}^N \chi_{[-\infty, \varphi]}(\Phi(\bar{u}, x^\nu)) \geq \frac{1}{N} \sum_{\nu \in K} \bar{\delta}_\nu = \frac{1}{N} \sum_{\nu=1}^N \bar{\delta}_\nu = \bar{\alpha}^*, \qquad (24)$$

where $\bar{\alpha}^*$ is the optimal objective value of problem (16). We notice that inequality (24) is written for the fixed realization $(x^1, x^2, \ldots x^N)$ of the sample.

Now, let $\tilde{u}$ be an optimal solution to problem (14). Let

$$\tilde{K} := \left\{ \nu \mid \mathcal{Y}^*(\tilde{u}, x^\nu) \neq \emptyset, \nu = \overline{1, N} \right\}. \qquad (25)$$

For each $\nu \in \tilde{K}$ let us choose $\tilde{y}^\nu \in \mathcal{Y}^*(\tilde{u}, x^\nu)$, $\tilde{\lambda}^\nu \in \Lambda^*$ in such a way that $\Phi(\tilde{u}, x^\nu) = \Psi(\tilde{u}, \tilde{y}^\nu, x^\nu)$. The existence of such values $\tilde{y}^\nu$ follows from the representation (12), because the minimum of the lower semicontinuous function $(y, \lambda) \mapsto \Psi(\tilde{u}, y, x^\nu) + \delta_\Lambda(\tilde{u}, y, \lambda, x^\nu)$ is attained on the compact set $Y^* \times \Lambda^*$. If $\nu \notin \tilde{K}$, then we take $\tilde{y}^\nu \in Y^*$, $\tilde{\lambda}^\nu \in \Lambda^*$ arbitrarily. If $\nu \in \tilde{K}$ and $\Phi(\tilde{u}, x^\nu) \leq \varphi$, then $\tilde{\delta}_\nu = 1$; otherwise $\tilde{\delta}_\nu = 0$. Let $\tilde{\zeta}_i^\nu = 1$ if $\tilde{y}_i^\nu = 0$ and $\tilde{\zeta}_i^\nu = 0$ if $\tilde{y}_i^\nu > 0$, $i = \overline{1, s}$; $\tilde{\eta}_j^\nu = 1$ if $\tilde{\lambda}_j^\nu = 0$ and $\tilde{\eta}_j^\nu = 0$ if $\tilde{\lambda}_j^\nu > 0$, $j = \overline{1, k}$. All the constraints (17)–(23) are satisfied for the solution $\left(\tilde{u}, (\tilde{y}^\nu), (\tilde{\lambda}^\nu), \tilde{\delta}, (\tilde{\eta}^\nu), (\tilde{\zeta}^\nu)\right)$. Thus,

$$P_\varphi^{(N)}(\tilde{u}) = \frac{1}{N} \sum_{\nu=1}^N \chi_{[-\infty, \varphi]}(\Phi(\tilde{u}, x^\nu)) = \frac{1}{N} \sum_{\nu=1}^N \tilde{\delta}_\nu \leq \bar{\alpha}^*. \qquad (26)$$

Taking into account the optimality of $\tilde{u}$, we obtain from inequality (24) that

$$P_\varphi^{(N)}(\tilde{u}) \geq P_\varphi^{(N)}(\bar{u}) \geq \bar{\alpha}^*. \qquad (27)$$

Hence, $\frac{1}{N} \sum_{\nu=1}^N \tilde{\delta}_\nu = \bar{\alpha}^*$. This proves the third assertion of the theorem. Combining (26) and (27), we get

$$P_\varphi^{(N)}(\bar{u}) = P_\varphi^{(N)}(\tilde{u}) = \bar{\alpha}^*. \qquad (28)$$

This implies the first assertion of the theorem. By definition, $\bar{\alpha}^* = \frac{1}{N} \sum_{\nu=1}^N \bar{\delta}_\nu$. We conclude from (28) that

$$\alpha_N = P_\varphi^{(N)}(\tilde{u}) = \bar{\alpha}^* = \frac{1}{N} \sum_{\nu=1}^N \bar{\delta}_\nu.$$

This equality proves the second assertion. All the assertions of Theorem 2 are proved.

Remark 1. Let us consider the model from Example 2. If the sets of feasible values of leader's and follower's variables are bounded then the function γ_i can be taken constant. In this case, constraints (17)–(23) are linear or convex depending on the properties of the function f.

The quantile minimization problem (15) reduces to a mixed integer programming problem:

$$\varphi \rightarrow \min_{\varphi \in \mathbb{R}^*,\, u \in U,\, y^\nu \in Y,\, \lambda^\nu \in \mathbb{R}^k,\, \delta \in \{0,1\}^N,\, \eta^\nu \in \{0,1\}^k,\, \zeta^\nu \in \{0,1\}^s,\, \nu = \overline{1,N}} \tag{29}$$

subject to

$$\frac{1}{N} \sum_{\nu=1}^{N} \delta_\nu \geq \alpha. \tag{30}$$

and (17)–(23).

Theorem 3. *Let the conditions of Theorem 1 hold. Suppose that Λ^* satisfies condition (iv) of Theorem 1. Then,*

1. *if $\bar{u}$ is an optimal value of the variable u in problem (29), then $\bar{u} \in V^{(N)}$;*
2. *if $\bar{\varphi}$ is the optimal value of the variable φ in problem (29), then $\varphi_N = \bar{\varphi}$;*
3. *for any optimal $\tilde{u} \in V^{(N)}$ there exist values $\tilde{y}^\nu \in Y$, $\tilde{\lambda}^\nu \in \mathbb{R}^k$, $\tilde{\delta} \in \{0,1\}^N$, $\tilde{\eta}^\nu \in \{0,1\}^k$, $\tilde{\zeta}^\nu \in \{0,1\}^s$, $\nu = \overline{1,N}$ such that*

$$\left(\varphi_N, \tilde{u}, (\tilde{y}^\nu), (\tilde{\lambda}^\nu), \tilde{\delta}, (\tilde{\eta}^\nu), (\tilde{\zeta}^\nu) \right)$$

is an optimal solution to problem (29).

Proof. Let $\left(\bar{\varphi}, \bar{u}, \bar{y}^\nu, \bar{\lambda}^\nu, \bar{\delta}, \bar{\eta}^\nu, \bar{\zeta}^\nu \right)$ be an optimal solution to problem (16). Due to constraint (30),

$$\frac{1}{N} \sum_{\nu=1}^{N} \bar{\delta}_\nu \geq \alpha. \tag{31}$$

It follows from inequalities (18)–(23) that

$$\Phi(\bar{u}, x^\nu) \leq \Psi(\bar{u}, \bar{y}^\nu, x^\nu) \leq \bar{\varphi}. \tag{32}$$

if $\bar{\delta}_\nu = 1$ (see the proof of Theorem 2). Since inequalities (31) and (32) hold,

$$\varphi_\alpha^{(N)}(\bar{u}) = \min \left\{ \varphi \mid \frac{1}{N} \sum_{\nu=1}^{N} \chi_{[-\infty,\varphi]}(\Phi(\bar{u}, x^\nu)) \geq \alpha \right\} \leq \bar{\varphi}. \tag{33}$$

Now, let $\tilde{u}$ be an optimal solution to problem (15). For each $\nu \in \tilde{K}$ let us choose $\tilde{y}^\nu \in \mathcal{Y}^*(\tilde{u}, x^\nu)$, $\tilde{\lambda}^\nu \in \Lambda^*$ in such a way that $\Phi(\tilde{u}, x^\nu) = \Psi(\tilde{u}, \tilde{y}^\nu, x^\nu)$, where $\tilde{K}$ is defined in (25). If $\nu \notin \tilde{K}$, then we take $\tilde{y}^\nu \in Y^*$, $\tilde{\lambda}^\nu \in \Lambda^*$ arbitrarily. Let $\tilde{\zeta}_i^\nu = 1$ if $\tilde{y}_i^\nu = 0$ and $\tilde{\zeta}_i^\nu = 0$ if $\tilde{y}_i^\nu > 0$, $i = \overline{1,s}$; $\tilde{\eta}_j^\nu = 1$ if $\tilde{\lambda}_j^\nu = 0$ and $\tilde{\eta}_j^\nu = 0$ if $\tilde{\lambda}_j^\nu > 0$, $j = \overline{1,k}$. If $\nu \in \tilde{K}$ and $\Phi(\tilde{u}, x^\nu) \leq \varphi_N$, then $\tilde{\delta}_\nu = 1$; otherwise

$\tilde{\delta}_\nu = 0$. Taking into account the definition of the sample quantile function, we get $\sum_{\nu=1}^N \tilde{\delta}_\nu \geq \alpha$. Thus, all the constraints (17)–(23) and (30) are satisfied for the solution $\left(\varphi_N, \tilde{u}, (\tilde{y}^\nu), (\tilde{\lambda}^\nu), \tilde{\delta}, (\tilde{\eta}^\nu), (\tilde{\zeta}^\nu)\right)$. Hence $\bar{\varphi} \leq \varphi_N$.

Since φ_N is the optimal objective value in (15), $\varphi_N \leq \varphi_\alpha^{(N)}(\bar{u})$. Due to (33), we obtain

$$\varphi_N \leq \varphi_\alpha^{(N)}(\bar{u}) \leq \bar{\varphi} \leq \varphi_N.$$

Thus, $\varphi_\alpha^{(N)}(\bar{u}) = \bar{\varphi} = \varphi_N$. This proves assertions 1 and 2 of Theorem 3. Assertion 3 follows from the existence of the solution $\left(\varphi_N, \tilde{u}, (\tilde{y}^\nu), (\tilde{\lambda}^\nu), \tilde{\delta}, (\tilde{\eta}^\nu), (\tilde{\zeta}^\nu)\right)$ such that $\bar{\varphi} = \varphi_N$. Theorem 3 is proved.

5   Convergence of the Sample Approximations

The convergence of sample approximations of stochastic programming problems with probabilistic criterion is studied in [13], where it was proved that $\lim_{N\to\infty} \alpha_N = \alpha^*$ almost surely (a.s.) (with respect to the probability measure $\mathbf{P}'$) if the function $(u, x) \mapsto \Phi(u, x)$ is a normal integrand and U is nonempty and compact. The sufficient conditions guaranteeing that the function $(u, x) \mapsto \Phi(u, x)$ is a normal integrand are given in Theorem 1 and Corollary 1. Let us formulate the theorem on the convergence of the sample approximations of the bilevel stochastic programming problem with probabilistic criterion. Denote by

$$D(S, T) := \sup_{s \in S} \inf_{t \subset T} \|s - t\|$$

the deviation of the set $S \subset \mathbb{R}^r$ from the set $T \subset \mathbb{R}^r$.

Theorem 4. *Suppose that the function $(u, x) \mapsto \Phi(u, x)$ is a normal integrand. Let the set U be nonempty and compact. Then $\lim_{N\to\infty} \alpha_N = \alpha^*$ a.s. and $\lim_{N\to\infty} D\left(U^{(N)}, U^*\right) = 0$ a.s.*

Proof. The convergence of α_N to α^* a.s. is proved in [13, Theorem 7]. Also, it was proved that, under the conditions of the theorem, that every limit point $\bar{u}$ of the sequence (u^N), where $u^N \in U^{(N)}$, is optimal in problem (6) a.s., i.e., $\bar{u} \in U^*$ a.s. To prove the set convergence, suppose that $\limsup_{N\to\infty} D\left(U^{(N)}, U^*\right) > 0$ with nonzero probability. This implies that there exists $\epsilon > 0$ such that $\limsup_{N\to\infty} D\left(U^{(N)}, U^*\right) > \epsilon$ with probability $\beta > 0$. Then, with probability β, we can find a sequence u^N such that

$$\limsup_{N\to\infty} \inf_{u \in U^*} \|u^N - u\| > \epsilon.$$

From the compactness of the set U and the continuity of the function $v \mapsto \inf_{u \in U^*} \|v - u\|$ it follows that there exists a limit point $\bar{u}$ of the sequence (u^N) such that $\inf_{u \in U^*} \|\bar{u} - u\| \geq \epsilon$. Therefore, with probability $\beta > 0$ there exists a limit point $\bar{u}$ (depending on the realization of the sample) such that $\bar{u} \notin U^*$. But $\bar{u} \in U^*$ a.s. This contradiction proves the theorem.

In applied problems, it is important to know the sample size guaranteeing a given accuracy of the approximation. This question is studied in [18, Theorem 1]. According to this result, if the set U is finite, then

$$\mathbf{P}'\{U^{(N)} \subset U_\epsilon\} \geq \beta$$

for

$$N \geq 2\frac{\ln |U| - \ln(1 - \beta)}{|\ln(1 - \epsilon^2)|},$$

where $\beta \in (0, 1)$, $\epsilon \in (0, 1)$, $U_\epsilon := \{u \in U \mid P_\varphi(u) \geq \alpha^* - \epsilon\}$ is the set of ϵ-optimal solutions to problem (6). This estimation was obtained for arbitrary functions $(u, x) \to \Phi(u, x)$ being normal integrands and taking values from $\mathbb{R}$. Replacing infinite values of $\Phi(u, x)$ in definition (3) by finite values that have the same sign does not change $P_\varphi(u)$ and $P_\varphi^{(N)}(u)$. Thus, the given estimation is valid for the considered bilevel problems (of course, if the conditions of Theorem 1 or Corollary 1 are satisfied).

Sufficient conditions of the convergence of problems with quantile criteria are given in [13, 14].

Theorem 5 ([14, Theorem 10]). *Suppose that*

(i) *The set U is compact and nonempty.*
(ii) *The function $(u, x) \mapsto \Phi(u, x)$ is a normal integrand, and $\Phi(u, x) > -\infty$ for all $(u, x) \in U \times \mathcal{X}$.*
(iii) *If $\varphi^* \neq +\infty$, then for all $\epsilon > 0$ there exists a pair $(\tilde{u}, \tilde{\varphi}) \in U \times \mathbb{R}$ such that $|\tilde{\varphi} - \varphi^*| \leq \epsilon$ and $P_{\tilde{\varphi}}(\tilde{u}) > \alpha$.*

Then $\lim_{N \to \infty} \varphi_N = \varphi^*$ a.s. and every limit point of the sequence (v^N), where $v^N \in V^{(N)}$, is optimal in problem (7) a.s.

Theorem 5 was proved for arbitrary functions $(u, x) \to \Phi(u, x)$ being normal integrands and taking values from $(-\infty, +\infty]$. Thus, it holds for the considered functions Φ in bilevel problems. Due to condition (ii), Theorem 5 is not applied to functions Φ taking value $-\infty$.

In the same manner as in the proof of Theorem 4, it can be proved that the assertion on the optimality of limit points in Theorem 5 can be replaced by

$$\lim_{N \to \infty} D\left(V^{(N)}, V^*\right) = 0 \text{ a.s.}$$

The most difficult point in applying Theorem 5 is to check assumption (iii). It is hard to describe sufficient conditions for this, because the dependence $(u, x) \mapsto \Phi(u, x)$ must be known. However, in some cases (for example, in the case of linear follower's problem [19]) this dependence can be found. It is easy to check that assumption (iii) of Theorem 5 holds if the function $x \mapsto \Phi(u, x)$ is strictly increasing and X has a positive on $\mathbb{R}$ density.

6   Conclusion

In this paper, sample approximations of the stochastic optimistic bilevel programming problems with probabilistic and quantile criteria were studied. The sample approximations reduced to deterministic optimization problems. These problems can be solved by using special software for nonlinear optimization. Conditions ensuring the convergence of the sample approximations were given. Since these conditions require that the leader's loss function is a normal integrand, some classes of the considered problems with such leader's loss functions were described. Although the convergence was proved, the sufficient sample size for the infinite set of the leader's variables (and for the quantile minimization problem even when the set of the variable is finite) is still unknown. This question can be studied in future research.

Acknowledgements. The reported study was funded by Russian Foundation for Basic Research (RFBR) according to the research project № 20-37-70022.

References

1. Bard, J.F.: Practical Bilevel Optimization: Algorithms and Applications. Kluwer Academie Publishers, Dordrecht (1998)
2. Dempe, S.: Foundations of Bilevel Programming. Kluwer Academie Publishers, Dordrecht (2002)
3. Dempe, S., Kalashnikov, V., Pérez-Valdés, G.A., Kalashnykova, N.: Bilevel Programming Problems – Theory, Algorithms and Applications to Energy Network. Springer Verlag, Heidelberg (2015). https://doi.org/10.1007/978-3-662-45827-3
4. Dempe, S.: Bilevel optimization: theory, algorithms, applications and a bibliography. In: Dempe, S., Zemkoho, A. (eds.) Bilevel Optimization. SOIA, vol. 161, pp. 581–672. Springer, Cham (2020). https://doi.org/10.1007/978-3-030-52119-6_20
5. Kibzun, A.I., Kan, Y.S.: Stochastic Programming Problems with Probability and Quantile Functions. John Wiley & Sons, Chichester (1996)
6. Sakawa, M., Katagiri, H., Matsui, T.: Stackelberg solutions for fuzzy random bilevel linear programming through level sets and probability maximization. Oper. Res. Int. J. **12**(3), 271–286 (2012). https://doi.org/10.1007/s12351-010-0090-2
7. Dempe, S., Ivanov, S., Naumov, A.: Reduction of the bilevel stochastic optimization problem with quantile objective function to a mixed-integer problem. Appl. Stoch. Models Bus. Ind. **33**(5), 544–554 (2017). https://doi.org/10.1002/asmb.2254
8. Shapiro, A., Dentcheva, D., Ruszczyński, A.: Lectures on Stochastic Programming. Society for Industrial and Applied Mathematics (SIAM), Philadelphia, Modeling and Theory (2014)
9. Burtscheidt, J., Claus, M., Dempe, S.: Risk-Averse models in bilevel stochastic linear programming. SIAM J. Optim. **30**(1), 377–406 (2020). https://doi.org/10.1137/19M1242240
10. Burtscheidt, J., Claus, M.: Bilevel linear optimization under uncertainty. In: Dempe, S., Zemkoho, A. (eds.) Bilevel Optimization. SOIA, vol. 161, pp. 485–511. Springer, Cham (2020). https://doi.org/10.1007/978-3-030-52119-6_17
11. Artstein, Z., Wets, R.J.-B.: Consistency of minimizers and the SLLN for stochastic programs. J. Convex Anal. **2**, 1–17 (1996)

12. Pagnoncelli, B.K., Ahmed, S., Shapiro, A.: Sample average approximation method for chance constrained programming: theory and applications. J. Optim. Theory Appl. **142**, 399–416 (2009). https://doi.org/10.1007/s10957-009-9523-6

13. Ivanov, S.V., Kibzun, A.I.: On the convergence of sample approximations for stochastic programming problems with probabilistic criteria. Autom. Remote Control **79**(2), 216–228 (2018). https://doi.org/10.1134/S0005117918020029

14. Ivanov, S.V., Kibzun, A.I.: General properties of two-stage stochastic programming problems with probabilistic criteria. Autom. Remote Control 80(6), 1041—1057 (2019). https://doi.org/10.1134/S0005117919060043

15. Rockafellar, R.T., Wets, R.J.-B.: Variational Analysis. Springer, Heidelberg (2009). https://doi.org/10.1007/978-3-642-02431-3

16. Norkin, V.I., Kibzun, A.I., Naumov, A.V.: Reducing two-stage probabilistic optimization problems with discrete distribution of random data to mixed-integer programming problems *. Cybern. Syst. Anal. **50**(5), 679–692 (2014). https://doi.org/10.1007/s10559-014-9658-9

17. Kibzun, A.I., Naumov, A.V., Norkin, V.I.: On reducing a quantile optimization problem with discrete distribution to a mixed integer programming problem. Autom. Remote Control **74**(6), 951–967 (2013). https://doi.org/10.1134/S0005117913060064

18. Ivanov, S.V., Zhenevskaya, I.D.: Estimation of the necessary sample size for approximation of stochastic optimization problems with probabilistic criteria. In: Khachay, M., Kochetov, Y., Pardalos, P. (eds.) MOTOR 2019. LNCS, vol. 11548, pp. 552–564. Springer, Cham (2019). https://doi.org/10.1007/978-3-030-22629-9_39

19. Ivanov, S.V.: A bilevel stochastic programming problem with random parameters in the follower's objective function. J. Appl. Ind. Math. **12**(4), 658–667 (2018). https://doi.org/10.1134/S1990478918040063

On Solving Bilevel Optimization Problems with a Nonconvex Lower Level: The Case of a Bimatrix Game

A. V. Orlov$^{(\boxtimes)}$ ⓘ

Matrosov Institute for System Dynamics and Control Theory of SB of RAS,
Irkutsk, Russia
anor@icc.ru

Abstract. This paper addresses the optimistic statement of one class of bilevel optimization problems (BOPs) with a nonconvex lower level. Namely, we study BOPs with a convex quadratic objective function at the upper level and with a bimatrix game at the lower level. It is known that the problem of finding a Nash equilibrium point in a bimatrix game is equivalent to the special nonconvex optimization problem with a bilinear structure. Nevertheless, we can replace such a lower level with its optimality conditions and transform the original bilevel problem into a single-level nonconvex optimization problem. Then we apply the original Global Search Theory (GST) for general D.C. optimization problems and the Exact Penalization Theory to the resulting problem. After that, a special method of local search, which takes into account the structure of the problem under consideration, is developed.

Keywords: Bilevel optimization · Bilevel problems with a nonconvex lower level · Optimistic solution · Bimatrix game · Nash equilibrium · Reduction theorem · Problem with D.C. constraints · Global search theory · Exact penalization theory · Local search

1 Introduction

Nowadays, bilevel optimization is a highly developing area of optimization [1, 2]. Bilevel optimization problems (BOPs) have a hierarchical structure involving two decision-makers: a leader (an upper level) and a follower (a lower level). These problems have a big number of practical applications, which are characterized by the unequal status of the participants (see, for example, [3–5], as well as Chapters 5, 6, and 20 in [2]).

The year 2019 marked the 85th anniversary of the famous book by Heinrich von Stackelberg [6] who can be considered as one of the founders of bilevel optimization. In 2020, Springer published a large volume dedicated to this event [2]. It largely consists of surveys on different topics of the modern bilevel optimization.

© Springer Nature Switzerland AG 2021
P. Pardalos et al. (Eds.): MOTOR 2021, LNCS 12755, pp. 235–249, 2021.
https://doi.org/10.1007/978-3-030-77876-7_16

The up-to-date challenge in studying bilevel optimization problems is finding a global solution to the bilevel problems with a nonconvex lower level, because the majority of existing solution algorithms require a preliminary transformation of the problem in question to a standard single-level optimization problem (e.g. by the KKT-conditions) [1,2]. This can be easily done only if the lower level of the original bilevel problem is convex with respect to the follower variable.

There are some interesting works concerning attempts to tackle bilevel problems with a nonconvex lower level. For example, in [7] a method based on the bounding technique with heuristics is suggested, in [8] G.-H. Lin et al. studied a special reformulation of simple bilevel problems (when the constraint set of the lower level does not depend on the variable of the upper level) based on the value function of the lower level problem, and in [9] four special types of bilevel problems with min–max optimization problems at the lower level are considered.

It can be noted that at the moment the range of publications related to the study of bilevel problems with a nonconvex lower level is rather small. It can be explained by the extreme complexity of this type of problems. Researchers are forced to apply various tricks and simplifications in order to somehow arrive at the development of methods for solving bilevel problems with a nonconvex lower level. In particular, it seems reasonable to distinguish some classes of problems with a similar structure and develop special methods for each class that uses this structure and special properties of the problems under study.

Simultaneously, the study of bilevel problems with many players at the lower (Single-Leader-Multi-Follower-Problem (SLMFP)) or at the upper level (Multi-Leader-Single-Follower-Problem (MLSFP)), or even Multi-Leader-Follower-Problems (MLFPs), with one or more Nash games at each level, is gaining popularity (see [10–12] and Chapter 3 in [2]). The research in this field is essentially motivated by real-life applications, but at the present time, there is no standard approach to constructing numerical methods for such problems, because, in general, a problem of finding a Nash equilibrium is nonconvex from the optimization point of view (whenever it is possible to make a corresponding transformation).

In this paper, we propose to approach this issue by studying one of the classes of bilevel problems with a so-called non-normalized parametric bimatrix game at the lower level [13,14], where the leader's objective function is convex quadratic subject to linear constraints.

In our recent work [15], an attempt was made to address a bilevel problem with a matrix game at the lower level. It is known that a matrix game is a convex problem because it is equivalent to a pair of linear programming problems [13,14]. From a practical perspective, such a model does not seem to be general enough, since the conflict between the players of a matrix game has an antagonistic character. In the present work, we are trying to extend the developed approach to a more difficult non-antagonistic conflict between two players at the lower level. So, we consider a bilevel problem with a special bimatrix game at the lower level (which is already a nonconvex problem from the optimization point of view).

In order to construct numerical methods for solving BOPs with such equilibrium at the lower level, we transform it, just as before, to a standard optimization problem using the special optimality conditions. The resulting problem is a global optimization problem with D.C. constraints (see, e.g., [16–18]). It is known that classical convex optimization methods do not allow to solve nonconvex optimization problems globally [18–20]. Therefore, to find global solutions to the reduced single-level problem, we employ an original Global Search Theory (GST) developed by A.S. Strekalovsky for D.C. optimization problems [18,21,22]. Lately, the GST has proven to be a powerful tool for developing numerical methods to solve different nonconvex problems of Operations Research (including problems with hierarchical and equilibrium structures), see [14,18,23–31].

In this connection, the structure of the paper is the following. Section 2 deals with the problem statement and properties of the non-normalized bimatrix game, which is formulated in Sect. 3 at the lower level of the bilevel problem. Then the transformation of the original bilevel problem to the single-level one is carried out. Section 4 addresses a D.C. decomposition for each nonconvex function from the latter formulation. In Sect. 5, Exact Penalization and the Global Optimality Conditions (GOCs) are formulated in terms of the obtained nonconvex problem. Section 6 is devoted to the Special Local Search Method. Section 7 presents concluding remarks.

2   Non-normalized Bimatrix Game and Its Properties

First of all, we need to study a so-called non-normalized 2-players bimatrix game and its basic properties. In particular, we are interested in such optimality conditions for the game (in a sense of a Nash equilibrium) that can be represented as a finite set of equalities and inequalities. Note, if a term like yB_1 is used, it means that y is a row vector, whereas the expression B_1z implies that z is a column vector.

Let us formulate the non-normalized bimatrix game in mixed strategies, which differs from a classical bimatrix game (see, e.g. [13,14]) in scalar parameters ξ_1, ξ_2 (y is the variable of Player 1, z is the variable of Player 2):

$$\left.\begin{array}{l} \langle y, B_1 z \rangle \uparrow \max_{y}, \quad y \in Y = \{y \in I\!\!R^{n_1} \mid y \geq 0, \ \langle e_{n_1}, y \rangle = \xi_1 > 0\}, \\ \langle y, B_2 z \rangle \uparrow \max_{z}, \quad z \in Z = \{z \in I\!\!R^{n_2} \mid z \geq 0, \ \langle e_{n_2}, z \rangle = \xi_2 > 0\}, \end{array}\right\} \quad (\Gamma B)$$

where B_1, B_2 are $(n_1 \times n_2)$-matrices, $e_{n_1} = (1, ..., 1)$, $e_{n_2} = (1, ..., 1)$ are vectors of appropriate dimension. It can be readily seen that we consider the game with simplexes depending on the parameters ξ_1 and ξ_2, instead of canonical simplexes.

A definition of a Nash equilibrium point in the game (ΓB) is the following.

Definition 1. A pair $(y^*, z^*) \in Y \times Z$ satisfying the inequalities

$$\begin{array}{ll} \alpha_* \stackrel{\triangle}{=} \langle y^*, B_1 z^* \rangle \geq \langle y, B_1 z^* \rangle & \forall y \in Y, \\ \beta_* \stackrel{\triangle}{=} \langle y^*, B_2 z^* \rangle \geq \langle y^*, B_2 z \rangle & \forall z \in Z. \end{array} \tag{1}$$

where α_* and β_* are the payoffs of Player 1 and Player 2 in the situation (y^*, z^*), respectively, is said to be a *Nash equilibrium point* (or a *Nash equilibrium*) of the game (ΓB) $((y^*, z^*) \in NE(\Gamma B))$.

The following optimality conditions for the non-normalized bimatrix game (ΓB) (in the sense of finding a Nash equilibrium) are a generalization of known optimality conditions for a classical bimatrix game [13, 14].

Theorem 1. *The situation* $(y^*, z^*) \in NE(\Gamma B)$, *if and only if the following inequalities and equalities hold:*

$$\left. \begin{array}{l} \xi_1(B_1 z^*)_i \leq \langle y^*, B_1 z^* \rangle \ \forall i = 1, ..., n_1, \quad z^* \geq 0, \ \langle e_{n_2}, z^* \rangle = \xi_2; \\ \xi_2(y^* B_2)_j \leq \langle y^*, B_2 z^* \rangle \ \forall j = 1, ..., n_2, \quad y^* \geq 0, \ \langle e_{n_1}, y^* \rangle = \xi_1. \end{array} \right\} \quad (2)$$

Proof. *Necessity.* Let $y_i = (0, ..., \overset{i}{\xi_1}, ..., 0) \in Y \ \forall i = 1, ..., n_1$ and $z_j = (0, ..., \overset{j}{\xi_2}, ..., 0) \in Z \ \forall j = 1, ..., n_2$ in (1). Then we arrive at (2).
Sufficiency. Scalarly multiplying the first inequality in (2) by an arbitrary $y \in Y$ and the first inequality in the second line of (2) by an arbitrary $z \in Z$, we obtain:

$$\xi_1 \langle y, B_1 z^* \rangle \leq \langle y^*, B_1 z^* \rangle \sum_{i=1}^{n_1} y_i \quad \forall y \in Y;$$

$$\xi_2 \langle y^*, B_2 z \rangle \leq \langle y^*, B_2 z^* \rangle \sum_{j=1}^{n_2} z_j \quad \forall z \in Z.$$

Hence, we obtain (1), because of $\sum\limits_{i=1}^{n_1} y_i = \xi_1$ and $\sum\limits_{j=1}^{n_2} z_j = \xi_2$. $\quad\square$

On the basis of Theorem 1, we can prove the following result.

Theorem 2. *The pair* $(y^*, z^*) \in NE(\Gamma B)$, *if and only if there exist numbers* α_* *and* β_*, *such that the following system takes place:*

$$\left. \begin{array}{l} \xi_1(B_1 z^*) \leq \alpha_* e_{n_1}, \ z^* \in Z, \quad \xi_2(y^* B_2) \leq \beta_* e_{n_2}, \ y^* \in Y, \\ \langle y^*, (B_1 + B_2) z^* \rangle = \alpha_* + \beta_*. \end{array} \right\} \quad (3)$$

Proof. *Necessity.* Let $(y^*, z^*) \in NE(\Gamma B)$. Set $\alpha_* := \langle y^*, B_1 z^* \rangle, \beta_* := \langle y^*, B_2 z^* \rangle$. Then, obviously, the last equality in the system (3) holds. Inequalities in the system (3) follow from the optimality conditions (2) written in vector form.
Sufficiency. Now, let a 4-tuple $(y^*, z^*, \alpha_*, \beta_*) \in Y \times Z \times \mathbb{R} \times \mathbb{R}$ satisfy the system (3). Scalarly multiplying the first inequality in (3) by an arbitrary $y \in Y$ and the second inequality in (3) by an arbitrary $z \in Z$, we get:

$$\left. \begin{array}{l} \xi_1 \langle y, B_1 z^* \rangle \leq \alpha_* \langle e_{n_1}, y \rangle, \\ \xi_2 \langle y^*, B_2 z \rangle \leq \beta_* \langle e_{n_2}, z \rangle. \end{array} \right\} \quad (4)$$

Therefore, when $y = y^*$ and $z = z^*$, we obtain:

$$\langle y^*, B_1 z^* \rangle \leq \alpha_*, \quad \langle y^*, B_2 z^* \rangle \leq \beta_*.$$

At the same time, according to the system (3), it should be $\alpha_* + \beta_* = \langle y^*, (B_1 + B_2)z^* \rangle$. This means that the last inequalities can be fulfilled only as equalities: $\alpha_* = \langle y^*, B_1 z^* \rangle$, $\beta_* = \langle y^*, B_2 z^* \rangle$. Substituting these values of α_* and β_* to the system (4), we obtain Definition 1 of a Nash equilibrium point in the game (ΓB). □

This theorem is a generalization of the known theorem of Mangasaryan-Stone [32]. The latter is used for solving bimatrix games by their transformation to a problem of mathematical optimization [14, 24]. So, it seems that the system (3) fits our purposes better than the system (2) as optimality conditions for the non-normalized bimatrix game (ΓB). It is easy to see that the last equality with respect to a couple of variables y and z is nonconvex.

Now, let us formulate the bilevel optimization problem with the nonconvex non-normalized bimatrix game at the lower level.

3 Bilevel Problem Formulation and Transformation

We will consider the BOPs with equilibrium at the lower level in the following formulation:

$$\left. \begin{array}{r} \langle x, Cx \rangle + \langle c, x \rangle + \langle y, D_1 y \rangle + \langle d_1, y \rangle + \langle z, D_2 z \rangle + \langle d_2, z \rangle \uparrow \max_{x,y,z}, \\ x \subset X, \ (y, z) \subset NE(\Gamma B(x)), \end{array} \right\} \quad (\mathcal{BP}_{\Gamma B})$$

where $X = \{x \in \mathbb{R}^m \mid Ax \leq a, \ x \geq 0, \ \langle b_1, x \rangle + \langle b_2, x \rangle - 1\}$, $NE(\Gamma B(x))$ is a set of Nash equilibrium points of the game

$$\left. \begin{array}{ll} \langle y, B_1 z \rangle \uparrow \max_{y}, & y \in Y(x) = \{y \mid y \geq 0, \ \langle e_{n_1}, y \rangle - \langle b_1, x \rangle \}, \\ \langle y, B_2 z \rangle \uparrow \max_{z}, & z \in Z(x) = \{z \mid z \geq 0, \ \langle e_{n_2}, z \rangle = \langle b_2, x \rangle \}; \end{array} \right\} \quad (\Gamma B(x))$$

$c, \ b_1, \ b_2 \in \mathbb{R}^m$; $y, \ d_1 \in \mathbb{R}^{n_1}$; $z, \ d_2 \in \mathbb{R}^{n_2}$; $a \in \mathbb{R}^{m_1}$; $b_1 > 0$, $b_1 \neq 0$, $b_2 \geq 0$, $b_2 \neq 0$; A, B_1, B_2, C, D_1, D_2 are matrices of appropriate dimension. $C = C^T$, $D_1 = D_1^T$, $D_2 = D_2^T$ are positive semidefinite matrices, so, the objective function of the leader is convex.

It is easy to see that at the lower level we formulate a non-normalized bimatrix game with mixed strategies from the previous section, where $\xi_1 = \langle b_1, x \rangle$, $\xi_2 = \langle b_2, x \rangle$. The equality $\langle b_1, x \rangle + \langle b_2, x \rangle = 1$ can be explained as some resource, which should be distributed by the leader among the followers.

Note that the problem $(\mathcal{BP}_{\Gamma B})$ is written in the so-called optimistic formulation when the interests of the leader can be agreed with the actions of the followers [1, 2]. In order to study the conditions guaranteeing the existence of a global solution in such formulation, one can use the corresponding theoretical results of bilevel optimization [1, 2].

As for developing numerical methods for finding solutions to the bilevel problem $(\mathcal{BP}_{\Gamma B})$, we need to rewrite it as a single-level problem. For this purpose, we can employ Theorem 2 and replace the lower level of $(\mathcal{BP}_{\Gamma B})$ with the system (3) where $\xi_1 := \langle b_1, x \rangle$, $\xi_2 := \langle b_2, x \rangle$ (x is fixed).

Therefore, it is possible to formulate the following single-level optimization problem which is equivalent to the bilevel problem $(\mathcal{BP}_{\Gamma B})$, from the global solutions point of view:

$$
\left.
\begin{aligned}
-f_0(x,y,z) &\overset{\triangle}{=} \langle x, Cx \rangle + \langle c, x \rangle + \langle y, D_1 y \rangle + \langle d_1, y \rangle \\
&+ \langle z, D_2 z \rangle + \langle d_2, z \rangle \uparrow \max_{x,y,z,\alpha,\beta}, \\
(x,y,z) \in S &\overset{\triangle}{=} \{ x,y,z \mid Ax \le a, \ x \ge 0, \ \langle b_1, x \rangle + \langle b_2, x \rangle = 1, \\
y \ge 0, \ &\langle e_{n_1}, y \rangle = \langle b_1, x \rangle, \quad z \ge 0, \ \langle e_{n_2}, z \rangle = \langle b_2, x \rangle \}, \\
&\langle b_1, x \rangle (B_1 z) \le \alpha e_{n_1}, \quad \langle b_2, x \rangle (y B_2) \le \beta e_{n_2}, \\
&\langle y, (B_1 + B_2) z \rangle = \alpha + \beta.
\end{aligned}
\right\} \quad (\mathcal{PB})
$$

More precisely, the following result is valid.

Theorem 3. *The 3-tuple (x^*, y^*, z^*) is a global (optimistic) solution to the bilevel problem $(\mathcal{BP}_{\Gamma B})$ $((x^*, y^*, z^*) \in \mathrm{Sol}(\mathcal{BP}_{\Gamma B}))$, if and only if there exist numbers α_* and β_* such that the 5-tuple $(x^*, y^*, z^*, \alpha_*, \beta_*)$ is a global solution to the problem $(\mathcal{PB})$.*

Proof. *Necessity.* Let the 3-tuple $(x^*, y^*, z^*) \in \mathrm{Sol}(\mathcal{BP}_{\Gamma B})$. Then, obviously, $(y^*, z^*) \in NE(\Gamma B(x^*))$ and Theorem 2 takes place. Hence, there exist α_* and β_* such that the conditions (3) are fulfilled under $\xi_1 = \langle b_1, x^* \rangle$ and $\xi_2 = \langle b_2, x^* \rangle$. So far as $x^* \in X$, then the 5-tuple $(x^*, y^*, z^*, \alpha_*, \beta_*)$ is feasible in the problem $(\mathcal{PB})$.

Let, on the contrary, $(x^*, y^*, z^*, \alpha_*, \beta_*) \notin \mathrm{Sol}(\mathcal{PB})$. Then there exists a 5-tuple $(\bar{x}, \bar{y}, \bar{z}, \bar{\alpha}, \bar{\beta})$ which is feasible in the problem $(\mathcal{PB})$ and

$$
\begin{aligned}
\langle \bar{x}, C\bar{x} \rangle &+ \langle c, \bar{x} \rangle + \langle \bar{y}, D_1 \bar{y} \rangle + \langle d_1, \bar{y} \rangle + \langle \bar{z}, D_2 \bar{z} \rangle + \langle d_2, \bar{z} \rangle \\
&> \langle x^*, Cx^* \rangle + \langle c, x^* \rangle + \langle y^*, D_1 y^* \rangle + \langle d_1, y^* \rangle + \langle z^*, D_2 z^* \rangle + \langle d_2, z^* \rangle.
\end{aligned} \quad (5)
$$

Simultaneously, the conditions (3) take place for the 5-tuple $(\bar{x}, \bar{y}, \bar{z}, \bar{\alpha}, \bar{\beta})$ (where $\xi_1 = \langle b_1, \bar{x} \rangle$, $\xi_2 = \langle b_2, \bar{x} \rangle$, and $(y^*, z^*, \alpha_*, \beta_*) = (\bar{y}, \bar{z}, \bar{\alpha}, \bar{\beta})$) since this 5-tuple is feasible in the problem $(\mathcal{PB})$. Therefore, by Theorem 2 $(\bar{y}, \bar{z}) \in NE(\Gamma B(\bar{x}))$, and the 3-tuple $(\bar{x}, \bar{y}, \bar{z})$ is feasible in the problem $(\mathcal{BP}_{\Gamma B})$, because $\bar{x} \in X$. So far as the objective functions of the problems $(\mathcal{BP}_{\Gamma B})$ and $(\mathcal{PB})$ coincide, the inequality (5) contradicts the fact that $(x^*, y^*, z^*) \in \mathrm{Sol}(\mathcal{BP}_{\Gamma B})$.

Sufficiency. Now let the 5-tuple $(x^*, y^*, z^*, \alpha_*, \beta_*) \in \mathrm{Sol}(\mathcal{PB})$. Then $x_* \in X$ and the conditions (3) are fulfilled for the 5-tuple $(x^*, y^*, z^*, \alpha_*, \beta_*)$ (where $\xi_1 = \langle b_1, x^* \rangle$ and $\xi_2 = \langle b_2, x^* \rangle$). Then by Theorem 2, $(y^*, z^*) \in NE(\Gamma B(x^*))$ and the 3-tuple (x^*, y^*, z^*) is feasible in the problem $(\mathcal{BP}_{\Gamma B})$.

Next, suppose that there exists a 3-tuple $(\tilde{x}, \tilde{y}, \tilde{z})$ which is feasible in the problem $(\mathcal{BP}_{\Gamma B})$ and

$$
-f_0(\tilde{x}, \tilde{y}, \tilde{z}) > -f_0(x^*, y^*, z^*). \quad (6)
$$

According to Theorem 2, again there exist numbers $\tilde{\alpha}$ and $\tilde{\beta}$ such that the optimality conditions (2) are valid for the 5-tuple $(\tilde{x}, \tilde{y}, \tilde{z}, \tilde{\alpha}, \tilde{\beta})$ (where $\xi_1 = \langle b_1, \tilde{x} \rangle$, $\xi_2 = \langle b_2, \tilde{x} \rangle$, and $(y^*, z^*, \alpha_*, \beta_*) = (\tilde{y}, \tilde{z}, \tilde{\alpha}, \tilde{\beta})$). In that case, the

5-tuple $(\tilde{x}, \tilde{y}, \tilde{z}, \tilde{\alpha}, \tilde{\beta})$ is feasible in the problem $(\mathcal{PB})$ and the inequality (6) takes place. As above, this contradicts the fact that $(x^*, y^*, z^*, \alpha_*, \beta_*) \in \mathrm{Sol}(\mathcal{PB})$. □

It is easy to see that the optimization problem $(\mathcal{PB})$ has a nonconvex feasible set (see, e.g., [16–18]). The structure of this problem differs from the problem arising in the study of bilevel problems with a matrix game at the lower level (see [15]) in the presence of an additional equality bilinear constraint. So, a nonconvexity in the problem $(\mathcal{PB})$ is produced by the $(n_1 + n_2)$ inequality bilinear constraints and by the single equality bilinear constraint. All of these constraints have arisen from the optimality conditions for a followers' game $(\Gamma B(x))$. As well-known, a bilinear function is a D.C. function, i.e. it can be decomposed into a difference of two convex functions [14, 25]. In order to solve the problem $(\mathcal{PB})$ with D.C. constraints [18, 21, 22, 33], we will employ the Global Search Theory (GST) mentioned above.

In this connection, first of all, we need to construct explicit representations of all nonconvex functions from the problem formulation as a difference of two convex functions.

4   D.C. Representation

First, let us find an explicit D.C. decomposition of the i-th scalar constraint in the first group of n_1 inequality constraints:

$$f_i(x, z, \alpha) = \langle b_1, x \rangle \langle (B_1)_i, z \rangle - \alpha \leq 0, \qquad i = 1, \ldots, n_1, \tag{7}$$

where $(B_1)_i$ is an i-th row of the matrix B_1

Let $Q_i^T = (b_1^{(1)}(B_1)_i; \; b_1^{(2)}(B_1)_i; \; \ldots; \; b_1^{(m)}(B_1)_i)$, where $b_1^{(1)}, b_1^{(2)} \ldots, b_1^{(m)}$ are components of the vector b_1, Q is a $(m \times n_2)$-matrix. Then we can transform bilinear inequalities (7) to a vector form $f_i(x, z, \alpha) = \langle x Q_i, z \rangle - \alpha \leq 0, i = 1, \ldots, n_1$. In that case, functions f_i have the following decomposition based on the known feature of a scalar product [14, 25]:

$$f_i(x, z, \alpha) = g_i(x, z, \alpha) - h_i(x, z), \tag{8}$$

where $g_i(x, z, \alpha) = \frac{1}{4}\|x Q_i + z\|^2 - \alpha, \; h_i(x, z) = \frac{1}{4}\|x Q_i - z\|^2$.

In the same way, if we take the matrix $R_{j-n_1}^T = (b_2^{(1)}(B_2)_{j-n_1}^T; \; b_2^{(2)}(B_2)_{j-n_1}^T; \; \ldots; b_2^{(m)}(B_2)_{j-n_1}^T)$ $((B_2)_{j-n_1}$ is a $(j - n_1)$-th column of the matrix B_2, R is a $(m \times n_1)$-matrix), we get a D.C. decompositions of n_2 inequality constraints from the second group:

$$f_j(x, y, \beta) = \langle b_2, x \rangle \langle y, (B_2)_j \rangle - \beta = g_j(x, y, \beta) - h_j(x, y), \; j = n_1 + 1, \ldots, n_1 + n_2, \tag{9}$$

where $g_j(x, y, \beta) = \frac{1}{4}\|x R_{j-n_1} + y\|^2 - \beta, \; h_j(x, y) = \frac{1}{4}\|x R_{j-n_1} - y\|^2$.

Afterwards, we can write the D.C. representation of the last equality bilinear constraint based on the same property of a scalar product:

$$f_{n_1+n_2+1}(y, z, \alpha, \beta) = g_{n_1+n_2+1}(y, z, \alpha, \beta) - h_{n_1+n_2+1}(y, z), \tag{10}$$

where $g_{n_1+n_2+1}(y,z,\alpha,\beta) = \frac{1}{4}\|y + B_1z\|^2 + \frac{1}{4}\|yB_2 + z\|^2 - \alpha - \beta$,

$h_{n_1+n_2+1}(y,z) = \frac{1}{4}\|y - B_1z\|^2 + \frac{1}{4}\|yB_2 - z\|^2$.

Note that in every group of constraints the functions $h(\cdot)$ which generate the so-called *basic nonconvexity* of the problem $(\mathcal{PB})$ depend on various groups of variables. Moreover, in each case the number of variables in functions $h(\cdot)$ is less than the number of variables in functions $g(\cdot)$. This means that the problem $(\mathcal{PB})$ has the property of the so-called incomplete-sized nonconvexity [14,24,28].

So, we can formulate the problem $(\mathcal{PB})$ as the minimization problem with a convex quadratic objective function and $(n_1 + n_2 + 1)$ D.C. constraints:

$$\left.\begin{array}{l} f_0(x,y,z) \downarrow \min_{x,y,z,v}, \quad (x,y,z) \in S, \\ f_i(x,z,\alpha) := g_i(x,z,\alpha) - h_i(x,z) \leq 0, \quad i \in \{1,\ldots,n_1\} =: \mathcal{I}, \\ f_j(x,y,\beta) := g_j(x,y,\beta) - h_j(x,y) \leq 0, \quad j \in \{n_1+1,\ldots,n_1+n_2\} =: \mathcal{J}, \\ f_{n_1+n_2+1}(y,z,\alpha,\beta) := g_{n_1+n_2+1}(y,z,\alpha,\beta) - h_{n_1+n_2+1}(y,z) = 0, \end{array}\right\} \quad (\mathcal{DCC})$$

where the functions f_0; y_i, h_i $\forall i \in \mathcal{I} = \{1,\ldots,n_1\}$; y_j, h_j $\forall j \in \mathcal{J} = \{n_1+1,\ldots,n_1+n_2\}$; $g_{n_1+n_2+1}$, and $h_{n_1+n_2+1}$, are convex with respect to the aggregate of all their variables; and the set

$$S = \{x,y,z \geq 0 \mid Ax \leq a, \langle b_1, x\rangle + \langle b_2, x\rangle = 1, \langle e_{n_1}, y\rangle = \langle b_1, x\rangle, \langle e_{n_2}, z\rangle = \langle b_2, x\rangle\},$$

is, obviously, convex too.

Let $\mathcal{F}$ be a feasible set of the problem $(\mathcal{DCC})$ $(N := n_1 + n_2 + 1)$:

$$\mathcal{F} := \{(x,y,z,\alpha,\beta) \mid (x,y,z) \in S; \ f_i(x,z,\alpha) \leq 0, \ i \in \mathcal{I}; \\ f_j(x,y,\beta) \leq 0, \ j \in \mathcal{J}; \ f_N(y,z,\alpha,\beta) = 0\}.$$

Note that the basic nonconvexity of the problem $(\mathcal{DCC})$ is generated, in particularly, by the functions $h_i(x,z)$, $i \in \mathcal{I}$ and $h_j(x,y)$, $j \in \mathcal{J}$. As for the last equality constraints, it is easy to see that, in general, both functions $g_N(\cdot)$ and $h_N(\cdot)$ make a nonconvexity because they are not affine. And whatever function $(g_N(\cdot)$ or $h_N(\cdot))$ we put hypothetically equal to zero, we get the nonconvex equality constraint. Nevertheless, it is possible to prove that under certain regularity conditions, when $h_N(y,z) \equiv 0$, we can reduce the equality nonconvex constraint to a convex inequality constraint (see, e.g., [23]). In other words, we can assume that it is the function $h_N(y,z)$ that contributes to the basic nonconvexity of the problem $(\mathcal{DCC})$. For convenience, here and in what follows, we use the term "nonconvex constraint" when the constraint defines a nonconvex feasible set in the problem under consideration, and the term "convex constraint" is specified similarly.

Of course, like any D.C. decomposition, the obtained D.C. decomposition is not unique [18]. In this case, the inclusion of the linear components of the representations to the functions $g(\cdot)$ is due to the need to "lighten" the functions $h(\cdot)$, which generate the nonconvexities, as much as possible. The simplest possible structure of functions $h(\cdot)$ allows more convenient work with the approximation of the level surface at the Global Search phase [18,21,22]. Now let us focus on how to apply the GST to the problem $(\mathcal{DCC})$.

In order to solve the problem $(\mathcal{DCC})$, we should develop the Global Search Algorithm (GSA) based on the Global Search Theory (GST) [18,21,22] using the D.C. decomposition (7)–(10) constructed above. According to the GST, this GSA consists of two main phases:

1) a special Local Search Method (LSM), which addresses the formulation of the problem under study [18,33];

2) the procedure of improving the point obtained at the Local Search phase, based on the Global Optimality Conditions (GOCs) [18,21,22].

In contrast to the nonconvex problem constructed in the study of a bilevel problem with a matrix game at the lower level [15], the problem $(\mathcal{DCC})$ has an additional nonconvex equality constraint. This means that it is impossible to apply the local search method from [33] to it, since this method was developed only for problems with inequality constraints. Instead, we propose the further transformation of the problem using the Exact Penalization Theory.

5   Exact Penalty and Global Optimality Conditions

Consider the penalized problem $(\theta := (x, y, z, \alpha, \beta))$:

$$\Phi_\sigma(\theta) := f_0(x, y, z) + \sigma W(\theta) \downarrow \min_\theta, \quad (x, y, z) \in S, \qquad (\mathcal{DC}(\sigma))$$

where $\sigma > 0$ is a penalty parameter, and the function $W(\cdot)$ is the penalty function for the problem $(\mathcal{DCC})$:

$$W(x, y, z, \alpha, \beta) := \max\{0, f_1(x, z, \alpha), \dots, f_{n_1}(x, z, \alpha),$$
$$f_{n_1+1}(x, y, \beta), \dots, f_{n_1+n_2}(x, y, \beta)\} + |f_N(y, z, \alpha, \beta)|.$$

It can be readily seen that the objective function of the problem $(\mathcal{DC}(\sigma))$ can be represented as a difference of two convex functions. Therefore, this problem belongs to the class of D.C. minimization problems [18] with a convex feasible set when σ is fixed.

From the Classical Penalty Theory [19,20] we know that if for some value of the parameter σ the 5-tuple $(x(\sigma), y(\sigma), z(\sigma), \alpha(\sigma), \beta(\sigma)) =: \theta(\sigma)$ is a solution to the problem $(\mathcal{DC}(\sigma))$ $(\theta(\sigma) \in \mathrm{Sol}(\mathcal{DC}(\sigma)))$, and $\theta(\sigma)$ is feasible in the problem $(\mathcal{DCC})$ $(\theta(\sigma) \in \mathcal{F})$, i.e. $W(\theta(\sigma)) = 0$, then $\theta(\sigma)$ is a global solution to the problem $(\mathcal{DCC})$ [19–22].

It is also well-known that if the equality $W(\theta(\sigma)) = 0$ takes place for some $\sigma := \hat{\sigma}$ at a solution $\theta(\sigma)$, then this solution to the problem $(\mathcal{DC}(\sigma))$ is a solution to the problem $(\mathcal{DCC})$ for all $\sigma \geq \hat{\sigma}$.

The key point of the Exact Penalization Theory is the existence of a threshold value $\hat{\sigma} > 0$ of the penalty parameter $\sigma : W(\theta(\sigma)) = 0 \ \forall \sigma \geq \hat{\sigma}$. Let us assume that the certain regularity conditions ensuring the existence of this value are fulfilled (see [21,22,30]) Then we can prove the following result concerning problems $(\mathcal{DCC})$ and $(\mathcal{DC}(\sigma))$.

Proposition 1 [19–22]. *Suppose, the 5-tuple* $(x_*, y_*, z_*, \alpha_*, \beta_*) =: \theta_*$ *is a global solution to the problem* $(\mathcal{DCC})$. *In that case, there exists* $\hat{\sigma} > 0$ *such that* θ_* *is a global solution to the problem* $(\mathcal{DC}(\hat{\sigma}))$. *Furthermore,* $\forall \sigma > \hat{\sigma}$ *any solution* $\theta(\sigma)$ *to the problem* $(\mathcal{DC}(\sigma))$ *is feasible in the problem* $(\mathcal{DCC})$, *i.e.* $W(\theta(\sigma)) = 0$, *and, therefore,* $\theta(\sigma)$ *is a solution to the problem* $(\mathcal{DCC})$, *so that* $\mathrm{Sol}(\mathcal{DCC}) \subset \mathrm{Sol}(\mathcal{DC}(\sigma))$. *The latter inclusion ensures the equality*

$$\mathrm{Sol}(\mathcal{DCC}) = \mathrm{Sol}(\mathcal{DC}(\sigma)) \quad \forall \sigma > \hat{\sigma}, \tag{11}$$

so that the problems $(\mathcal{DCC})$ *and* $(\mathcal{DC}(\sigma))$ *turn out to be equivalent (in the sense of (11)).*

Therefore, combining Proposition 1 with Theorem 3, we can infer that the declared interconnection between the problems $(\mathcal{DC}(\sigma))$ and $(\mathcal{DCC})$ allows us to find an optimistic solution to the problem $(\mathcal{BP}_{\Gamma B})$ by solving the single problem $(\mathcal{DC}(\sigma))$ (where $\sigma > \hat{\sigma}$ is fixed) instead of the problem $(\mathcal{DCC})$.

Next, we need to construct an explicit D.C. decomposition of the objective function $\Phi_\sigma(\theta)$. Using the well-known properties of "max" (see [16,18], and [30]), it is easy to prove that

$$\Phi_\sigma(\theta) \stackrel{\triangle}{=} f_0(x, y, z) + \sigma \max\{0, f_i(x, z, \alpha), i \in \mathcal{I}; \; f_j(x, y, \beta), j \in \mathcal{J}\} \\ + \sigma |f_N(y, z, \alpha, \beta)| = G_\sigma(\theta) - H_\sigma(\theta), \tag{12}$$

where

$$G_\sigma(\theta) := f_0(x, y, z) + \sigma \max\Big\{ \sum_{k \in \mathcal{I}} h_k(x, z, \alpha) + \sum_{k \in \mathcal{J}} h_k(x, y, \beta); \\ \Big[g_l(\cdot) + \sum_{\substack{k \in \mathcal{I} \cup \mathcal{J} \\ k \neq l}} h_k(\cdot) \Big], l \in \mathcal{I} \cup \mathcal{J} \Big\} + 2\sigma \max\{g_N(y, z, \alpha, \beta); h_N(y, z)\}, \tag{13}$$

$$H_\sigma(\theta) := \sigma \Big[\sum_{i \in \mathcal{I}} h_i(x, z) + \sum_{j \in \mathcal{J}} h_j(x, y) + g_N(y, z, \alpha, \beta) + h_N(y, z) \Big]. \tag{14}$$

Note that in some cases, for simplicity, here we have to use the expressions $g_l(\cdot)$ and $h_k(\cdot)$ without indicating the variables. It is explained by the fact that $g_l(\cdot)$ and $h_k(\cdot)$ when $k, l \in \mathcal{I}$ and $k, l \in \mathcal{J}$ depend on different groups of variables.

It can be readily seen that functions $G_\sigma(\cdot)$ and $H_\sigma(\cdot)$ are both convex functions. Now, denote $S' := \{(x, y, z, \alpha, \beta) \in \mathbb{R}^{m+n_1+n_2+2} \mid (x, y, z) \in S\}$ and formulate the necessary GOCs in terms of the problem $(\mathcal{DC}(\sigma))$ that represent the foundation of the Global Search Theory.

Theorem 4 [21,22]. *Suppose, a feasible point* $\theta_* \in \mathcal{F} \subset \mathbb{R}^{m+n_1+n_2+2}$, $\zeta := f_0(x_*, y_*, z_*)$ *is a global solution to the problem* $(\mathcal{DCC})$, *and a number* $\sigma : \sigma \geq \hat{\sigma} > 0$ *is fixed, where* $\hat{\sigma}$ *is a threshold value of the penalty parameter, such that* $\mathrm{Sol}(\mathcal{DCC}) = \mathrm{Sol}(\mathcal{DC}_\sigma) \; \forall \sigma \geq \hat{\sigma}$.

Then $\forall (\eta, \gamma) \in \mathbb{R}^{m+n_1+n_2+2} \times \mathbb{R}$, *such that*

$$H_\sigma(\eta) = \gamma - \zeta, \tag{15}$$

the inequality

$$G_\sigma(\theta) - \gamma \geq \langle \nabla H_\sigma(\eta), \theta - \eta \rangle \qquad \forall \theta \in S' \tag{16}$$

holds.   □

The conditions (15)–(16) possess the so-called constructive (algorithmic) property (see [18,21,22], and [30]). It means that if the main inequality (16) of the GOCs are violated, we can construct a feasible point that will be better than the current point.

Also, Theorem 4 produces the following convex (linearized) problems

$$\Psi_{\sigma\eta}(\theta) := G_\sigma(\theta) - \langle \nabla H_\sigma(\eta), \theta \rangle \downarrow \min_\theta, \quad \theta \in S', \qquad (\mathcal{P}_\sigma\mathcal{L}(\eta))$$

depending on $\eta \in \mathbb{R}^{m+n_1+n_2+2}$ which satisfy the equality (15). The linearization is realized here with respect to the "unified" nonconvexity of the problem $(\mathcal{DC}(\sigma))$ incorporated to the function $H_\sigma(\cdot)$.

And our task is to vary the parameters (η, γ) with the aim of violating the inequality (16). It is known [18,21,22] that it is convenient to carry this out together with a local search. Therefore, constructing a special Local Search Method that takes into account special properties of the problem under consideration is the primary task before developing a Global Search Algorithm [14,18,23,28].

6   Local Search Scheme

If a value of the penalty parameter $\sigma := \bar{\sigma} > 0$ is fixed, then the problem $(\mathcal{DC}(\sigma))$ belongs to one of the canonical nonconvex optimization classes, namely, D.C. minimization (on a convex feasible set). Therefore, in order to implement a local search in this problem, we can apply the well-known Special Local Search Method (SLSM) (DCA) [18,34]. This method consists in a consecutive solution of the linearized problems (see $(\mathcal{P}_\sigma\mathcal{L}(\eta))$), where the linearization is realized with respect to the function $H_\sigma(\cdot)$ that accumulated all of the nonconvexities of the problem $(\mathcal{DC}(\sigma))$ and, counsequently, $(\mathcal{DCC})$. Similar to the study of the nonlinear bilevel problem [30], in that case, a question about a threshold value of the penalty parameter should be solved in advance, before implementing a local search.

In this connection, unlike [15], in this work we suggest seeking a threshold value of the penalty parameter at the stage of a local search (as well as in [30]). To this end, we will use the Special Penalty Local Search Method (SPLSM) [30,35]. Keeping the ideology of linearization, it contains some steps for dynamic update of the penalty parameter.

Let there be given a starting point $(x_0, y_0, z_0) \in S$ and an initial value $\sigma^0 > 0$ of the penalty parameter σ. Suppose, at the iteration s of the SPLSM, we have obtained the triple $(x^s, y^s, z^s) \in S$ and the value $\sigma_s \geq \sigma^0$ of the penalty parameter. Calculate $\alpha_s := \langle y^s, B_1 z^s \rangle$, $\beta_s := \langle y^s, B_2 z^s \rangle$. Introduce the following notations: $G_s(\cdot) := G_{\sigma_s}(\cdot)$, $H_s(\cdot) := H_{\sigma_s}(\cdot)$, $\theta^s := (x^s, y^s, z^s, \alpha_s, \beta_s)$.

The linearized problem $(\mathcal{P}_s\mathcal{L}) = (\mathcal{P}_{\sigma_s}\mathcal{L}(\theta^s))$ at the iteration s has the following statement:

$$\Psi_s(\theta) := G_s(\theta) - \langle \nabla H_s(\theta^s), \theta \rangle \downarrow \min_\theta, \quad \theta \in S'. \qquad (\mathcal{P}_s\mathcal{L})$$

According to the scheme of the SPLSM [30, 35], we need to represent the penalty function $W(\theta)$ as a difference of two convex functions using the D.C. decomposition (12)–(14):

$$W(\theta) = G_W(\theta) - H_W(\theta),$$

where

$$G_W(\theta) := \frac{1}{\sigma}[G_\sigma(\theta) - f_0(x, y, z)], \ H_W(\theta) := \frac{1}{\sigma}[H_\sigma(\theta)].$$

Then introduce the following auxiliary convex problem, related to minimization of the penalty function $W(\theta)$:

$$\Psi_W(\theta) := G_W(\theta) - \langle \nabla H_W(\theta(\sigma_s)), \theta \rangle \downarrow \min_\theta, \quad \theta \in S', \qquad (\mathcal{APW}\mathcal{L}_s)$$

where $\theta(\sigma_s)$ is a solution to the problem $(\mathcal{P}_s\mathcal{L})$.

The scheme of the SPLSM in the problem $(\mathcal{DC}(\sigma))$ is the following.

Let there be given two scalar parameters $\mu_1, \mu_2 \in]0, 1[$ of the method.

Step 0. Set $s := 0$, $\sigma_s := \sigma^0$; $(x^s, y^s, z^s) := (x_0, y_0, z_0)$, $\alpha_s := \langle y^s, B_1 z^s \rangle$, $\beta_s := \langle y^s, B_2 z^s \rangle$, and $\theta_s := (x^s, y^s, z^s, \alpha_s, \beta_s)$.

Step 1. Solve the linearized problem $(\mathcal{P}_s\mathcal{L})$ to obtain $\theta(\sigma_s) \in \mathrm{Sol}(\mathcal{P}_s\mathcal{L})$.

Step 2. If $W(\theta(\sigma_s)) = 0$ then set $\sigma_+ := \sigma_s$, $\theta(\sigma_+) := \theta(\sigma_s)$ and move to **Step 7**.

Step 3. Else (if $W(\theta(\sigma_s)) > 0$), by solving the subproblems $(\mathcal{APW}\mathcal{L}_s)$ find $\theta_W^s \in \mathrm{Sol}(\mathcal{APW}\mathcal{L}_s)$.

Step 4. If $W(\theta_W^s) = 0$ then solve several problems $(\mathcal{P}_\sigma\mathcal{L}(\theta_W^s))$ (by increasing, if necessary, σ_s), trying to obtain $\sigma_+ > \sigma_s$ and the vector $\theta(\sigma_+) \in \mathrm{Sol}(\mathcal{P}_{\sigma_+}\mathcal{L}(\theta_W^s))$, such that $W(\theta(\sigma_+)) = 0$ and move to **Step 7**.

Step 5. Else, if $W(\theta_W^s) > 0$, or the value $\sigma_+ > \sigma_s$ such that $W(\theta(\sigma_+)) = 0$ is not found at the previous step, then find $\sigma_+ > \sigma_s$ satisfying the inequality

$$W(\theta(\sigma_s)) - W(\theta(\sigma_+)) \geq \mu_1[W(\theta(\sigma_s)) - W(\theta_W^s)]. \qquad (17)$$

Step 6. Increase σ_+, if necessary, to fulfil the inequality

$$\Psi_s(\theta(\sigma_s)) - \Psi_{\sigma_+}(\theta(\sigma_+)) \geq \mu_2\sigma_+[W(\theta(\sigma_s)) - W(\theta(\sigma_+))]. \qquad (18)$$

Step 7. $\sigma_{s+1} := \sigma_+$, $\theta^{s+1} := \theta(\sigma_+)$, $s := s + 1$ and loop to **Step 1**.   □

Note that the convex linearized problems $(\mathcal{P}_s\mathcal{L})$ and $(\mathcal{APW}\mathcal{L}_s)$ have non-differentiable objective functions. To solve these problems, we can reformulate them in order to eliminate non-smoothness (see, e.g. [35]) or use one of the appropriate methods of convex non-differentiable optimization [19, 36].

In order to apply the presented SPLSM in practice, we also should take into account the possibility of approximate solution of the linearized problems $(\mathcal{P}_s\mathcal{L})$ and $(\mathcal{APW}\mathcal{L}_s)$, as well as elaborate the stopping criteria. It should be noted that if we use only the obvious criterion $W(\theta(\sigma_+)) = 0$ (or $W(\theta(\sigma_+)) \leq \varepsilon$), it will not be enough for the local search goals [35].

7 Concluding Remarks

In this paper, we proposed a new approach to finding optimistic solutions to one class of bilevel optimization problems (BOPs) with a nonconvex lower level. The BOPs with an equilibrium (namely, with a parametric bimatrix game) at the lower level were studied. The approach is based on the original A.S. Strekalovsky's Global Search Theory for general D.C. optimization problems using the Exact Penalization Theory.

We presented and substantiated the reduction of the original bilevel problem to a nonconvex single-level problem, obtained an explicit D.C. representation of all functions from the problem statement, described the Global Optimality Conditions and Special Penalty Local Search Method in terms of the problem under consideration.

The present paper is the first phase of studying one class of bilevel optimization problems with a nonconvex equilibrium problem at the lower level. In our future research, we will build a range of test examples that belong to this class of bilevel problems, elaborate and test the described local search scheme, as well as construct and test a global search method for such bilevel problems, using the considered theoretical foundations. Based on our recent computational experience (see, for example, the results on the solution of other bilevel problems [26,27,29,31]), we hope that the proposed approach can also be used for the efficient numerical solution of bilevel problems with a nonconvex equilibrium problem at the lower level.

Acknowledgement. The research was funded by the Ministry of Education and Science of the Russian Federation within the framework of the project "Theoretical foundations, methods and high-performance algorithms for continuous and discrete optimization to support interdisciplinary research" (No. of state registration: 121041300065-9).

References

1. Dempe, S.: Foundations of Bilevel Programming. Kluwer Academic Publishers, Dordrecht (2002)
2. Dempe, S., Zemkoho, A. (eds.): Bilevel Optimization: Advances and Next Challenges. Springer International Publishing, New York (2020). https://doi.org/10.1007/978-3-030-52119-6
3. Colson, B., Marcotte, P., Savard, G.: An overview of bilevel optimization. Ann. Oper. Res. **153**, 235–256 (2007)
4. Dempe, S.: Bilevel programming. In: Audet, C., Hansen, P., Savard, G. (eds.) Essays and Surveys in Global Optimization, pp. 165–193. Springer, Boston (2005). https://doi.org/10.1007/0-387-25570-2_6
5. Dempe, S., Kalashnikov, V.V., Perez-Valdes, G.A., Kalashnykova, N.: Bilevel Programming Problems: Theory, Algorithms and Applications to Energy Networks. Springer-Verlag, Berlin-Heidelberg (2015). https://doi.org/10.1007/978-3-662-45827-3

6. Stackelberg, H.F.V.: Marktform und Gleichgewicht. Springer, Wien (1934). (in german)
7. Mitsos, A., Lemonidis, P., Barton, P.I.: Global solution of bilevel programs with a nonconvex inner program. J. Global Optim. **42**, 475–513 (2008)
8. Lin, G.-H., Xu, M., Ye, J.J.: On solving simple bilevel programs with a nonconvex lower level program. Math. Program. **144**, 277–305 (2013). https://doi.org/10.1007/s10107-013-0633-4
9. Zhu, X., Guo, P.: Approaches to four types of bilevel programming problems with nonconvex nonsmooth lower level programs and their applications to newsvendor problems. Math. Methods Oper. Res. **86**(2), 255–275 (2017). https://doi.org/10.1007/s00186-017-0592-2
10. Hu, M., Fukushima, M.: Existence, uniqueness, and computation of robust Nash equilibria in a class of multi-leader-follower games. SIAM J. Optim. **23**(2), 894–916 (2013)
11. Ramos, M., Boix, M., Aussel, D., Montastruc, L., Domenech, S.: Water integration in eco-industrial parks using a multi-leader-follower approach. Comput. Chem. Eng. **87**, 190–207 (2016)
12. Yang, Z., Ju, Y.: Existence and generic stability of cooperative equilibria for multi-leader-multi-follower games. J. Global Optim. **65**(3), 563–573 (2015). https://doi.org/10.1007/s10898-015-0393-1
13. Mazalov, V.: Mathematical Game Theory and Applications. John Wiley & Sons, New York (2014)
14. Strekalovsky, A.S., Orlov, A.V.: Bimatrix Games and Bilinear Programming. Fiz-MatLit, Moscow (2007). (in russian)
15. Orlov, A.V., Gruzdeva, T.V.: The local and global searches in bilevel problems with a matrix game at the lower level. In: Khachay, M., Kochetov, Y., Pardalos, P. (eds.) MOTOR 2019. LNCS, vol. 11548, pp. 172–183. Springer, Cham (2019). https://doi.org/10.1007/978-3-030-22629-9_13
16. Törn, A., Žilinskas, A. (eds.): Global Optimization. LNCS, vol. 350. Springer, Heidelberg (1989). https://doi.org/10.1007/3-540-50871-6
17. Strongin, R.G., Sergeyev, Ya.D.: Global Optimization with Non-convex Constraints. Sequential and Parallel Algorithms. Springer-Verlag, New York (2000). https://doi.org/10.1007/978-1-4615-4677-1
18. Strekalovsky, A.S.: Elements of Nonconvex Optimization. Nauka, Novosibirsk (2003). (in Russian)
19. Nocedal, J., Wright, S.J.: Numerical Optimization. Springer-Verlag, New York (2000). https://doi.org/10.1007/978-0-387-40065-5
20. Bonnans, J.-F., Gilbert, J.C., Lemarechal, C., Sagastizabal, C.A.: Numerical Optimization: Theoretical and Practical Aspects. Springer, Berlin-Heidelberg (2006). https://doi.org/10.1007/978-3-540-35447-5
21. Strekalovsky, A.S.: Global optimality conditions and exact penalization. Optim. Lett. **13**(3), 597–615 (2017). https://doi.org/10.1007/s11590-017-1214-x
22. Strekalovsky, A.S.: On a global search in D.C. optimization problems. In: Jaćimović, M., Khachay, M., Malkova, V., Posypkin, M. (eds.) OPTIMA 2019. CCIS, vol. 1145, pp. 222–236. Springer, Cham (2020). https://doi.org/10.1007/978-3-030-38603-0_17
23. Strekalovsky, A.S., Orlov, A.V.: Linear and Quadratic-linear Problems of Bilevel Optimization. SB RAS Publishing, Novosibirsk (2019). (in russian)
24. Orlov, A.V., Strekalovsky, A.S.: Numerical search for equilibria in bimatrix games. Comput. Math. Math. Phys. **45**, 947–960 (2005)

25. Orlov, A.V.: Numerical solution of bilinear programming problems. Comput. Math. Math. Phys. **48**, 225–241 (2008)
26. Gruzdeva, T.V., Petrova, E.G.: Numerical solution of a linear bilevel problem. Comput. Math. Math. Phys. **50**, 1631–1641 (2010)
27. Strekalovsky, A.S., Orlov, A.V., Malyshev, A.V.: On computational search for optimistic solutions in bilevel problems. J. Global Optim. **48**(1), 159–172 (2010)
28. Orlov, A.V., Strekalovsky, A.S., Batbileg, S.: On computational search for Nash equilibrium in hexamatrix games. Optim. Lett. **10**(2), 369–381 (2014). https://doi.org/10.1007/s11590-014-0833-8
29. Orlov, A.V.: The global search theory approach to the bilevel pricing problem in telecommunication networks. In: Kalyagin, V.A., Pardalos, P.M., Prokopyev, O., Utkina, I. (eds.) NET 2016. SPMS, vol. 247, pp. 57–73. Springer, Cham (2018). https://doi.org/10.1007/978-3-319-96247-4_5
30. Orlov, A.V.: On a solving bilevel D.C.-convex optimization problems. In: Kochetov, Y., Bykadorov, I., Gruzdeva, T. (eds.) MOTOR 2020. CCIS, vol. 1275, pp. 179–191. Springer, Cham (2020). https://doi.org/10.1007/978-3-030-58657-7_16
31. Strekalovsky, A.S., Orlov, A.V.: Global search for bilevel optimization with quadratic data. In: Dempe, S., Zemkoho, A. (eds.) Bilevel Optimization. SOIA, vol. 161, pp. 313–334. Springer, Cham (2020). https://doi.org/10.1007/978-3-030-52119-6_11
32. Mangasarian, O.L., Stone, H.: Two-person nonzero games and quadratic programming. J. Math. Anal. Appl. **9**, 348–355 (1964)
33. Strekalovsky, A.S.: On local search in D.C. optimization problems. Appl. Math. Comput. **255**, 73–83 (2015)
34. Tao, P.D., Souad, L.B.: Algorithms for solving a class of non convex optimization. Methods of subgradients. In: Hiriart-Urruty J.-B. (ed.) Fermat Days 85, pp. 249–271. Elservier Sience Publishers B.V., North Holland (1986)
35. Strekalovsky, A.S.: Local search for nonsmooth DC optimization with DC equality and inequality constraints. Accepted for publication. In: Bagirov, A. et al. (Eds.) Numerical Nonsmooth Optimization - State of the Art Algorithms (2020)
36. Ben-Tal, A., Nemirovski, A.: Non-Euclidean restricted memory level method for large-scale convex optimization. Math. Program. **102**, 407–456 (2005)

Scheduling Problems

Two-Machine Routing Open Shop: How Long Is the Optimal Makespan?

Ilya Chernykh[1,2]($\boxtimes$) (iD)

[1] Sobolev Institute of Mathematics, Koptyug Avenue 4, Novosibirsk 630090, Russia
[2] Novosibirsk State University, Pirogova Street 2, Novosibirsk 630090, Russia
idchern@math.nsc.ru

Abstract. We consider a routing open shop problem being a natural generalization of the metric TSP and the classic open shop scheduling problem. The maximal possible ratio of the optimal makespan and the standard lower bound for the routing open shop has already been investigated in the last few years. The two-machine case is mostly covered. It is constructively proven in 2013 that the ratio mentioned above cannot be greater than 4/3, however, we do not know of any problem instance with the value of that ratio greater than 6/5. The latter ratio is achievable for a simplest case with two nodes. On the other hand, it is known that optimal makespan is at most 6/5 of the standard lower bound for at least a few special cases of the transportation network: one is with at most three nodes, and another is a tree.

In this paper, we introduce an ultimate instance reduction technique, which allows reducing the general problem into a case with at most four nodes and at most six jobs. As a by-product, we propose a new polynomially solvable case of the two-machine routing open shop problem.

Keywords: Routing open shop · Standard lower bound · Optima localization · Instance reduction · Efficient algorithm

1 Introduction

One of important directions in the research of scheduling problems is establishing useful structural properties of optimal solutions. One classic example is the investigation of the properties of an optimal permutation for two-machine flow shop problem [14]. Those properties were used to describe an efficient exact algorithm for the problem. As other examples, important properties for preemptive *flow shop* and *job shop* problems were investigated in [16] and [17], respectively. These properties allow to describe finite (although obviously not efficient) algorithms to solve those problems to the optimum—a task, trivial for problems without preemption, is however difficult for the cases, when preemption is allowed.

This research was supported by the program of fundamental scientific researches of the SB RAS No I.5.1., project No 0314-2019-0014, and by the Russian Foundation for Basic Research, projects 20-01-00045 and 20-07-00458.

P. Pardalos et al. (Eds.): MOTOR 2021, LNCS 12755, pp. 253–266, 2021.
https://doi.org/10.1007/978-3-030-77876-7_17

Fig. 1. An illustration for the instance transformation concept.

In this paper we want to address another approach to establish some properties of any problem instance by means of an *instance reduction*. Assume $\mathcal{I}$ is the set of instances of the given problem, and we want to verify that for some property $\mathcal{P}$

$$\forall I \in \mathcal{I} \ (\mathcal{P}(I) \text{ holds}).$$

Suppose we have some transformation $\varphi : \mathcal{I} \to \mathcal{I}$ such that

$$\forall I \subset \mathcal{I} \ (\mathcal{P}(\varphi(I)) \ \rightarrow \ \mathcal{P}(I)).$$

In this case, it is sufficient to establish the property $\mathcal{P}$ for the *image* $\varphi(\mathcal{I})$ of the transformation φ in order to obtain the global result. We refer to the set of instances $\varphi(\mathcal{I})$ as to a *kernel* of $\mathcal{I}$ with respect to $\mathcal{P}$.

Further we aim to establish properties for the optimal schedules for instances from $\mathcal{I}$ in the following form:

$$\forall I \in \mathcal{I} \ \exists \text{ feasible schedule } S(I) \text{ such that } \mathcal{P}(S(I)) \text{ holds}.$$

We call a transformation φ *reversible* for the property $\mathcal{P}$, if there exists a transformation procedure φ^* for schedules, such that for any feasible schedule $S(\varphi(I))$ schedule $\varphi^*(S(\varphi(I)))$ is feasible for I, and φ^* preserves the property $\mathcal{P}$. Evidently, for any reversible φ its image $\varphi(\mathcal{I})$ is a kernel of $\mathcal{I}$ with respect to $\mathcal{P}$.

Naturally, we want the property $\mathcal{P}$ for the kernel to be provable easier, than for the whole set $\mathcal{I}$. If the kernel contains constrained instances, obeying some strict properties, making those instances simpler. For example, for scheduling problems transformation might drastically reduce the number of jobs (hence the name *instance reduction*). So, our goal is **to choose as restricted kernel as possible**.

That approach can also be useful for designing either exact or approximation efficient algorithms for the problem under consideration, providing that transformations φ and φ^* are doable in polynomial time. Then one can take any instance I, reduce it to a simplified instance I', build an optimal (or good enough) schedule S' for I' and use it to restore an approximate solution S for the initial instance I. This idea is pictured in Fig. 1. The approximation ratio for such an approach depends on the property $\mathcal{P}$, as well as on the quality of the schedule S'.

In this paper we demonstrate this approach on the following *routing open shop* problem (introduced in [1,2]), which is a natural combination of two classic discrete optimization problems: well-known metric traveling salesman problem [12] and open shop scheduling problem [13].

In the routing open shop problem a fleet of mobile machines has to process a set of immovable jobs, located at the nodes of some transportation network, described by an undirected edge-weighted graph $G = \langle V; E \rangle$. Each node contains at least one job, and weight $\text{dist}(u, v)$ represents travel time of any machine between nodes u and v. Each machine M_i has to perform an operation O_{ji} on each job J_j, providing that operations of the same jobs are not performed simultaneously. The processing times p_{ji} are given for each operation O_{ji} in advance. All the machines start from the same node v_0 referred to as *the depot*, and have to return to the depot after processing all the job. No restriction on the machines' traveling are in order: any number of machines can travel over the same edge of the network simultaneously in any direction, machines are allowed to visit each node multiple times, or to bypass a node without performing any operation at that node. However, machine has to reach a node prior to be able to process jobs located there. Without loss of generality we assume, that triangle inequality for distances holds.

The goal is to minimize the *makespan* $R_{\max}$, *i.e.* the completion time of the last machine's activity (either traveling back to the depot or performing an operation on a job, located at the depot). The problem is clearly a generalization of the metric traveling salesman problem and therefore is NP-hard in strong sense even in a single machine case. On the other hand, it generalizes the classical open shop problem, which is well-known to be NP-hard for the case of three and more machines, and is polynomially solvable for the two-machine case [13]. Surprisingly, the routing open shop is NP-hard even in the two-machine case on the transportation network consisting of at least two nodes (including the depot) [2]. We use notation $ROm||R_{\max}$ for the routing open shop with m machines. Optional notation $G = X$ in the second field is used in case we want to specify the structure of the transportation network, with X being the name of the structure (*e.g.* K_p or *tree*). A set of instances of the $ROm|G = X|R_{\max}$ problem is denoted by $\mathcal{I}_m^X$ (or $\mathcal{I}_m$ for a general case of unspecified X).

The property we aim to establish is tightly connected with the following *standard lower bound* on the optimal makespan, introduced in [1]:

$$\bar{R} = \max\left\{\ell_{\max} + T^*, \max_{v \in V}(d_{\max}(v) + 2\text{dist}(v_0, v))\right\}. \tag{1}$$

Here $\ell_{\max} = \max_i \ell_i = \max_i \sum_{j=1}^{n} p_{ji}$ is the *maximum machine load*, $d_{\max}(v) = \max_{j \in \mathcal{J}(v)} d_j = \max_{j \in \mathcal{J}(v)} \left(\sum_{i=1}^{m} p_{ji}\right)$ is the *maximum length of job* from node v, with $\mathcal{J}(v)$ being the set of jobs located at v, while T^* is the TSP optimum on G. Values ℓ_i and d_j are called the *machine load* of M_i and *job length* of J_j, respectively. The property under research is so-called *optima localization* and can be described as follows: how much (by which factor) can optimal makespan differ from the standard lower bound $\bar{R}$ for a given class of instances $\mathcal{K}$? More precisely, for some class $\mathcal{K}$ we want to find

$$\alpha\left(\mathcal{K}\right) = \sup_{I \in \mathcal{K}} \alpha(I) = \sup_{I \in \mathcal{K}} \frac{R^*_{\max}(I)}{\bar{R}(I)}.$$

Here $R^*_{\max}(I)$ and $\bar{R}(I)$ denote optimal makespan and the value of $\bar{R}$ for I, respectively, and $\alpha(I)$ is referred to as the *abnormality* of instance I.

Not only the optima localization gives us an estimation on how tight the standard lower bound is for a given class of instances, it also can be used for the design of approximation algorithms (using the concept, showed in Fig. 1) with the best theoretically possible approximation ratio with respect to the standard lower bound. Moreover, often the underlying procedures φ and φ^* can be done in linear time, which is smallest theoretically possible running time for any algorithm, which needs to specify a starting time for each operation.

It is known that for the classical two-machine open shop (which can be denoted as $RO2|G = K_1|R_{\max}$ for consistency) optimal makespan always coincides with the standard lower bound, therefore $\alpha\left(\mathcal{I}_2^{K_1}\right) = 1$ [13]. It is not the case for the three-machine problem, where optimal makespan can reach as much as $\frac{4}{3}\bar{R}$. It was actually proved in [18] that $\alpha\left(\mathcal{I}_3^{K_1}\right) = \frac{4}{3}$. This result was recently generalized in [7] for $G = K_2$ case: $\alpha\left(\mathcal{I}_3^{K_2}\right) = \frac{4}{3}$. As for the classical open shop, the value $\alpha\left(\mathcal{I}_4^{K_1}\right)$ is still an open question, however we have no evidence that it is greater than $\frac{4}{3}$.

Optima localization research for $m = 2$ is now completed only for small number of nodes and for the *tree* structure of the transportation network. All the results, known up to the moment, are shown in Table 1.

Table 1. Known optima localization results for the routing open shop problem.

Problem	Abnormality	Year	Reference
$RO2|G = K_1|R_{\max}$	1	1976	[13]
$RO3|G = K_1|R_{\max}$	4/3	1998	[18]
$RO2|G = K_2|R_{\max}$	6/5	2005	[1]
$RO2|G = K_3|R_{\max}$	6/5	2016	[8]
$RO3|G = K_2|R_{\max}$	4/3	2020	[7]
$RO2|G = tree|R_{\max}$	6/5	To appear	[5]

In this paper we focus on the two-machine case. The general question is, what is the value $\alpha\left(\mathcal{I}_2\right)$? We have a lower bound $\frac{6}{5}$ from [1]. On the other hand, there is an algorithm [4] that provides a schedule with makespan not greater that $\frac{4}{3}\bar{R}(I)$ for any problem instance $I \in \mathcal{I}_2$, therefore $\frac{6}{5} \leqslant \alpha\left(\mathcal{I}_2\right) \leqslant \frac{4}{3}$. (It should be mentioned that this algorithm has polynomial running time only under the assumption, that the optimal solution to the underlying TSP is known. Otherwise, obviously, the minimal approximation factor we could hope to achieve for the $RO2\|R_{\max}$ problem is $\frac{3}{2}$, as soon as the best approximation algorithm for the metric TSP known up to date is $\frac{3}{2}$-approximation by Christofides and Serdyukov [3,11,15]).

Although we have no evidence that $\alpha\left(\mathcal{I}_2\right)$ is greater than $\frac{6}{5}$, how this bound can be proved? There is still a possibility, that the abnormality depends on the structure of the transportation network, and/or on the number of nodes. The proofs from [1,5,8] are heavily utilizing the specific graph structure in each case. One could try to continue this line of work to find the values $\alpha\left(\mathcal{I}_2^{K_4}\right)$, or $\alpha\left(\mathcal{I}_2^{pseudotree}\right)$ for instance, but how to stop this infinite series of incremental results and still reach the ultimate goal of discovering the general value $\alpha\left(\mathcal{I}_2\right)$? This is exactly the question we address in this paper. The answer is: we need the ultimate instance reduction procedure, allowing us to describe a relatively small and simple kernel of $\mathcal{I}_2$ with respect to the desired property.

The structure of this paper is as follows. In Sect. 2 we discuss known instance simplification techniques and suggest some new ones. In Sect. 3 we provide a new polynomially solvable subcase of $RO2\|R_{\max}$ (Corollary 1) and prove the main result, describing a kernel of $\mathcal{I}_2$ with respect to the optima localization (Theorem 2). Each instance from that kernel consists of at most six jobs, located at four or less nodes. Conclusions and open questions for the future research are given in Sect. 4.

2   Instance Transformations

The property $\mathcal{P}$, connected with optima localization for some class of instances $\mathcal{K}$ is the following: is it true that for any instance $I \in \mathcal{K}$ its optimal makespan belongs to the interval $[\bar{R}(I), \rho^*\bar{R}(I)]$? Here $\rho^* = \alpha(\mathcal{K})$. This condition is equivalent to the following

$$\forall I \in \mathcal{K} \; \exists \text{ feasible schedule } S \text{ for } I \mid R_{\max}(S) \leqslant \rho^*\bar{R}(I).$$

The following proposition gives sufficient conditions for a transformation φ to produce a kernel with respect to $\mathcal{P}$.

Proposition 1. *Let φ be a transformation of $\mathcal{K}$, and I' denotes $\varphi(I)$. Suppose φ obeys the following conditions:*

$$\forall I \in \mathcal{K} \; \bar{R}(I') = \bar{R}(I), \tag{2}$$

$$\forall \text{ feasible schedule } S(I') \; R_{\max}(\varphi^*(S(I'))) = R_{\max}(S(I')). \tag{3}$$

Then $\varphi(\mathcal{K})$ is a kernel of $\mathcal{K}$ with respect to $\mathcal{P}$, and φ is reversible for the property $\mathcal{P}$.

Proof. Conditions (2) and (3) imply that $\alpha(I') \geqslant \alpha(I)$, therefore if $\mathcal{P}$ holds for I', then $\mathcal{P}$ holds for I. $\square$

Transformations, obeying the conditions of the Proposition 1 will be referred to as *valid* ones. In this section we describe several valid instance transformation operations, which can be used for our purposes of describing optima localization kernels.

The first operation worth mentioning is a well-known transformation, which reduces the number of jobs, referred to as *job aggregation* or *job grouping*. The idea is to combine a set of jobs from the same node into a single one, adding up the processing times independently for each machine. Such a procedure was used, *e.g.*, in [18] for the classic open shop problem, and in [8] for the two-machine routing open shop. While the procedure is clearly reversible (one can treat an operation of an aggregated job as a sequence of the initial operation), its validity has to be maintained explicitly. More precisely, let v be some node of the transportation network, and $\mathcal{O}$ be a subset of jobs from $\mathcal{J}(v)$. Then (1) implies, that job aggregation of the set $\mathcal{O}$ is valid if and only if

$$\sum_{j \in \mathcal{O}} d_j \leqslant \bar{R} - 2\mathrm{dist}(v_0, v). \tag{4}$$

As for the classical open shop, it is possible to perform valid job aggregations for any instance of $Om||C_{\max}$ so that the resulting instance would contain at most $2m - 1$ jobs [18]. As for $RO2||R_{\max}$, one can aggregate jobs in such a valid manner that every node (except for at most one) has a single job, and the "exceptional" one (if any) contains at most 3 jobs [8]. Such an exceptional node v is referred to as *overloaded*, as the total load $\Delta(v)$ of the node v should be big enough:

$$\Delta(v) = \sum_{j \in \mathcal{J}(v)} d_j > \bar{R} - 2\mathrm{dist}(v_0, v).$$

It follows from [8], that any instance of the $RO2||R_{\max}$ problem contains at most one overloaded node. As for the general $ROm||R_{\max}$, it was shown in [7], that any instance of the $ROm||R_{\max}$ problem contains at most $m - 1$ overloaded nodes, and a valid aggregation can be performed in such a way, that the total number of jobs in all the overloaded nodes together doesn't exceed $2m - 1$. All the described aggregations can be done in linear time [7,8,18].

However, it would be of the most interest to describe some valid transformations to simplify the structure of G. An example of such a reduction is so-called *terminal edge contraction*, which can be described as follows. Suppose G contains a terminal node $v \neq v_0$ with a single job J_j in $\mathcal{J}(v)$. Let u be the node adjacent to v, and $\tau = \mathrm{dist}(u, v)$. The transformation is the following: we translate the job J_j to the node u, increase its operations' processing times by 2τ each, and eliminate the obsolete node v. Such a transformation is reversible, as one can treat the processing of a new operation O_{ji} as a concatenation of traveling of M_i from u to v, processing of the initial operation and traveling back to u (see Fig. 2). It is proved in [9] that for any instance $I \in \mathcal{I}_2$ one can perform a valid transformation $I \to I'$ such that the transportation network in I' contains at most two terminal nodes. This helps to efficiently reduce any tree to a chain.

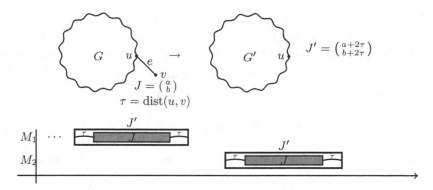

Fig. 2. The reversibility of a terminal edge contraction.

Similar operation called *terminal cycle contraction* was introduced in [6]. It can be described as follows. Consider a cycle C in graph G. Node $v \in C$ is called *a gate*, if its degree if either greater than 2, or $v = v_0$. Any cycle with a single gate is referred to as *terminal* one.

Let C be a terminal cycle with gate u, T is the length of C (total weight of its edges), and $J(C \backslash \{u\})$ is the set of all jobs from $C \backslash \{u\}$. Let us replace all the jobs from C, except the ones in u, with a new single job J' with operations processing times $p'_i = T + \sum_{j \in J(C \backslash \{u\})} p_{ji}$, and locate J' at u. Obsolete nodes of C (all except u) can now be removed from G. Such a transformation is reversible, as soon as we can treat the processing of operations of job J' as traveling along the cycle C and processing the jobs on the way. Actually, terminal edge contraction can be seen as a special case of terminal cycle contraction, as edge can be replaced with a 2-cycle.

Let Q be some terminal element (edge or cycle) with gate u, and $W(Q)$ is its weight, defined as follows:

– if Q is an edge $[u, v]$ of length τ with single job J_j in v, $W(Q) = d_j + 2m\tau$,
– if Q is a cycle C of length T, $W(Q) = mT + \sum_{j \in J(C \backslash \{u\})} d_j$.

In other words, $W(Q)$ is the length of a new job J' constructed by the contraction of Q.

Proposition 2. *Let Q be a terminal element with gate u. Contraction of Q is valid if and only if*

$$W(Q) \leqslant \bar{R} - 2\text{dist}(v_0, u). \tag{5}$$

Proof. Straightforward from (1). □

On the other hand, a graph might have a complex structure even without terminal elements. Below we describe a new approach to the instance reduction which allows to significantly simplify the structure of a transportation network in a valid manner. From now on we focus on a case $m = 2$.

Let cycle σ be an optimal solution of the underlying TSP. Any edge $e \notin \sigma$ is referred to as *chord*. A chord e is referred to as *critical* if removing it from G increases the standard lower bound $\bar{R}$. The distance between nodes u, v along the cycle σ (using only edges from σ) will be denoted as $\widehat{\text{dist}}(u, v)$. Note that removing any chord $[u, v]$ doesn't affect $\text{dist}(u, v)$ and

$$\text{dist}(u, v) \leqslant \widehat{\text{dist}}(u, v) \leqslant \frac{T^*}{2}. \tag{6}$$

Remark 1. It is possible, that some chord e only becomes critical after removing another chord e'.

Lemma 1. *Let $I \in \mathcal{I}_2$ and the transportation network G of I is a complete graph with metric weights of edges. Then removing all the chords which are not incident to the depot v_0 doesn't increase the standard lower bound $\bar{R}$.*

Proof. Evidently, removing all of those chords cannot affect distances $\text{dist}(v_0, v)$ for any $v \in V$. Now it is sufficient to note, that the standard lower bound (1) doesn't depend on any distance between any two non-depot node.   □

Consider an instance $I \in \mathcal{I}_2$. Let $\Delta = \sum_j d_j$ be the *total load* if I. Note that (1) implies

$$\Delta \leqslant 2\ell_{\max} \leqslant 2(\bar{R} - T^*). \tag{7}$$

Lemma 2. *Let instance $I \in \mathcal{I}_2$ contain a critical chord $[v_0, v]$. Then removing all other chords doesn't affect the standard lower bound $\bar{R}$.*

Proof. Chord $e = [v_0, v]$ is critical, therefore $\bar{R} < 2\widehat{\text{dist}}(v_0, v) + d_{\max}(v) \leqslant T^* + d_{\max}(v)$. Assuming we have another critical chord $[v_0, u]$, we have $\bar{R} < T^* + d_{\max}(u)$. Combining those two inequalities we obtain $2\bar{R} < d_{\max}(u) + d_{\max}(v) + 2T^* \leqslant \Delta + 2T^*$. Lemma is proved by contradiction with (7).   □

Lemma 3. *Let $I \in \mathcal{I}_2$, node v is overloaded and chord $[v_0, u]$ is critical. Then $u = v$.*

Proof. We have $\Delta(v) > \bar{R} - 2\text{dist}(v_0, v) \geqslant \bar{R} - T^*$ and $d_{\max}(u) > \bar{R} - T^*$. Assume $u \neq v$, then $\Delta \geqslant \Delta(v) + d_{\max}(u) > 2(\bar{R} - T^*)$. Lemma is proved by contradiction with (7).   □

Combining Lemmas 1–3 we can describe the following valid transformation:

1. Consider the transitive closure of graph G.
2. Choose some optimal tour σ.
3. Remove all chords which are not incident to the depot v_0.
4. Remove all the other chords, except at most one critical.

This transformation is valid (although not always polynomial, due to Step 2), and helps to reduce the order of magnitude of the number of edges. However, it doesn't affect the number of nodes. In the next section we are going to deal with that problem.

3 The Main Result

The first minor result is the following

Theorem 1. *Let $I \in \mathcal{I}_2$ such that the depot v_0 is overloaded. Then $\alpha(I) = 1$.*

Proof. Note that $\Delta(v_0) > \bar{R}$. It follows from Lemma 3 that I contains no critical chords, therefore eliminating all the chords is a valid transformation of I. Now the transportation network is cyclic, therefore the cycle σ is terminal (with single gate v_0). Note that $W(\sigma) = 2T^* + \Delta - \Delta(v_0)$. Let us prove that contraction of σ is valid. Suppose otherwise, then by Proposition 2 $W(\sigma) > \bar{R}$. Then

$$2\bar{R} < W(\sigma) + \Delta(v_0) = \Delta + 2T^*,$$

which contradicts (7).

Performing the contraction of σ, we obtain a single-node instance I', which actually is an instance of a classical $O2||C_{\max}$ problem, therefore $\alpha(I') = 1$. Validity of the transformation $I \to I'$ confirms that $\alpha(I) = \alpha(I') = 1$. □

Note that the transformation described in the proof of Theorem 1 can be done in $O(n)$ time, providing that the have the knowledge of an optimal Hamiltonian walk on graph G. Therefore we have the following

Corollary 1. *Let $I \in \mathcal{I}_2$ such that the depot v_0 is overloaded. Then a feasible schedule of makespan $\bar{R}$ can be constructed in $O(n + t_{TSP})$ time, where t_{TSP} is time needed to solve the TSP on $G(I)$.*

This corollary actually describes a polynomially solvable subcase of $RO2||R_{\max}$ on the subclass of instances, obeying the following properties:

1. The TSP on graph G is solvable in polynomial time (for instance, due to the simple structure of G, or special properties of the matrix of distances);
2. $\Delta(v_0) > \bar{R}$.

This can be considered as an improvement on the similar polynomially solvable subclass of $RO2||R_{\max}$, described in [10], which requires stronger conditions on the load of the depot $\Delta(v_0)$.

Now we describe a *chain contraction* transformation. Suppose G contains a chain $C = (v_1 - v_2 - \cdots - v_k)$ consisting of non-depot nodes, such that all the nodes $v_2, \ldots, v_{k-1}$ are of degree 2. Let τ be the length of chain (the distance between v and u along C) and $\mathcal{J}(C)$ is the set of jobs from nodes $v_1, \ldots, v_k$. The transformation is the following: we replace the chain $(v_1 - \cdots - v_k)$ with a new *special* node v_C containing single job J_C with processing times $p_{Ci} = \tau + \sum_{J_j \in \mathcal{J}(C)} p_{ji}$. Processing of operation of J_C can be treated as traveling of the machine along C and processing operations of its jobs on the way (Fig. 3).

Such a transformation is not reversible in general. To make it reversible we need to apply certain restriction on schedules for the transformed instance:

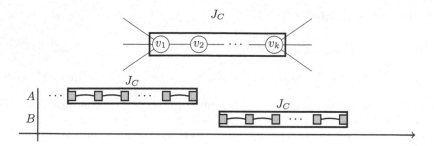

Fig. 3. An illustration of chain contraction transformation. Rectangle denotes the new special node.

1. If machine arrives at J_C from one end, the machine is considered to be at another end after the completion of operation of job J_C.
2. Any machine can bypass the node v_C, but this takes τ time units.

We say that the chain contraction transformation is *conditionally reversible*, meaning that a schedule is reversible under condition of a special treatment (described above) of nodes, obtained by chain contractions. Such nodes will be called *special* and denoted in figures as rectangles.

Consider the contraction of a chain $C = (v_1, \ldots, v_k)$ of length τ. We use the following notation:

- $W(C) = \sum_{t=1}^{k} \Delta(v_t) + 2\tau$—the *weight* of chain C (again, it coincides with the length of new job J_C),
- $\text{dist}(v, C) = \min\{\text{dist}(v, v_1), \text{dist}(v, v_k)\}$—the distance from v to chain C.

Remark 2. The chain contraction, as described above, is (conditionally) valid, if and only if chain C belongs to the optimal Hamiltonian walk over G and

$$2\text{dist}(v_0, C) + W(C) \leqslant \bar{R}. \tag{8}$$

Indeed, by chain contraction, the load of each machine is increased by τ, but on the other hand, the length of optimal Hamiltonian walk (TSP optimum) is decreased by the same value. Now it is sufficient to note that it is possible to process job J_C in time $2\text{dist}(v_0, C) + W(C)$ while obeying the special restrictions, described above. To that end, machine M_1 travels to the farthest end of C (say, v_k) and waits until machine M_2 processes jobs starting from another end (Fig. 4).

The main result of this paper is the following description of the kernel of $\mathcal{I}_2$ with respect to the optima localization property.

Theorem 2. *For any instance $I \in \mathcal{I}_2$ there exists a combination of job aggregations, valid chord eliminations and chain/cycle contractions $I \to I'$, such that I' contains at most 6 jobs and at most four nodes, from which at most two are special.*

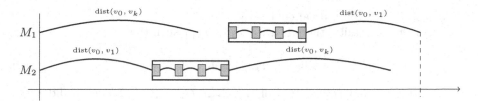

Fig. 4. How to process contracted chain in time.

Proof. Assume instance I contains more than four nodes. The main goal of the transformation is to reduce the number of nodes of the transportation network to at most four. After that, we can apply job aggregation, described in [8] to reduce the number of jobs to at most six: at most three at the possible overloaded node, and exactly one at each of other nodes, which gives a total of at most six.

Let σ be an optimal TSP solution. We start with elimination of non-critical chords, as described in the previous section.

Enumerate the nodes $v_0, v_1, \ldots, v_p$ along the cycle σ, and denote $\widehat{\text{dist}}(v_i, v_{i+1})$ by τ_i (for consistency, $v_{p+1} = v_0$).

Case 1. All the chords are removed (there is no critical one).

Let s be maximal index such that

$$\sum_{t=1}^{s} (2\tau_{t-1} + \Delta(v_t)) \leqslant \bar{R}.$$

Now let w be such a minimal index, greater that s, that

$$\sum_{t=w}^{p} (2\tau_t + \Delta(v_t)) \leqslant \bar{R}.$$

Due to remark 2, contractions of both chains $C_1 = (v_1, \ldots, v_s)$ and $C_2 = (v_w, \ldots, v_p)$ are valid. Indeed, $\text{dist}(v_0, C_1) = \tau_0$ and $\text{dist}(v_0, C_2) = \tau_p$, thus (8) follows.

Let us prove that $w \leqslant s + 2$. Suppose otherwise. By the choice of s and w we have

$$\sum_{t=1}^{s+1} (2\tau_{t-1} + \Delta(v_t)) > \bar{R},$$

$$\sum_{t=w-1}^{p} (2\tau_t + \Delta(v_t)) > \bar{R}.$$

Combining those inequalities, we have a contradiction with (7).

There is at most one non-depot node not covered by chains C_1 and C_2. Therefore, we reduced the transportation network into a cycle of at most four nodes (might be less, if one of the chains is empty, or $w = s+1$). All the possible results are pictured in Fig. 5a)–d).

Case 2. There is a critical chord $[v_0, v_{s+1}]$.
Due to the criticality, there is a job J_j at v_{s+1} such that

$$\bar{R} < 2\widehat{\text{dist}}(v_0, v_{s+1}) + d_j \leqslant T^* + d_j. \tag{9}$$

Let us prove, that contraction of the chain $C_1 = (v_1, \ldots, v_s)$ is valid. Note that

$$\text{dist}(v_0, C_1) + \sum_{t=1}^{s-1} \tau_t \leqslant \sum_{t=0}^{s} \tau_t \leqslant T^* - \widehat{\text{dist}}(v_0, v_{s+1}). \tag{10}$$

Indeed, either $\sum_{t=0}^{s} \tau_t = \widehat{\text{dist}}(v_0, v_{s+1}) \leqslant \frac{1}{2}T^*$, or $\widehat{\text{dist}}(v_0, v_{s+1}) = \sum_{t=s+1}^{p} \tau_t$ and $T^* = \sum_{t=0}^{s} \tau_t + \widehat{\text{dist}}(v_0, v_{s+1})$.

Assume (8) does not hold for C_1:

$$2\text{dist}(v_0, C_1) + W(C) = 2(\text{dist}(v_0, C_1) + \sum_{t=1}^{s-1} \tau_t) + \sum_{t=1}^{s} \Delta(v_t) > \bar{R},$$

and by (10) $\bar{R} < 2(T^* - \widehat{\text{dist}}(v_0, v_{s+1}) + \sum_{t=1}^{s} \Delta(v_t)$. Combining this with (9) we have a contradiction with (7). Similar consideration works for the chain $C_2 = (v_{s+2}, \ldots, v_p)$:

The result of the transformation is shown in Fig. 5e). □

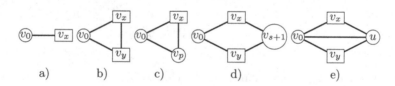

a)   b)   c)   d)   e)

Fig. 5. Possible results of instance reduction, Theorem 2.

4   Conclusions

Although we still don't know the value $\alpha(\mathcal{I}_2)$, Theorem 2 allows to finally perform that research. One just need to consider special case with at most four nodes. Moreover, the structure of the resulting instances from the kernel is not arbitrary: all the possible variants are shown in Fig. 5. Case a) is trivial: for such an instance a schedule of makespan $\bar{R}$ can be easily constructed. For the triangular network (cases b) and c)) the research is partly performed in [8], however, the proof has to be revised in order to confirm special treatment of special nodes. Our working conjecture is that $\alpha(\mathcal{I}_2) = \frac{6}{5}$.

We should note that instance transformation, described in Theorem 2 relies on the knowledge of the optimal solution for the underlying TSP and, therefore, cannot be done in polynomial time in general case. This doesn't break the theoretical result, but can serve as a basis for a approximation algorithm (supposedly $\frac{6}{5}$-approximation) only for the special cases with simple structure of the transportation network, allowing polynomial-time solution of the TSP.

We suggest the following directions for the future research.

1. Complete the investigation of $\alpha(\mathcal{I}_2)$.
2. Generalize the main result for the $ROm||R_{\max}$ problem.
3. Perform the similar research for a generalization $RO2|Qtt|R_{\max}$, in which each machine travels with its own speed (and therefore, travel times are proportional).
4. Note that this idea doesn't work for the general $RO2|Rtt|R_{\max}$ problem, in which travel times of machines are unrelated, because in this case optimal routes for two machines might be different. This makes the research of optima localization in this case even more interesting.

References

1. Averbakh, I., Berman, O., Chernykh, I.: A 6/5-approximation algorithm for the two-machine routing open-shop problem on a two-node network. Euro. J. Oper. Res. **166**(1), 3–24 (2005). https://doi.org/10.1016/j.ejor.2003.06.050
2. Averbakh, I., Berman, O., Chernykh, I.: The routing open-shop problem on a network: complexity and approximation. Eur. J. Oper. Res. **173**(2), 531–539 (2006). https://doi.org/10.1016/j.ejor.2005.01.034
3. van Bevern, R., Slugina, V.A.: A historical note on the 3/2-approximation algorithm for the metric traveling salesman problem. Hist. Math. **53**, 118–127 (2020). https://doi.org/10.1016/j.hm.2020.04.003
4. Chernykh, I., Kononov, A.V., Sevastyanov, S.: Efficient approximation algorithms for the routing open shop problem. Comput. Oper. Res. **40**(3), 841–847 (2013). https://doi.org/10.1016/j.cor.2012.01.006
5. Chernykh, I., Krivonogiva, O.: Optima localization for the two-machine routing open shop on a tree (in Russian) (2021), submitted to Diskretnyj Analiz i Issledovanie Operacij
6. Chernykh, I., Krivonogova, O.: Efficient algorithms for the routing open shop with unrelated travel times on cacti. In: Jaćimović, M., Khachay, M., Malkova, V., Posypkin, M. (eds.) OPTIMA 2019. CCIS, vol. 1145, pp. 1–15. Springer, Cham (2020). https://doi.org/10.1007/978-3-030-38603-0_1
7. Chernykh, I., Krivonogova, O.: On the optima localization for the three-machine routing open shop. In: Kononov, A., Khachay, M., Kalyagin, V.A., Pardalos, P. (eds.) MOTOR 2020. LNCS, vol. 12095, pp. 274–288. Springer, Cham (2020). https://doi.org/10.1007/978-3-030-49988-4_19
8. Chernykh, I., Lgotina, E.: The 2-machine routing open shop on a triangular transportation network. In: Kochetov, Y., Khachay, M., Beresnev, V., Nurminski, E., Pardalos, P. (eds.) DOOR 2016. LNCS, vol. 9869, pp. 284–297. Springer, Cham (2016). https://doi.org/10.1007/978-3-319-44914-2_23

9. Chernykh, I., Lgotina, E.: Two-machine routing open shop on a tree: instance reduction and efficiently solvable subclass. Optim. Methods Softw. (2020). https://doi.org/10.1080/10556788.2020.1734802

10. Chernykh, I., Pyatkin, A.: Irreducible bin packing: complexity, solvability and application to the routing open shop. In: Matsatsinis, N.F., Marinakis, Y., Pardalos, P. (eds.) LION 2019. LNCS, vol. 11968, pp. 106–120. Springer, Cham (2020). https://doi.org/10.1007/978-3-030-38629-0_9

11. Christofides, N.: Worst-Case Analysis of a New Heuristic for the Travelling Salesman Problem. Report 388, Graduate School of Industrial Administration, Carnegie-Mellon University, Pittsburg, PA (1976)

12. Christofides, N.: Graph Theory: An Algorithmic Approach. Academic Press, New York (1975)

13. Gonzalez, T.F., Sahni, S.: Open shop scheduling to minimize finish time. J. ACM 23(4), 665–679 (1976). https://doi.org/10.1145/321978.321985

14. Johnson, S.M.: Optimal two- and three-stage production schedules with setup times included. Naval Res. Logist. Q. 1(1), 61–68 (1954). https://doi.org/10.1002/nav.3800010110

15. Serdyukov, A.: On some extremal routes on graphs (in Russian). Upravlyaemye Syst. 17, 76–79 (1978)

16. Sevastyanov, S.V., Chemisova, D.A., Chernykh, I.D.: Some properties of optimal schedules for the Johnson problem with preemption. J. Appl. Ind. Math. 1(3), 386–397 (2007). https://doi.org/10.1134/s1990478907030143

17. Sevastyanov, S.V., Chemisova, D.A., Chernykh, I.D.: On some properties of optimal schedules in the job shop problem with preemption and an arbitrary regular criterion. Ann. Oper. Res. 213(1), 253–270 (2012). https://doi.org/10.1007/s10479-012-1290-3

18. Sevastianov, S.V., Tchernykh, I.D.: Computer-aided way to prove theorems in scheduling. In: Bilardi, G., Italiano, G.F., Pietracaprina, A., Pucci, G. (eds.) ESA 1998. LNCS, vol. 1461, pp. 502–513. Springer, Heidelberg (1998). https://doi.org/10.1007/3-540-68530-8_42

Minimizing Total Completion Time in Multiprocessor Job Systems with Energy Constraint

Alexander Kononov[ID] and Yulia Kovalenko[(✉)][ID]

Sobolev Institute of Mathematics, Novosibirsk, Russia
alvenko@math.nsc.ru, kovalenko@ofim.oscsbras.ru

Abstract. We consider the problem of scheduling multiprocessor jobs to minimize the total completion time under the given energy budget. Each multiprocessor job requires more than one processor at the same moment of time. Processors may operate at variable speeds. Running a job at a slower speed is more energy efficient, however it takes longer time and affects the performance. The complexity of both parallel and dedicated versions of the problem is investigated. We propose approximation algorithms for various particular cases. In our algorithms, initially a sequence of jobs and their processing times are calculated and then a feasible solution is constructed using list-type scheduling rule.

Keywords: Multiprocessor job · Speed scaling · Scheduling · Approximation algorithm · NP-hardness

1 Introduction

We investigate the problem of non-preemptive scheduling a set of jobs $\mathcal{J} = \{1, \ldots, n\}$ on m speed scalable parallel processors. Each job $j \in \mathcal{J}$ is characterized by processing volume (work) V_j and the number $size_j$ or the set fix_j of required processors. Note that parameter $size_j$ for job $j \in \mathcal{J}$ indicates that the job can be processed on any subset of parallel processors of the given size. Such jobs are called rigid jobs [5]. Parameter fix_j states that the job uses the prespecified subset of dedicated processors. Such jobs are called single mode multiprocessor jobs [5]. We also consider moldable jobs [5]. In contrast to the previous job types, a moldable job j may be performed on any number of processors lower or equal to the given upper bound δ_j.

The standard homogeneous model in speed-scaling is considered. When a processor runs at a speed s, then the rate with which the energy is consumed (the *power*) is s^α, where $\alpha > 1$ is a constant (usually, $\alpha \approx 3$). Each of m processors may operate at variable speed. However, we assume that the total work V_j of a job $j \in \mathcal{J}$ should be uniformly divided between the utilized processors, i.e. if job j uses m_j processors, then processing volumes are the same for all m_j processors

© Springer Nature Switzerland AG 2021
P. Pardalos et al. (Eds.): MOTOR 2021, LNCS 12755, pp. 267–279, 2021.
https://doi.org/10.1007/978-3-030-77876-7_18

(denoted by $W_j := \frac{V_j}{m_j}$), and all these processors run at the same speed. It is supposed that a continuous spectrum of processor speeds is available.

The aim is to find a feasible schedule with the minimum sum of completion times $\sum C_j$ so that the energy consumption is not greater than a given energy budget E. This is a natural assumption in the case when the energy of a battery is fixed, i.e. the problem finds applications in computer devices whose lifetime depends on a limited battery efficiency (for example, multi-core laptops). Moreover, the bicriteria problems of minimizing energy consumption and a scheduling metric arise in real practice. The most obvious approach is to bound one of the objective functions and optimize the other. The energy of the battery may reasonably be estimated, so we bound the energy used, and optimize the regular timing criterion.

The non-preemptive rigid, moldable and single-mode variants of the speed-scaling scheduling subject to bound on energy consumption are denoted by $P|size_j, energy|\sum C_j$, $P|var, \delta_j, energy|\sum C_j$ and $P|fix_j, energy|\sum C_j$, respectively.

2   Previous Research

Pruhs et al. [14] investigated the single-processor problem of minimizing the average flow time of jobs, given a fixed amount of energy and release times of jobs. For unit-work jobs, they proposed a polynomial-time algorithm that simultaneously computes, for each possible energy level, the schedule with smallest average flow time. Bunde [3] adopted the approach to multiple processors. $O(1)$-approximation algorithm, allowing an additional factor of $(1 + \varepsilon)$ energy, has been proposed for scheduling arbitrary work jobs on single processor. Albers and Fujiwara [1] have investigated online and offline versions of single-processor scheduling to minimize energy consumption plus job flow times. A deterministic constant competitive online algorithm and offline dynamic programming algorithm with polynomial time complexity were proposed for unit-work jobs.

Shabtay et al. [16] analyzed a closely related problem of scheduling single-processor jobs on identical parallel processors, where job-processing times p_j are controllable through the allocation of a nonrenewable common limited resource as $p_j(R_j) = \left(\frac{W_j}{R_j}\right)^\kappa$. Here W_j is the workload of job j, R_j is the amount of resource allocated to processing job j and $0 < \kappa \leq 1$ is a positive constant. Exact polynomial time algorithm was proposed for the multiprocessor non-preemptive instances of minimizing the sum of completion times. The algorithm can be adopted to the speed scaling scheduling of single-processor jobs.

The speed scaling scheduling with makespan criterion has been widely investigated. Various approaches to construct approximation algorithms for single-processor and multiprocessor jobs were proposed (see, e.g., [3,11,14]).

Now we review the known results for the the classic problem of scheduling multiprocessor jobs with given durations and without energy constraint. The non-preemptive problem with rigid jobs is strongly NP-hard even in the case of two processors [13], the preemptive one is NP-hard when the number of processors is a part of the input [6]. For the non-preemptive instances constant factor approximation algorithms were proposed. These algorithms use list-type scheduling [17] and scheduling to minimize average response time (SMART) [15]. The non-preemptive single-mode problem is NP-hard in the case of two processors [10] and strongly NP-hard in the case of two-processor jobs [12]. The strategy from preemptive schedule to the non-preemptive one gives a 2-approximation algorithm for two-processor problem [4], and First Fit Coloring strategy allows to obtain a 2-approximate solution for unit-work two-processor jobs [7].

Our Results. We prove NP-hardness of problems $P|size_j \leq \frac{m}{2}, energy| \sum C_j$ and $P|fix_j, |fix_j| = 2, energy| \sum C_j$, and develop two-stage approximation algorithms for the following particular cases:

- rigid jobs requiring at most $\frac{m}{2}$ processors,
- moldable jobs,
- two-processor dedicated instances.

At the first stage, we obtain a lower bound on the total completion time and calculate a sequence and processing times of jobs using an auxiliary convex program. Then, at the second stage, we transform our problem to the classic scheduling problem without speed scaling, and we use "list-scheduling" algorithms to obtain feasible solutions. Whenever a subset of processors falls idle, a "list-scheduling" algorithm schedules from a given priority list the first job that does not require more processors than are available.

3   Rigid Jobs

In this section we consider rigid jobs. Firstly, we prove that the problem is NP-hard. Secondly, 2-approximation algorithm is presented for jobs, which require at most $\frac{m}{2}$ processors and have identical workloads on utilized processors (i.e. $V_j = W \cdot size_j$).

3.1   NP-Hardness

Theorem 1. *Problem* $P|size_j \leq \frac{m}{2}, W_j = 1, energy| \sum C_j$ *is NP-hard in the strong sense.*

Proof. We show that the strongly NP-complete 3-PARTITION problem polynomially transforms to the decision version of scheduling problem $P|size_j \leq \frac{m}{2}, W_j = 1, energy| \sum C_j$.

We consider an instance of the 3-PARTITION problem: Given a set of $3q$ elements with weights a_j, $j = 1, \ldots, 3q$, where $\sum_{j=1}^{3q} a_j = Bq$ and $\frac{B}{4} \leq a_j \leq \frac{B}{2}$. Could the set be partitioned into q subsets $A_1, \ldots, A_q$ such that $\sum_{j \in A_i} a_j = B$?

An instance of $P|size_j, energy| \sum C_j$ is constructed as follows. Put the number of jobs $n = 3q$, the number of processors $m = B$, and the energy budget $E = Bq$. For every a_j we generate a job j, $j = 1, \ldots, 3q$. We set $W_j = 1$, $size_j = a_j$, $V_j = a_j$ for $j \in \mathcal{J}$. In the decision version of $P|size_j, W_j = 1, energy| \sum C_j$ it is required to answer the question: Is there a schedule with $\sum C_j$ value not greater than a given threshold T?

In order to determine the value of T we solve an auxiliary problem with $\sum_{j=1}^{3q} a_j$ single-processor jobs of unit works, i.e. each rigid job is replaced by $size_j$ single-processor jobs. Such problem has the unique optimal solution (with the accuracy of placing jobs on processors and permuting jobs on each processor), where each processor executes q jobs and uses energy budget q. Now we find optimal durations of jobs on each processor, solving the following convex program:

$$\sum_{j=1}^{q} p_j(q - j + 1) \to \min, \tag{1}$$

$$\sum_{j=1}^{q} p_j^{1-\alpha} = q, \tag{2}$$

$$p_j \geq 0, \; j = 1, \ldots, q. \tag{3}$$

Here p_j is the execution time of j-th job on a processor, $j = 1, \ldots, q$.

We compose the Lagrangian function $L(p_j, \lambda) = \sum_{j=1}^{q} p_j(q - j + 1) + \lambda \left(\sum_{j=1}^{q} p_j^{1-\alpha} - q \right)$ and calculate the optimal solution by equating to zero partial derivatives:

$$p_j^* = \frac{\left(\sum_{j=1}^{q}(n - j + 1)^{\frac{\alpha-1}{\alpha}} \right)^{\frac{1}{\alpha-1}}}{q^{\frac{1}{\alpha-1}}(q - j + 1)^{\frac{1}{\alpha}}}, \; j = 1, \ldots, q,$$

$$\sum C_j^* = \sum_{j=1}^{q} p_j(q - j + 1) = \left(\sum_{j=1}^{q}(n - j + 1)^{\frac{\alpha-1}{\alpha}} \right)^{\frac{\alpha}{\alpha-1}} q^{\frac{1}{1-\alpha}}.$$

Note that each next job has more duration than the previous one. The optimal schedule for each processor does not have idle times. The optimal total completion time for all processors is equal to $m \sum C_j^*$.

Set the threshold $T := 3 \sum C_j^*$, since at most 3 rigid jobs can be executed in parallel. We show that a positive answer (a negative answer) to 3-PARTITION implies a positive answer (a negative answer) to the constructed $P|size_j, energy| \sum C_j$ with $\sum C_j \leq T$.

Firstly, we assume that the answer to 3-Partition is positive. Then there is a feasible schedule with $\sum C_j \leq T$, where three jobs, corresponding to three elements forming set A_i, $i = 1, \ldots, q$ such that $\sum_{j \in A_i} a_j = B$, are executed in parallel. This schedule is similar to the optimal schedule of the corresponding problem with single-processor jobs. The value of criterion is equal to $3 \sum C_j^*$.

Secondly, we show that the negative answer to 3-Partition implies the negative answer to our speed scaling scheduling problem. Indeed, in this case we can not construct a schedule, which is identical to the optimal schedule for the corresponding single-processor jobs (with the accuracy of placing jobs on processors and permuting jobs on each processor). In other words, in any feasible schedule the sum of completion times is greater than $3 \sum C_j^*$ as this schedule has idle times.

The presented transformation is polynomial. So, problem $P|size_j \leq \frac{m}{2}, W_j = 1, energy| \sum C_j$ is strongly NP-hard.   □

In the next subsection, we present a polynomial time approximation algorithm with constant factor approximation guarantee. Processing times of jobs are calculated using convex program and approximate schedule is constructed by "list-scheduling" algorithm.

3.2   Approximation Algorithm

The sequence of jobs and the completion time of each job are important in the problems with criterion $\sum C_j$. Now we compute a lower bound for the case when a jobs sequence is given. Suppose that the jobs are started in accordance with permutation $\pi = (\pi_1, \ldots, \pi_n)$. Using the lower bound on the total completion time presented in [17] for rigid jobs with the given durations, we formulate the following convex program:

$$\frac{1}{m} \sum_{j=1}^{n} \sum_{i=1}^{j} size_{\pi_i} p_{\pi_i} + \frac{1}{2} \sum_{j=1}^{n} p_{\pi_j} - \frac{1}{2m} \sum_{j=1}^{n} size_{\pi_j} p_{\pi_j} \rightarrow \min, \tag{4}$$

$$\sum_{j=1}^{n} W_j^{\alpha} p_j^{1-\alpha} size_j \leq E. \tag{5}$$

$$p_j \geq 0, \ j \in \mathcal{J}. \tag{6}$$

We solve the program by means of the Lagrangian method. Define the Lagrangian function $L(p_{\pi_j}, \lambda)$ as

$$L(p_{\pi_j}, \lambda) = \frac{1}{m} \sum_{j=1}^{n} \sum_{i=1}^{j} size_{\pi_i} p_{\pi_i} + \frac{1}{2} \sum_{j=1}^{n} \left(1 - \frac{size_{\pi_j}}{m}\right) p_{\pi_j}$$

$$+ \lambda \left(\sum_{j=1}^{n} W_{\pi_j}^{\alpha} p_{\pi_j}^{1-\alpha} size_{\pi_j} - E \right).$$

The necessary and sufficient conditions for an optimal solution are (partial derivatives are equal to zero):

$$\frac{\partial L}{\partial p_{\pi_i}} = \frac{1}{m} size_{\pi_i}(n-i+1) + \frac{1}{2}\left(1 - \frac{size_{\pi_i}}{m}\right) + \lambda W_{\pi_i}^{\alpha} size_{\pi_i}(1-\alpha)p_{\pi_i}^{-\alpha} = 0,$$

$$i = 1, \ldots, n.$$

Rewriting the expressions, we obtain

$$p_{\pi_i} = ((\alpha-1)\lambda m)^{\frac{1}{\alpha}} \frac{W_{\pi_i} size_{\pi_i}^{\frac{1}{\alpha}}}{(size_{\pi_i}(n-i+0.5)+0.5m)^{\frac{1}{\alpha}}},$$

$$i = 1, \ldots, n.$$

The processing times are placed into equation

$$\frac{\partial L}{\partial \lambda} = \sum_{j=1}^{n} W_j^{\alpha} p_j^{1-\alpha} size_j - E = 0.$$

As a result we calculate the durations of jobs

$$p_{\pi_i} = \frac{E^{\frac{1}{1-\alpha}} W_{\pi_i} size_{\pi_i}^{\frac{1}{\alpha}}}{(size_{\pi_i}(n-i+0.5)+0.5m)^{\frac{1}{\alpha}}} \cdot \left(\sum_{j=1}^{n} \frac{W_{\pi_j} size_{\pi_j}^{\frac{1}{\alpha}}}{(size_{\pi_j}(n-j+0.5)+0.5m)^{\frac{1-\alpha}{\alpha}}}\right)^{\frac{1}{\alpha-1}},$$

$$i = 1, \ldots, n.$$

The obtained values for execution times are placed in expression (4) and the lower bound on $\sum C_j$ in the general case for an arbitrary permutation π of jobs is calculated as follows

$$LB(\pi) = \frac{E^{\frac{1}{1-\alpha}}}{m}\left(\sum_{i=1}^{n} W_{\pi_i} size_{\pi_i}^{\frac{1}{\alpha}}(size_{\pi_i}(n-i+0.5)+0.5m)^{\frac{\alpha-1}{\alpha}}\right)^{\frac{\alpha}{\alpha-1}}. \quad (7)$$

Here π_i is the i-th job in accordance with permutation π. So, it is required to find permutation, that gives $\min_{\pi} LB(\pi)$.

From now on, we suppose that the processing works of jobs on processors are identical, i.e. $W_j = W$, $j \in \mathcal{J}$. Then the minimization of (7) is equivalent to the minimization of

$$G(\pi) = \sum_{i=1}^{n} W size_{\pi_i}\left(n-i+0.5+\frac{0.5m}{size_{\pi_i}}\right)^{\frac{\alpha-1}{\alpha}}.$$

We define vectors $WS_{\pi} = (W size_{\pi_1}, W size_{\pi_2}, \ldots, W size_{\pi_n})$ and $NS_{\pi} = \left(\left(n-0.5+\frac{0.5m}{size_{\pi_1}}\right)^{\frac{\alpha-1}{\alpha}}, \left(n-1.5+\frac{0.5m}{size_{\pi_2}}\right)^{\frac{\alpha-1}{\alpha}}, \ldots, \left(0.5+\frac{0.5m}{size_{\pi_n}}\right)^{\frac{\alpha-1}{\alpha}}\right)$. Then $G(\pi)$ can be expressed as the following scalar product $WS_{\pi} \cdot (NS_{\pi})^{\mathrm{T}}$. It is easy to see that the minimum of $WS_{\pi} \cdot (NS_{\pi})^{\mathrm{T}}$ is reached on the permutation, where the jobs are ordered by non-decreasing of the required processors numbers $size_j$, $j \in \mathcal{J}$. Indeed, in this case the first vector is ordered nondecreasingly, the second one is ordered nonincreasingly and the scalar product is minimal (the proof is based on the permutation approach).

Let the jobs are ordered by non-decreasing of $size_j$. We denote by $\bar{p}_j$ the durations of jobs, and by $LB(\bar{p}_j) = \frac{1}{m} \sum_{j=1}^{n} \sum_{i=1}^{j-1} size_j \bar{p}_j + \frac{1}{2} \sum_{j=1}^{n} \bar{p}_j + \frac{1}{2m} \sum_{j=1}^{n} size_j \bar{p}_j$ the lower bound corresponding to the optimal solution of problem (4)–(6). We construct the schedule using "list-scheduling" algorithm with processing times $\bar{p}_j$, $j \in \mathcal{J}$: The first job is scheduled at time 0. The next job is scheduled at the earliest time such that there are enough processors to execute it. Recall that each job requires at most $\frac{m}{2}$ processors and $W_j = W$.

"List-scheduling" algorithm has time complexity $O(n^2)$ and allows to construct 2-approximate schedule. Indeed, the starting time of the j-th job is not greater than $2\left(\frac{1}{m} \sum_{i=1}^{j-1} size_i \bar{p}_i\right)$, as at least $\frac{m}{2}$ processors are busy at each time moment in the schedule. The completion time of the j-th job satisfies condition $C_j \leq \frac{2}{m} \sum_{i=1}^{j-1} size_i \bar{p}_i + \bar{p}_j$. In the sum we have

$$\sum_{j \in \mathcal{J}} C_j \leq 2\left(\frac{1}{m} \sum_{j=1}^{n} \sum_{i=1}^{j-1} size_i \bar{p}_i + \frac{1}{2} \sum_{j=1}^{n} \bar{p}_j\right).$$

We compare this value with the lower bound $LB(\bar{p}_j)$ and conclude that $\sum_{j \in \mathcal{J}} C_j \leq 2LB(\bar{p}_j)$.

Therefore, the following theorem takes place.

Theorem 2. *A 2-approximate schedule can be found in polynomial time for problem* $P|size_j, W_j = W, size_j \leq \frac{m}{2}, energy| \sum C_j$.

4   Moldable Jobs

Now we provide 2-approximation algorithm for moldable jobs with identical works. Recall that V_j denote the total processing volume of job j, i.e. the execution time of this job on one processor with unit speed. Let $\delta_j \leq m$ be the maximal possible number of processors, that may be utilized by job j.

In order to obtain a lower bound on the sum of completion times, we formulate the following convex model in the case of the given sequence π of jobs

$$\frac{1}{m} \sum_{j=1}^{n} \sum_{i=1}^{j} p_{\pi_i} + \frac{1}{2} \sum_{j=1}^{n} \frac{p_{\pi_j}}{\delta_{\pi_j}} - \frac{1}{2m} \sum_{j=1}^{n} p_{\pi_j} \rightarrow \min, \tag{8}$$

$$\sum_{j=1}^{n} V_j^{\alpha} p_j^{1-\alpha} \leq E. \tag{9}$$

Here p_j is the execution time of job j on one processor in the total volume V_j.

Using the same arguments as in Subsect. 3.2, we can show that the minimum of (8), say

$$LB := \frac{E^{1/1-\alpha}}{m} \left(\sum_{j=1}^{n} V_{\pi_j} \left(n - j + 0.5 + \frac{0.5m}{\delta_{\pi_j}} \right)^{\alpha-1/\alpha} \right)^{\alpha/\alpha-1},$$

is reached on the permutation π, where the jobs are ordered by non-decreasing of the maximum processors numbers δ_j, $j \in \mathcal{J}$, in the case when works of jobs V_j are identical or the non-decreasing order of V_j corresponds to the non-decreasing order of δ_j (i.e. $V_i < V_j$ implies $\delta_i \leq \delta_j$). The corresponding durations of jobs will be denoted by $\bar{p}_j$, $j \in \mathcal{J}$.

Now we assign the number of processors m_j for jobs as follows

$$m_j = \begin{cases} \delta_j & \text{if } \delta_j < \lceil \frac{m}{2} \rceil, \\ \lceil \frac{m}{2} \rceil & \text{if } \delta_j \geq \lceil \frac{m}{2} \rceil, \end{cases}$$

and construct the schedule using "list-scheduling" algorithm based on the order of jobs in non-decreasing of δ_j, $j \in \mathcal{J}$. Let us prove that the total completion time $\sum C_j(\bar{p}_j) \leq 2LB$. Indeed,

$$\sum C_j(\bar{p}_j) \leq \frac{2}{m} \sum_{j=1}^{n} \sum_{i=1}^{j-1} \bar{p}_i + \sum_{j=1}^{n} \frac{\bar{p}_j}{m_j} = \frac{2}{m} \sum_{j=1}^{n} \sum_{i=1}^{j} \bar{p}_i + \sum_{j=1}^{n} \frac{\bar{p}_j}{m_j} - \frac{2}{m} \sum_{j=1}^{n} \bar{p}_j$$

$$= \frac{2}{m} \sum_{j=1}^{n} \sum_{i=1}^{j} \bar{p}_i + \sum_{j=1}^{n} \bar{p}_j \left(\frac{1}{m_j} - \frac{2}{m} \right) \leq \frac{2}{m} \sum_{j=1}^{n} \sum_{i=1}^{j} \bar{p}_i + \sum_{j=1}^{n} \bar{p}_j \left(\frac{1}{\delta_j} - \frac{1}{m} \right) \leq 2LB.$$

Therefore, we have

Theorem 3. *A 2-approximate schedule can be found in polynomial time for problem* $P|any, V_j = V, \delta_j, energy| \sum C_j$.

We note here, that the complexity status of speed scaling scheduling problem $P|any, V_j = V, \delta_j, energy| \sum C_j$ is open.

5   Single Mode Multiprocessor Jobs

In this section we consider single-mode multiprocessor jobs. Firstly, we prove that the problem is NP-hard. Then we provide a polynomial time algorithm for two-processor instances.

5.1 NP-Hardness

Theorem 4. *Problem* $P|fix_j, |fix_j| = 2, V_j = 2, energy| \sum C_j$ *is NP-hard in the strong sense.*

Proof. The proof is similar to the proof of Theorem 1, but it is based on the polynomial reduction of the strongly NP-complete Chromatic Index problem for cubic graphs [9].

The chromatic index of a graph is the minimum number of colors required to color the edges of the graph in such a way that no two adjacent edges have the same color. Consider an instance of the Chromatic Index problem on a cubic graph $G = (V, A)$, which asks whether the chromatic index $\chi'(G)$ is three. It is well-known that $\chi'(G) = 3$ or 4, and $|V|$ is even. Moreover, $\chi'(G) = 3$ if and only if each color class has exactly $\frac{1}{2}|V|$ edges.

We construct an instance of $P|fix_j, energy| \sum C_j$ as follows. Put the number of jobs $n = |A|$, the number of processors $m = |V|$ and the energy budget $E = 2|A| = 3|V|$. Vertices correspond to processors. For every edge $\{u_j, v_j\}$ we generate a job j with $fix_j = \{u_j, v_j\}$, $j = 1, \ldots, |A|$. We set $V_j = 2$ and $W_j = \frac{V_j}{|fix_j|} = 1$ for all $j = 1, \ldots, n$. In the decision version of $P|fix_j, W_j = 1, energy| \sum C_j$ it is required to answer the question: Is there a schedule, in which the total completion time is not greater than a given threshold T?

In order to define the value of T we solve auxiliary problem with $2n$ single-processor jobs, i.e. each two processor job is replaced by two single-processor jobs. Such problem has the unique optimal solution (with the accuracy of permuting jobs on processors), where each processor execute tree jobs and uses energy budget 3. Now we find optimal durations of jobs on each processor, solving the following convex program:

$$p_1 + 2p_2 + p_3 \to \min, \tag{10}$$

$$\sum_{j=1}^{3} p_j^{1-\alpha} = 3, \tag{11}$$

$$p_j \geq 0, \ j = 1, \ldots, n. \tag{12}$$

Here p_j is the execution time of j-th job on a processor.

We compose the Lagrangian function

$$L(p_j, \lambda) = (3p_1 + 2p_2 + p_3) + \lambda \left(p_1^{1-\alpha} + p_2^{1-\alpha} + p_3^{1-\alpha} - 3 \right)$$

and calculate

$$p_j^* = \frac{3^{1/1-\alpha}}{(4-j)^{1/\alpha}} \left(3^{\alpha-1/\alpha} + 2^{\alpha-1/\alpha} + 1 \right)^{1/\alpha-1}, \ j = 1, 2, 3, \tag{13}$$

$$\sum C_j^* = \left(\frac{\left(3^{\alpha-1/\alpha} + 2^{\alpha-1/\alpha} + 1 \right)^{\alpha}}{3} \right)^{1/\alpha-1}. \tag{14}$$

The sum of job completion times on all processors is equal to $m \sum C_j^*$. The optimal schedule does not have idle times. Set the threshold $T := \frac{m}{2} \sum C_j^*$ because at most $\frac{m}{2}$ two processor jobs can be executed in parallel. We prove that a positive answer (a negative answer) to Chromatic Index problem corresponds a positive answer (a negative answer) to the constructed decision version of $P|fix_j, W_j = 1, energy| \sum C_j$.

Now we assume that the answer to Chromatic Index problem is positive. Then there is a feasible schedule, where $\frac{m}{2}$ jobs, corresponding to $\frac{m}{2}$ edges forming one coloring class, are executed in parallel. This schedule is similar to the optimal schedule of the corresponding problem with single-processor jobs. The value of criterion is equal to $\frac{m}{2} \sum C_j^*$.

It is easy to see that the negative answer to Chromatic Index problem implies the negative answer to our scheduling problem. Indeed, in this case we can not construct a schedule, which is identical to the optimal schedule for the corresponding single-processor jobs (with the accuracy of permuting jobs on processors). In other words, any feasible schedule has idle times, and, therefore, the total sum of completion times is greater than $\frac{m}{2} \sum C_j^*$.

The presented reduction is polynomial. So, speed scaling scheduling problem $P|fix_j, W_j = 1, energy| \sum C_j$ is strongly NP-hard. $\qquad\square$

5.2   Two-Processor Instances

Now we consider the non-preemptive problem with two processors and propose polynomial time algorithm with constant-factor approximation guarantee. Denote by $\mathcal{J}_i$ the set of jobs using only processor $i = 1, 2$ and by $\mathcal{J}_{12}$ the set of two-processors jobs, $\mathcal{J} = \mathcal{J}_1 \cup \mathcal{J}_2 \cup \mathcal{J}_{12}$. In order to obtain a lower bound on the total completion time we identify two subproblems: the first one schedules only jobs from $\mathcal{J}' = \mathcal{J}_1 \cup \mathcal{J}_{12}$, $|J'| = n'$, the second one schedules only jobs from $\mathcal{J}'' = \mathcal{J}_2 \cup \mathcal{J}_{12}$, $|J''| = n''$.

The optimal solution of the first subproblem in the case of the given sequence π of jobs can be found by solving the following convex program:

$$\sum_{i=1}^{n'} (n' - i + 1) p_{\pi_i} \rightarrow \min,$$

$$\sum_{i \in \mathcal{J}'} |fix_i|(p_i)^{1-\alpha} W_i^\alpha \leq E.$$

Solving this subproblem via KKT-conditions, we obtain durations of jobs

$$p_{\pi_i} = \left(\frac{E}{\sum_{j=1}^{n'} W_{\pi_j} |fix_{\pi_j}|^{1/\alpha} (n' - j + 1)^{\alpha - 1/\alpha}} \right)^{1/1-\alpha} \frac{W_{\pi_i} |fix_{\pi_i}|^{1/\alpha}}{(n' - i + 1)^{1/\alpha}},$$

$$i = 1, \ldots, n',$$

and the sum of completion times

$$\sum C_j^1(\pi) = (E)^{1/1-\alpha} \left(\sum_{j=1}^{n'} W_{\pi_j} |fix_{\pi_j}|^{1/\alpha} (n'-j+1)^{\alpha-1/\alpha} \right)^{\alpha/\alpha-1}.$$

So, the minimum sum of completion times is reached on the permutation π', where the jobs are ordered by non-decreasing of $W_i |fix_i|^{1/\alpha}$, $i \in \mathcal{J}'$, since values $(n'-j+1)$ decrease.

Using the same approach for the second subproblem we conclude that the minimum sum of completion times $\sum C_j^2$ is reached on the permutation π'', where the jobs are ordered by non-decreasing of $W_i |fix_i|^{1/\alpha}$, $i \in \mathcal{J}''$. So, the lower bound LB for the general problem can be calculated as $\max\{\sum C_j^1(\pi'), \sum C_j^2(\pi'')\}$. Note that subsequences of two-processor jobs are identical in optimal solutions of both subproblems.

Decreasing the energy budget in both subproblems in two times, we obtain $2^{1/\alpha-1}$-approximate solutions S' and S'' for them with the same sequences of jobs as in case of energy budget E. Let C_j' and C_j'' (p_j' and p_j'') denote the completion times (the processing times) of two processor job j in S' and S'', respectively. Now we construct a preemptive schedule for the general problem, and then transform this schedule to the non-preemptive one.

In the constructed preemptive schedule each two-processor job j is executed without preemptions in interval $(\max\{C_j', C_j''\} - \min\{p_j', p_j''\}, \max\{C_j', C_j''\}]$. Note that execution intervals of two-processor jobs do not intersect each other. Single-processor jobs are performed in the same order and with the same durations as in S' and S'', but may be preempted by two processor jobs (idle times between single-processor jobs are not allowed). It is easy to see that the schedule is feasible, and has the total completion time $\sum C_j \leq 2 \cdot 2^{1/\alpha-1} LB$ as the completion times of single-processor jobs are no later than in S' and S'' by construction.

Now we go to calculate a non-preemptive feasible schedule. The obtained preemptive schedule may be reconstructed without increasing the completion times of jobs such that at most one single-processor job is preempted by each two-processor job (if a two-processor job j preempts two single-processor jobs, then moving j slightly earlier will lower the completion time of j without affecting the completion times of any other jobs). Let $S'(j)$ and $C'(j)$ denote the starting time and completion time, respectively, of any job j in the reconstructed preemptive schedule. Identify single-processor jobs $j_{i_1}, j_{i_2}, \ldots, j_{i_k}$ that are preempted by some two-processor jobs. Suppose that these jobs are ordered by increasing of starting times $S'(j_{i_1}) < \cdots < S'(j_{i_k})$, and therefore completion times satisfy $C'(j_{i_1}) < \cdots < C'(j_{i_k})$. Let $F(j_{i_l})$ be the last two-processor job that preempts j_{i_l}, $g(j_{i_l})$ be the amount of processing time of j_{i_l} scheduled before the starting time of $F(j_{i_l})$, $h(j_{i_l})$ be the number of jobs that complete later than $F(j_{i_l})$, $l = 1, \ldots, k$. The construction procedure consists of k steps. At step l we insert an idle time period of length $g(j_{i_l})$ on both processors immediately after the completion of $F(j_{i_l})$. Change the start time of j_{i_l} to the completion time of $F(j_{i_l})$. So, at step l, inserting the idle time period will increase the total completion time of the schedule by $g(j_{i_l}) \cdot h(j_{i_l})$, and in total after all steps the non-preemptive schedule has the objective value

$$\sum C_j^{npr} \leq \sum C_j + \sum_{l=1}^{k} g(j_{i_l}) \cdot h(j_{i_l}).$$

Since each two-processor job preempts at most one single-processor job, the time intervals $(S'(j_{i_1}), C'(F(j_{i_1}))], (S'(j_{i_2}), C'(F(j_{i_2}))], \ldots, (S'(j_{i_k}), C'(F(j_{i_k}))]$ do not overlap with each other. Therefore,

$$\sum C_j \geq \sum_{l=1}^{k} (C'(F(j_{i_l})) - S'(j_{i_l})) \cdot h(j_{i_l}) > \sum_{l=1}^{k} g(j_{i_l}) \cdot h(j_{i_l}).$$

As a result we have the following bound on the total completion time

$$\sum C_j^{npr} \leq \sum C_j + \sum_{l=1}^{k} g(j_{i_l}) \cdot h(j_{i_l}) < 2 \sum C_j \leq 2^{2\alpha-1/\alpha-1} LB.$$

Theorem 5. *A $2^{2\alpha-1/\alpha-1}$-approximate schedule can be found in polynomial time for problem $P2|fix_j, energy| \sum C_j$.*

Corollary 1. *A $2^{\alpha/\alpha-1}$-approximate schedule can be found in polynomial time for problem $P2|fix_j, pmtn, energy| \sum C_j$.*

The complexity status of both preemptive and non-preemptive problems with two processors is open.

Conclusion

NP-hardness of both parallel and dedicated versions of the speed scaling problem with the total completion time criterion under the given energy budget is proved. We propose an approach to construct approximation algorithms for various particular cases of the problem. In our algorithms, initially a sequence of jobs and their processing times are calculated and then a feasible solution is constructed using list-type scheduling rule.

Further research might address the approaches to the problems with more complex structure, where processors are heterogeneous and jobs have alternative execution modes with various characteristics. Open questions are the complexity status of the problem with moldable jobs and two-processor dedicated problem, and constant factor approximation guarantee for two-processor jobs in the system with arbitrary number of processors.

Acknowledgements. The reported study was funded by RFBR, project number 20-07-00458.

References

1. Albers, S., Fujiwara, H.: Energy-efficient algorithms for flow time minimization. ACM Trans. Algorithms **3**(4), 17 (2007)
2. Bampis, E., Letsios, D., Lucarelli, G.: A note on multiprocessor speed scaling with precedence constraints. In: Proceedings of the 26th ACM Symposium on Parallelism in Algorithms and Architectures (SPAA 2014), pp. 138–142 (2014)
3. Bunde, D.P.: Power-aware scheduling for makespan and flow. J. Sched. **12**, 489–500 (2009)
4. Cai, X., Lee, C.-Y., Li, C.-L.: Minimizing total completion time in two-processor task systems with prespecified processor allocations. Naval Res. Logisti. (NRL) **45**(2), 231–242 (1998)
5. Drozdowski, M.: Scheduling for Parallel Processing. Springer-Verlag, London (2009). https://doi.org/10.1007/978-1-84882-310-5
6. Drozdowski, M., DellOlmo, P.: Scheduling multiprocessor tasks for mean flow time criterion. Comput. Oper. Res. **27**(6), 571–585 (2000)
7. Giaro, K., Kubale, M., Maiafiejski, M., Piwakowski, K.: Chromatic scheduling of dedicated 2-processor UET tasks to minimize mean flow time. In: Proceedings of 7th IEEE International Conference on Emerging Technologies and Factory Automation (ETFA 1999), pp. 343–347 (1999)
8. Glasgow, J., Shachnai, H.: Minimizing the Flow Time for Parallelizable Task Systems. Tech, Report (1993)
9. Holyer, I.: The NP-completeness of edge-coloring. SIAM J. Comput. **10**, 718–720 (1981)
10. Hoogeveen, J.A., van de Velde, S.L., Veltman, B.: Complexity of scheduling multiprocessor tasks with prespecified processor allocations. Discrete Appl. Math. **55**(3), 259–272 (1994)
11. Kononov, A., Kovalenko, Y.: Makespan minimization for parallel jobs with energy constraint. In: Kononov, A., Khachay, M., Kalyagin, V.A., Pardalos, P. (eds.) MOTOR 2020. LNCS, vol. 12095, pp. 289–300. Springer, Cham (2020). https://doi.org/10.1007/978-3-030-49988-4_20
12. Kubale, M.: Preemptive versus nonpreemtive scheduling of biprocessor tasks on dedicated processors. Eur. J. Oper. Res. **94**(2), 242–251 (1996)
13. Lee, C.-Y., Cai, X.: Scheduling one and two-processor tasks on two parallel processors. IEE Trans. **31**(5), 445–455 (1999)
14. Pruhs, K., Uthaisombut, P., Woeginger, G.: Getting the best response for your erg. ACM Trans. Algorithms **4**(3), 17 (2008)
15. Schwiegelshohn, U., Ludwig, W., Wolf, J.L., Turek, J., Yu, P.S.: Smart SMART bounds for weighted response time scheduling. SIAM J. Comput. **28**, 237–253 (1998)
16. Shabtay, D., Kaspi, M.: Parallel machine scheduling with a convex resource consumption function. Eur. J. Oper. Res. **173**, 92–107 (2006)
17. Turek, J., et al.: Scheduling parallelizable tasks to minimize average response time. In: Proceedings of the Sixth Annual ACM Symposium on Parallel Algorithms and Architectures, pp. 200–209 (1994)

Maximising the Total Weight of On-Time Jobs on Parallel Machines Subject to a Conflict Graph

Yakov Zinder[1] , Joanna Berlińska[2(✉)] , and Charlie Peter[1]

[1] School of Mathematical and Physical Sciences, University of Technology Sydney, Ultimo, Australia
Yakov.Zinder@uts.edu.au, 13240039@student.uts.edu.au
[2] Faculty of Mathematics and Computer Science, Adam Mickiewicz University, Poznań, Poland
Joanna.Berlinska@amu.edu.pl

Abstract. The paper considers scheduling on parallel machines under the constraint that some pairs of jobs cannot be processed concurrently. Each job has an associated weight, and all jobs have the same deadline. The objective is to maximise the total weight of on-time jobs. The problem is known to be strongly NP-hard in general. A polynomial-time algorithm for scheduling unit execution time jobs on two machines is proposed. The performance of a broad family of approximation algorithms for scheduling unit execution time jobs on more than two machines is analysed. For the case of arbitrary job processing times, two integer linear programming formulations are proposed and compared with two formulations known from the earlier literature. An iterated variable neighborhood search algorithm is also proposed and evaluated by means of computational experiments.

Keywords: Scheduling · Parallel machines · Total weight of on-time jobs · Conflict graph

1 Introduction

The paper is concerned with scheduling a set N of n jobs on $m > 1$ identical parallel machines under the restriction that some pairs of jobs cannot be processed concurrently. The jobs are numbered from 1 to n and are referred to by these numbers, i.e. $N = \{1, ..., n\}$. In order to be completed, a job j should be processed during p_j units of time, where the processing time p_j is integer. Each job can be processed by only one of the machines, and if a machine starts processing a job j, then it should process it without interruptions (without preemptions) for the entire job's processing time p_j. At any point in time a machine can process at most one job. The only exceptions are the points in time when one job finishes processing and another commences its processing.

© Springer Nature Switzerland AG 2021
P. Pardalos et al. (Eds.): MOTOR 2021, LNCS 12755, pp. 280–295, 2021.
https://doi.org/10.1007/978-3-030-77876-7_19

The undirected graph $G(N, E)$, where the set of jobs N is the set of nodes and the set of edges E is the set of all pairs of jobs that cannot be processed concurrently, usually is referred to as the conflict graph. The complement of the conflict graph $\overline{G}(N, E)$, that is the graph with the same set of nodes N and with the set of edges that is comprised of all edges that are not in E, will be referred to as the agreement graph. So, jobs can be processed concurrently only if they induce a complete subgraph (a clique) of the agreement graph.

The processing of jobs commences at time $t = 0$. A schedule σ specifies for each $j \in N$ its starting time $S_j(\sigma)$. Since preemptions are not allowed, the completion time of job j in this schedule is

$$C_j(\sigma) = S_j(\sigma) + p_j.$$

Each job j has an associated positive weight w_j and all jobs have the same deadline D. The objective is to find a schedule with the largest total weight of on-time jobs (also referred to as the weighted number of on-time jobs)

$$F(\sigma) = \sum_{j \in J} w_j(1 - U(C_j(\sigma))), \tag{1}$$

where

$$U(t) = \begin{cases} 1 & \text{if } t > D \\ 0 & \text{otherwise} \end{cases}.$$

The situation when some jobs cannot be processed concurrently due to technological restrictions and when it may not be possible to complete all jobs during the given time period, arises in planning of maintenance. Thus, an operator of a communication network faces such a situation when it is needed to execute on the parallel computers the so called change requests that modify some parts of this network [19]. Each computer can execute only one program (change request) at a time, and each change request can be assigned to only one computer. Any two change requests which affect overlapping parts of the network, cannot be executed concurrently. Every change request is initiated by a technician who remains involved during the entire period of the request's execution, and it may not be possible to execute all change requests during one shift. The change requests have different importance, which is modelled by associating with each change request a certain positive number (weight). The goal is to maximise the total weight of the change requests that are executed during the current shift.

The considered scheduling problem also arises in various make-to-order systems [1], where D producers are to be assigned to jobs, each of which is a certain production process during the time interval specified by this job. Each job can be allocated to at most one producer, and each producer can be assigned to any job subject to the following two restrictions: jobs cannot be allocated to the same producer if their time intervals overlap and each producer cannot be assigned to more than $m > 1$ jobs, where m is an integer. Each job has a weight, for example, the associated profit. The goal is to maximise the total weight of all allocated jobs. The problem of assigning the producers to jobs in such a make-to-order system is equivalent to the problem of scheduling jobs on m parallel identical machines under the restrictions imposed by a conflict graph. Indeed, associate with each job in the make-to-order system a job which can be processed on any of these m parallel machines and which requires one unit of processing time. In the literature on scheduling, jobs with unit processing (execution) time are referred to as UET jobs. Let N be the set of all UET jobs and let $G(N, E)$ be the conflict graph where the set of edges is the set of all pairs of UET jobs for which the corresponding jobs in the make-to-order system overlap. Then, the two problems are equivalent if all UET jobs have the same deadline D and each UET job has the same weight as the corresponding job in the make-to-order system. Observe that in the make-to-order systems the conflict graphs have a special structure. Such graphs are called interval graphs [3].

The scheduling problems with parallel machines and the restrictions imposed by a conflict graph, are an area of intensive research. The main focus in this research was on the minimisation of the makespan

$$C_{max}(\sigma) = \max_{j \in N} C_j(\sigma), \tag{2}$$

and despite of various applications and the challenging mathematical nature, the maximisation of (1) has attracted much less attention than it deserves. The paper addresses this gap in the knowledge by establishing the existence of a polynomial-time algorithm, analysing the performance of approximation algorithms, and by presenting integer linear programming formulations and heuristics together with their comparison by means of computational experiments.

2   Related Work

As has been mentioned above, the majority of publications on scheduling on parallel identical machines under the restrictions imposed by a conflict graph, pertain to the makespan minimisation [2,4–7,9,13–15,20,21,24]. The publications on the scheduling problems with UET jobs constitute considerable part of the generated literature, including, in particular, [2,13–15,20] from the list above. As has been shown in [2], the problem of scheduling UET jobs on parallel identical machines with a conflict graph and the objective function (2) arises in balancing the load on parallel processors when partial differential equations are solved using the domain decomposition. In this application, each region of the domain is viewed as a job and each pair of regions which have common points is viewed as an edge in the conflict graph. Another application of the UET case of the makespan minimisation, mentioned in [2], is the exams timetabling problem, where two exams cannot be scheduled concurrently if some students must sit for both of them.

The makespan minimisation problem with parallel identical machines, UET jobs and a conflict graph closely relates to two problems that are among the central in the graph theory: the graph coloring problem [10] and the maximum matching problem [22]. In the graph coloring problem, each color class is the set of jobs that are processed concurrently, and therefore, this problem has an additional restriction that the cardinality of each color class cannot exceed the limit imposed by the number of machines. Since the problem of partitioning a graph into triangles is NP-complete in the strong sense [16], the graph coloring problem (and the equivalent makespan minimisation problem) is NP-hard in the strong sense even in the case when each color class cannot have more than three nodes. Therefore, the research on the graph coloring with a limit on the size of each color class was focused on various particular classes of graphs [8]. The relevance of the maximum matching problem follows from the observation that any two jobs, which can be processed concurrently, correspond to an edge in the agreement graph and the minimisation of the makespan for $m = 2$ is equivalent to finding the largest number of edges in the agreement graph that do not have common nodes [2]. Observe that the maximum matching problem as well as the maximum weight matching problem can be solved in polynomial time [11,12].

If all jobs can be completed by the deadline D, then the corresponding schedule is optimal for (1). Therefore, NP-hardness results for the makespan imply the NP-hardness for the total weight of on-time jobs. Furthermore, the well known result of [28] established that, for all $\varepsilon > 0$, it is NP-hard to find an approximation for the maximum clique problem even within $n^{1-\varepsilon}$. The latter implies the inapproximability of the maximisation of (1) in the case when $m = n$, $D = 1$, and the jobs are UET jobs with equal weights.

3   Scheduling UET Jobs

This section assumes that the optimal makespan is greater than D, because otherwise the maximisation of (1) is equivalent to the makespan minimisation.

3.1   Polynomial-Time Algorithm for $m = 2$

As has been mentioned above, if $m = 2$, a schedule that minimises the makespan can be found by constructing a maximum matching in the agreement graph. This can be done in $O(\sqrt{\kappa}\chi)$ by the algorithm of Micali and Vazirani, where κ and χ are the number of nodes and the number of edges. In contrast, it can be shown that a maximum weight matching may contain no edges which represent pairs of jobs executed concurrently in any schedule that maximises (1). As will be shown below, this scheduling problem can be solved in polynomial time by using the idea suggested in [25] for the generalisation of the maximum weight matching called in [25] the constrained matching.

Let $G(N, E^c)$ be the complement of the conflict graph $G(N, E)$. This agreement graph is transformed in two steps. According to the first step, each edge $e = \{j, g\}$ in E^c is assigned the weight $u_e = w_j + w_g$ and, for each job $j \in N$, a new node j' and the edge $e = \{j, j'\}$ with the weight $u_e = w_j$ are introduced. This doubles the number of nodes and increases the number of edges by n. Denote the set of new nodes by N' and the set of new edges by E'. The second step is based on the idea in [25]: the introduction of the set Q of $2(n - D)$

additional nodes together with the set E_Q of $2(n-D)|N \cup N'|$ edges, all of the same weight $u = 0$, that link each node in Q with every node in $N \cup N'$.

Any optimal schedule σ induces a perfect matching M in $G(N \cup N' \cup Q, E^c \cup E' \cup E_Q)$, where each pair of jobs j and g, processed concurrently and completed on time, induces the edge $\{j, g\}$; each job j, which is completed on time and is not processed concurrently with any other job, induces the edge in E' that covers this job; and the remaining $2(n-D)$ edges are from E_Q, each covering a distinct node in $N \cup N'$, which is not covered by the induced edges from $E^c \cup E'$, with a distinct node in Q.

Conversely, any maximum weight perfect matching M in $G(N \cup N' \cup Q, E^c \cup E' \cup E_Q)$ induces a schedule σ, where two jobs j and g are processed in σ concurrently only if $\{j, g\} \in M$; the jobs, covered by the edges in E_Q, are tardy (being perfect, M must contain $2(n-D)$ edges from E_Q) and the remaining jobs, covered by the remaining D edges (all these edges are from $E^c \cup E'$) are completed on time.

In the discussion above, the weight of the matching M induced by σ, is equal to $F(\sigma)$, and $F(\sigma)$ for the schedule σ induced by the matching M, is equal to the weight of M. Therefore, any maximum weight perfect matching (constructed by Gabow's algorithm in $O(\xi(\zeta + \xi \log \xi))$ where $\xi = |N \cup N' \cup Q|$ and $\zeta = |E^c \cup E' \cup E_Q|$) induces an optimal schedule.

3.2   Approximation Algorithms for $m > 2$

This subsection is concerned with the performance of the algorithms that construct a schedule for arbitrary m, using as a subroutine the method of constructing an optimal schedule for $m = 2$. For any algorithm A, denote by σ^A a schedule constructed according to this algorithm. For any schedule σ, let $J(\sigma)$ be the set of jobs that are completed on time in this schedule. Denote by σ^* an optimal schedule, i.e. a schedule with the largest value of (1), and by σ_2 a schedule with the largest total weight of on-time jobs for the problem, obtained from the original problem by replacing the original number of machines m by 2. Observe that σ_2 can be constructed in polynomial time, using the method described in the previous subsection. Let $\mathfrak{A}$ be the set of all algorithms A such that $J(\sigma_2) \subseteq J(\sigma^A)$.

Theorem 1. *For each $A \in \mathfrak{A}$,*

$$F(\sigma^*) \leq \frac{m}{2} F(\sigma^A)$$

and this performance guarantee is tight.

Proof. For each integer $1 \leq t \leq D$, denote by $k(t)$ the number of jobs processed in schedule σ^* in the time slot $[t-1, t]$, i.e. the number of jobs j such that $C_j(\sigma^*) = t$. By virtue of the assumption that the makespan is greater than D, for all considered integers, $k(t) \geq 1$. Let $j_{t,1}, ..., j_{t,k(t)}$ be the jobs, processed in the time slot $[t-1, t]$ and numbered in a nonincreasing order of their weights, i.e.

$$w_{t,1} \geq ... \geq w_{t,k(t)}, \tag{3}$$

and denote
$$u(t) = \begin{cases} 0 & \text{if } k(t) \leq 2 \\ w_{t,3} & \text{otherwise} \end{cases}.$$

Let σ_2^* be any schedule for the agreement graph $G(N, E^c)$ and two machines such that, for any integer $1 \leq t \leq D$ and each integer $1 \leq i \leq \min[2, k(t)]$, $C_{j_{t,i}}(\sigma_2^*) = t$. Then, taking into account (3), for each integer $1 \leq t \leq D$,

$$u(t) \leq \frac{1}{2} \sum_{\{j:\, C_j(\sigma_2^*)=t\}} w_j$$

and consequently, using $J(\sigma_2) \subseteq J(\sigma^A)$,

$$F(\sigma^*) = \sum_{1 \leq t \leq D} \sum_{\{j:\, C_j(\sigma^*)=t\}} w_j \leq \sum_{1 \leq t \leq D} \left[\sum_{\{j:\, C_j(\sigma_2^*)=t\}} w_j + (m-2)u(t) \right]$$
$$\leq \frac{m}{2} \sum_{1 \leq t \leq D} \sum_{\{j:\, C_j(\sigma_2^*)=t\}} w_j \leq \frac{m}{2} F(\sigma_2) \leq \frac{m}{2} F(\sigma^A).$$

The performance guarantee $\frac{m}{2}$ is tight for each algorithm $A \in \mathfrak{A}$, because, as will be shown below, for any $\varepsilon > 0$, there exists an instance for which

$$F(\sigma^*) > \left(\frac{m}{2} - \varepsilon \right) F(\sigma^A). \tag{4}$$

Indeed, assume that $\varepsilon \leq \frac{1}{2}$ and consider the instance where the agreement graph is comprised of a complete graph of mD nodes and D disjoint edges. All jobs, constituting the complete graph, have the same weight w, whereas all jobs, constituting the disjoint edges, have the same weight $w + \delta$, where

$$0 < \delta < \frac{2\varepsilon w}{m - 2\varepsilon}. \tag{5}$$

Since each job, covered by the D disjoint edges, has a weight greater than the weight of any job that is not covered by these D edges, the set $J(\sigma_2)$ is the set of all jobs covered by the D disjoint edges, and

$$F(\sigma_2) = 2D(w + \delta).$$

On the other hand, any job covered by one of the D disjoint edges, can be processed concurrently only with one job, namely the job covered by the same edge. Furthermore, since, for all $m \geq 3$,

$$\frac{2\varepsilon w}{m - 2\varepsilon} \leq \frac{(m-2)w}{2},$$

by virtue of (5),

$$mw > 2(w + \delta),$$

and therefore, $F(\sigma^*) = mwD$. Consequently, taking into account (5),

$$\frac{F(\sigma^*)}{F(\sigma^A)} = \frac{mwD}{2D(w+\delta)} = \frac{m(w+\delta)}{2(w+\delta)} - \frac{m\delta}{2(w+\delta)} > \frac{m}{2} - \varepsilon,$$

which implies (4). □

4   Scheduling Jobs with Arbitrary Processing Times

4.1   Integer Linear Programming Formulations

Two integer linear programming formulations for the case of arbitrary processing times were proposed in [19]. This section presents two new formulations, based on the approach from [26]. Let $M = \{1, \ldots, m\}$ be the set of available machines. Note that the maximum position on which an on-time job may be scheduled on a machine is at most $k_{max} = \min\{n, \lfloor D/\min_{j=1}^{n}\{p_j\}\rfloor\}$. Let $K = \{1, \ldots, k_{max}\}$ be the set of available positions of on-time jobs. Let e denote the number of conflicting pairs of jobs.

For the first formulation, denoted by ILP1, define for all $j \in N$, $k \in K$ and $l \in M$ binary variables u_{jkl} such that $u_{jkl} = 1$ if job j is not tardy and is scheduled at position k on machine l, and $u_{jkl} = 0$ otherwise. For any $j \in N$, let $U_j = 1$ if job j is tardy, and $U_j = 0$ if it is completed on time. Moreover, for any two conflicting jobs $j, g \in N$, let $y_{jg} = 0$ if job j precedes job g, i.e., job j finishes before g starts, and $y_{jg} = 1$ otherwise. Finally, let t_{kl} denote the starting time of the job at position k on machine l, and let τ_j be the starting time of job j. The considered problem can be stated as follows.

$$\text{(ILP1)   minimise} \quad \sum_{j=1}^{n} w_j U_j \tag{6}$$

$$\text{s.t.} \quad \sum_{k=1}^{n}\sum_{l=1}^{m} u_{jkl} + U_j = 1 \quad \forall\, j \in N \tag{7}$$

$$\sum_{j=1}^{n} u_{jkl} \leq 1 \quad \forall\, k \in K, l \in M \tag{8}$$

$$t_{kl} + \sum_{j=1}^{n} p_j u_{jkl} \leq t_{k+1,l} \quad \forall\, k \in K \setminus \{k_{max}\}, l \in M \tag{9}$$

$$t_{kl} + \sum_{j=1}^{n} p_j u_{jkl} \leq D \quad \forall\, k \in K, l \in M \tag{10}$$

$$\tau_j + D(1 - u_{jkl}) \geq t_{kl} \quad \forall\, j \in N, k \in K, l \in M \tag{11}$$

$$t_{kl} + D(1 - u_{jkl}) \geq \tau_j \quad \forall\, j \in N, k \in K, l \in M \tag{12}$$

$$\tau_j + p_j(1 - U_j) - Dy_{jg} \leq \tau_g \quad \forall\, (j, g) \in E \tag{13}$$

$$y_{jg} + y_{gj} \leq 1 \quad \forall \; (j,g) \in E, j < g \tag{14}$$

$$t_{kl} \geq 0 \quad \forall \; k \in K, l \in M \tag{15}$$

$$\tau_j \geq 0 \quad \forall \; j \in N \tag{16}$$

$$U_j \in \{0,1\} \quad \forall \; j \in N \tag{17}$$

$$u_{jkl} \in \{0,1\} \quad \forall \; j \in N, k \in K, l \in M \tag{18}$$

$$y_{jg} \in \{0,1\} \quad \forall \; (j,g) \in E \tag{19}$$

The objective in the above program is to minimise the weighted number of tardy jobs (6), which is equivalent to maximising the weighted number of on-time jobs. Equations (7) guarantee that each job is either tardy or scheduled at exactly one position on one machine. By (8), at most one job is scheduled at each position on each machine. Constraints (9) ensure that no two jobs are executed at the same time on the same machine. According to (10), a job that is scheduled at position k on machine l is completed by time D. Inequalities (11) and (12) ensure that if job j is scheduled at position k on machine l, then $\tau_j = t_{kl}$. Constraints (13) guarantee that for any two conflicting jobs j and g, if j is scheduled and $y_{jg} = 0$, then j is finished before job g starts. By (14), no two conflicting jobs are executed at the same time. The program (6)–(19) contains $O(nmk_{max} + e)$ binary variables, $O(mk_{max} + n)$ continuous variables, and $O(nmk_{max} + e)$ constraints.

To construct the second formulation, denoted by ILP2, let binary variables U_j and y_{jg} be defined as in ILP1. Additionally, for each $j \in N$ and $t \in \{0, \ldots, D - p_j\}$, let $v_{jt} = 1$ if job j starts at time t (on an arbitrary machine), and $v_{jt} = 0$ otherwise. The optimal schedule can be found in the following way.

$$\text{(ILP2)} \quad \text{minimise} \quad \sum_{j=1}^{n} w_j U_j \tag{20}$$

$$\text{s.t.} \quad \sum_{t=0}^{D-p_j} v_{jt} + U_j = 1 \quad \forall \; j \in N \tag{21}$$

$$\sum_{j=1}^{n} \sum_{s=\max\{0,t-p_j+1\}}^{\min\{t,D-p_j\}} v_{js} \leq m \quad \forall \; t = 0, \ldots, D - 1 \tag{22}$$

$$\sum_{t=0}^{D-p_j} t v_{jt} + p_j(1 - U_j) - D y_{jg} \leq \sum_{t=0}^{D-p_g} t v_{gt} \quad \forall \; (j,g) \in E \tag{23}$$

$$y_{jg} + y_{gj} \leq 1 \quad \forall \; (j,g) \in E, j < g \tag{24}$$

$$v_{jt} \in \{0,1\} \quad \forall \; j \in N, t = 0, \ldots, D - p_j \tag{25}$$

$$U_j \in \{0,1\} \quad \forall \; j \in N \tag{26}$$

Once again, the weighted number of tardy jobs is minimised (20). Constraints (21) guarantee that each job j is either tardy or scheduled to start at exactly one moment $t \leq D - p_j$. At most m jobs are executed at any time t by (22). Inequalities (23) ensure that for any two conflicting jobs j and g, if j is scheduled and $y_{jg} = 0$, then j is finished before job g starts. Conflicting jobs are not executed at the same time by (24). The program (20)–(26) contains $O(nD + e)$ binary variables and $O(nD + e)$ constraints.

4.2 Heuristic Algorithms

This section presents heuristic algorithms for the analysed problem. In the schedule representation used, a list of assigned jobs and their starting times is stored for each machine. Additionally, a separate list of tardy jobs is maintained.

Firstly, a variable neighborhood search algorithm VNS is proposed. Variable neighborhood search is a metaheuristic for solving combinatorial optimisation problems, introduced by [23]. It consists in a systematic change of neighborhood within a local search algorithm. There exist many variants of variable neighborhood search, which have been successfully applied in many areas [17]. In this work, the variable neighborhood descent method is used. Consider k_{max} neighborhoods $N_1, \ldots, N_{k_{max}}$. Let x be the initial solution passed to the algorithm. At the beginning, the current neighborhood number is $k = 1$. In each iteration of the algorithm, the best solution x' in neighborhood $N_k(x)$ is found. If an improvement is obtained, i.e. x' is better than x, then x is updated to x' and k is set to 1. Otherwise, k is increased by 1 and the next neighborhood $N_k(x)$ is considered. The process continues until k exceeds k_{max}.

In the proposed variable neighborhood search, the initial schedule is delivered by a list scheduling algorithm which processes the jobs in the weighted shortest processing time order. Thus, jobs are first sorted in such a way that $p_1/w_1 \leq \cdots \leq p_n/w_n$. Afterwards, each consecutive job j is assigned to the machine that is the first to finish processing assigned jobs in the current schedule. The earliest possible starting time of job j is then computed, taking into account the time when processing on the selected machine finishes, as well as the processing intervals of already scheduled jobs conflicting with j. If it is not possible to finish job j by time D, this job is removed from the machine's list and added to the list of tardy jobs. After the initial schedule is constructed, the following six neighborhoods are considered in the search procedure.

- $N_1(\sigma)$ consists of all schedules obtained from σ by swapping a single pair of jobs on one machine;
- $N_2(\sigma)$ consists of all schedules obtained from σ by moving one job to a different position on the machine it is assigned to;
- $N_3(\sigma)$ consists of all schedules obtained from σ by swapping a pair of jobs assigned to different machines;
- $N_4(\sigma)$ consists of all schedules obtained from σ by moving a job scheduled on any machine i to an arbitrary position on a different machine;
- $N_5(\sigma)$ consists of all schedules obtained from σ by replacing an arbitrary scheduled job by an arbitrary tardy job;
- $N_6(\sigma)$ consists of all schedules obtained from σ by moving an arbitrary tardy job to an arbitrary position on any machine.

The order of the neighborhoods was selected on the basis of preliminary computational experiments, which suggested that the best results are produced when the neighborhoods obtained by changes on a single machine are considered first, and the neighborhoods obtained by assigning to machines the jobs from the tardy list are considered last. Naturally, after changing the assignment of jobs to machines and positions, the starting times of the jobs have to be recomputed, taking into account job conflicts. If due to the changes made, a job that was assigned to a machine becomes tardy, it is removed from the machine's list of jobs and added to the list of tardy jobs. Moreover, after executing the changes, the list of tardy jobs is scanned in the weighted shortest processing time order, and additional jobs from this list are assigned to the least loaded machines, if possible.

Secondly, the proposed VNS algorithm is embedded in the iterated search framework. The iterated variable neighborhood search algorithm IVNS starts with executing the VNS. The obtained schedule is then modified by making $\lceil 0.05n \rceil$ changes consisting in moving a random (scheduled or tardy) job to a random position on a random machine. Job starting times are recomputed, and additional jobs are scheduled if possible, as explained in the description of the VNS algorithm. After this shaking step, the variable neighborhood search is run again. The whole procedure is repeated $r = 10$ times. The number of changes made in the shaking procedure was selected on the basis of preliminary computational experiments. The number of iterations r was chosen as a compromise between solution quality and the running time of the algorithm.

5   Computational Experiments

In this section, the results of computational experiments on the quality of the proposed algorithms are presented. The algorithms were implemented in C++, and integer linear programs were solved using Gurobi. In addition to the algorithms presented in Sect. 4, integer linear programming formulations F1 and F2 proposed in [19] were implemented. The experiments were performed on an Intel Core i7-7820HK CPU @ 2.90 GHz with 32 GB RAM.

In the generated test instances, the number of machines was $m \in [2, 10]$, and the number of jobs was $n \in \{5m, 10m\}$. The job processing times p_j were chosen randomly from the interval $[50, 150]$, and the job weights w_j from the interval $[1, 5]$. A parameter $\delta > 0$ was used to control the ratio between the available time window and the expected amount of work per machine. Since the expected duration of a job was 100, the common deadline of all jobs was set to $D = 100\delta n/m$. The number of conflicting jobs was controlled by a parameter $c \in (0, 1)$. For a given value of c, the conflict graph G contained $c\binom{n}{2}$ randomly chosen edges. In the experiments presented in this paper, $c = 0.1$ was used. For each analysed combination of instance parameters, 30 tests were generated.

Many instances could not be solved to optimality in reasonable time using the integer linear programs. Therefore, a one hour time limit was imposed on algorithms ILP1, ILP2, F1 and F2. Since the optimum solutions were not known in all cases, the total weights of on-time jobs returned by the respective algorithms were compared to an upper bound UB defined as the smallest upper bound obtained by Gurobi while solving the integer linear programs. Schedule quality was measured by the relative percentage error with respect to UB.

Table 1. Performance of the integer linear programs.

Algorithm	$m = 2, n = 10$		$m = 2, n = 20$		$m = 4, n = 40$	
	Error (%)	Time (s)	Error (%)	Time (s)	Error (%)	Time (s)
ILP1	0.00	4.22E−2	0.00	2.12E−1	0.00	1.50E+2
ILP2	0.00	2.36E+0	0.00	1.55E+1	0.00	2.45E+2
F1 [19]	0.00	1.71E+2	0.00	3.61E+3	1.09	3.61E+3
F2 [19]	0.00	2.14E+1	0.00	3.61E+3	1.68	3.61E+3

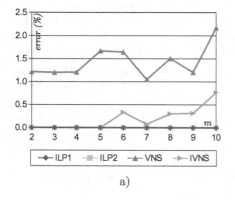

a)

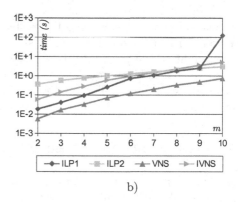

b)

Fig. 1. Algorithm performance vs. m for $n = 5m$, $\delta = 0.3$. a) Average quality, b) average execution time.

In the first experiment, the performance of integer linear programs ILP1, ILP2, F1 and F2 was compared on instances of various sizes with $\delta = 0.7$. The obtained results are presented in Table 1. All analysed algorithms were able to solve all the instances with $m = 2$ and $n = 10$ within the imposed time limit. Still, the running times of the algorithms differed significantly. ILP1 had the best average running time, followed by ILP2, then F2, and finally F1, which was four orders of magnitude slower than ILP1. The optimum schedules were also found by all algorithms for all instances with $m = 2$ and $n = 20$. However, F1 and F2 did not finish their computations within an hour for any such tests. Although the optimum solutions were found, the upper bounds computed during one hour using these formulations were larger. All generated instances with $m = 4$, $n = 40$ were solved to optimality by ILP1 and ILP2, but F1 and F2 did not find the best schedules for some of these tests. Their average errors were 1.09% and 1.68%, correspondingly. It can be concluded that the formulations ILP1 and ILP2 are more efficient than F1 and F2 proposed in [19]. As the running times of F1 and F2 were very long even for small instances, these two algorithms were not used in the remaining experiments.

Figure 1 presents the results obtained for instances with $m = 2, \ldots, 10$, $n = 5m$ and $\delta = 0.3$. All tests in this group were solved to optimality by both ILP1 and ILP2 (see Fig. 1a). VNS produced good schedules, its average error is below 2% for all $m \leq 9$, and only reaches 2.17% for $m = 10$. Using IVNS leads to obtaining substantially better results, although at a higher computational cost. All tests with $m \leq 5$ were solved to optimality by this algorithm, and the largest average error, obtained for $m = 10$, is only 0.77%. Naturally, the running times of all algorithms increase with growing m (cf. Fig. 1b). VNS is the fastest among the analysed algorithms. ILP1 is faster than IVNS and ILP2 for $m \leq 9$, which means that it is very suitable for solving small instances with a small δ. However, its average running time rapidly increases when $m = 10$. This is caused by the fact that ILP1 reached the time limit of 1 h for one instance with $m = 10$. Contrarily, ILP2 is slower than IVNS for $m \leq 7$, but faster than IVNS for $m \geq 8$.

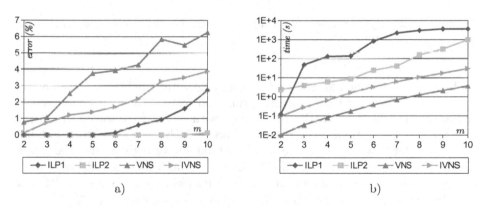

Fig. 2. Algorithm performance vs. m for $n = 5m$, $\delta = 0.7$. a) Average quality, b) average execution time.

The results obtained for tests with $m = 2, \ldots, 10$, $n = 5m$ and $\delta = 0.7$ are shown in Fig. 2. ILP1 found optimal solutions for all instances with $m \leq 5$ but was unsuccessful at some tests for each larger m. In particular, it was not able to finish computations within an hour for any instances with $m \geq 9$. The average ILP1 error is below 1% for $m \leq 8$ and reaches 2.72% for $m = 10$. ILP2 performed much better, finding the optimum schedules for all instances with $m \leq 9$. Even for $m = 10$, the average error of ILP2 is only 0.13%. The distance between VNS and IVNS schedules is small for $m \in \{2,3\}$, but increases for larger m. The errors obtained by VNS are below 6.5%, and the IVNS errors are smaller than 4%. Thus, instances with $\delta = 0.7$ are in general more demanding than those with $\delta = 0.3$. In the group of tests with $\delta = 0.7$, the running time of ILP1 is close to that of IVNS only for $m = 2$. For $m \geq 3$, the running time of ILP1 significantly increases, and ILP1 becomes the slowest of all analysed algorithms. In particular, for $m \geq 9$, the average running time of ILP1 is one hour, since it solved no instances within the imposed time limit. In contrast, the average execution time of ILP2 is 985 s for $m = 10$.

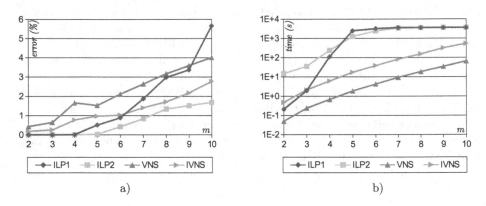

a)   b)

Fig. 3. Algorithm performance vs. m for $n = 10m$, $\delta = 0.7$. a) Average quality, b) average execution time.

The results obtained for the largest instances, with variable m, $n = 10m$ and $\delta = 0.7$, are shown in Fig. 3. In this group, no tests with $m \geq 7$ were solved within the time limit by ILP1, and no tests with $m \geq 8$ were solved within the time limit by ILP2. Hence, the running times of both these algorithms stabilise at one hour for large m. There are significant differences between the qualities of solutions obtained by ILP1 and ILP2. For $m \leq 4$, both these algorithms always find optimum solutions, but for larger m, ILP2 clearly outperforms ILP1. For $m = 10$, the average distance from UB is 1.68% in the case of ILP2 solutions, and 5.66% in the case of ILP1 schedules. Moreover, ILP1 is outperformed by both VNS and IVNS for such m. Thus, ILP1 is not recommended for solving large instances. It is interesting that VNS and IVNS obtain here better results than for the tests with $n = 5m$ and $\delta = 0.7$. The average errors of VNS reach 4.01% for $m = 10$ and the average distance of IVNS schedules from the upper bound is below 3% for all m.

In addition to the above experiments, a preliminary comparison of the methods in this paper with the ones in communications [27] and [18] is presented. The communications [27] and [18] appeared after this paper has been completed. The algorithms ILP2 and IVNS are compared with the integer linear programming formulation F3 from [18], and the integer linear program FT and heuristic HE from [27]. The comparison is preliminary because of the use of different computers and the possible differences between Gurobi and CPLEX.

Algorithms F3, ILP2 and IVNS were run on the set of 432 instances used in [18], obtained from the authors. Algorithm ILP2 solved optimally 430 instances, and its average error was 0.02%. Formulation F3 delivered optimal solutions for 394 tests, and had an average error of 0.58%. The average execution time was 149 s in the case of ILP2 and 489 s in the case of F3. Thus, it seems that ILP2 is more efficient than F3. Heuristic IVNS obtained 339 optimal solutions, average error 0.70% and average execution time 1.42 s.

Furthermore, algorithms ILP2 and IVNS were executed on the set of 3840 large instances used in [27], which are publicly available. Algorithm ILP2 solved all of them optimally in the average time of 1.76 s, while FT is reported in [27] to solve all but one instance optimally, in the average time of 35 s. These results cannot be compared directly because different computers and different solvers were used to run both algorithms. However, the difference between the execution times seems significant. Algorithm IVNS solved optimally 2430 instances, and HE found optimum solutions for 1851 tests [27]. The average error of HE on instances not solved optimally is reported to be 1.91%, which gives the average error of 0.99% overall, while the average error of IVNS was 0.56%. The average computation time of HE is 0.21 s according to [27], and the average running time of IVNS was 4.77 s. Again, the time results cannot be compared directly. Still, it seems that IVNS obtains better solutions than HE, but at a higher computational cost.

It is worth noting that the integer linear programs ILP2, F3 and FT, which use time-indexed variables, significantly outperform formulations F1 and F2 (as shown here and in [18,27]), which do not use such variables. Thus, it seems that the time-indexed approach to the considered problem is particularly efficient.

6   Conclusions

For the problem of maximising the total weight of on-time jobs with a common deadline, scheduled on parallel machines subject to a conflict graph, the paper presents a polynomial-time algorithm for the case of two parallel machines, and a performance guarantee which is tight for a broad family of algorithms for an arbitrary number of machines. Both these results were obtained for UET jobs. For jobs with general processing times, the paper presents two new integer linear programming formulations and a variable neighborhood search algorithm, which is embedded in an iterated search framework. Computational experiments show that both proposed integer linear programs obtain better results than those in [19], ILP2 is particularly efficient for large instances, and the heuristic algorithm IVNS obtains good results in a reasonable time. Further development of such optimisation procedures, and their experimental evaluation, should be one of the main directions of future research. In particular, the experiments should include a more detailed comparison of the proposed algorithms with the methods recently announced in [18,27].

References

1. Arkin, E.M., Silverberg, E.B.: Scheduling jobs with fixed start and end times. Discrete Appl. Math. **18**(1), 1–8 (1987)
2. Baker, B.S., Coffman, E.G.: Mutual exclusion scheduling. Theor. Comput. Sci. **162**(2), 225–243 (1996)
3. Balakrishnan, R., Ranganathan, K.: A Textbook of Graph Theory. Springer, New York (2012). https://doi.org/10.1007/978-1-4614-4529-6
4. Bampis, E., Kononov, A., Lucarelli, G., Milis, I.: Bounded max-colorings of graphs. J. Discrete Algorithms **26**, 56–68 (2014)
5. Bendraouche, M., Boudhar, M.: Scheduling jobs on identical machines with agreement graph. Comput. Oper. Res. **39**(2), 382–390 (2012)
6. Bendraouche, M., Boudhar, M.: Scheduling with agreements: new results. Int. J. Prod. Res. **54**(12), 3508–3522 (2016)
7. Bendraouche, M., Boudhar, M., Oulamara, A.: Scheduling: agreement graph vs resource constraints. Eur. J. Oper. Res. **240**(2), 355–360 (2015)
8. Bodlaender, H.L., Jansen, K.: Restrictions of graph partition problems. Part I. Theor. Comput. Sci. **148**(1), 93–109 (1995)
9. Demange, M., De Werra, D., Monnot, J., Paschos, V.T.: Time slot scheduling of compatible jobs. J. Sched. **10**(2), 111–127 (2007). https://doi.org/10.1007/s10951-006-0003-7
10. Diestel, R.: Graph Theory, Volume 173. Graduate Texts in Mathematics, p. 7 (2012)
11. Edmonds, J.: Maximum matching and a polyhedron with 0, 1-vertices. J. Res. Natl. Bureau Stand. B **69**(125–130), 55–56 (1965)
12. Edmonds, J.: Paths, trees, and flowers. Can. J. Math. **17**, 449–467 (1965)
13. Even, G., Halldórsson, M.M., Kaplan, L., Ron, D.: Scheduling with conflicts: online and offline algorithms. J. Sched. **12**(2), 199–224 (2009). https://doi.org/10.1007/s10951-008-0089-1
14. Gardi, F.: Mutual exclusion scheduling with interval graphs or related classes. Part II. Discrete Appl. Math. **156**(5), 794–812 (2008)
15. Gardi, F.: Mutual exclusion scheduling with interval graphs or related classes, part I. Discrete Appl. Math. **157**(1), 19–35 (2009)
16. Garey, M.R., Johnson, D.S.: Computers and Intractability, vol. 174. Freeman, San Francisco (1979)
17. Gendreau, M., Potvin, J.Y.: Handbook of Metaheuristics. Springer, Boston (2010). https://doi.org/10.1007/978-1-4419-1665-5
18. Hà, M.H., Ta, D.Q., Nguyen, T.T.: Exact algorithms for scheduling problems on parallel identical machines with conflict jobs (2021). https://arxiv.org/abs/2102.06043v1
19. Hà, M.H., Vu, D.M., Zinder, Y., Thanh, T.: On the capacitated scheduling problem with conflict jobs. In: 2019 11th International Conference on Knowledge and Systems Engineering (KSE), pp. 1–5. IEEE (2019)
20. Jansen, K.: The mutual exclusion scheduling problem for permutation and comparability graphs. Inf. Comput. **180**(2), 71–81 (2003)
21. Kowalczyk, D., Leus, R.: An exact algorithm for parallel machine scheduling with conflicts. J. Sched. **20**(4), 355–372 (2016). https://doi.org/10.1007/s10951-016-0482-0
22. Lovász, L., Plummer, M.D.: Matching Theory, vol. 367. American Mathematical Society, Providence (2009)

23. Mladenović, N., Hansen, P.: Variable neighborhood search. Comput. Oper. Res. **24**(11), 1097–1100 (1997)
24. Mohabeddine, A., Boudhar, M.: New results in two identical machines scheduling with agreement graphs. Theor. Comput. Sci. **779**, 37–46 (2019)
25. Plesník, J.: Constrained weighted matchings and edge coverings in graphs. Discrete. Appl. Math. **92**(2–3), 229–241 (1999)
26. Sevaux, M., Thomin, P.: Heuristics and metaheuristics for a parallel machine scheduling problem: a computational evaluation. Technical report 01-2-SP, University of Valenciennes (2001)
27. Tresoldi, E.: Solution approaches for the capacitated scheduling problem with conflict jobs (2021). https://doi.org/10.13140/RG.2.2.11756.18566
28. Zuckerman, D.: Linear degree extractors and the inapproximability of max clique and chromatic number. In: Proceedings of the Thirty-Eighth Annual ACM Symposium on Theory of Computing, pp. 681–690 (2006)

Game Theory and Optimal Control

Optimal Boundary Control of String Vibrations with Given Shape of Deflection at a Certain Moment of Time

V. Barseghyan[1,2]($^{(\boxtimes)}$) and S. Solodusha[3]($^{(\boxtimes)}$)

[1] Institute of Mechanics NAS RA, Yerevan, Armenia
[2] Yerevan State University, Yerevan, Armenia
barseghyan@sci.am
[3] Melentiev Energy Systems Institute SB RAS, Irkutsk, Russia
solodusha@isem.irk.ru

Abstract. We consider the problem of optimal boundary control of string vibrations with given initial and final conditions and a given value of the string deflection function at some intermediate time moment and with a quality criterion given over the entire time interval. It is controlled by the displacement of one end while the other end is fixed. A constructive approach to constructing the optimal boundary control action is proposed. A computational experiment was carried out with the construction of the corresponding graphs and their comparative analysis, which confirm the results obtained.

Keywords: Boundary control · Optimal control of vibrations · Intermediate conditions · Separation of variables

1 Introduction

Controlled vibration systems are widely used in various theoretical and applied fields of science. The need for control and optimal control of vibrational processes with both distributed and boundary actions is an urgent problem, the solution of which is paid attention to by many researchers [1–13]. In practice, problems of boundary control and optimal control often arise, in particular, when it is necessary to generate vibrations with predetermined (desired) intermediate parameters (deflection shape, speed of string points, etc.). Modeling and control of dynamical systems described by both ordinary differential equations and partial differential equations with intermediate conditions are an actively developing area in modern control theory. In particular, papers [8–15] are devoted to the study of such problems.

The research of S. Solodusha was carried out within the state assignment of Ministry of Science and Higher Education of the Russian Federation (Project FWEU-2021-0006, theme No. AAAA-A21-121012090034-3).

P. Pardalos et al. (Eds.): MOTOR 2021, LNCS 12755, pp. 299–313, 2021.
https://doi.org/10.1007/978-3-030-77876-7_20

The purpose of this work is to develop a constructive approach to building a function of optimal boundary control of string vibrations by displacement of one end while the other end is fixed with a given shape of the string deflection at a certain moment in time and with a quality criterion set over the entire time interval.

2   Statement of the Problem and Its Reduction to a Problem with Zero Boundary Conditions

Consider small transverse vibrations of a stretched homogeneous string. They are described by the function $Q(x,t)$, $0 \leq x \leq l$, $0 \leq t \leq T$, which satisfies the wave equation

$$\frac{\partial^2 Q}{\partial t^2} = a^2 \frac{\partial^2 Q}{\partial x^2}, \ 0 < x < l, \ t > 0 \tag{1}$$

with the boundary conditions

$$Q(0,t) = u(t), \ Q(l,t) = 0, \ \ 0 \leq t \leq T. \tag{2}$$

Let the initial and final conditions be given

$$Q(x,0) = \varphi_0(x), \ \left.\frac{\partial Q}{\partial t}\right|_{t=0} = \psi_0(x), \ 0 \leq x \leq l, \tag{3}$$

$$Q(x,T) = \varphi_T(x) = \varphi_2(x), \ \left.\frac{\partial Q}{\partial t}\right|_{t=T} = \psi_T(x) = \psi_2(x), \ 0 \leq x \leq l, \tag{4}$$

where T is some given finite moment of time. In Eq. (1) $a^2 = \frac{T_0}{\rho}$, where T_0 is the string tension, ρ is the density of the homogeneous string, and function $u(t)$ is the boundary control. It is assumed that the function $Q(x,t) \in C^2(\Omega_T)$, where the set $\Omega_T = \{(x,t) : x \in [0,l], \ t \in [0,T]\}$. Let at some intermediate moment of time t_1 $(0 = t_0 < t_1 < t_2 = T)$ the values of the velocities of the string points are given in the form:

$$Q(x,t_1) = \varphi_1(x), \ 0 \leq x \leq l. \tag{5}$$

We formulate the following boundary control problem of string vibrations with a given value of the deflection function at an intermediate point in time.

Among the possible boundary controls for $u(t)$, $0 \leq t \leq T$, (2) it is required to find such optimal control $u^0(t)$ under the influence of which the vibrational motion of system (1) from a given initial state (3) goes into the final state (4), ensuring that condition (5) is satisfied and minimizing functional

$$\int_0^T u^2(t)dt. \tag{6}$$

We assume that the functions $\varphi_i(x)$ ($i = 0, 1, 2$) belong to the $C^2[0, l]$, and functions $\psi_0(x)$ and $\psi_T(x)$ belong to the $C^1[0, l]$. It is also assumed that all functions are such that the compatibility conditions below are satisfied:

$$u(0) = \varphi_0(0), \ u'(0) = \psi_0(0), \ \varphi_0(l) = \psi_0(l) = 0, \tag{7}$$

$$u(t_1) = \varphi_1(0), \ \varphi_1(l) = 0, \tag{8}$$

$$u(T) = \varphi_T(0), \ u'(T) = \psi_T(0), \ \varphi_T(l) = \psi_T(l) = 0. \tag{9}$$

As the boundary conditions (2) are non-homogeneous, the solution of the stated problem is reduced to a problem with zero boundary conditions. Thus, following [16], the solution to Eq. (1) is sought in the form of a sum

$$Q(x, t) = V(x, t) + W(x, t), \tag{10}$$

where $V(x, t)$ is an unknown function to be determined with the homogeneous boundary conditions

$$V(0, t) = V(l, t) = 0, \tag{11}$$

and the function $W(x, t)$ is solution to Eq. (1) with the non-homogeneous boundary conditions

$$W(0, t) = u(t), \ W(l, t) = 0.$$

The function $W(x, t)$ has the form

$$W(x, t) = \left(1 - \frac{x}{l}\right) u(t). \tag{12}$$

Substituting (10) into (1) and taking into account (12), we obtain the following equation for determining the function $V(x, t)$:

$$\frac{\partial^2 V}{\partial t^2} = a^2 \frac{\partial^2 V}{\partial x^2} + F(x, t), \ F(x, t) = \left(\frac{x}{l} - 1\right) u''(t). \tag{13}$$

The function $V(x, t)$ by virtue of conditions (2)–(5) and the compatibility conditions (7)–(9) must satisfy the initial, intermediate and final conditions

$$V(x, 0) = \varphi_0(x) + \left(\frac{x}{l} - 1\right) \varphi_0(0), \ \left.\frac{\partial V}{\partial t}\right|_{t=0} = \psi_0(x) + \left(\frac{x}{l} - 1\right) \psi_0(0), \tag{14}$$

$$V(x, t_1) = \varphi_1(x) + \left(\frac{x}{l} - 1\right) \varphi_1(0), \tag{15}$$

$$V(x, T) = \varphi_T(x) + \left(\frac{x}{l} - 1\right) \varphi_T(0), \ \left.\frac{\partial V}{\partial t}\right|_{t=T} = \psi_T(x) + \left(\frac{x}{l} - 1\right) \psi_T(0). \tag{16}$$

Thus, the solution of the problem of optimal boundary control of string vibrations with a given shape of the points deflection at an intermediate time instant is reduced to the problem of controlling vibrational motion (13) with boundary conditions (11) and minimized functional (6), which is formulated as follows: it is required to find such optimal boundary control $u^0(t)$, $0 \le t \le T$ that transfers the vibration described by Eq. (13) with boundary conditions (11) from the given initial state (14) through the intermediate state (15) to the final state (16) and minimizes functional (6).

3 Reduction of the Solution of a Problem with Zero Boundary Conditions to the Problem of Moments

Taking into account that the boundary conditions (11) in problem (13) are homogeneous and the compatibility conditions are satisfied (7)–(9), according to the theory of Fourier series, we seek the solution to Eq. (13) in the form

$$V(x,t) = \sum_{k=1}^{\infty} V_k(t) \sin \frac{\pi k}{l} x, \quad V_k(t) = \frac{2}{l} \int_0^l V(x,t) \sin \frac{\pi k}{l} x \, dx. \qquad (17)$$

We represent the functions $F(x,t)$, $\varphi_i(x)$ $(i = 0, 1, 2)$ and $\psi_i(x)$ $(i = 0, 2)$ in the form of Fourier series and, substituting their values together with $V(x,t)$ into Eq. (13) and conditions (7)–(9), we obtain

$$\ddot{V}_k(t) + \lambda_k^2 V_k(t) = F_k(t), \quad \lambda_k^2 = \left(\frac{a\pi k}{l}\right)^2, \quad F_k(t) = -\frac{2a}{\lambda_k l} u''(t), \qquad (18)$$

$$V_k(0) = \varphi_k^{(0)} - \frac{2a}{\lambda_k l} \varphi_0(0), \quad \dot{V}_k(0) = \psi_k^{(0)} - \frac{2a}{\lambda_k l} \psi_0(0), \qquad (19)$$

$$V_k(t_1) = \varphi_k^{(1)} - \frac{2a}{\lambda_k l} \varphi_1(0), \qquad (20)$$

$$V_k(T) = \varphi_k^{(T)} - \frac{2a}{\lambda_k l} \varphi_T(0), \quad \dot{V}_k(T) = \psi_k^{(T)} - \frac{2a}{\lambda_k l} \psi_T(0), \qquad (21)$$

where $F_k(t)$, $\varphi_k^{(i)}$ $(i = 0, 1, 2)$ and $\psi_k^{(i)}$ $(i = 0, 2)$ denote the Fourier coefficients, corresponding to the functions $F(x,t)$, $\varphi_i(x)$ $(i = 0, 1, 2)$ and $\psi_i(x)$ $(i = 0, 2)$. The general solution to Eq. (18) with initial conditions (19) has the form

$$V_k(t) = V_k(0) \cos \lambda_k t + \frac{1}{\lambda_k} \dot{V}_k(0) \sin \lambda_k t + \frac{1}{\lambda_k} \int_0^t F_k(\tau) \sin \lambda_k (t - \tau) d\tau. \qquad (22)$$

Now, taking into account the intermediate (20) and final (21) conditions, from (22) with conditions (14)–(16), we obtain that the function $u(\tau)$ for each k must satisfy the following system of equalities:

$$\int_0^T u(\tau) \sin \lambda_k (T - \tau) d\tau = C_{1k}(T), \quad \int_0^T u(\tau) \cos \lambda_k (T - \tau) d\tau = C_{2k}(T), \qquad (23)$$

$$\int_0^T u(\tau) h_k^{(1)} (\tau) d\tau = C_{1k}(t_1), \quad k = 1, 2, \ldots,$$

where

$$h_k^{(1)}(\tau) = \begin{cases} \sin \lambda_k (t_1 - \tau), & 0 \leq \tau \leq t_1, \\ 0, & t_1 < \tau \leq T, \end{cases}$$

$$C_{1k}(T) = \frac{1}{\lambda_k^2} \left[\frac{\lambda_k l}{2a} \tilde{C}_{1k}(T) + X_{1k} \right],$$

$$\tilde{C}_{1k}(T) = \lambda_k V_k(T) - \lambda_k V_k(0) \cos \lambda_k T - \dot{V}_k(0) \sin \lambda_k T,$$

$$C_{2k}(T) = \frac{1}{\lambda_k^2} \left[\frac{\lambda_k l}{2a} \tilde{C}_{2k}(T) + X_{2k} \right],$$

$$\tilde{C}_{2k}(T) = \dot{V}_k(T) + \lambda_k V_k(0) \sin \lambda_k T - \dot{V}_k(0) \cos \lambda_k T, \tag{24}$$

$$C_{1k}(t_1) = \frac{1}{\lambda_k^2} \left[\frac{\lambda_k l}{2a} \tilde{C}_{1k}(t_1) + X_{1k}^{(1)} \right],$$

$$\tilde{C}_{1k}(t_1) = \lambda_k V_k(t_1) - \lambda_k V_k(0) \cos \lambda_k t_1 - \dot{V}_k(0) \sin \lambda_k t_1,$$

$$X_{1k} = \lambda_k \varphi_T(0) - \psi_0(0) \sin \lambda_k T - \lambda_k \varphi_0(0) \cos \lambda_k T,$$

$$X_{2k} = \psi_T(0) - \psi_0(0) \cos \lambda_k T + \lambda_k \varphi_0(0) \sin \lambda_k T,$$

$$X_{1k}^{(1)} = \lambda_k \varphi_1(0) - \psi_0(0) \sin \lambda_k t_1 - \lambda_k \varphi_0(0) \cos \lambda_k t_1.$$

Thus, the solution of the posed problem of optimal control is reduced to finding the optimal boundary control $u^0(t)$, $0 \leq t \leq T$, which for each $k = 1, 2, \ldots$ satisfies the integral relations (23) and minimizes functional (6). Since functional (6) is the square of the norm of a linear normed space, and integral relations (23), generated by the function $u(t)$, are linear, the problem of determining the optimal control for each $k = 1, 2, \ldots$ can be regarded as a moment problem [1,14,17]. Therefore, a solution can be constructed using an algorithm for solving the problem of moments.

4   The Problem Solution

In practice, the first n harmonics of the vibrations are chosen and the control synthesis problem is solved using the methods of control theory for finite-dimensional systems. Therefore, we construct a solution of problem (6) and (23) for $k = 1, 2, \ldots, n$ using the algorithm for solving the problem of moments. To solve the finite-dimensional (for $k = 1, 2, \ldots, n$) moment problem (6) and (23), following [17], it is necessary to find the values p_k, q_k, γ_k, $k = 1, \ldots, n$, related by the condition

$$\sum_{k=1}^{n} [p_k C_{1k}(T) + q_k C_{2k}(T) + \gamma_k C_{1k}(t_1)] = 1, \tag{25}$$

for which

$$(\rho_n^0)^2 = \min_{(25)} \int_0^T h_{1n}^2(\tau) d\tau, \tag{26}$$

where

$$h_{1n}(\tau) = \sum_{k=1}^{n} \left[p_k \sin \lambda_k (T - \tau) + q_k \cos \lambda_k (T - \tau) + \gamma_k h_k^{(1)}(\tau) \right]. \tag{27}$$

To determine the values p_k^0, q_k^0, γ_k^0, $k = 1, ..., n$ minimizing (26), we apply the Lagrange's method of undetermined multipliers. For this purpose, we introduce the function

$$f_n = \int_0^T (h_{1n}(\tau))^2 \, d\tau + \beta_n \left[\sum_{k=1}^{n} (p_k C_{1k}(T) + q_k C_{2k}(T) + \gamma_k C_{1k}(t_1)) - 1 \right],$$

where β_n—undetermined Lagrange multiplier. Based on this method, calculating the derivatives of the function f_n with respect to p_k, q_k, γ_k, $k = 1, ..., n$ and equating them to zero with notation (27), we obtain the following system of integral relations

$$\sum_{j=1}^{n} \int_0^T [p_j \sin \lambda_j (T - \tau) + q_j \cos \lambda_j (T - \tau)$$
$$+ \gamma_j h_j^{(1)}(\tau)] \sin \lambda_k (T - \tau) \, d\tau = -\frac{\beta_n}{2} C_{1k}(T),$$

$$\sum_{j=1}^{n} \int_0^T [p_j \sin \lambda_j (T - \tau) + q_j \cos \lambda_j (T - \tau)$$
$$+ \gamma_j h_j^{(1)}(\tau)] \cos \lambda_k (T - \tau) \, d\tau = -\frac{\beta_n}{2} C_{2k}(T), \tag{28}$$

$$\sum_{j=1}^{n} \int_0^T [p_j \sin \lambda_j (T - \tau) + q_j \cos \lambda_j (T - \tau)$$
$$+ \gamma_j h_j^{(1)}(\tau)] h_k^{(1)}(\tau) d\tau = -\frac{\beta_n}{2} C_{1k}(t_1), \ k = 1, ..., n.$$

Calculating the integrals on the left sides of Eqs. (28), with notation (24) and adding condition (25) to the resulting equations, we obtain closed system of $3n + 1$ linear algebraic equations in as many unknowns p_k, q_k, γ_k, $k = 1, ..., n$ and β_n:

$$\sum_{j=1}^{n} (a_{jk}p_j + b_{jk}q_j + c_{jk}\gamma_j) = -\frac{\beta_n}{2} C_{1k}(T),$$

$$\sum_{j=1}^{n} (d_{jk}p_j + e_{jk}q_j + f_{jk}\gamma_j) = -\frac{\beta_n}{2} C_{2k}(T), \tag{29}$$

$$\sum_{j=1}^{n} \left(a_{jk}^{(1)} p_j + b_{jk}^{(1)} q_j + g_{jk}\gamma_j \right) = -\frac{\beta_n}{2} C_{1k}(t_1), \ k = 1, ..., n,$$

$$\sum_{k=1}^{n} [p_k C_{1k}(T) + q_k C_{2k}(T) + \gamma_k C_{1k}(t_1)] = 1,$$

where

$$a_{jk} = \int_0^T \sin \lambda_j \, (T - \tau) \sin \lambda_k \, (T - \tau) \, d\tau, \ b_{jk} = \int_0^T \cos \lambda_j \, (T - \tau) \sin \lambda_k \, (T - \tau) \, d\tau,$$

$$c_{jk} = \int_0^T h_j^{(1)} \, (\tau) \sin \lambda_k \, (T - \tau) \, d\tau, \ d_{jk} = \int_0^T \sin \lambda_j \, (T - \tau) \cos \lambda_k \, (T - \tau) \, d\tau,$$

$$e_{jk} = \int_0^T \cos \lambda_j \, (T - \tau) \cos \lambda_k \, (T - \tau) \, d\tau, \ f_{jk} = \int_0^T h_j^{(1)} \, (\tau) \cos \lambda_k \, (T - \tau) \, d\tau,$$

$$a_{jk}^{(1)} = \int_0^T \sin \lambda_j \, (T - \tau) \, h_k^{(1)} \, (\tau) \, d\tau, \ b_{jk}^{(1)} = \int_0^T \cos \lambda_j \, (T - \tau) \, h_k^{(1)} \, (\tau) \, d\tau,$$

$$g_{jk} = \int_0^T h_j^{(1)} \, (\tau) \, h_k^{(1)} \, (\tau) \, d\tau.$$

Let the values p_k^0, q_k^0, γ_k^0, $k = 1, ..., n$ and β_n^0, are the solution to system (29). Then, according to (27), (26) we have

$$h_{1n}^0(\tau) = \sum_{k=1}^n \left[p_k^0 \sin \lambda_k \, (T - \tau) + q_k^0 \cos \lambda_k \, (T - \tau) + \gamma_k^0 h_k^{(1)}(\tau) \right], \qquad (30)$$

where

$$(\rho_n^0)^2 = \int_0^T \left(h_{1n}^0(\tau) \right)^2 \, d\tau.$$

Following [17], the optimal boundary controls $u_n^0 \, (\tau)$, according to formulas (24) and (30), for any $n = 1, 2, ...$ will be represented as follows:

$$u_n^0(\tau) = \begin{cases} \frac{1}{(\rho_n^0)^2} \sum_{k=1}^n \left[p_k^0 \sin \lambda_k (T - \tau) + q_k^0 \cos \lambda_k (T - \tau) \right. \\ \qquad\qquad \left. + \gamma_k^0 \sin \lambda_k (t_1 - \tau) \right], \ 0 \le \tau \le t_1, \\ \frac{1}{(\rho_n^0)^2} \sum_{k=1}^n \left[p_k^0 \sin \lambda_k \, (T - \tau) + q_k^0 \cos \lambda_k (T - \tau) \right], \ t_1 < \tau \le t_2 = T. \end{cases}$$

$$(31)$$

It should be noted that the values of the optimal control $u_n^0(\tau)$ at the end of the interval $[0, t_1]$ coincide with the values at the beginning of the interval $(t_1, T]$, and this value has the following form:

$$u_n^0 \, (t_1) = \frac{1}{(\rho_n^0)^2} \sum_{k=1}^n \left[p_k^0 \sin \lambda_k \, (T - t_1) + q_k^0 \cos \lambda_k \, (T - t_1) \right].$$

Thus, the constructed optimal boundary controls $u_n^0(\tau)$, as functions of time, are continuous on the interval $[0, T]$.

Now we construct the deflection function corresponding to the optimal control $u_n^0(\tau)$. Substituting the obtained expressions for the optimal controls $u_n^0(\tau)$ from (31) into (18), and the expression obtain for $F_k^0(t)$ into (22), we obtain the function $V_k^0(t)$, $t \in [0, T]$, $k = 1, ..., n$. Further, from formula (17), we have

$$V_n^0(x,t) = \sum_{k=1}^{n} V_k^0(t) \sin \frac{\pi k}{l} x, \tag{32}$$

and from (12) the function $W_n^0(x, t)$ has the form

$$W_n^0(x, t) = \left(1 - \frac{x}{l}\right) u_n^0(t). \tag{33}$$

Thus, according to (10), for the first n harmonics, the optimal string deflection function $Q_n^0(x, t)$ can be written in the form

$$Q_n^0(x, t) = V_n^0(x, t) + W_n^0(x, t). \tag{34}$$

5   Constructing the Solution for Case $n = 1$

Applying the above proposed approach, we construct a boundary control for $n = 1$ (i.e., $k = 1$). Then, according to (29), we have

$$a_{11}p_1 + b_{11}q_1 + c_{11}\gamma_1 = -\frac{\beta_1}{2}C_{11}(T), \quad d_{11}p_1 + e_{11}q_1 + f_{11}\gamma_1 = -\frac{\beta_1}{2}C_{21}(T), \tag{35}$$

$$a_{11}^{(1)}p_1 + b_{11}^{(1)}q_1 + g_{11}\gamma_1 = -\frac{\beta_1}{2}C_{11}(t_1), \quad p_1 C_{11}(T) + q_1 C_{21}(T) + \gamma_1 C_{11}(t_1) = 1,$$

where

$$a_{11} = \frac{T}{2} - \frac{1}{2\lambda_1} \sin \lambda_1 T \cos \lambda_1 T, \quad b_{11} = d_{11} = \frac{1}{2\lambda_1} \sin^2 \lambda_1 T,$$

$$c_{11} = a_{11}^{(1)} = \frac{t_1}{2} \cos \lambda_1 (T - t_1) - \frac{1}{2\lambda_1} \sin \lambda_1 t_1 \cos \lambda_1 T,$$

$$e_{11} = \frac{T}{2} + \frac{1}{2\lambda_1} \sin \lambda_1 T \cos \lambda_1 T, \quad g_{11} = \frac{t_1}{2} - \frac{1}{2\lambda_1} \sin \lambda_1 t_1 \cos \lambda_1 t_1,$$

$$f_{11} = b_{11}^{(1)} = \frac{1}{2\lambda_1} \sin \lambda_1 t_1 \sin \lambda_1 T - \frac{t_1}{2} \sin \lambda_1 (T - t_1), \tag{36}$$

$$C_{11}(T) = \frac{l}{2a\lambda_1}\left[\lambda_1 V_1(T) - \lambda_1 V_1(0) \cos \lambda_1 T - \dot{V}_1(0) \sin \lambda_1 T\right] \\ + \frac{1}{\lambda_1^2}\left[\lambda_1 \varphi_T(0) - \psi_0(0) \sin \lambda_1 T - \lambda_1 \varphi_0(0) \cos \lambda_1 T\right],$$

$$C_{21}(T) = \frac{l}{2a\lambda_1} \left[\dot{V}_1(T) + \lambda_1 V_1(0) \sin \lambda_1 T - \dot{V}_1(0) \cos \lambda_1 T \right]$$
$$+ \frac{1}{\lambda_1^2} \left[\psi_T(0) - \psi_0(0) \cos \lambda_1 T + \lambda_1 \varphi_0(0) \sin \lambda_1 T \right],$$

$$C_{11}(t_1) = \frac{l}{2a\lambda_1} \left[\lambda_1 V_1(t_1) - \lambda_1 V_1(0) \cos \lambda_1 t_1 - \dot{V}_1(0) \sin \lambda_1 t_1 \right]$$
$$+ \frac{1}{\lambda_1^2} \left[\lambda_1 \varphi_1(0) - \psi_0(0) \sin \lambda_1 t_1 - \lambda_1 \varphi_0(0) \cos \lambda_1 t_1 \right].$$

Next, we find a solution to system (35), i.e. values p_1^0, q_1^0, γ_1^0 and β_1^0:

$$p_1^0 = \frac{1}{2\Delta} \left(\left(f_{11}^2 - e_{11}g_{11} \right) C_{11}(T) + (b_{11}g_{11} - c_{11}f_{11}) C_{21}(T) \right.$$
$$\left. + (e_{11}c_{11} - b_{11}f_{11}) C_{11}(t_1) \right),$$

$$q_1^0 = \frac{1}{2\Delta} \left((b_{11}g_{11} - c_{11}f_{11}) C_{11}(T) + \left(c_{11}^2 - a_{11}g_{11} \right) C_{21}(T) \right. \qquad (37)$$
$$\left. + (a_{11}f_{11} - b_{11}c_{11}) C_{11}(t_1) \right),$$

$$\gamma_1^0 = \frac{1}{2\Delta} \left((e_{11}c_{11} - b_{11}f_{11}) C_{11}(T) + (a_{11}f_{11} - b_{11}c_{11}) C_{21}(T) \right.$$
$$\left. + \left(b_{11}^2 - a_{11}e_{11} \right) C_{11}(t_1) \right),$$

$$\beta_1^0 = \frac{1}{\Delta} \left(a_{11}g_{11}e_{11} + 2c_{11}b_{11}f_{11} - a_{11}f_{11}^2 - e_{11}c_{11}^2 - g_{11}b_{11}^2 \right),$$

where

$$\Delta = \frac{1}{2} \left[\left(f_{11}^2 - e_{11}g_{11} \right) C_{11}^2(T) + \left(c_{11}^2 - a_{11}g_{11} \right) C_{21}^2(T) \right.$$
$$\left. + \left(b_{11}^2 - a_{11}e_{11} \right) C_{11}^2(t_1) \right]$$
$$+ (e_{11}c_{11} - b_{11}f_{11}) C_{11}(t_1)C_{11}(T) + (b_{11}g_{11} - c_{11}f_{11}) C_{21}(T)C_{11}(T)$$
$$+ (a_{11}f_{11} - b_{11}c_{11}) C_{11}(t_1)C_{21}(T).$$

Following (31), we obtain

$$u_1^0(\tau) = \begin{cases} \frac{1}{(\rho_1^0)^2} \left(p_1^0 \sin \lambda_1(T - \tau) + q_1^0 \cos \lambda_1(T - \tau) + \right. \\ \qquad \left. + \gamma_1^0 \sin \lambda_1(t_1 - \tau) \right), \qquad 0 \leq \tau \leq t_1, \\ \frac{1}{(\rho_1^0)^2} \left(p_1^0 \sin \lambda_1 (T - \tau) + q_1^0 \cos \lambda_1 (T - \tau) \right), \qquad t_1 < \tau \leq t_2 = T, \end{cases} \qquad (38)$$

where

$$\left(\rho_1^0 \right)^2 = \frac{T}{2} \left(\left(q_1^0 \right)^2 + \left(p_1^0 \right)^2 \right) + \frac{t_1}{2} \left((\gamma_1^0)^2 + 2\gamma_1^0 \left[p_1^0 \cos \lambda_1(T - t_1) \right. \right.$$
$$\left. - q_1^0 \sin \lambda_1(T - t_1) \right] \right) + \frac{1}{\lambda_1} \left(p_1^0 q_1^0 \sin^2 \lambda_1 T - \frac{(\gamma_1^0)^2}{2} \sin \lambda_1 t_1 \cos \lambda_1 t_1 \right.$$
$$\left. + \gamma_1^0 \left(q_1^0 \sin \lambda_1 T - p_1^0 \cos \lambda_1 T \right) \sin \lambda_1 t_1 + \frac{(q_1^0)^2 - (p_1^0)^2}{2} \sin \lambda_1 T \cos \lambda_1 T \right).$$

The optimal function of the string deflection $Q_n^0(x,t)$ according to formula (34) will have the form:

$$Q_1^0(x,t) = V_1^0(x,t) + W_1^0(x,t) = V_1^0(t) \sin \frac{\pi}{l} x + \left(1 - \frac{x}{l} \right) u_1^0(t). \qquad (39)$$

6   Numerical Experiment

We present the results of a computational experiment for a given initial, intermediate, and final state of the string. For simplicity, we assume that

$$t_1 = 4\frac{l}{a},\ T = 8\frac{l}{a},\ \lambda_1 = \frac{a\pi}{l},$$

so that

$$t_1\lambda_1 = 4\pi,\ T\lambda_1 = 8\pi,\ \lambda_1(T - t_1) = 4\pi.$$

Then, according to (36) and (37), we obtain

$$p_1^0 = \frac{2\left(C_{11}(T) - C_{11}(t_1)\right)}{\delta_1},\ q_1^0 = \frac{C_{21}(T)}{\delta_1},\ \gamma_1^0 = \frac{2\left(2C_{11}(t_1) - C_{11}(T)\right)}{\delta_1},$$

$$\beta_1^0 = -\frac{8l}{a\delta_1},\ \delta_1 = 4C_{11}^2(t_1) + C_{21}^2(T) - 4C_{11}(t_1)C_{11}(T) + 2C_{11}^2(T).$$

Thus, according to (38), we have

$$u_1^0(\tau) = \begin{cases} \frac{a}{4l}C_{21}(T)\cos\lambda_1\tau - \frac{a}{2l}C_{11}(t_1)\sin\lambda_1\tau, & 0 \leq \tau \leq t_1, \\ \frac{a}{4l}C_{21}(T)\cos\lambda_1\tau - \frac{a}{2l}\left(C_{11}(T) - C_{11}(t_1)\right)\sin\lambda_1\tau, & t_1 < \tau \leq t_2 = T. \end{cases}$$

$$(40)$$

Let us present the results of a numerical experiment for a given initial, intermediate, and final state of a string, under the assumption that $a = \frac{1}{4}$, $l = 1$ and compare the behavior of the string deflection function with the given initial functions. For the chosen values a and l we have

$$t_1 = \frac{4l}{a} = 16,\ T = \frac{8l}{a} = 32,\ \lambda_1 = \frac{\pi}{4}.$$

Let the following initial state be given for $t_0 = 0$:

$$\varphi_0(x) = -\frac{1}{2}x^3 + \frac{3x^2}{10} + \frac{1}{5}x,\ \psi_0(x) = x^2 - x,$$

for $t_1 = 16$ an intermediate state is given:

$$\varphi_1(x) = x^2 - \frac{9x}{10} - \frac{1}{10},$$

and for $T = 32$ the final state is given:

$$\varphi_T(x) = 0,\ \psi_T(x) = 0.$$

The Fourier coefficients for the functions $\varphi_0(x)$, $\psi_0(x)$, $\varphi_1(x)$, $\varphi_T(x)$, $\psi_T(x)$ are equal:

$$\varphi_1^{(0)} = \frac{18}{5\pi^3},\ \psi_1^{(0)} = -\frac{8}{\pi^3},\ \varphi_1^{(1)} = -\frac{8}{\pi^3} - \frac{1}{5\pi},\ \varphi_1^{(T)} = 0, \psi_1^{(T)} = 0,$$

respectively. The values of these functions at the edges of the string are as follows:

$$\varphi_1(0) = -\tfrac{1}{10}, \; \varphi_0(0) = \varphi_T(0) = \psi_T(0) = \psi_0(0) = \varphi_0(1)$$
$$= \varphi_1(1) = \varphi_T(1) = \psi_T(1) = \psi_0(1) = 0.$$

From (19)–(21) we have

$$V_1(0) = \frac{18}{5\pi^3}, \, \dot{V}_1(0) = -\frac{8}{\pi^3}, V_1(t_1) = -\frac{8}{\pi^3}, V_1(T) = 0, \dot{V}_1(T) = 0.$$

And according to (36) we have

$$C_{11}(T) = \frac{l}{2a}\left(V_1(T) - V_1(0)\right) + \frac{\varphi_T(0) - \varphi_0(0)}{\lambda_1} = -\frac{36}{5\pi^3},$$

$$C_{21}(T) = \frac{l}{2a\lambda_1}\left(\dot{V}_1(T) - \dot{V}_1(0)\right) + \frac{\psi_T(0) - \psi_0(0)}{\lambda_1^2} = \frac{64}{\pi^4}, \tag{41}$$

$$C_{11}(t_1) = \frac{l}{2a}\left(V_1(t_1) - V_1(0)\right) + \frac{\varphi_1(0) - \varphi_0(0)}{\lambda_1} = -\frac{116}{5\pi^3} - \frac{2}{5\pi}.$$

From formulas (40), (41) we have

$$u_1^0(t) = \begin{cases} \frac{4}{\pi^4}\cos\frac{\pi}{4}t + \frac{\pi^3+58\pi}{20\pi^4}\sin\frac{\pi}{4}t, & 0 < t < t_1, \\ \frac{4}{\pi^4}\cos\frac{\pi}{4}t - \frac{\pi^3+40\pi}{20\pi^4}\sin\frac{\pi}{4}t, & t_1 < t \le t_2 = T. \end{cases} \tag{42}$$

Graph of function $u_1^0(t)$ is shown in Fig. 1. Note that

$$u_1^0(t_1) = \frac{4}{\pi^4}, \; \max_{0 \le t \le T}\left|u_1^0(t)\right| \approx 0.1169.$$

Taking into account (42) for the function $V_1^0(t)$ we obtain:

$$V_1^0(t) = \begin{cases} \left(\frac{144-29t}{40\pi^3} - \frac{t}{80\pi}\right)\cos\frac{\pi}{4}t + \left(\frac{10t-291}{10\pi^4} + \frac{1}{20\pi^2}\right)\sin\frac{\pi}{4}t, & 0 \le t \le t_1, \\ \left(\frac{t-32}{2\pi^3} + \frac{t-32}{80\pi}\right)\cos\frac{\pi}{4}t + \left(\frac{t-34}{\pi^4} - \frac{1}{20\pi^2}\right)\sin\frac{\pi}{4}t, & t_1 < t \le t_2 = T. \end{cases} \tag{43}$$

Graph of function $V_1^0(t)$ is illustrated in Fig. 2. According to formula (39), using (42) and (43) we obtain at $t \in [0, 16]$

$$Q_1^0(x,t) = \left(\left(\frac{144-29t}{40\pi^3} - \frac{t}{80\pi}\right)\cos\frac{\pi}{4}t + \left(\frac{10t-291}{10\pi^4} + \frac{1}{20\pi^2}\right)\sin\frac{\pi}{4}t\right)\sin\pi x$$
$$+ (1-x)\left(\frac{4}{\pi^4}\cos\frac{\pi}{4}t + \frac{\pi^3+58\pi}{20\pi^4}\sin\frac{\pi}{4}t\right),$$

at $t \in (16, 32]$

$$Q_1^0(x,t) = \left(\left(\frac{t-32}{2\pi^3} + \frac{t-32}{80\pi}\right)\cos\frac{\pi}{4}t + \left(\frac{t-34}{\pi^4} - \frac{1}{20\pi^2}\right)\sin\frac{\pi}{4}t\right)\sin\pi x$$
$$+ (1-x)\left(\frac{4}{\pi^4}\cos\frac{\pi}{4}t - \frac{\pi^3+40\pi}{20\pi^4}\sin\frac{\pi}{4}t\right).$$

Fig. 1. Graph of $u_1^0(t)$.

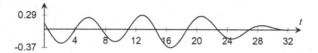

Fig. 2. Graph of $V_1^0(t)$.

Let us give the form of the functions $Q_1^0(x,t)$, $\dot{Q}_1^0(x,t)$, at fixed times $t = 0$, 16, 32:

$$Q_1^0(x,0) = \frac{18}{5\pi^3}\sin\pi x + \frac{4}{\pi^4}(1-x),$$

$$\left.\frac{\partial Q_1^0(x,t)}{\partial t}\right|_{t=0} = \dot{Q}_1^0(x,0) = -\frac{8}{\pi^3}\sin\pi x + \frac{58+\pi^2}{80\pi^2}(1-x),$$

$$Q_1^0(x,16) = -\left(\frac{8}{\pi^3}+\frac{1}{5\pi}\right)\sin\pi x + \frac{4}{\pi^4}(1-x),$$

$$Q_1^0(x,32) = \frac{4}{\pi^4}(1-x),\quad \left.\frac{\partial Q_1^0(x,t)}{\partial t}\right|_{t=32} = \dot{Q}_1^0(x,32) = -\frac{40+\pi^2}{80\pi^2}(1-x).$$

Note that the value of the deflection function $Q_1^0(x,16)$ at the end moment of the first interval coincides with the value at the beginning of the second interval.

Graphical representations of functions $Q_1^0(x,0)$ and $\varphi_0(x)$, $\dot{Q}_1^0(x,0)$ and $\psi_0(x)$, $Q_1^0(x,16)$ and $\varphi_1(x)$, $Q_1^0(x,32)$ and $\dot{Q}_1^0(x,32)$ are illustrated in Fig. 3 respectively. Let us introduce the following notation

$$\varepsilon_1(x,t_j) = \left|Q_1^0(x,t_j)-\varphi_j(x)\right|,\ \widehat{\varepsilon}_1(x,t_m) = \left|\dot{Q}_1^0(x,t_j)-\psi_m(x)\right|,$$

where $j = \overline{0,2}$, $m = 0,2$ ($j = m = 2$ correspond to the moment in time $t_2 = T$). Then

$$\max_{0\leq x\leq 1}\varepsilon_1(x,0)\approx 0.0411,\ \int_0^1\varepsilon_1(x,0)\,dx\approx 0.0293,$$

$$\max_{0\leq x\leq 1}\widehat{\varepsilon}_1(x,0)\approx 0.0860,\ \int_0^1\widehat{\varepsilon}_1(x,0)\,dx\approx 0.0454,$$

$$\max_{0\leq x\leq 1}\varepsilon_1(x,16)\approx 0.0411,\ \int_0^1\varepsilon_1(x,16)\,dx\approx 0.0365,$$

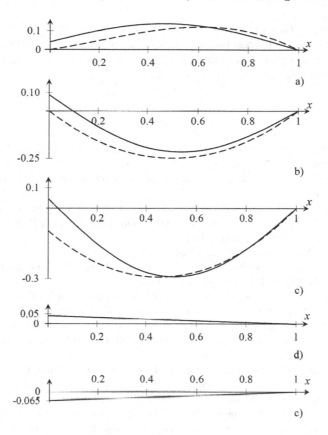

Fig. 3. Graphs of functions: a) $Q_1^0(x, 0)$ (solid line) and $\varphi_0(x)$ (dashed line); b) $\dot{Q}_1^0(x, 0)$ (solid line) and $\psi_0(x)$ (dashed line); c) $Q_1^0(x, 16)$ (solid line) and $\varphi_1(x)$ (dashed line); d) $Q_1^0(x, 32)$ (solid line); e) $\dot{Q}_1^0(x, 32)$ (solid line).

$$\max_{0 \leq x \leq 1} \varepsilon_1(x, 32) \approx 0.0411, \quad \int_0^1 \varepsilon_1(x, 32) \, dx \approx 0.0205,$$

$$\max_{0 \leq x \leq 1} \widehat{\varepsilon}_1(x, 32) \approx 0.0632, \quad \int_0^1 \widehat{\varepsilon}_1(x, 32) \, dx \approx 0.0316.$$

Thus, the results of the analysis showed that under the influence of the constructed control, the behavior of the string deflection function and its derivative are sufficiently close to the given initial functions.

7   Conclusion

A constructive method is proposed for constructing the optimal boundary control of vibrations of a homogeneous string by displacement of one end at a fixed other end with a given shape of deflection and speed of points at different intermediate times. The results can be used in designing the optimal boundary control of the vibration processes. The proposed method can be extended to other multidimensional vibrating systems.

References

1. Butkovskii, A.G.: Control Methods for Systems with Distributed Parameters. Nauka, Moscow (1975). (in Russian)
2. Znamenskaya, L.N.: Control of Elastic Vibrations. Fizmatlit, Moscow (2004). (in Russian)
3. Abdukarimov, M.F.: On optimal boundary control of displacements in the process of forced vibrationson both ends of a string. Dokl. Akad. Nauk Resp. Tadzhikistan **56**(8), 612–618 (2013). (in Russian)
4. Gibkina, N.V., Sidorov, M.V., Stadnikova, A.V.: Optimal boundary control of vibrations of uniform string. Radioelektronika i informatika Nauchno-tekhnicheskij zhurnal HNURE **2**, 3–11 (2016). (in Russian)
5. Il'in, V.A., Moiseev, E.I.: Optimization of boundary controls of string vibrations. Russ. Math. Surv. **60**(6), 1093–1119 (2005). https://doi.org/10.4213/rm1678
6. Moiseev, E.I., Kholomeeva, A.A.: On an optimal boundary control problem with a dynamic boundary condition. Differ. Eqn. **49**(5), 640–644 (2013). https://doi.org/10.1134/S0012266113050133
7. Kopets, M.M.: The problem of optimal control of the string vibration process. In: The Theory of Optimal Solutions, pp. 32–38. V. M. Glushkov Institute of Cybernetics NAS of Ukraine, Kiev (2014). (in Russian)
8. Barseghyan, V.R.: Optimal control of string vibrations with nonseparate state function conditions at given intermediate instants. Autom. Remote.Control **81**(2), 226–235 (2020). https://doi.org/10.1134/S0005117920020034
9. Barseghyan, V.R.: About one problem of optimal control of string oscillations with non-separated multipoint conditions at intermediate moments of time. In: Tarasyev, Alexander, Maksimov, Vyacheslav, Filippova, Tatiana (eds.) Stability, Control and Differential Games. LNCISP, pp. 13–25. Springer, Cham (2020). https://doi.org/10.1007/978-3-030-42831-0_2
10. Barsegyan, V.R.: The problem of optimal control of string vibrations. Int. Appl. Mech. **56**(4), 471–480 (2020). https://doi.org/10.1007/s10778-020-01030-w
11. Barseghyan, V.R., Saakyan, M.A.: The optimal control of wire vibration in the states of the given intermediate periods of time. Proc. NAS RA Mech. **61**, 52–60 (2008). (in Russian)
12. Barseghyan, V.R., Solodusha, S.V.: The problem of boundary control of string vibrations by displacement of the left end when the right end is fixed with the given values of the deflection function at intermediate times. Russian Universities Reports. Mathematics **25**(130), 131–146 (2020). https://doi.org/10.20310/2686-9667-2020-25-130-131-146. (in Russian, Abstr. in Engl.)
13. Barseghyan, V.R., Movsisyan, L.A.: Optimal control of the vibration of elastic systems described by the wave equation. Int. Appl. Mech. **48**(2), 234–239 (2012). https://doi.org/10.1007/s10778-012-0519-9
14. Barseghyan, V.R.: Control of Composite Dynamic Systems and Systems with Multipoint in Termediate Conditions. Nauka, Moscow (2016). (in Russian)

15. Korzyuk, V.I., Kozlovskaya, I.S.: Two-point boundary problem for string oscillation equation with given velocity in arbitrary point of time. II. Tr. In-ta mat. NAN Belarusi **19**(1), 62–70 (2011). (in Russian)
16. Tikhonov, A.N., Samarskii, A.A.: Equations of Mathematical Physics. Nauka, Moscow (1977). (in Russian)
17. Krasovsky, N.N.: The Theory of Motion Control. Nauka, Moscow (1968). (in Russian)

A Discrete Game Problem
with a Non-convex Terminal Set
and a Possible Breakdown in Dynamics

Igor' V. Izmest'ev[1,2]([envelope]) [iD] and Viktor I. Ukhobotov[1,2] [iD]

[1] N.N. Krasovskii Institute of Mathematics and Mechanics, S. Kovalevskaya Street,
16, 620108 Yekaterinburg, Russia
ukh@csu.ru
[2] Chelyabinsk State University, Br. Kashirinykh Street, 129,
454001 Chelyabinsk, Russia

Abstract. A one-dimensional discrete game problem with a given end-point is considered. A terminal set is a union of an infinite number of disjoint segments of equal length. This terminal set has the meaning of the neighborhood of the desired state of the system, taking into account the periodicity. It is believed that one breakdown is possible, which leads to a change in the dynamics of the controlled process. The breakdown time is not known in advance. The first player's control is based on the principle of minimizing the guaranteed result. The opposite side is the second player and the moment of the breakdown. In this paper, we have found necessary and sufficient termination conditions and constructed the corresponding controls of the players. As an example, we consider the problem of controlling a rotational mechanical system with disturbance and possible breakdown.

Keywords: Game · Control · Breakdown

1 Introduction

This article deals with a discrete conflict-controlled process (see as example [5, p. 44–63], [4,11]). Discrete control processes arise, as a rule, when solving applied problems. This is because it is possible to obtain information about the state of real controlled systems and adjust the controls in them only at discrete time moments.

A control problem with possible changes in the dynamics as a result of a breakdown can be analyzed within an approach based on the principle of optimization of the guaranteed result [9]. Such an approach is natural if we know only a time interval when a breakdown may happen. The paper [10] is among the first studies devoted to control problems with a breakdown in this formulation.

This work was funded by the Russian Science Foundation (project no. 19-11-00105).

P. Pardalos et al. (Eds.): MOTOR 2021, LNCS 12755, pp. 314–325, 2021.
https://doi.org/10.1007/978-3-030-77876-7_21

In the present paper, continuing the research begun in [7,8], we consider a discrete game problem, in which the terminal set is the union of an infinite number of disjoint segments of equal length. This terminal set has the meaning of ε-neighborhood of the target position of the system, taking into account the periodicity. The goal of the first player is to lead the phase variable at a given time to the terminal set. The goal of the second player is the opposite. Also, we assume the possibility of one breakdown, which results in a change in the dynamics of the first player. The time of the breakdown is not known in advance. The control of the first player is constructed based on the principle of minimization of the guaranteed result. Necessary and sufficient conditions for the possibility of termination are found.

The obtained results can find application in solving problems of controlling rotational mechanical systems (see as example [1,2,6–8,12]) with uncontrolled disturbance, in which the control goal in the original problem acquires the meaning of minimizing the modulus of deviation of the angle from the desired state.

In this paper, as an example of illustrating the theory, we consider the problem of controlling the rod, which is attached to the rotor of the electric motor. The control is the value of the voltage applied to the electric motor. The goal of control is to bring the rod to ε-neighborhood of the target position at a given time, taking into account the periodicity.

2   Problem Statement

Consider discrete game problem

$$z(k+1) = z(k) - a(k)u(k) + b(k)v(k), \tag{1}$$

where $z \in \mathbb{R}$, $a(k) \geq 0$, $b(k) \geq 0$, $u(k) \in S$, $v(k) \in S$, $k = \overline{0, N-1}$. Here, $u(k)$ is control of the first player, $v(k)$ is control of the second player, $S = [-1, 1]$.

Rule of transition from $z(k)$ to $z(k+1)$, $k = \overline{0, N-1}$. At the moment of time k, knowing the value of $z(k)$, the first player chooses control $u(k) \in S$ and informs the second player about his choice. After that, the second player, knowing $z(k)$ and the chosen control of the first player $u(k)$, chooses control $v(k) \in S$. Then, for the selected pair of controls by the formula (1) $z(k+1)$ is realized.

The sequences of controls $u(k)$ and $v(k)$, $k = \overline{0, N-1}$ formed according to this rule will be called admissible strategies of the first and second players, respectively.

The numbers α, $\varepsilon \in \mathbb{R}$ are given such that $0 \leq 2\varepsilon < \alpha$. The goal of the first player is to lead the point z at the time moment N to the terminal set Z:

$$z(N) \in \bigcup_{i \in I} [i\alpha - \varepsilon, i\alpha + \varepsilon].$$

Here, $I = 0, \pm 1, \pm 2, \pm 3, \ldots$. The goal of the second player is the opposite.

3   Necessary and Sufficient Conditions of Termination

Define function

$$f(k) = \sum_{j=k}^{N-1} (a(j) - b(j)) \text{ for } 0 \le k \le N - 1 \text{ and } f(N) = 0. \tag{2}$$

Denote

$$q_1 = \min\{k \in \overline{0,N} : \varepsilon + f(j) < \alpha - \varepsilon - f(j) \text{ for all } j \in \overline{k,N}\}, \tag{3}$$

$$q_2 = \min\{k \in \overline{0,N-1} : 0 \le \varepsilon + f(j+1) - b(j) \text{ for all } j \in \overline{k,N-1}\} \tag{4}$$

for $\varepsilon - b(N-1) \ge 0$ and $q_2 = N$ for $\varepsilon - b(N-1) < 0$.

Define sets $W(k)$ for $k \in \overline{0,N}$ as follows:

$$W(k) = \begin{cases} \bigcup_{i \in I}[i\alpha - \varepsilon - f(k), i\alpha + \varepsilon + f(k)] & \text{for } \max(q_1, q_2) \le k \le N, \\ \mathbb{R} & \text{for } k < q_1, q_2 < q_1, \\ \emptyset & \text{for } k < q_2, q_1 \le q_2. \end{cases} \tag{5}$$

Here, $\emptyset$ denotes empty set.

Theorem 1. *Let $z(k) \in W(k)$, then there exists a control of the first player $u(k) \in S$ that guarantees the fulfilment of the inclusion $z(k+1) \in W(k+1)$ for any control of the second player $v(k) \in S$.*

Proof. Case 1. If $W(k) \ne \mathbb{R}$, then condition $z(k) \in W(k)$ is equivalent to the following (see (5)): there exists $i_* \in I$ such that

$$z(k) \in [\alpha i_* - f(k) - \varepsilon, \alpha i_* + f(k) + \varepsilon].$$

This inclusion can be written as the inequality

$$|z(k) - \alpha i_*| \le \varepsilon + f(k). \tag{6}$$

Using (1), we obtain

$$z(k+1) - \alpha i_* = z(k) - \alpha i_* - a(k)u(k) + b(k)v(k). \tag{7}$$

Case 1.1. Let

$$|z(k) - \alpha i_*| > a(k),$$

then the first player chooses control

$$u(k) = \text{sign}(z(k) - \alpha i_*). \tag{8}$$

Substituting control (8) in (7), we obtain

$$|z(k+1) - \alpha i_*| \le |z(k) - \alpha i_*| - a(k) + b(k) \le f(k+1). \tag{9}$$

Here, we use formulas (2) and (6).

From (9) we obtain inclusion

$$z(k+1) \in [\alpha i_* - f(k+1) - \varepsilon, \alpha i_* + f(k+1) + \varepsilon].$$

Hence $z(k+1) \in W(k+1)$.

Case 1.2. Let

$$|z(k) - \alpha i_*| \le a(k),$$

then the first player chooses control

$$u(k) = \frac{z(k) - \alpha i_*}{a(k)} \text{ for } a(k) > 0. \tag{10}$$

and any $u(k) \in S$ for $a(k) = 0$.

Substituting control (10) in (7), we obtain

$$|z(k+1) - \alpha i_*| \le b(k) \le \varepsilon + f(k+1).$$

Last inequality holds because $W(k) \ne \emptyset$ and $W(k) \ne \mathbb{R}$ that imply $q_2 \le k$ and, therefore, $0 \le \varepsilon + f(k+1) - b(k)$ (see (4)).

Case 2. Let $W(k) = \mathbb{R}$ and $W(k+1) \ne \mathbb{R}$. From this and formulas (3) and (5), we obtain the inequality

$$\varepsilon + f(k) \ge \alpha - \varepsilon - f(k).$$

Hence

$$\frac{\alpha}{2} \le \varepsilon + f(k).$$

On the other hand, there exists $i_* \in I$ such that

$$|z(k) - i_* \alpha| \le \frac{\alpha}{2}.$$

Therefore, the inequality $|z(k) - i_* \alpha| \le \varepsilon + f(k)$ holds. Thus, we go to the case 1 of the proof.

Case 3. If $W(k+1) = \mathbb{R}$, then the first player can choose any control $u(k) \in S$.

Theorem 2. *Let $z(k) \notin W(k)$, then for any control of the first player $u(k) \in S$ there exists a control of the second player $v(k) \in S$, which guarantees the fulfilment of the condition $z(k+1) \notin W(k+1)$.*

Proof. Case 1. If $W(k) \ne \emptyset$, then condition $z(k) \notin W(k)$ is equivalent to the following: there exists $i_* \in I$ such that

$$z(k) \in (\alpha i_* + f(k) + \varepsilon, \alpha(i_* + 1) - f(k) - \varepsilon).$$

This inclusion can be written as the inequality

$$|z(k) - \alpha(i_* + 0.5)| < 0.5\alpha - \varepsilon - f(k). \tag{11}$$

Using (1), we obtain

$$z(k+1) - \alpha(i_* + 0.5) = z(k) - \alpha(i_* + 0.5) - a(k)u(k) + b(k)v(k). \qquad (12)$$

Case 1.1. Let

$$|z(k) - \alpha(i_* + 0.5) - a(k)u(k)| > b(k),$$

then the second player chooses control

$$v(k) = -\text{sign}(z(k) - \alpha(i_* + 0.5) - a(k)u(k)).$$

Substituting this control in (12), we obtain

$$|z(k+1) - \alpha i_*(i_* + 0.5)| \le |z(k) - \alpha(i_* + 0.5) - a(k)| - b(k)$$
$$\le |z(k) - \alpha(i_* + 0.5)| + a(k) - b(k) < 0.5\alpha - \varepsilon - f(k+1). \qquad (13)$$

Here, we use formulas (2) and (11).

From (13) we obtain inclusion

$$z(k+1) \in (\alpha i_* + f(k+1) \mid \varepsilon, \alpha(i_* + 1) - f(k+1) - \varepsilon).$$

Hence $z(k+1) \notin W(k+1)$.

Case 1.2. Let

$$|z(k) - \alpha(i_* + 0.5) - a(k)u(k)| \le b(k),$$

then the second player chooses control

$$v(k) = -\frac{z(k) - \alpha(i_* + 0.5) - a(k)u(k)}{b(k)} \quad \text{for } b(k) > 0$$

and any $v(k) \in S$ for $b(k) = 0$.

Substituting this control in (12), we obtain

$$|z(k+1) - \alpha(i_* + 0.5)| = 0.$$

Note that $W(k+1) \ne \mathbb{R}$, therefore $0 < 0.5\alpha - \varepsilon - f(k+1)$. Thus, we obtain

$$|z(k+1) - \alpha(i_* + 0.5)| < 0.5\alpha - \varepsilon - f(k+1).$$

Case 2. Let $W(k) = \emptyset$ and $W(k+1) \ne \emptyset$. These equality imply that

$$\varepsilon + f(k) < 0.$$

On the other hand, there exists $i_* \in I$ such that

$$|z(k) - \alpha(i_* + 0.5)| \le \frac{\alpha}{2}.$$

From here we obtain

$$|z(k) - \alpha(i_* + 0.5)| < \frac{\alpha}{2} - \varepsilon - f(k).$$

Thus, we go to case 1 of the proof.

Case 3. If $W(k+1) = \emptyset$, then the second player can choose any control $v(k) \in S$.

4   Problem with an Unknown Moment of Change in the Dynamics of the First Player

Consider a modification of the original problem in which

$$a(k,\tau) = a_1(k) \text{ for } k < \tau, \quad a(k,\tau) = 0 \text{ for } \tau \le k < \tau + \Delta, \qquad (14)$$

$$a(k,\tau) = a_1(k) \text{ for } \tau + \Delta \le k, \qquad (15)$$

where $a_1(k,\tau) \ge 0$, $k = \overline{0, N-1}$.

The reason for this change in dynamics may be a breakdown. The moment of breakdown $\tau \in \overline{0, N}$ is not known to the first player in advance. It takes $\Delta > 0$ time (Δ is integer number) to fix the breakdown. According to (14), for moments $\tau \le k < \min(N; \tau + \Delta)$, the right-hand side of the equation of motion (1) does not depend on the control $u(k)$ of the first player. The first player receives information about whether a breakdown has occurred or not at the beginning of each move before choosing his control $u(k)$. If a breakdown occurs, the first player remembers this moment in time.

Consider the case when the moment of change in the dynamics of the first player τ is chosen by the second player once before the start of the game process.

Define for $\tau \in \overline{0, N}$ the following function:

$$f(k,\tau) = \sum_{j=k}^{N-1} (a(j,\tau) - b(j)) \text{ for } 0 \le k \le N-1 \text{ and } f(N,\tau) = 0.$$

Denote

$$q_1(\tau) = \min\{k \in \overline{0, N} : \varepsilon + f(j,\tau) < \alpha - \varepsilon - f(j,\tau) \text{ for all } j \in \overline{k, N}\},$$

$$q_2(\tau) = \min\{k \in \overline{0, N-1} : 0 \le \varepsilon + f(j+1,\tau)) - b(j) \text{ for all } j \in \overline{k, N-1}\}$$

for $\varepsilon - b(N-1) \ge 0$ and $q_2(\tau) = N$ for $\varepsilon - b(N-1) < 0$.

Define sets $W(k,\tau)$ for $k \in \overline{0, N}$ and fixed $\tau \in \overline{0, N}$ as follows:

$$W(k,\tau) = \begin{cases} \bigcup_{i \in I}[i\alpha - \varepsilon - f(k,\tau), i\alpha + \varepsilon + f(k,\tau)] & \text{for } \max(q_1(\tau), q_2(\tau)) \le k, \\ \mathbb{R} & \text{for } k < q_1(\tau), q_2(\tau) < q_1(\tau), \\ \emptyset & \text{for } k < q_2(\tau), q_1(\tau) \le q_2(\tau). \end{cases}$$

Furthermore, we define sets $W^*(k)$ for $k \in \overline{0, N}$:
$W^*(k) = \emptyset$, if there exists $\tau \in \overline{k, N}$ such that $k < q_2(\tau)$, $q_1(\tau) \le q_2(\tau)$;
$W^*(k) = \mathbb{R}$, if $k < q_1(\tau)$, $q_2(\tau) < q_1(\tau)$ for all $\tau \in \overline{k, N}$;

$$W^*(k) = \bigcup_{i \in I}[i\alpha - \varepsilon - \min_{\tau \in T(k)} f(k,\tau), i\alpha + \varepsilon + \min_{\tau \in T(k)} f(k,\tau)],$$

if $q_2(\tau) \le k$ or $q_2(\tau) < q_1(\tau)$ for all $\tau \in \overline{k, N}$, and, in addition, $T(k) \ne \emptyset$. Here,

$$T(k) = \{\tau \in \overline{k, N} : q_1(\tau) \le k\}.$$

Theorem 3. *Let $z(0) \in W^*(0)$, then there exists an admissible strategy of the first player that guarantees the fulfillment of the inclusion $z(N) \in Z$ for any breakdown moment $\tau \in \overline{0, N}$ and any admissible strategy of the second player.*

Proof. Assume that inclusion $z(k) \in W^*(k)$ holds for some $k \in \overline{0, N-1}$, and until moment k breakdown has not occurred.

Case 1. Let the breakdown not occur at the moment k.

Case 1.1. Let $W^*(k) \neq \mathbb{R}$ and $W^*(k+1) \neq \mathbb{R}$. Then we can choose $i_* \in I$ such that

$$|z(k) - \alpha i_*| \leq \varepsilon + \min_{\tau \in T(k)} f(k, \tau). \tag{16}$$

Also, it can be shown that inequality

$$\min_{\tau \in T(k)} f(k, \tau) \leq \min_{\tau \in T(k+1)} f(k+1, \tau) + a(k) - b(k) \tag{17}$$

holds.

Case 1.1.1. Let $|z(k) - \alpha i_*| > a_1(k)$. Then using (7), (16) and (17), by analogy with case 1.1 of the proof of Theorem 1, we can proof that the control of the first player (8) guarantees the inequality

$$|z(k+1) - \alpha i_*| \leq \varepsilon + \min_{\tau \in T(k+1)} f(k+1, \tau). \tag{18}$$

Therefore, $z(k+1) \in W^*(k+1)$.

Case 1.1.2. Let $|z(k) - \alpha i_*| \leq a_1(k)$. Then using (7), by analogy with case 1.2 of the proof of Theorem 1, we can show that the control of the first player (10) (in the definition of which $a(k)$ is replaced by $a_1(k)$) guarantees the inequality

$$|z(k+1) - \alpha i_*| \leq b(k).$$

It can be shown that if $W(k) \neq \emptyset$ and $W(k) \neq \mathbb{R}$, then

$$b(k) \leq \varepsilon + \min_{\tau \in T(k+1)} f(k+1, \tau).$$

From this, we obtain inequality (18) and, consequently, the required inclusion.

Case 1.2. Let $W^*(k) = \mathbb{R}$ and $W^*(k+1) \neq \mathbb{R}$. These formulas imply that

$$W^*(k+1) = \bigcup_{i \in I} [\alpha i - \varepsilon - \min_{\tau \in T(k+1)} f(k+1, \tau), \alpha i + \varepsilon + \min_{\tau \in T(k+1)} f(k+1, \tau)].$$

There exists $\tau_* \in T(k+1)$ such that

$$W^*(k+1) = \bigcup_{i \in I} [\alpha i - \varepsilon - f(k+1, \tau_*), \alpha i + \varepsilon + f(k+1, \tau_*)] = W(k+1, \tau_*).$$

On the other hand, if $W^*(k) = \mathbb{R}$, then $W(k, \tau_*) = \mathbb{R}$. Therefore, we can construct the control of the first player $u(k)$ (see case 1.2 of the proof of Theorem 1), which guarantees inclusion $z(k+1) \in W(k+1, \tau_*)$. Therefore $z(k+1) \in W^*(k+1)$

Case 1.3. If $W^*(k+1) = \mathbb{R}$, then the first player can choose any control $u(k) \in S$.

Case 2. Let a breakdown occur at time k. Show that $W^*(k) \subseteq W(k,k)$. Indeed, if $W(k,k) = \mathbb{R}$, then this inclusion is obvious. If $W(k,k) \neq \mathbb{R}$ and $W(k,k) \neq \emptyset$, then $q_1(k) \leq k$. Therefore $k \in T(k)$ and

$$\min_{\tau \in T(k)} f(k,\tau) \leq f(k,k).$$

From this, we obtain the required inclusion.

Thus, $z(k) \in W(k,k)$. Next, the first player constructs his controls $u(j)$, $j = \overline{k, N-1}$ as described in the proof of Theorem 1.

Theorem 4. *Let $z(0) \notin W^*(0)$, then there exists a breakdown moment $\tau \in \overline{0,N}$ such that for any admissible strategy of the first player there exists an admissible strategy of the second player that guarantees the fulfillment of the condition $z(N) \notin Z$.*

Proof. Let the second player takes a moment of breakdown $\tau_* \in \overline{0,N}$ such that $W^*(0) = W(0,\tau_*)$. Let us show that such τ_* moment exists. Indeed, if $W^*(0) = \emptyset$ then there exists $\tau_* \in \overline{0,N}$ such that $0 < q_2(\tau_*)$, $q_1(\tau_*) \leq q_2(\tau_*)$. For this τ_* $W(0,\tau_*) = \emptyset$. If $W^*(0) \neq \emptyset$ and $W^*(0) \neq \mathbb{R}$. Then $T(0) \neq \emptyset$, and there exists $\tau_* \in T(0)$ such that

$$\min_{\tau \in T(0)} f(0,\tau) = f(0,\tau_*).$$

Thus, $z(0) \notin W(0,\tau_*)$. Next, the second player constructs his controls $v(k)$, $k = \overline{0, N-1}$ as described in the proof of Theorem 2.

5   Example

The rotor axis of the electric motor passes through the point O perpendicular to the plane of the figure (see Fig. 1). One end of the rod OA is rigidly attached to the axis of the rotor so that it can rotate together with the rotor about its axis in the plane of the figure.

The rotation angle of the rod is denoted by ϕ. Mass of the rod is equal to m. The moment of inertia of a system consisting of the rotor and the rod, concerning the rotor axis, is denoted by J.

Neglecting the inductance in the motor rotor circuit, we assume [2,12] that the moment of electromagnetic forces applied to the rotor on the side of the stator is equal to $c_1\xi - c_2\dot{\phi}$, $c_i > 0$, $i = 1,2$. Here, ξ is voltage applied to the motor. Geometric constraint $|\xi| \leq \sigma$ is imposed on ξ, where the number $\sigma > 0$ is given. Product $c_2\dot{\phi}$ describes moment of forces, which arise because of counter-emf.

Denote by l distance from point O to the center of mass of the rod and write down the Lagrange equation, which describes the motion of the system

$$J\ddot{\phi} = -mgl\sin\phi - c_2\dot{\phi} + c_1\xi.$$

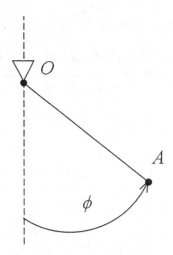

Fig. 1. The problem of controlling the rod using the rotor of the electric motor.

Angle ϕ_* and time moment $p > 0$ are given. The goal of choice of applied voltage $\xi(t)$ is to implement the inclusion from a given initial position $\phi(0)$

$$\phi(p) - \phi_* \in \bigcup_{i \in I} [2\pi i - \varepsilon, 2\pi i + \varepsilon].$$

Here, $0 \leq \varepsilon < \pi$.

Following [3], we take nonlinear addend in the Lagrange equation as disturbance

$$\eta = -\frac{mgl}{J} \sin \phi.$$

It is believed that for distance l we know only the estimate of its value $0 < l \leq l_*$. Then for disturbance constraint

$$|\eta| \leq \delta, \quad \delta = \frac{mgl_*}{J}$$

holds.

Denote

$$x_1 = \phi, \quad x_2 = \dot{\phi}, \quad \gamma = \frac{c_2}{J}, \quad \beta = \frac{c_1}{J}.$$

Write down the considered problem as a control system under disturbance

$$\begin{pmatrix} \dot{x}_1 \\ \dot{x}_2 \end{pmatrix} = \begin{pmatrix} 0 & 1 \\ 0 & -\gamma \end{pmatrix} \begin{pmatrix} x_1 \\ x_2 \end{pmatrix} + \begin{pmatrix} 0 \\ \beta \end{pmatrix} \xi + \begin{pmatrix} 0 \\ 1 \end{pmatrix} \eta, \quad |\xi| \leq \sigma, \quad |\eta| \leq \delta,$$

in which the terminal condition takes the form

$$x_1(p) - \phi_* \in \bigcup_{i \in I} [2\pi i - \varepsilon, 2\pi i + \varepsilon].$$

Introduce a variable

$$z = x_1 + \frac{1}{\gamma}\left(1 - e^{(t-p)\gamma}\right)x_2 - \phi_*.$$

Then

$$\dot{z} = -\frac{\beta\sigma}{\gamma}\left(1 - e^{(t-p)\gamma}\right)u + \frac{\delta}{\gamma}\left(1 - e^{(t-p)\gamma}\right)\widehat{v}, \quad u = -\frac{\xi}{\sigma}, \quad \widehat{v} = \frac{\eta}{\delta}, \quad |u| \le 1, \quad |\widehat{v}| \le 1. \tag{19}$$

Since $z(p) = x_1(p) - \phi_*$, then the terminal condition takes the form

$$z(p) \in \bigcup_{i \in I}[2\pi i - \varepsilon, 2\pi i + \varepsilon]. \tag{20}$$

Let's fix the sequence of numbers $\overline{0, N}$ so that $N = p$, otherwise we scale the time.

Define

$$a_1(k) = \frac{\beta\sigma}{\gamma}\int_k^{k+1}\left(1 - e^{(r-N)\gamma}\right)dr = \frac{\beta\sigma}{\gamma}\left(1 + \frac{1}{\gamma}\left(e^{(k-N)\gamma} - e^{(k+1-N)\gamma}\right)\right),$$

$$b(k) = \frac{\delta}{\gamma}\int_k^{k+1}\left(1 - e^{(r-N)\gamma}\right)dr = \frac{\delta}{\gamma}\left(1 + \frac{1}{\gamma}\left(e^{(k-N)\gamma} - e^{(k+1-N)\gamma}\right)\right).$$

We assume that controls u can be chosen only piecewise constant:

$$u(t) = u(k) = \text{const for } t \in [k, k+1).$$

It can be shown that there exists $v(k) \in S$ such that

$$\frac{\delta}{\gamma}\int_k^{k+1}\left(1 - e^{(r-N)\gamma}\right)\widehat{v}(r)dr - b(k)v(k).$$

We assume that at some time moment $\tau \in \overline{0, N}$ a breakdown may occur in the dynamics as defined in formulas (14), (15).

Thus, taking into account the previous assumptions and notation, the equation of motion of the system (19) takes the form

$$z(k+1) = z(k) - a(k, \tau)u(k) + b(k)v(k), \quad u(k) \in S, \quad v(k) \in S.$$

Taking disturbance $v(k)$ as the control of the second player, we can construct the set $W^*(0)$.

Assume that $z(0) \in W^*(0)$. Let us write down the values of the applied voltages $\xi(k)$, $k = \overline{0, N-1}$, which guarantees the fulfilment of the terminal condition (20).

Case 1. Let $a(t, \tau) > 0$ and the following condition is satisfied: $W^*(k+1) \ne \mathbb{R}$, if no breakdown occurred before moment k; $W(k+1, \tau) \ne \mathbb{R}$, if a breakdown occurs at moment $\tau \le k$.

Case 1.1. Let $|z(k) - 2\pi i_*| > a_1(k)$, then

$$\xi(k) = -\sigma \operatorname{sign}(z(k) - 2\pi i_*).$$

Case 1.2. Let $|z(k) - 2\pi i_*| \leq a_1(k)$, then

$$\xi(k) = -\sigma \frac{z(k) - 2\pi i_*}{a_1(k)}.$$

Case 2. Let $a(t, \tau) = 0$ or the following condition is satisfied: $W^*(k+1) = \mathbb{R}$, if no breakdown occurred before moment k; $W(k+1, \tau) = \mathbb{R}$, if a breakdown occurs at moment $\tau \leq k$. Then we can choose any voltage $\xi(k) \in \sigma S$.

6 Conclusion

In this paper, we consider a one-dimensional discrete game of pursuit. The terminal set in this game is the union of an infinite number of disjoint segments. This terminal set has the meaning of ε–neighborhood of the target position of the system, taking into account the periodicity. For this problem, we find the necessary and sufficient conditions for the possibility of termination and constructed the corresponding controls of the players. Also, a modification of the original problem is considered, in which a change (breakdown) occurs in the dynamics of the first player at an unknown time moment.

As an example, we consider the problem of controlling the rod, which is attached to the rotor of the electric motor. The theoretical results obtained can also find application in solving other problems of controlling rotational mechanical systems.

In the future, this problem can be considered in the case, when the moment of change in dynamics does not depend on the choice of the second player.

References

1. Andrievsky, B.R.: Global stabilization of the unstable reaction-wheel pendulum. Control of Big Syst. **24**, 258–280 (2009). (in Russian)
2. Beznos, A.V., Grishin, A.A., Lenskiy, A.V., Okhozimskiy, D.E., Formalskiy, A.M.: The control of pendulum using flywheel. In: Workshop on Theoretical and Applied Mechanics, pp. 170–195. Publishing of Moscow State University, Moscow (2009). (in Russian)
3. Chernous'ko, F.L.: Decomposition and synthesis of control in nonlinear dynamical systems. Proc. Steklov Inst. Math. **211**, 414–428 (1995)
4. Ibragimov, G.I: Problems of linear discrete games of pursuit. Math. Notes **77**(5), 653–662 (2005)
5. Isaacs, R.: Differential games: A Mathematical Theory with Applications to Warfare and Pursuit, Control and Optimization. Wiley, New York (1965)
6. Izmest'ev, I.V., Ukhobotov, V.I.: On a linear control problem under interference with a payoff depending on the modulus of a linear function and an integral. In: 2018 IX International Conference on Optimization and Applications (OPTIMA 2018) (Supplementary Volume), DEStech, pp. 163–173. DEStech Publications, Lancaster (2019). https://doi.org/10.12783/dtcse/optim2018/27930

7. Izmest'ev, I.V., Ukhobotov, V.I.: On a single-type differential game with a non-convex terminal set. In: Khachay, M., Kochetov, Y., Pardalos, P. (eds.) MOTOR 2019. LNCS, vol. 11548, pp. 595–606. Springer, Cham (2019). https://doi.org/10.1007/978-3-030-22629-9_42

8. Izmest'ev, I.V., Ukhobotov, V.I.: On a one-dimensional differential game with a non-convex terminal payoff. In: Kononov, A., Khachay, M., Kalyagin, V.A., Pardalos, P. (eds.) MOTOR 2020. LNCS, vol. 12095, pp. 200–211. Springer, Cham (2020). https://doi.org/10.1007/978-3-030-49988-4_14

9. Krasovskii, N.N.: Control of a Dynamical System. Nauka Publishing, Moscow (1985). (in Russian)

10. Nikol'skii, M.S.: The crossing problem with possible engine shutoff. Diff. Eq. **29**(11), 1681–1684 (1993)

11. Shorikov, A.F.: An algorithm of adaptive minimax control for the pursuit-evasion process in discrete-time dynamical system. Proc. Steklov Inst. Math. **2**, 173–190 (2000)

12. Ushakov, V.N., Ukhobotov, V.I., Ushakov, A.V., Parshikov, G.V.: On solution of control problems for nonlinear systems on finite time interval. IFAC-PapersOnLine **49**(18), 380–385 (2016). https://doi.org/10.1070/10.1016/j.ifacol.2016.10.195

Altruistic-Like Equilibrium in a Differential Game of Renewable Resource Extraction

Vladimir Mazalov[1,2] ⓘ, Elena Parilina[2(✉)] ⓘ, and Jiangjing Zhou[3]

[1] Institute of Applied Mathematical Research, Karelian Research Center of the Russian Academy of Sciences, 11, Pushkinskaya Street, Petrozavodsk 185910, Russia
vmazalov@krc.karelia.ru
[2] Saint Petersburg State University, 7/9 Universitetskaya nab., Saint Petersburg 199034, Russia
e.parilina@spbu.ru
[3] School of Mathematics and Statistics, Qingdao University, Qingdao 266071, People's Republic of China
st092028@student.spbu.ru

Abstract. We consider a model of renewable resource extraction described by a differential game with infinite horizon. The environmental problems are often considered from cooperative prospective as selfish behavior of the players may negatively affects not only on other players' profits, but also on the environment. The reason is the joint stock of resource which is influenced by all players. We characterize the Berge and altruistic equilibrium in a differential game of renewable resource extraction and compare them with the Nash equilibrium. According to the concept of altruistic equilibrium players can choose the part of the other players' payoffs they support and summarize with the part of their own profit. This equilibrium can be considered as an intermediate between Berge and Nash equilibria. We make numerical simulations and demonstrate theoretical results for the case of n symmetric players.

Keywords: Dynamic games · Berge equilibrium · Altruistic equilibrium · Renewable resources

1 Introduction

We consider a game-theoretical model of renewable resource extraction played by many players in continuous time. The players (countries or companies) extract some resource on a joint territory (harvest fish from the lake, cut the trees at the forest). Traditionally, this problem is modeled from a selfish behavior prospective when the players try to maximize their own profits in a competitive environment. Another approach in the literature is to examine a full cooperative situation when all players form a grand coalition and maximize their joint

The work is supported by the Russian Science Foundation (grant no. 17-11-01079).

profit. Here we consider an altruistic approach when the players want to apply "positive" behavior with respect to other players and maximize the part of the other players' profits summarized with their own costs. This approach recently has appeared in the literature on cooperative economic behavior and it can be compared with the Berge equilibrium.

In 1957, Claude Berge proposed a new concept of an equilibrium, according to which in a coalition game, members of a coalition can work together to maximize the profits of the players in other coalition [5]. On the basis of the notion of equilibrium for a coalitional game introduced by Berge, Zhukovskii introduced the Berge equilibrium for non-cooperative games [21]. This equilibrium can be used as an alternative solution when there is no Nash equilibrium (see [7]) or when there are many of them. In this equilibrium, each player obtains his maximal payoff if the situation is favorable for him: by obligation or willingness, the other players choose strategies favorable for him. The Berge equilibrium concept formalizes mutual support among players motivated by the altruistic social value orientation in the games. The relationship between mutually beneficial practices like creation of teams and Berge equilibria is examined in [8].

In [1–4], Abalo and Kostreva introduce a more general definition of the Berge equilibrium. They also provide a theorem for existence of this equilibrium [1,2], based on an earlier Radjef's theorem [19]. Nessah et al. in [17] describe a simple game that verifies the assumptions of Abalo and Kostreva's theorem without Berge equilibrium in the sense of Zhukovskii, which is a particular case of Berge equilibrium in the sense of Abalo and Kostreva. Colman et al. prove some basic results for Berge equilibria and its connection with the Nash equilibria, and provide a straightforward method for finding the Berge equilibria in n-person games [7]. They explain how the Berge equilibrium provides a compelling model of cooperation in social dilemmas. Kudryatsev et al. proposed a concept of a weak Berge equilibrium (WBE) in a non-cooperative game $(N, f_i(x), i \in N)$ according to which a profile of strategies x^* is the WBE if this profile is the Nash equilibrium in game $(N, \sum_{j \neq i} f_j(x), i \in N)$ (see [12]). The Berge equilibrium concept is described in details in the review of Larbani and Zhukovskii [13] and the book of Salukvadze and Zhukovskiy [20].

Nevertheless, the Berge equilibrium concept has some drawbacks. One of these drawbacks is that the Berge equilibrium rarely exists in pure strategies. Pukacz et al. found an example where in n-person game ($n \geq 3$) with a finite set of strategies, Berge equilibrium may not exist in the class of mixed strategies [18]. Another concern about the Berge equilibrium is that the players need to cooperate to maximize any player's payoff. But this assumption takes additional requirements on the concept which are not considered in the definition of the Berge equilibrium.

We use the concepts of altruistic equilibrium in a dynamic game of renewable resource extraction. Cooperation in environmental problems leads to lower values of pollution stock [11] or higher levels of renewable resource [14]. The cooperation in non-renewable resource extraction with random initial time is examined in [9]. However, cooperation can be difficult to realize (see [6] for stability of coalitions in a pollution game). In the altruistic equilibrium examined in our paper it is supposed that the players are interested in maximizing the benefits of other players and minimizing their own costs at the same time. Therefore, the scheme of altruistic equilibrium is an intermediate one between the Nash equilibrium (selfish behavior) and Berge equilibrium (full altruistic behavior). The altruistic equilibrium concept can be applied to other environmental games including the fish wars ([15], see [10] for a survey on dynamic games in environmental sciences).

In the paper, we characterize the altruistic equilibrium in a differential game of renewable resource extraction and compare the equilibrium strategies, state trajectories and player's payoffs with the Nash and Berge equilibria. As expected, the altruistic equilibrium lies in between the Nash and Berge equilibria. We present comparative analysis of all equilibria on the numerical simulations.

The remainder of the paper is organized as follows. Section 2 introduces the model. In Sect. 2.1 we define the equilibrium concepts including the Nash, Berge and altruistic equilibria. Section 3 contains the main results about equilibria in a differential game of renewable resource extraction. In Sect. 4 we represent the results of numerical simulations, while in Sect. 5 we briefly conclude.

2   The Model

We consider an economy of n players (countries) over an infinite planning horizon in continuous time. The set of players is $N = \{1, \ldots, n\}$. We denote by $u_i(t) \geq 0$ a player i's extraction level at time $t \in [0, \infty)$. We assume that the profit of a player is proportional to the extraction level, that is $q_i u_i(t)$ with $q_i > 0$ for any $i \in N$. Therefore, the revenue of player i at time t is defined as

$$R_i(u_i(t)) = q_i u_i(t). \tag{1}$$

We also assume that each player bears the costs extracting the resource, that is the convex increasing function of strategy:

$$C_i(u_i(t)) = c_i u_i^2(t) \tag{2}$$

with $c_i > 0$.

Denote by $u(t) = (u_1(t), \ldots, u_n(t))$ the vector of players' strategies at time t, and by $x(t) \in X \subset \mathbb{R}^+$ the resource stock at this time. We assume that the stock growth is affected by the players' strategies. The evolution of this stock is defined by the following differential equation:

$$\dot{x}(t) = \alpha x(t) - \sum_{i=1}^{n} u_i(t), \tag{3}$$

with a given initial stock $x(0) = x_0$, where $\alpha \geq 1$ is the natural growth rate.

The instant profit of any player i is given by function

$$\Pi_i(t, u, x) = R_i(u_i(t)) - C_i(u_i(t)). \tag{4}$$

The player i's profit in a differential game is

$$J_i = \int_0^\infty e^{-rt} \Pi_i(t, u(t), x(t)) dt = \int_0^\infty e^{-rt} \left\{ R_i(u_i(t)) - C_i(u_i(t)) \right\} dt \qquad (5)$$

$$= \int_0^\infty e^{-rt} \left\{ q_i u_i(t) - c_i u_i^2(t) \right\} dt$$

subject to state dynamics (3) with initial state x_0 given.

We consider feedback information structure when any player i's strategy ϕ_i is a function of time and state, i.e., $\phi_i = \phi_i(t, x(t))$. We denote the profile of feedback strategies at time t by $\phi(t, x) = (\phi_1(t, x), \ldots, \phi_n(t, x))$.

2.1   Equilibrium Concepts

We consider the Nash equilibrium as a basic equilibrium concept in the non-cooperative game [16].

Definition 1. *The Nash equilibrium (in feedback strategies) is the profile of strategies $\phi^*(t, x) = (\phi_1^*(t, x), \ldots, \phi_n^*(t, x))$ the following inequality holds:*

$$J_i(\phi^*(\cdot)) \geq J_i(\phi_i(\cdot), \phi_{-i}^*(\cdot))$$

for any admissible feedback strategy $\phi_i(\cdot)$ of player $i \in N$. Here $\phi_{-i}^*(\cdot)$ is a vector of feedback strategies of the players from set $N \backslash i$.

We also consider two more solution concepts which involve assumptions about altruistic behavior of the players. First, we define the Berge equilibrium in a differential game according to which any player is supported by all other players in the sense that all other players maximize his payoff in a differential game.

Definition 2. *The Berge equilibrium (in feedback strategies) is the profile of strategies $\phi^b(t, x) = (\phi_1^b(t, x), \ldots, \phi_n^b(t, x))$ such that for any $i \in N$ and any $\phi_{-i}(t, x)$ the following inequality holds:*

$$J_i(\phi^b(\cdot)) \geq J_i(\phi_i^b(\cdot), \phi_{-i}(\cdot)).$$

In the Berge equilibrium, the payoff of player i is maximized over the product set of other players' strategies[1], i.e. the players from the set $N \backslash i$ solve the following maximization problem:

$$\max_{\phi_{-i}(\cdot)} J_i(\phi_i(\cdot), \phi_{-i}(\cdot))$$

for any player i.

In the so-called *altruistic* equilibrium any player maximizes the sum of the profits of other players with the exception of their costs, summarized with his personal costs, i.e., he/she maximizes

$$J_i' = \int_0^\infty e^{-rt} \left\{ \sum_{j \in N, j \neq i} R_j(t, u(t), x(t)) - C_i(u_i(t)) \right\} dt. \qquad (6)$$

---

[1] When $n = 2$, to find the Berge equilibrium one needs to find the Nash equilibrium in the game when players exchange their payoff functions.

Definition 3. *The altruistic equilibrium (in feedback strategies) is the profile of strategies $\phi^a(t,x) = (\phi_1^a(t,x), \ldots, \phi_n^a(t,x))$ such that for any $i \in N$ and any feasible feedback strategy $\phi_i(t,x)$ the following inequality holds:*

$$J_i'(\phi^a(\cdot)) \geq J_i'(\phi_i(\cdot), \phi_{-i}^a(\cdot)).$$

We refer the Berge and altruistic equilibria as altruistic-like equilibria as they reflect the idea of maximizing the payoffs of other players in different manners.

3   Altruistic-Like Equilibria in a Differential Game of Renewable Resource Extraction

In this section, we characterize the altruistic and Berge equilibria for the defined differential game. Then we make a comparison of the equilibrium strategies and trajectories in these equilibria with the Nash equilibrium ones.

We consider symmetric players, i.e., $q_i = q$, $c_i = c$ for any $i \in N$. The main tool to find the equilibria from Definitions 1–3 in feedback strategies is the Hamilton-Jacobi-Bellman (HJB) equation. To use HJB equation for the value function $V_i(x,t)$ we specify the linear-quadratic form of the value function for any problem:

$$V_i(x,t) = V_i(x) = Ax^2 + Bx + D.$$

The next two propositions characterize the altruistic and Berge equilibria defined in Sect. 2.1.

Proposition 1. *The altruistic equilibria in dynamic game with symmetric players and payoff functions (6), and state dynamics given by (3) with initial condition $x(0) = x_0$, is the profile of strategies*

$$u_i^a(x(t)) = \frac{2\alpha - r}{2n - 1}x - \frac{q(n-1)(2\alpha - r)}{2\alpha c(2n - 1)}, \quad i \in N,$$

and the corresponding state trajectory for altruistic equilibrium is

$$x^a(t) = \left(x_0 - \frac{qn(n-1)(2\alpha - r)}{2\alpha c(\alpha - nr)}\right)e^{-\frac{\alpha - nr}{2n - 1}t} + \frac{qn(n-1)(2\alpha - r)}{2\alpha c(\alpha - nr)}. \tag{7}$$

The value function of any player $i \in N$ is

$$V_i^a(x) = -\frac{(2\alpha - r)}{4\alpha^2 c(2n - 1)}(2\alpha cx - q(n-1))^2.$$

Proof. We assume a linear-quadratic form of the value function $V_i(x,t) = Ax^2 + Bx + D$. The HJB equation for player i is

$$rV_i(x,t) = \max_{u_i \geq 0}\left\{\sum_{j \in N, j \neq i} q_j u_j - c_i u_i^2 + \frac{\partial V_i(x,t)}{\partial x}\left(\alpha x - \sum_{j=1}^{n} u_j\right)\right\}. \tag{8}$$

Taking into account the form of the value function and the symmetric form of the game, we substitute an expression of the value function into (8) and find the maximum of RHS in (8) and obtain u_i:

$$u_i^a(x) = -\frac{A}{c}x - \frac{B}{2c} \tag{9}$$

for any player $i \in N$.

From Eq. (8) we obtain the system to find coefficients A, B, D in the value function:

$$rA = \frac{2n-1}{c}A^2 + 2A\alpha,$$

$$rB = \frac{2n-1}{c}AB - \frac{(n-1)q}{c}A + \alpha B,$$

$$rD = \frac{(2n-1)B^2 - 2(n-1)qB}{4c}.$$

Simplifying this system we obtain the solution:

$$A = -\frac{c(2\alpha - r)}{2n-1},$$

$$B = \frac{q(n-1)(2\alpha - r)}{(2n-1)\alpha},$$

$$D = -\frac{q^2(n-1)^2(2\alpha - r)}{4\alpha^2 c(2n-1)}.$$

Notice that A is negative when $\alpha > r/2$. The equilibrium strategy has the form

$$u_i^a(x) = \frac{2\alpha - r}{2n-1}x - \frac{q(n-1)(2\alpha - r)}{2\alpha c(2n-1)}, \quad i \in N, \tag{10}$$

and the value function is

$$V_i^a(x) = V^a(x) = -\frac{(2\alpha - r)}{4\alpha^2 c(2n-1)}(2\alpha cx - q(n-1))^2.$$

The value function of any player in the altruistic equilibrium is

$$V_i^a(x_0) = -\frac{(2\alpha - r)}{4\alpha^2 c(2n-1)}(2\alpha cx_0 - q(n-1))^2.$$

We should notice that the payoff of any player given by function (5) can be calculated substituting the equilibrium state and strategies into (5).

The corresponding state trajectory can be found by substituting the altruistic equilibrium strategies given by (9) into state equation (3), which yields

$$\dot{x}(t) = \frac{nr - \alpha}{2n-1}x(t) + \frac{qn(n-1)(2\alpha - r)}{2\alpha c(2n-1)}, \quad x(0) = x_0. \tag{11}$$

The solution of Eq. (11) is

$$x^a(t) = \left(x_0 - \frac{qn(n-1)(2\alpha - r)}{2\alpha c(\alpha - nr)}\right) e^{-\frac{\alpha - nr}{2n-1}t} + \frac{qn(n-1)(2\alpha - r)}{2\alpha c(\alpha - nr)}. \tag{12}$$

This finishes the proof.

Remark 1. The altruistic equilibrium state trajectory is defined by Eq. (12). The game state dynamics have a globally asymptotically stable steady state if $\alpha > nr$. When $t \to \infty$, the state x^a tends to the steady state

$$x^a(\infty) = \frac{qn(n-1)(2\alpha - r)}{2\alpha c(\alpha - nr)},$$

which is an increasing function of n.

Proposition 2. *The Berge equilibrium in dynamic game with symmetric players, payoff functions defined by (5), and state dynamics given by (3) with initial condition $x(0) = x_0$ is represented by the strategy*

$$u_i^b(x,t) = 0, \quad i \in N,$$

and the corresponding state trajectory for the Berge equilibrium is $x^b(t) = x_0 e^{\alpha t}$, $t \geq 0$. The Berge equilibrium payoff of any player $i \in N$ is $J_i^b = 0$.

Proof. The HJB equation for player i is

$$rV_i(x,t) = \max_{u_j \geq 0, j \neq i} \left\{ q_i u_i - c_i u_i^2 + \frac{\partial V_i(x,t)}{\partial x}\left(\alpha x - \sum_{k=1}^{n} u_k\right) \right\}. \tag{13}$$

Taking into account the form of the value function and the symmetric form of the game, we substitute an expression of the value function $V_i(x,t) = Ax^2 + Bx + D$ into (13) and find the maximum of RHS in (13) and obtain u_i:

$$u_i^b(x,t) = 0 \tag{14}$$

for any player $i \in N$ assuming $\frac{\partial V_i(x,t)}{\partial x} > 0$.

From Eq. (13) we obtain that the coefficients A, B, D are all equal to zero. The corresponding state trajectory is defined as a solution of equation

$$\dot{x}(t) = \alpha x(t), \quad x(0) = x_0, \tag{15}$$

which is

$$x^B(t) = x_0 e^{\alpha t}. \tag{16}$$

The payoff of any player in the Berge equilibrium is zero.

In the following proposition we determine the Nash equilibrium strategies and corresponding state trajectory. We provide this result for further comparison with the Berge and altruistic equilibria.

Proposition 3. *The Nash equilibrium in the dynamic game with symmetric players, payoff functions defined by (5), and state dynamics given by (3) with initial condition $x(0) = x_0$, is the profile of strategies*

$$u_i^*(x) = \frac{2\alpha - r}{2n - 1} x - \frac{q(\alpha - nr)}{2c\alpha(2n - 1)}, \quad i \in N,$$

and the corresponding state trajectory is

$$x^*(t) = \left(x_0 - \frac{nq}{2c\alpha}\right) e^{-\frac{\alpha - nr}{2n-1}t} + \frac{nq}{2c\alpha}.$$

The Nash equilibrium payoff of any player $i \in N$ is equal to

$$J_i^* = -\frac{(2\alpha - r)r(2\alpha c x_0 - nq)^2 - \alpha^2(2n - 1)}{4\alpha^2 cr(2n - 1)}.$$

The proof is given in Appendix A.

Remark 2. The game state dynamics have a globally asymptotically stable steady state if $\alpha > nr$. When $t \to \infty$, the state x^* tends to a steady state

$$x^*(\infty) = \frac{nq}{2c\alpha}.$$

In the steady state the strategy of any player is

$$u^*(\infty) = \frac{(2\alpha - r)q}{2\alpha c(2n - 1)} \left(\frac{nr(2\alpha - 1) - \alpha}{2n - 1} - n + 1\right).$$

Remark 3. The following properties immediately follow from Propositions 1–3:

1. The resource stock is larger with the altruistic equilibrium than with the Nash equilibrium:
$$x^a(t) > x^*(t)$$

for any $t > 0$. This can be easily proved considering the difference

$$x^a(t) - x^*(t) = \left[\frac{qn(n-1)(2\alpha - r)}{2ac(\alpha - nr)} - \frac{nq}{2c\alpha}\right]\left(1 - e^{-\frac{\alpha-nr}{2n-1}t}\right)$$

$$= \frac{qn(n-1)(\alpha(2n - 3) + r)}{2ac(\alpha - nr)}\left(1 - e^{-\frac{\alpha-nr}{2n-1}t}\right)$$

$$> 0$$

for any $t > 0$ and $n > 1$.

2. At the initial state $x(0) = x_0$, the equilibrium strategy in the Nash equilibrium is larger than in the altruistic equilibrium which is positive,

$$u_i^*(0) > u_i^a(0) > u_i^b(0) = 0$$

for any player $i \in N$. As the stock in the altruistic equilibrium is larger than in the Nash one, then for some time instant $t > 0$ it can appear that $u_i^*(t) < u_i^a(t)$.

3. The difference in the steady states in the altruistic and Nash equilibria is

$$x^a(t) - x^*(t) = \frac{qn(n-1)(\alpha(2n-3)+r)}{2\alpha c(\alpha - nr)},$$

which is positive and an increasing function of n.

For the players, the Nash equilibrium behavior is preferable in comparison with the altruistic one if we take into account only their payoffs. But from the Nature's prospective, the altruistic equilibrium is better than the Nash equilibrium as the resource stock is larger in the altruistic equilibrium in comparison with the Nash equilibrium. We demonstrate the properties of the equilibria and the dependence of equilibrium strategies, states and profits on the number of players in the next section on the numerical examples.

4   Numerical Simulations

Example 1. Let the parameters of the game be $n = 2$, $x_0 = 4$, $\alpha = 2.0$, $c = 0.65$, $q = 2.9$, $r = 0.4$. The altruistic and Nash equilibrium state trajectories are

$$x^a(t) = 6.69231 - 2.69231e^{-0.4t},$$
$$x^*(t) = 2.23077 + 1.76923e^{-0.4t}.$$

The altruistic and Nash equilibrium strategies are

$$u^a(x) = -1.33846 + 1.2x^a(t),$$
$$u^*(x) = -0.446154 + 1.2x^*(t).$$

The equilibrium state trajectory and equilibrium strategies are represented in Fig. 1. As expected, the resource level in altruistic equilibrium is higher than in the Nash equilibrium. We can also observe the steady state for both equilibria, $x^a(\infty) = 6.69231$, $x^*(\infty) = 2.23077$.

The equilibrium payoffs of any player in the altruistic and the Nash equilibria are $J_i^a = -6.49038$, $J_i^* = 5.645$ respectively. We should notice that in the Berge equilibrium, the strategies are zero and the players' payoffs are also zero. As expected the real payoff of any player in the altruistic equilibrium is less than in the Nash equilibrium. The player's payoff in the altruistic equilibrium is not equal to the value function given in Proposition 1 because any player maximizes the value of function (6) but not his own profit.

Example 2. Now we increase the number of players up to $n = 3$ while other parameters are kept at the same level. The state trajectories are

$$x^a(t) = 30.1154 - 26.1154e^{-0.14t},$$
$$x^*(t) = 3.34615 + 0.653846e^{-0.14t}.$$

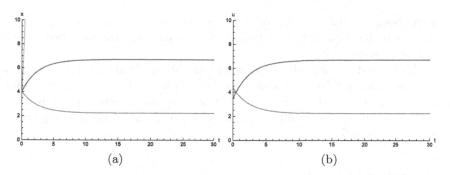

Fig. 1. (a) State $x(t)$ and (b) strategy $u(t)$ in the Berge (green), altruistic (blue), Nash (red) equilibrium in the game from Example 1 when $n = 2$. (Color figure online)

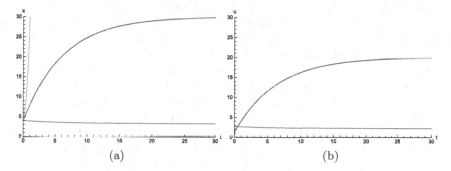

Fig. 2. (a) State $x(t)$ and (b) strategy $u(t)$ in the Berge (green), altruistic (blue), Nash (red) equilibrium in the game from Example 2 when $n - 3$. (Color figure online)

The altruistic and Nash equilibrium strategies are

$$u^a(x) = -1.60615 + 0.72x^a(t),$$
$$u^*(x) = 0.178462 + 0.72x^*(t).$$

The equilibrium state trajectory and equilibrium strategy are represented in Fig. 2. We can also see the steady state for both equilibria, $x^a(\infty) = 30.1154$, $x^*(\infty) = 3.34615$. In the altruistic equilibrium, the resource stock is larger with three players than with two. And we should also notice that the equilibrium stock in the Nash equilibrium with three players is larger with two players.

The equilibrium payoffs of any player in the altruistic and the Nash equilibria with three players are $J_i^a = -49.6495$, $J_i^* = 7.88646$ respectively. It is interesting that the players in the altruistic equilibrium earns less with the larger number of players, and in the Nash equilibrium they have larger profits with three players than with two. The smaller payoffs in the altruistic equilibrium with the larger number of players can be explained by the larger number of players that each player supports which leads to higher costs for him.

If we compare the equilibrium strategies with two and three players, we conclude that the slope of both u functions becomes less with an increase of the number of players.

We have the following observations about the equilibrium stock under two equilibrium concepts: (i) the resource stock is larger with the altruistic equilibrium that with the Nash equilibrium; (ii) both equilibria tend to steady states when the horizon tends to infinity under the conditions given in the proofs of Propositions 1 and 3.

5   Conclusion

In the paper, we have considered the altruistic-like concepts of the solutions in dynamic games. The Berge equilibrium is a theoretical implication of the altruistic social value orientation, in which a player's utility function is maximized by all other players. We also find the altruistic equilibrium when players maximize the summarized benefits of other players minus his own costs.

From cooperative dynamic game prospective, the Berge equilibrium provides an attractive model of cooperation in the society. In cooperative dynamic game theory, one can usually compare cooperative and selfish behavior. In the first case, the optimal control problem is solved, and in the latter case, the Nash equilibrium is considered as a basic equilibrium concept. As a rule, in the optimal control problem, the target function is selected as the sum of the utilities of all players. This causes some criticism, because all players are different and this criterion is unfair. Generally speaking, cooperation involves some kind of discussion and negotiation. The Berge and other altruistic-like equilibrium concepts presented in the paper can serve as an argument for negotiations.

We demonstrate the altruistic equilibrium approach on the problem of renewable resource extraction when several countries or players exploit a common resource. The problem is modeled as a differential game with infinite horizon. We have shown that the altruistic equilibrium is better for the Nature to have higher resource stock. Thus, when modeling cooperative dynamic games, we can choose the Berge or other altruistic-like equilibrium as a cooperative solution and then use regularization to achieve compliance with this solution for all participants.

In the future, we plan to consider this approach for the dynamic games in finite horizon and different games in environmental management sciences. We can also extend the results for a general form of a linear-quadratic games with finite and infinite horizons (see also [22]).

Appendix A (Proof of Proposition 3)

Proof. We assume a linear-quadratic form of the value function $V_i(x,t) = Ax^2 + Bx + D$. The HJB equation for player i is

$$rV_i(x,t) = \max_{u_i \geq 0} \left\{ q_i u_i - c_i u_i^2 + \frac{\partial V_i(x,t)}{\partial x}\left(\alpha x - \sum_{j=1}^{n} u_j\right) \right\}. \quad (17)$$

Taking into account the form of the value function and the symmetric form of the game, we substitute an expression of the value function into (17) and find the maximum of RHS in (17) and obtain u_i:

$$u_i^*(x) = -\frac{A}{c}x + \frac{q-B}{2c} \quad (18)$$

for any player $i \in N$.

From Eq. (17) we obtain the system to find coefficients A, B, D in the value function:

$$rA = \frac{2n-1}{c}A^2 + 2A\alpha,$$

$$rB = \frac{2n-1}{c}AB - \frac{qn}{c}A + \alpha B,$$

$$rD = \frac{(2n-1)B^2 - 2nqB + q^2}{4c}.$$

Simplifying this system we obtain the solution:

$$A = -\frac{c(2\alpha - r)}{2n-1},$$

$$B = \frac{nq(2\alpha - r)}{(2n-1)\alpha},$$

$$D = \frac{q^2(\alpha - nr)((2n-1)\alpha - nr)}{4\alpha^2 cr(2n-1)}.$$

Coefficient A is negative if $\alpha \geq r/2$.

The equilibrium is defined by strategy

$$u_i(x) = \frac{2\alpha - r}{2n-1}x - \frac{q(\alpha - nr)}{2c\alpha(2n-1)}, \quad i = 1,\ldots,n, \quad (19)$$

and the value function is

$$V_i(x) = V(x) = -\frac{(2\alpha - r)r(2\alpha cx - nq)^2 - \alpha^2(2n-1)}{4\alpha^2 cr(2n-1)}.$$

The corresponding state trajectory can be found by substituting (7) into state equation (1), which yields

$$\dot{x}(t) = -\frac{\alpha - nr}{2n-1}x(t) + \frac{nq(\alpha - nr)}{2(2n-1)c\alpha}, \quad x(0) = x_0. \quad (20)$$

The solution of Eq. (20) is

$$x(t) = \left(x_0 - \frac{nq}{2c\alpha}\right) e^{-\frac{\alpha-nr}{2n-1}t} + \frac{nq}{2c\alpha}. \tag{21}$$

and the Nash equilibrium player's payoff is

$$J(u^*) = V(x_0) = -\frac{(2\alpha - r)r(2\alpha c x_0 - nq)^2 - \alpha^2(2n-1)}{4\alpha^2 cr(2n-1)}, \quad i \in N. \tag{22}$$

This finishes the proof.

References

1. Abalo, K.Y., Kostreva, M.M.: Berge equilibrium: some recent results from fixed-point theorems. Appl. Math. Comput. **169**, 624–638 (2005)
2. Abalo, K.Y., Kostreva, M.M.: Some existence theorems of Nash and Berge equilibria. Appl. Math. Lett. **17**, 569–573 (2004)
3. Abalo, K.Y., Kostreva, M.M.: Fixed points, Nash games and their organizations. Topol. Methods Nonlinear Anal. **8**, 205–215 (1996)
4. Abalo, K.Y., Kostreva, M.M.: Equi-well-posed games. J. Optim. Theory Appl. **89**, 89–99 (1996). https://doi.org/10.1007/BF02192642
5. Berge, C.: Théorie générale des jeux à -personnes. Gauthier Villars, Paris (1957)
6. Breton, M., Fredj, K., Zaccour, G.: International cooperation, coalitions stability and free riding in a game of pollution control. Manch. Sch. **74**, 103–122 (2006)
7. Colman, A.M., Körner, T.W., Musy, O., Tazdaït, T.: Mutual support in games: some properties of Berge equilibria. J. Math. Psychol. **55**(2), 166–175 (2011)
8. Crettez, B.: On Sugden's "mutually beneficial practice" and Berge equilibrium. Int. Rev. Econ. **64**(4), 357–366 (2017). https://doi.org/10.1007/s12232-017-0278-3
9. Gromova, E.V., López-Barrientos, J.D.: A differential game model for the extraction of nonrenewable resources with random initial times-the cooperative and competitive cases. Int. Game Theory Rev. **18**(2) (2016). Art. no. 1640004
10. Jørgensen, S., Martín-Herrán, G., Zaccour, G.: Dynamic games in the economics and management of pollution. Environ. Model. Assess. **15**, 433–467 (2010). https://doi.org/10.1007/s10666-010-9221-7
11. Kossioris, G., Plexousakis, M., Xepapadeas, A., de Zeeuw, A., Mäler, K.-G.: Feedback Nash equilibria for non-linear differential games in pollution control. J. Econ. Dyn. Control **32**(4), 1312–1331 (2008)
12. Kudryavtsev, K., Malkov, U., Zhukovskiy, V.: Weak Berge equilibrium. In: Kochetov, Y., Bykadorov, I., Gruzdeva, T. (eds.) MOTOR 2020. CCIS, vol. 1275, pp. 231–243. Springer, Cham (2020). https://doi.org/10.1007/978-3-030-58657-7_20
13. Larbani, M., Zhukovskii, V.I.: Berge equilibrium in normal form static games: a literature review. Izv. IMI UdGU **49**, 80–110 (2017). https://doi.org/10.20537/2226-3594-2017-49-04
14. Mazalov, V.V., Rettieva, A.N.: Fish wars and cooperation maintenance. Ecol. Model. **221**(12), 1545–1553 (2010)
15. Mazalov, V.V., Rettieva, A.N.: Fish wars with many players. Int. Game Theory Rev. **12**(4), 385–405 (2010)
16. Nash, J.F.: Noncooperative games. Ann. Math. **54**, 286–295 (1951)

17. Nessah, R., Larbani, M., Tazdait, T.: A note on Berge equilibrium. Appl. Math. Lett. **20**(8), 926–932 (2007)
18. Pykacz, J., Bytner, P., Frackiewicz, P.: Example of a finite game with no Berge equilibria at all. Games **10**(1), 7 (2019). https://doi.org/10.3390/g10010007
19. Radjef, M.S.: Sur l'existence d'un équilibre de Berge pour un jeu différentiel a-personnes. Cah. Math. Univ. d'Oran **1**, 89–93 (1988)
20. Salukvadze, M.E., Zhukovskiy, V.I.: The Berge Equilibrium: A Game-Theoretic Framework for the Golden Rule of Ethics. Static & Dynamic Game Theory: Foundations & Applications. Birkhäuser, Cham (2020)
21. Zhukovskii, V.I.: Linear Quadratic Differential Games. Naoukova Doumka, Kiev (1994)
22. Zhukovskiy, V.I., Gorbatov, A.S., Kudryavtsev, K.N.: Berge and Nash equilibria in a linear-quadratic differential game. Autom. Remote. Control. **81**(11), 2108–2131 (2020). https://doi.org/10.1134/S0005117920110119

Multicriteria Dynamic Games with Random Horizon

Anna Rettieva[1,2]([⊠]) [iD]

[1] Karelian Research Center of RAS, Institute of Applied Mathematical Research,
11 Pushkinskaya Street, Petrozavodsk 185910, Russia
[2] Saint Petersburg State University, 7/9 Universitetskaya nab.,
Saint Petersburg 199034, Russia
annaret@krc.karelia.ru

Abstract. We consider a dynamic, discrete-time, game model where the players use a common resource and have different criteria to optimize. Moreover, the planning horizon is assumed to be random. To construct a multicriteria Nash equilibrium the bargaining solution is adopted. To obtain a multicriteria cooperative equilibrium, a modified bargaining scheme that guarantees the fulfillment of rationality conditions is applied. To stabilize the multicriteria cooperative solution a time-consistent payoff distribution procedure is constructed. To illustrate the presented approaches, a dynamic bi-criteria bioresource management problem with many players and random planning horizon is investigated.

Keywords: Dynamic games · Multicriteria games · Nash bargaining solution · Random horizon

1 Introduction

Game-theoretic models that take into account the presence of several objective functions of participants [19] are closer to reality. Players often seek to achieve several goals simultaneously, which can be incomparable. For example, in bioresource management problems the players wish to maximize their exploitation rates and to minimize the harm to the environment. The multicriteria approach helps to determine an optimal behavior in such situations.

In static multicriteria games, the solution concepts are usually based on the Pareto set [2, 19] or some convolutions of the criteria [1]. Recently, some other approaches have been suggested to solve multicriteria games (for example, the ideal equilibrium [21], the E-equilibrium [13]). However, Pareto equilibrium is the most studied concept in static multicriteria game theory.

This work was supported by the Russian Science Foundation (No. 17-11-01079).

P. Pardalos et al. (Eds.): MOTOR 2021, LNCS 12755, pp. 340–355, 2021.
https://doi.org/10.1007/978-3-030-77876-7_23

The methods of static multicriteria games are not applicable to the dynamic statements. A key issue addressed to the research is to find optimal compromise solutions for game-theoretic models with vector payoffs. Hence, in the series of papers [14–17], new approaches to obtain players' optimal behavior in dynamic multicriteria games were suggested. The noncooperative equilibrium was constructed combining the ideas of multi-objective optimization (nadir points [23]) and the classical concept of Nash equilibrium. The Nash bargaining approach [8,9,20] was applied to determine the cooperative behavior of the players in [15]. To guarantee the rationality of cooperative behavior, a method to construct cooperative strategies and payoffs that combines compromise programming [23] and the Nash bargaining scheme was presented in [16]. It was shown [15,17] that the obtained cooperative behavior in dynamic multicriteria resource management problems with finite and infinite planning horizons is beneficial for the players and, that is more important, improves the ecological situation.

As is well known, the Nash bargaining scheme is not dynamically stable [4]. The concept of time-consistency (dynamic stability) was introduced by Petrosyan L.A. [11]. Petrosyan L.A. and Danilov N.N. [12] have developed the notion of time-consistent imputation distribution procedure. The designing of dynamically stable payoff distribution procedures for dynamic games with random horizon has been studied in a series of papers [3,10,22]. Different payoff distribution procedures, including the time-consistent ones, for multicriteria multistage games were presented in [5–7]. To stabilize the cooperative solution in dynamic multicriteria games with finite horizon the idea of payoff distribution procedure was applied in [16,17]. Moreover, in [17], the conditions for rational behavior were defined for dynamic multicriteria games.

The main purpose of this paper is to adopt the presented approaches to the multicriteria dynamic game with random horizon. This extension of the model is most appropriate to describe reality as external random factors can cause a game breach. For example, in management problems, negative factors include an economic crisis, variations in the rate of inflation, international economic and political situations. In bioresource exploitation process, the firms can go bankrupt, their exploitation tools can be damaged, etc. All these processes can possibly interrupt the game process, and the solution concept should capture the possibility of the players' leaving the game.

We consider a dynamic, discrete-time, game model where the players use a common resource and have different criteria to optimize. Moreover, the planning horizon is assumed to be random. To construct a multicriteria Nash equilibrium the bargaining solution is adopted [14]. To design a multicriteria cooperative equilibrium, a modified bargaining scheme [16] that guarantees the fulfillment of rationality conditions is applied. To stabilize the multicriteria cooperative solution a time-consistent payoff distribution procedure [17] is constructed. To illustrate the presented approaches, a dynamic bi-criteria bioresource management problem with many players and random planning horizon is investigated.

Further exposition has the following structure. Section 2 describes the non-cooperative and cooperative solution concepts for a multicriteria dynamic game with many players and random horizon. The time-consistent payoff distribution procedure is presented in Sect. 2.3. A bi-criteria discrete-time game-theoretic bioresource management model (harvesting problem) is treated in Sect. 3. Finally, Sect. 4 provides the basic results and their discussion.

2   Dynamic Multicriteria Game with Random Horizon

Consider a multicriteria dynamic game in discrete time. Let $N = \{1, \ldots, n\}$ players exploit a common resource and each of them wishes to optimize k different criteria. The state dynamics is in the form

$$x_{t+1} = f(x_t, u_{1t}, \ldots, u_{nt}), \quad x_1 = x, \tag{1}$$

where $x_t \geq 0$ is the resource size at time $t \geq 0$, $f(x_t, u_{1t}, \ldots, u_{nt})$ denotes the natural growth function, and $u_{it} \geq 0$ gives the exploitation rate of player i at time t, $i \in N$.

By assumption, the players stop exploitation at random time step as external stochastic processes can cause a game breach. In [18] two-players bi-criteria dynamic game with different random planning horizons was investigated. Here, the planning horizon is assumed to be identical for all the players. For example, participants possess similar economic or political conditions and their random changes simultaneously stop the game for all of the players.

Suppose that the players exploit a common resource during T steps. Here T represents a discrete random variable taking values $\{1, \ldots, m\}$ with the corresponding probabilities $\{\theta_1, \ldots, \theta_m\}$.

Each player has k goals to optimize. The vector payoffs of the players are determined via the expectation operator:

$$J_i = \begin{pmatrix} J_i^1 = E\Big\{ \sum_{t=1}^{T} \delta^t g_i^1(x_t, u_{1t}, \ldots, u_{nt}) \Big\} \\ \ldots \\ J_i^k = E\Big\{ \sum_{t=1}^{T} \delta^t g_i^k(x_t, u_{1t}, \ldots, u_{nt}) \Big\} \end{pmatrix}, \quad i \in N, \tag{2}$$

where $g_i^j(u_{1t}, \ldots, u_{nt}) \geq 0$ gives the instantaneous utility, $i \in N$, $j = 1, \ldots, k$, $\delta \in (0, 1)$ denotes the discount factor.

Assuming that the distribution function of the planning horizon T is known to the players we can rewrite the payoff function in the following form:

$$J_i = \begin{pmatrix} J_i^1 = \sum_{T=1}^{m} \theta_T \sum_{t=1}^{T} \delta^t g_i^1(x_t, u_{1t}, \dots, u_{nt}) \\ \dots \\ J_i^k = \sum_{T=1}^{m} \theta_T \sum_{t=1}^{T} \delta^t g_i^k(x_t, u_{1t}, \dots, u_{nt}) \end{pmatrix}, \ i \in N. \tag{3}$$

2.1 Multicriteria Nash Equilibrium

We design the noncooperative behavior in dynamic multicriteria game applying the bargaining products and the classical concept of Nash equilibrium [14,15]. Therefore, we begin with the construction of guaranteed payoffs which play the role of status quo points.

The possible concepts to determine the guaranteed payoffs for the game with two players were presented in [14]. As it was demonstrated, the variant where the guaranteed payoffs are obtained as the Nash equilibrium solutions is the best for the state of the exploited system and profitable for the players. Therefore, for the multicriteria game with random horizon we adopt this concept. Namely,

$G_1^1, \dots, G_n^1$ are the Nash equilibrium payoffs in the dynamic game with random horizon $\langle x, N, \{U_i\}_{i=1}^n, \{J_i^1\}_{i=1}^n \rangle$,

$\dots$

$G_1^k, \dots, G_n^k$ are the Nash equilibrium payoffs in the dynamic game with random horizon $\langle x, N, \{U_i\}_{i=1}^n, \{J_i^k\}_{i=1}^n \rangle$,
where the state dynamics is in the form (1).

To construct multicriteria payoff functions, we adopt the Nash products where the guaranteed payoffs play the role of the status quo points:

$$H_1(u_{1t}, \dots, u_{nt}) = (J_1^1(x_t, u_{1t}, \dots, u_{nt}) - G_1^1) \cdot \dots \cdot (J_1^k(x_t, u_{1t}, \dots, u_{nt}) - G_1^k),$$

$$\dots$$

$$H_n(u_{1t}, \dots, u_{nt}) = (J_n^1(x_t, u_{1t}, \dots, u_{nt}) - G_n^1) \cdot \dots \cdot (J_n^k(x_t, u_{1t}, \dots, u_{nt}) - G_n^k).$$

The multicriteria Nash equilibrium strategies are constructed in the feedback form $u_{it}^N = u_{it}^N(x_t)$, $i \in N$.

Definition 1. *A strategy profile $u_t^N = u_t^N(x_t) = (u_{1t}^N, \dots, u_{nt}^N)$ is called a multicriteria Nash equilibrium [14] of the problem (1), (3) if*

$$H_i(u_t^N) \geq H_i(u_{1t}^N, \dots, u_{i-1t}^N, u_{it}, u_{i+1t}^N, \dots, u_{nt}^N) \ \forall u_{it} \in U_i, \ i \in N. \tag{4}$$

For the duration of the game, the expected payoffs of the players have the form

$$V_i^j(1, x) = \sum_{T=1}^{m} \theta_T \sum_{t=1}^{T} \delta^t g_i^j(x_t, u_{1t}, \dots, u_{nt}), \ i \in N, \ j = 1, \dots, k. \tag{5}$$

Further the payoffs obtained by the players as the game reaches step τ, $\tau = 1, \ldots, m$, are considered. Note that the probabilities that a player continues exploitation for $\tau, \tau + 1, \ldots, m$ steps are

$$\frac{\theta_\tau}{\sum\limits_{l=\tau}^m \theta_l}, \frac{\theta_{\tau+1}}{\sum\limits_{l=\tau}^m \theta_l}, \ldots, \frac{\theta_m}{\sum\limits_{l=\tau}^m \theta_l}.$$

Hence, as step τ occurs, the expected payoffs $V_i^j(\tau, x_\tau)$, $i \in N$, $j = 1, \ldots, k$, of the players take the form

$$V_i^j(\tau, x_\tau) = \sum_{T=\tau}^m \frac{\theta_T}{\sum\limits_{l=\tau}^m \theta_l} \sum_{t=\tau}^T \delta^t g_i^j(x_t, u_{1t}, \ldots, u_{nt}). \tag{6}$$

Next, we obtain a relationship between the expected payoffs when steps τ and $\tau + 1$ occur in the game, $V_i^j(\tau, x_\tau)$ and $V_i^j(\tau + 1, x_{\tau+1})$, respectively. Let us rewrite the expected payoff of player i on j's criterion (6) as the game reaches step τ:

$$V_i^j(\tau, x_\tau) = \frac{\theta_\tau}{\sum\limits_{l=\tau}^m \theta_l} \delta^\tau g_i^j(x_\tau, u_\tau) + \sum_{T=\tau+1}^m \frac{\theta_T}{\sum\limits_{l=\tau}^m \theta_l} \sum_{t=\tau}^T \delta^t g_i^j(x_t, u_t)$$

$$= \frac{\theta_\tau}{\sum\limits_{l=\tau}^m \theta_l} \delta^\tau g_i^j(x_\tau, u_\tau) + \sum_{T=\tau+1}^m \frac{\theta_T}{\sum\limits_{l=\tau}^m \theta_l} \left[\sum_{t=\tau+1}^T \delta^t g_i^j(x_t, u_t) + \delta^\tau g_i^j(x_\tau, u_\tau) \right]$$

$$= \delta^\tau g_i^j(x_\tau, u_\tau) + \sum_{T=\tau+1}^m \frac{\theta_T}{\sum\limits_{l=\tau}^m \theta_l} \sum_{t=\tau+1}^T \delta^t g_i^j(x_t, u_t)$$

$$= \delta^\tau g_i^j(x_\tau, u_\tau) + \sum_{T=\tau+1}^m \frac{\theta_T}{\sum\limits_{l=\tau+1}^m \theta_l} \frac{\sum\limits_{l=\tau+1}^m \theta_l}{\sum\limits_{l=\tau}^m \theta_l} \sum_{t=\tau+1}^T \delta^t g_i^j(x_t, u_t)$$

$$= \delta^\tau g_i^j(x_\tau, u_\tau) + \Theta_\tau^{\tau+1} V_i^j(\tau + 1, x_{\tau+1}), \tag{7}$$

where

$$\Theta_\tau^{\tau+1} = \frac{\sum\limits_{l=\tau+1}^m \theta_l}{\sum\limits_{l=\tau}^m \theta_l}. \tag{8}$$

According to Definition 1 to construct the multicriteria Nash equilibrium strategies it is required to solve the problem (4), hence

$$(V_1^{1N}(1,x) - G_1^1) \cdot \ldots \cdot (V_1^{kN}(1,x) - G_1^k)$$

$$= \left(\sum_{T=1}^m \theta_T \sum_{t=1}^T \delta^t g_1^1(x_t^N, u_{1t}^N, \ldots, u_{nt}^N) - G_1^1 \right) \cdot \ldots \cdot$$

$$\cdot \left(\sum_{T=1}^m \theta_T \sum_{t=1}^T \delta^t g_1^k(x_t^N, u_{1t}^N, \ldots, u_{nt}^N) - G_1^k \right)$$

$$= \max_{u_{11}, \ldots, u_{1m}} \left\{ (V_{-1}^{1N}(1,x) - G_1^1) \cdot \ldots \cdot (V_{-1}^{kN}(1,x) - G_1^k) \right\},$$

$$\ldots$$

$$(V_n^{1N}(1,x) - G_n^1) \cdot \ldots \cdot (V_n^{kN}(1,x) - G_n^k)$$

$$= \left(\sum_{T=1}^m \theta_T \sum_{t=1}^T \delta^t g_n^1(x_t^N, u_{1t}^N, \ldots, u_{nt}^N) - G_n^1 \right) \cdot \ldots \cdot$$

$$\cdot \left(\sum_{T=1}^m \theta_T \sum_{t=1}^T \delta^t g_n^k(x_t^N, u_{1t}^N, \ldots, u_{nt}^N) - G_n^k \right)$$

$$= \max_{u_{n1}, \ldots, u_{nm}} \left\{ (V_{-n}^{1N}(1,x) - G_n^1) \cdot \ldots \cdot (V_{-n}^{kN}(1,x) - G_n^k) \right\},$$

where $V_{-i}^{jN}(\tau, x_\tau)$ take the form

$$V_{-i}^{jN}(\tau, x_\tau) = \sum_{T=\tau}^m \frac{\theta_T}{\sum_{l=\tau}^m \theta_l} \sum_{t=\tau}^T \delta^t g_i^j(x_t, u_{1t}^N, \ldots, u_{i-1t}^N, u_{it}, u_{i+1t}^N, \ldots, u_{nt}^N)$$

and satisfy the relations (7), G_i^j are the guaranteed payoff points, $i \in N$, $j = 1, \ldots, k$, noncooperative trajectory x_t^N is defined by (1) where $u_{it} = u_{it}^N$, $i \in N$.

2.2   Multicriteria Cooperative Equilibrium

The multicriteria cooperative equilibrium was obtained as a solution of the Nash bargaining scheme with the multicriteria Nash equilibrium payoffs acting as the status quo points in [15]. Namely, in accordance with this approach, the product of the distances from the sum of the players' payoffs to the sum of the noncooperative payoffs for all the criteria was applied to determine the cooperative ones. As the presented scheme does not guarantee the fulfillment of the individual rationality conditions that is the key point to maintain cooperative behavior a new approach to determine cooperative strategies in dynamic multicriteria game with asymmetric players was presented in [16]. This solution concept guarantees the rationality of cooperative behavior as the cooperative payoffs of the players are greater than or equal to the multicriteria Nash payoffs. Moreover, this approach is similar to the classical definition of cooperative behavior as the players seek to maximize the sum of their individual payoffs. The goal of each player is to maximize the distance to the noncooperative payoffs, and under cooperation the players prefer to do it jointly.

More specifically, the cooperative strategies and payoffs of the players are determined from the modified bargaining solution that combines compromise programming [23] and the Nash bargaining scheme [8,9]. The status quo points are the noncooperative payoffs obtained by the players using the multicriteria Nash equilibrium strategies u_t^N:

$$
J_i^N = \begin{pmatrix} J_i^{1N} = E\Big\{ \sum_{t=1}^{T} \delta^t g_i^1(x_t^N, u_t^N) \Big\} \\ \dots \\ J_i^{kN} = E\Big\{ \sum_{t=1}^{T} \delta^t g_i^k(x_t^N, u_t^N) \Big\} \end{pmatrix}, \quad i \in N. \tag{9}
$$

The cooperative strategies in the feedback form $u_{it}^c = u_{it}^c(x_t)$, $i \in N$, and payoffs are obtained from the following problem:

$$
(V_1^{1c}(1,x) - J_1^{1N}) \cdot \dots \cdot (V_1^{kc}(1,x) - J_1^{kN}) + \dots +
$$
$$
+ (V_n^{1c}(1,x) - J_n^{1N}) \cdot \dots \cdot (V_n^{kc}(1,x) - J_n^{kN})
$$

$$
= (E\Big\{ \sum_{t=1}^{T} \delta^t g_1^1(x_t^c, u_{1t}^c, \dots, u_{nt}^c) \Big\} - J_1^{1N}) \cdot \dots \cdot
$$

$$
\cdot (E\Big\{ \sum_{t=1}^{T} \delta^t g_1^k(x_t^c, u_{1t}^c, \dots, u_{nt}^c) \Big\} - J_1^{kN}) + \dots
$$

$$
+ (E\Big\{ \sum_{t=1}^{T} \delta^t g_n^1(x_t^c, u_{1t}^c, \dots, u_{nt}^c) \Big\} - J_n^{1N}) \cdot \dots \cdot
$$

$$
\cdot (E\Big\{ \sum_{t=1}^{T} \delta^t g_n^k(x_t^c, u_{1t}^c, \dots, u_{nt}^c) \Big\} - J_n^{kN})
$$

$$
= \max_{u_{1t},\dots,u_{nt}} \Big\{ (V_1^1(1,x) - J_1^{1N}) \cdot \dots \cdot (V_1^k(1,x) - J_1^{kN}) + \dots +
$$
$$
+ (V_n^1(1,x) - J_n^{1N}) \cdot \dots \cdot (V_n^k(1,x) - J_n^{kN}) \Big\}, \tag{10}
$$

where J_i^{jN} are the noncooperative payoffs given by (9), $V_i^j(1,x)$ have the forms (6) and satisfy the relations (7), $i \in N$, $j = 1, \dots, k$, cooperative trajectory x_t^c is defined by (1) where $u_{it} = u_{it}^c$, $i \in N$.

Definition 2. *A strategy profile $u_t^c = u_t^c(x_t) = (u_{1t}^c, \dots, u_{nt}^c)$ is a rational multicriteria cooperative equilibrium [16] of problem (1), (3) if it is the solution of problem (10).*

2.3   Time-Consistent Payoff Distribution Procedure

The players' expected cooperative payoffs for the duration of the game can be calculated as

$$
J_i^c(1, x) = \begin{pmatrix} J_i^{1c}(1, x) = E\Big\{ \sum_{t=1}^{T} \delta^t g_i^1(x, u_t^c(x)) \Big\} \\ \dots \\ J_i^{kc}(1, x) = E\Big\{ \sum_{t=1}^{T} \delta^t g_i^k(x, u_t^c(x)) \Big\} \end{pmatrix}, \; i \in N, \tag{11}
$$

where $u_t^c = (u_{1t}^c, \dots, u_{nt}^c)$ are the cooperative strategies obtained from (10).

Similarly we determine the cooperative payoffs $J_i^c(t, x_t^c)$, $i = 1, \dots, n$, for every subgame started from the state x_t^c at a time t.

To stabilize the cooperative solution in multicriteria dynamic game with random horizon we adopt the time-consistent payoff distribution procedure [5,11,12,17]. The main idea of this scheme is to distribute the cooperative gain along the game path. Dynamic stability guarantees that the players following the cooperative trajectory are guided by the same optimal behavior determination approach (10) at each current time and hence do not have any incentives to deviate from the cooperation agreement.

The payment to player i, $i \in N$, in all criteria at a time t is defined from the following definitions.

Definition 3. *A vector*

$$
\beta(t, x_t) = (\beta_1(t, x_t), \dots, \beta_n(t, x_t)),
$$

where

$$
\beta_i(t, x_t) = \begin{pmatrix} \beta_i^1(t, x_t) \\ \dots \\ \beta_i^k(t, x_t) \end{pmatrix}, \; i \in N,
$$

is a payoff distribution procedure (PDP) for the dynamic multicriteria game with random horizon (1), (3), if

$$
J_i^{jc}(1, x) = E\Big\{ \sum_{t=1}^{T} \delta^t \beta_i^j(t, x_t) \Big\}, \; i \in N, \; j = 1, \dots, k. \tag{12}
$$

Definition 4. *A vector $\beta(t, x_t) = (\beta_1(t, x_t), \dots, \beta_n(t, x_t))$ is a time-consistent [11, 12] PDP for dynamic multicriteria game with random horizon (1), (3), if for every $t \geq 0$*

$$
J_i^{jc}(1, x) = \sum_{l=1}^{t} \delta^l \beta_i^j(l, x_l) + J_i^{jc}(t + 1, x_{t+1}), \; i \in N, \; j = 1, \dots, k. \tag{13}
$$

Theorem 1. *A vector* $\beta(t, x_t) = (\beta_1(t, x_t), \ldots, \beta_n(t, x_t))$, *where*

$$\beta_i(t, x_t) = \frac{1}{\delta^t}[J_i^c(t, x_t) - \Theta_t^{t+1} J_i^c(t+1, x_{t+1})], \ i \in N, \tag{14}$$

is a time-consistent payoff distribution procedure for dynamic multicriteria game with random horizon (1), (3).

Proof. Conditions (12) of Definition 3 are satisfied:

$$E\left\{\sum_{t=1}^{T}\delta^t\beta_i^j(t,x_t)\right\} = \sum_{T=1}^{m}\theta_T\sum_{t=1}^{m}\delta^t\frac{1}{\delta^t}[J_i^{jc}(t,x_t) - \Theta_t^{t+1}J_i^{jc}(t+1,x_{t+1})]$$

$$= \sum_{t=1}^{m}\theta_T J_i^{jc}(1,x) - J_i^{jc}(2,x_2)\Theta_1^2\sum_{T=1}^{m}\theta_T + J_i^{jc}(2,x_2)\sum_{T=2}^{m}\theta_T - \ldots -$$

$$-J_i^{jc}(m,x_m)\Theta_{m-1}^m\sum_{T-m-1}^{m}\theta_T + J_i^{jc}(m,x_m)\theta_m - \Theta_m^{m+1}J_i^{jc}(m+1,x_{m+1})\theta_m$$

$$= J_i^{jc}(1,x)$$

as $J_i^{jc}(m+1, x_{m+1}) = 0$, $i \in N$, $j = 1, \ldots, k$.
Conditions (13) follow from the equalities

$$J_i^{jc}(1,x) - J_i^{jc}(t+1,x_{t+1}) = E\left\{\sum_{l=1}^{T}\delta^l\beta_i^j(l,x_l)\right\} - E\left\{\sum_{l=t+1}^{T}\delta^l\beta_i^j(l,x_l)\right\}$$

$$= E\left\{\sum_{l=1}^{t}\delta^l\beta_i^j(l,x_l)\right\} = \sum_{l=1}^{t}\delta^l\beta_i^j(l,x_l), \ i \in N, j = 1, \ldots, k.$$

As the non-negativity of PDP (14) is not guaranteed, the approaches to overcome this drawback such as [3,10] are planned to be applied in the future research.

Next, we consider a dynamic bi-criteria model with many players and random planning horizon related with the bioresource management problem (harvesting) to illustrate the suggested concepts.

3   Dynamic Bi-criteria Resource Management Problem with Random Horizon

Consider a bi-criteria discrete-time dynamic resource management model. Let n players (countries or firms) exploit a bioresource during T steps. T is a discrete random variable taking values $\{1, \ldots, m\}$ with the corresponding probabilities $\{\theta_1, \ldots, \theta_m\}$.

The bioresource evolves according to the equation

$$x_{t+1} = \varepsilon x_t - u_{1t} - \ldots - u_{nt}, \ x_1 = x, \tag{15}$$

where $x_t \geq 0$ is the resource size at a time $t \geq 0$, $\varepsilon \geq 1$ denotes the natural birth rate, and $u_{it} = u_{it}(x_t) \geq 0$ specifies the exploitation strategy of player i at a time $t \geq 0$, $i \in N = \{1, \ldots, n\}$.

Each player seeks to achieve two goals: to maximize the revenue from resource sales and to minimize the exploitation costs. Assume that the players have different market prices but the same costs that depend quadratically on the exploitation rate. The vector payoffs of the players take the form

$$
J_1 = \begin{pmatrix} J_1^1 = E\left\{ \sum_{t=1}^{T} \delta^t p_1 u_{1t}(x_t) \right\} \\ J_1^2 = -E\left\{ \sum_{t=1}^{m} \delta^t c u_{1t}^2(x_t) \right\} \end{pmatrix}, \ldots, J_n = \begin{pmatrix} J_n^1 = E\left\{ \sum_{t=1}^{m} \delta^t p_n u_{nt}(x_t) \right\} \\ J_n^2 = -E\left\{ \sum_{t=1}^{m} \delta^t c u_{nt}^2(x_t) \right\} \end{pmatrix},
$$

$$
(16)
$$

where $p_i \geq 0$ is the market price of the resource for player i, $i \in N$, $c \geq 0$ indicates the catching cost, and $\delta \in (0, 1)$ denotes the discount factor.

3.1 Multicriteria Nash Equilibrium

First, we construct the guaranteed payoffs adopting one of the modifications from [14]. The guaranteed payoff points $G_1^1, \ldots, G_n^1$ will be defined as the Nash equilibrium in the game $\langle x, N, \{U_i\}_{i=1}^n, \{J_i^1\}_{i=1}^n \rangle$.

Let $V_i(\tau, x)$ be a value function for player i, $i \in N$. Assume that the value functions have the linear forms $V_i(\tau, x) = A_i^\tau x + B_i^\tau$, $i \in N$, similarly to the relations (7) these functions satisfy

$$
A_i^\tau x + B_i^\tau = \max_{u_{i\tau}} \{ \delta^\tau p_i u_{i\tau} + \Theta_\tau^{\tau+1}(A_i^\tau(\varepsilon x - u_{1\tau} - \ldots - u_{n\tau}) + B_i^\tau) \} .
$$

Searching for the linear strategies, we obtain that the Nash equilibrium strategies coincide and take the form

$$
u_{1\tau} = \ldots = u_{n\tau} = \gamma_\tau x_\tau = \frac{\Theta_\tau^{\tau+1} \varepsilon - 1}{(n-1)\Theta_\tau^{\tau+1}} x_\tau ,
$$

and the guaranteed payoff points have the form

$$
G_1^1 = p_1 A x, \ldots, G_n^1 = p_n A x, \tag{17}
$$

where $A = \frac{\delta^{m-1}}{\Theta_{m-1}^m}$.

Similarly, determining the Nash equilibrium in the game with the second criteria of all players $\langle x, N, \{U_i\}_{i=1}^n, \{J_i^2\}_{i=1}^n \rangle$, yields the equilibrium strategies

$$
u_{1\tau} = \ldots = u_{n\tau} = \frac{\varepsilon \Theta_\tau^{\tau+1} G}{\delta^\tau - n\Theta_\tau^{\tau+1} G} x_\tau ,
$$

and n more guaranteed payoffs points

$$
G_1^2 = \ldots = G_n^2 = cG x^2 , \tag{18}
$$

where $G = -\frac{\delta^{m-1}}{2n^2\Theta_{m-1}^m}[2n - \varepsilon\Theta_{m-1}^m + \varepsilon(\Theta_{m-1}^m(\varepsilon^2\Theta_{m-1}^m + 4n(n-1)))^{1/2}]$.

According to Definition 1, to determine the multicriteria Nash equilibrium of the game (15), (16) the following problem has to be solved:

$$p_1 c(E\{\sum_{t=1}^{T}\delta^t u_{1t}(x_t)\} - Ax)(-E\{\sum_{t=1}^{T}\delta^t u_{1t}^2(x_t)\} - Gx^2) \to \max_{u_{11},\ldots,u_{1m}},$$

$$\cdots$$

$$p_n c(E\{\sum_{t=1}^{T}\delta^t u_{nt}(x_t)\} - Ax)(-E\{\sum_{t=1}^{T}\delta^t u_{nt}^2(x_t)\} - Gx^2) \to \max_{u_{n1},\ldots,u_{nm}}. \quad (19)$$

Proposition 1. *The multicriteria Nash equilibrium payoffs in the problem (15), (16) have the form*

$$V_i^{1N}(m-k,x) = p_i(\delta^{m-k}\gamma_{m-k}^N + (\varepsilon - n\gamma_{m-k}^N)B_{m-k+1}^N)x$$
$$= p_i\tilde{V}^{1N}(m-k,x)x, \; i \in N,$$
$$V_i^{2N}(m-k,x) = -c(\delta^{m-k}(\gamma_{m-k}^N)^2 + (\varepsilon - n\gamma_{m-k}^N)^2 D_{m-k+1}^N)x^2$$
$$= -c\tilde{V}^{2N}(m-k,x)x^2, \; i \in N, \; k = 1,\ldots,m-1, \quad (20)$$

where $\tilde{V}_i^{1N}(\tau,x) = \frac{1}{p_i}V_i^{1N}(\tau,x),\; \tilde{V}_i^{2N}(\tau,x) = \frac{1}{c}V_i^{1N}(\tau,x),$

$$B_{m-k+1}^N = \sum_{j=m-k+1}^{m-1}\frac{\delta^j\gamma_j^N}{\varepsilon - n\gamma_j^N}\prod_{l=m-k+1}^{j}\Theta_{l-1}^l(\varepsilon - n\gamma_i^N) + A\prod_{l=m-k}^{m-1}\Theta_l^{l+1}(\varepsilon - \gamma_i^N),$$

$$D_{m-k+1}^N = -\sum_{j=m-k+1}^{m-1}\frac{\delta^j(\gamma_j^N)^2}{(\varepsilon - n\gamma_j^N)^2}\prod_{l=m-k+1}^{j}\Theta_{l-1}^l(\varepsilon - n\gamma_i^N)^2$$
$$+ G\prod_{l=m-k}^{m-1}\Theta_1^{l+1}(\varepsilon - n\gamma_i^N)^2.$$

The multicriteria Nash equilibrium strategies take the form

$$u_{1m-k}^N = \ldots = u_{nm-k}^N = \gamma_{m-k}^N x_{m-k}, \; k = 2,\ldots,m-1,$$

and are related by

$$P(\gamma_{m-1}^N)\prod_{j=2}^{m-k}(\varepsilon - n\gamma_{m-j}^N)$$
$$= \frac{\gamma_{m-k}^N - \delta\Theta_{m-k}^{m-k+1}\gamma_{m-k+1}^N(\varepsilon - n\gamma_{m-k}^N)(\varepsilon - (n+1)\gamma_{m-k+1}^N)}{1 - \delta\Theta_{m-k}^{m-k+1}(\varepsilon - (n+1)\gamma_{m-k+1}^N)}, \quad (21)$$

where

$$P(\gamma_{m-1}^N) = \frac{\delta^{m-1}\gamma_{m-1}^N + \Theta_{m-1}^m G(\varepsilon - n\gamma_{m-1}^N)}{\delta^{m-1} - \Theta_{m-1}^m A}.$$

The strategy of the players when the penult step occurs (the quantity γ_{m-1}^N) is evaluated through the following equation

$$(\tilde{V}^{2N}(1,x) - G) = P(\gamma_{m-1}^N)(\tilde{V}^{1N}(1,x) - A).$$

Proof. Let us denote $\tilde{V}_{-i}^1(\tau,x) = \frac{1}{p_i}V_{-i}^1(\tau,x)$, $\tilde{V}_{-i}^2(\tau,x) = \frac{1}{c}V_{-i}^1(\tau,x)$.

The analysis begins with step m occurs in the game. The players have zero payoffs at the next step $m+1$. Hence, the optimal Nash strategies coincide with the guaranteed ones, and the payoffs have the form

$$\tilde{V}_i^{1N}(m,x) = Ax, \tilde{V}_i^{2N}(m,x) = Gx^2, \ i \in N.$$

Now, suppose that the game reaches step $m-1$. The problem (19) takes the form

$$(\tilde{V}_{-1}^{1N}(m-1,x) - Ax)(\tilde{V}_{-1}^{2N}(m-1,x) - Gx^2) \to \max_{u_{1m-1}},$$

$$\ldots$$

$$(\tilde{V}_{-n}^{1N}(m-1,x) - Ax)(\tilde{V}_{-n}^{2N}(m-1,x) - Gx^2) \to \max_{u_{nm-1}},$$

where

$$\tilde{V}_{-i}^{1N}(m-1,x) = \delta^{m-1}u_{im-1}$$

$$+\Theta_{m-1}^m\tilde{V}_i^{1N}(m,\varepsilon x - u_{1\,m-1}^N - \cdots - u_{i-1\,m-1}^N - u_{im-1} - u_{i+1\,m-1}^N - \cdots - u_{nm-1}^N),$$

$$\tilde{V}_{-i}^{2N}(m-1,x) = -\delta^{m-1}(u_{im-1})^2$$

$$+\Theta_{m-1}^m\tilde{V}_i^{2N}(m,\varepsilon x - u_{1\,m-1}^N - \cdots - u_{i-1\,m-1}^N - u_{im-1} - u_{i+1\,m-1}^N - \cdots - u_{nm-1}^N).$$

As usual, we seek for the strategies of the linear form $u_{im-1}^N = \gamma_{im-1}^N x$, $i \in N$. Notice that all the strategies coincide $\gamma_{1m-1}^N = \ldots = \gamma_{nm-1}^N = \gamma_{m-1}^N$ and can be obtained from the first-order condition

$$(\delta^{m-1} - A\Theta_{m-1}^m)[\ \delta^{m-1}(\gamma_{m-1}^N)^2 + \Theta_{m-1}^m G(\varepsilon - n\gamma_{m-1}^N) - G]$$

$$= 2(\delta^{m-1}\gamma_{m-1}^N - \Theta_{m-1}^m G(\varepsilon - n\gamma_{m-1}^N))[\delta^{m-1}\gamma_{m-1}^N + \Theta_{m-1}^m A(\varepsilon - n\gamma_{m-1}^N) - A].$$

Now, we study the situation when step $m-2$ occurs in the game. Then the problem (19) takes the form

$$(\tilde{V}_{-1}^{1N}(m-2,x) - Ax)(\tilde{V}_{-1}^{2N}(m-2,x) - Gx^2) \to \max_{u_{1m-1},u_{1m-2}},$$

$$\ldots$$

$$(\tilde{V}_{-n}^{1N}(m-2,x) - Ax)(\tilde{V}_{-n}^{2N}(m-2,x) - Gx^2) \to \max_{u_{1m-1},u_{nm-2}},$$

where

$$\tilde{V}_{-i}^{1N}(m-2,x) = \delta^{m-2}u_{im-2}$$

$$+\Theta_{m-2}^{m-1}\tilde{V}_{-i}^{1N}(m-1,\varepsilon x - u_{1\,m-2}^N - \cdots - u_{i-1\,m-2}^N - u_{im-2} - u_{i+1\,m-2}^N - \cdots - u_{nm-2}^N),$$

$$\tilde{V}_{-i}^{2N}(m-2,x) = -\delta^{m-2}(u_{im-2})^2$$

$$+\Theta_{m-2}^{m-1}\tilde{V}_{-i}^{2N}(m-1,\varepsilon x - u_{1\,m-2}^N - \cdots - u_{i-1\,m-2}^N - u_{im-2} - u_{i+1\,m-2}^N - \cdots - u_{nm-2}^N).$$

Searching for the linear strategies $u_{it}^N = \gamma_{it}^N x$, $i \in N$, $t = m - 2, m - 1$, again notice that all the strategies coincide $\gamma_{1t}^N = \ldots = \gamma_{nt}^N = \gamma_t^N$, $t = m - 2, m - 1$.

From the first-order optimality conditions we obtain the following relationship between the equilibrium strategies of the players when the game reaches steps $m - 2$ and $m - 1$:

$$(\varepsilon - n\gamma_{m-2}^N)(1 - \delta\Theta_{m-2}^{m-1}(\varepsilon - (n+1)\gamma_{m-1}^N))[\delta^{m-1}\gamma_{m-1}^N + \Theta_{m-1}^m G(\varepsilon - n\gamma_{m-1}^N)]$$
$$= (\gamma_{m-2}^N - \delta\Theta_{m-2}^{m-1}\gamma_{m-1}^N(\varepsilon - n\gamma_{m-2}^N)(\varepsilon - (n+1)\gamma_{m-1}^N))[\delta^{m-1} - \Theta_{m-1}^m A].$$

By continuing the process until the game reaches step k, we obtain the payoffs (20) and the relations to determine the Nash equilibrium strategies of the form (21).

3.2 Cooperative Equilibrium

To construct the cooperative payoffs and strategies the modified bargaining scheme [16] will be applied. First, we have to determine the noncooperative payoffs gained by the players using the multicriteria Nash strategies. Then, we construct the sum of the Nash products with the noncooperative payoffs of players acting as the status quo points.

In view of Proposition 1, the noncooperative payoffs have the form

$$J_i^{1N}(x) = p_i \tilde{V}^{1N}(1, x)x\,,$$
$$J_i^{2N}(x) = -c\tilde{V}^{2N}(1, x)x^2\,, \quad i \in N\,.$$

According to Definition 2, to construct the multicriteria cooperative equilibrium the following problem has to be solved:

$$p_1(E\left\{\sum_{t=1}^T \delta^t u_{1t}(x_t)\right\} - Mx)(-E\left\{\sum_{t=1}^T \delta^t u_{1t}^2(x_t)\right\} + Kx^2) + \ldots$$

$$+p_n(E\left\{\sum_{t=1}^T \delta^t u_{nt}(x_t)\right\} - Mx)(-E\left\{\sum_{t=1}^m \delta^t u_{nt}^2(x_t)\right\} + Kx^2) \to \max_{u_{1t},\ldots,u_{nt}}, \quad (22)$$

where $M = \tilde{V}^{1N}(1, x)$, $K = \tilde{V}^{2N}(1, x)$.

Considering the process starting when step m occurs in the game till the game reaches step k and seeking the strategies in linear form, we construct cooperative behavior.

Proposition 2. *The multicriteria cooperative equilibrium payoffs in the problem* (15), (16) *have the form*

$$V_i^{1c}(m - k, x)$$
$$= p_i(\delta^{m-k}\gamma_{im-k}^c + (\varepsilon - \gamma_{1m-k}^c - \ldots - \gamma_{nm-k}^c)B_{im-k+1}^c)x, k = 1, \ldots, m - 1,$$
$$V_i^{2c}(m - k, x)$$
$$= -c(\delta^{m-k}(\gamma_{im-k}^c)^2 + (\varepsilon - \gamma_{1m-k}^c - \ldots - \gamma_{nm-k}^c)^2 D_{im-k+1}^c)x^2, i \in N, (23)$$

where

$$B^c_{im-k+1} = \sum_{j=m-k+1}^{m-1} \frac{\delta^j \gamma^c_{ij}}{\varepsilon - \gamma^c_{1j} - \ldots - \gamma^c_{nj}} \prod_{l=m-k+1}^{j} \Theta^l_{l-1}(\varepsilon - \gamma^c_{1l} - \ldots - \gamma^c_{nl})$$

$$+ M \prod_{l=m-k}^{m-1} \Theta^{l+1}_l(\varepsilon - \gamma^c_{1l} - \ldots - \gamma^c_{nl}),$$

$$D^c_{im-k+1} = - \sum_{j=m-k+1}^{m-1} \frac{\delta^j (\gamma^c_{ij})^2}{(\varepsilon - \gamma^c_{1j} - \ldots - \gamma^c_{nj})^2} \prod_{l=m-k+1}^{j} \Theta^l_{l-1}(\varepsilon - \gamma^c_{1l} - \ldots - \gamma^c_{nl})^2$$

$$+ K \prod_{l=m-k}^{m-1} \Theta^{l+1}_l(\varepsilon - \gamma^c_{1l} - \ldots - \gamma^c_{nl})^2 .$$

The multicriteria cooperative equilibrium strategies take the form

$$u^c_{im-k} = \gamma^c_{im-k} x , \quad i \in N , \quad k = 2, \ldots, m-1 ,$$

and are related by

$$\varepsilon - \gamma^c_{1m-k} - \ldots - \gamma^c_{nm-k} = \frac{\gamma^c_{im-k} - \gamma^c_{jm-k}}{\gamma^o_{im-k+1} - \gamma^c_{jm-k+1}} , \quad i, j = 1, \ldots, n, \ i \neq j . \quad (24)$$

The strategies of the players when the penult step occurs (the quantities γ^c_{im-1}) are evaluated through the following system of equations

$$p_i(V^{2c}_i(1,x) - M) - p_j(V^{2c}_j(1,x) - M) =$$

$$= 2[p_i \gamma_{im-1}(V^{1c}_i(1,x) + K) - p_j \gamma_{jm-1}(V^{1c}_j(1,x) + K)], \quad i, j = 1, \ldots, n, \ i \neq j .$$

Proof. Similar to noncooperative case.

Proposition 3. *The time-consistent payoff distribution procedure in the problem (15), (16) takes the form*

$$\beta_i(t, x_t) = \begin{pmatrix} \beta^1_i(t, x_t) \\ \beta^2_i(t, x_t) \end{pmatrix} , \quad i \in N ,$$

where

$$\beta_i^1(t, x_t) = p_i \gamma_{im-t}^c x_t + \frac{p_i}{\delta^t} x_t \cdot$$

$$\cdot [\sum_{j=m-t}^{m-2} \frac{\delta^{j+1} \gamma_{ij+1}^c (\varepsilon - \gamma_{1j}^c - \ldots - \gamma_{nj}^c - 1)}{\varepsilon - \gamma_{1j}^c - \ldots - \gamma_{nj}^c} \prod_{l=m-t}^{j} \Theta_l^{l+1}(\varepsilon - \gamma_{1l}^c - \ldots - \gamma_{nl}^c)$$

$$+ M \prod_{l=m-t}^{m-2} \Theta_l^{l+1}(\varepsilon - \gamma_{1l}^c - \ldots - \gamma_{nl}^c)(\varepsilon - \gamma_{1m-1}^c - \ldots - \gamma_{nm-1}^c - 1)],$$

$$\beta_i^2(t, x) = -c(\gamma_{im-t}^c)^2 x_t^2 - \frac{c}{\delta^t} x_t^2 \cdot$$

$$\cdot [\sum_{j=m-t}^{m-2} \frac{\delta^{j+1} \gamma_{ij+1}^c ((\varepsilon - \gamma_{1j}^c - \ldots - \gamma_{nj}^c)^2 - 1)}{(\varepsilon - \gamma_{1j}^c - \ldots - \gamma_{nj}^c)^2} \prod_{l=m-t}^{j} \Theta_l^{l+1}(\varepsilon - \gamma_{1l}^c - \ldots - \gamma_{nl}^c)^2$$

$$+ K \prod_{l=m-t}^{m-2} \Theta_l^{l+1}(\varepsilon - \gamma_{1l}^c - \ldots - \gamma_{nl}^c)^2((\varepsilon - \gamma_{1m-1}^c - \ldots - \gamma_{nm-1}^c)^2 - 1)].$$

Proof. Follows from Theorem 1 and the form of cooperative payoffs given in Proposition 2.

4   Conclusions

The multicriteria dynamic games with random planning horizon has been investigated. First, the multicriteria Nash equilibrium has been obtained. Second, the multicriteria cooperative strategies and payoffs have been constructed via the modified bargaining scheme. We have adopted the concept of dynamic stability for multicriteria dynamic games with random horizon and have constructed the time-consistent payoff distribution procedure.

To illustrate the presented approaches, we have studied a bi-criteria discrete-time bioresource management problem with random planning horizon. Multicriteria Nash and cooperative equilibria strategies have been derived analytically in linear forms. As cooperative behavior improves the ecological situation, the dynamic stability concept has been applied to stabilize the cooperative agreement. The time-consistent payoff distribution procedure has been also derived analytically.

The results presented in this paper can be applied in biological, economic and other game-theoretic models with vector payoffs.

References

1. Ghose, D., Prasad, U.R.: Solution concepts in two-person multicriteria games. J. Optim. Theor Appl. **63**, 167–189 (1989)
2. Ghose, D.: A necessary and sufficient condition for Pareto-optimal security strategies in multicriteria matrix games. J. Optim. Theory Appl. **68**, 463–481 (1991)

3. Gromova, E.V., Plekhanova, T.M.: On the regularization of a cooperative solution in a multistage game with random time horizon. Discrete Appl. Math. **255**, 40–55 (2019)
4. Haurie, A.: A note on nonzero-sum differential games with bargaining solution. J. Optim. Theor. Appl. **18**, 31–39 (1976)
5. Kuzyutin, D., Nikitina, M.: Time consistent cooperative solutions for multistage games with vector payoffs. Oper. Res. Lett. **45**(3), 269–274 (2017)
6. Kuzyutin, D., Smirnova, N., Gromova, E.: Long-term implementation of the cooperative solution in a multistage multicriteria game. Oper. Res. Perspect. **6**, 100107 (2019)
7. Kuzyutin, D., Gromova, E., Smirnova, N.: On the cooperative behavior in multistage multicriteria game with chance moves. Lecture Notes in Computer Science, vol. 12095, pp. 184–199 (2020)
8. Marin-Solano, J.: Group inefficiency in a common property resource game with asymmetric players. Econ. Lett. **136**, 214–217 (2015)
9. Mazalov, V.V., Rettieva, A.N.: Asymmetry in a cooperative bioresource management problem. In: Game-Theoretic Models in Mathematical Ecology, pp. 113–152. Nova Science Publishers (2015)
10. Parilina, E.M., Zaccour, G.: Node-consistent shapley value for games played over event trees with random terminal time. J. Optim. Theory Appl. **175**, 236–254 (2017)
11. Petrosjan, L.A.: Stable solutions of differential games with many participants. Viestnik Leningrad Univ. **19**, 46–52 (1977)
12. Petrosjan, L.A., Danilov, N.N.: Stable solutions of nonantogonostic differential games with transferable utilities. Viestnik Leningrad Univ. **1**, 52–59 (1979)
13. Pusillo, L., Tijs, S.: E-equilibria for multicriteria games. Ann. ISDG. **12**, 217–228 (2013)
14. Rettieva, A.N.: Equilibria in dynamic multicriteria games. Int. Game Theory Rev. **19**(1), 1750002 (2017)
15. Rettieva, A.N.: Dynamic multicriteria games with finite horizon. Mathematics **6**(9), 156 (2018)
16. Rettieva, A.N.: Dynamic multicriteria games with asymmetric players. J. Glob. Optim. 1–17 (2020). 10.1007/s10898-020-00929-5
17. Rettieva, A.: Rational behavior in dynamic multicriteria games. Mathematics **8**, 1485 (2020)
18. Rettieva, A.N.: Cooperation in dynamic multicriteria games with random horizons. J. Glob. Optim. **76**, 455–470 (2020)
19. Shapley, L.S.: Equilibrium points in games with vector payoffs. Naval Res. Log. Q. **6**, 57–61 (1959)
20. Sorger, G.: Recursive Nash bargaining over a productive assert. J. Econ. Dyn. Control **30**, 2637–2659 (2006)
21. Voorneveld, M., Grahn, S., Dufwenberg, M.: Ideal equilibria in noncooperative multicriteria games. Math. Methods Oper. Res **52**, 65–77 (2000)
22. Yeung, D.W.K., Petrosyan, L.A.: Subgame consistent cooperative solutions for randomly furcating stochastic dynamic games with uncertain horizon. Int. Game Theory Rev. **16**(2), 1440012 (2014)
23. Zeleny, M.: Compromising Programming, Multiple Criteria Decision Making. University of South Carolina Press, Columbia (1973)

Feedback Maximum Principle for a Class of Linear Continuity Equations Inspired by Optimal Impulsive Control

Maxim Staritsyn[1]($\boxtimes$)(iD), Nikolay Pogodaev[1,2](iD), and Elena Goncharova[1](iD)

[1] Matrosov Institute for System Dynamics and Control Theory of the Siberian Branch of the Russian Academy of Sciences, Irkutsk, Russia
goncha@icc.ru

[2] Krasovskii Institute of Mathematics and Mechanics of the Ural Branch of the Russian Academy of Sciences, Ekaterinburg, Russia

Abstract. The paper deals with an optimal control problem for the simplest version of a "reduced" linear impulsive continuity equation. The latter is introduced in our recent papers as a model of dynamical ensembles enduring jumps, or impulsive ODEs having a probabilistic uncertainty in the initial data. The model, addressed in the manuscript, is, in fact, equivalent to the mentioned impulsive one, while it is stated within the usual, continuous setup. The price for this reduction is the appearance of an integral constraint on control, which makes the problem non-standard.

The main focus of our present study is on the theory of so-called feedback necessary optimality conditions, which are one of the recent achievements in the optimal control theory of ODEs. The paradigmatic version of such a condition, called the feedback maximum (or minimum) principle, is formulated with the use of standard constructions of the classical Pontryagin's Maximum Principle but is shown to strengthen the latter (in particular, for different pathological cases). One of the main advantages of the feedback maximum principle is due to its natural algorithmic property, which enables using it as an iterative numeric algorithm.

The paper presents a version of the feedback maximum principle for linear transport equations with integrally bounded controls.

Keywords: Optimal control · Ensemble control · Impulsive control · Linear transport equation · Pontryagin's Maximum Principle · Feedback controls · Iterative methods for optimal control

1 Introduction

In the paper, we consider a specific optimal control problem for a simple distributed system, namely, a linear transport equation on the space of probability measures. The problem (P) writes:

The second author is supported by the Russian science foundation (project No. 17-11-01093).

$$\text{Minimize } I[u] = \int_{\mathbb{R}^n} \ell(x)\, \mathrm{d}\mu_T(x) \tag{1}$$

subject to

$$\partial_t \mu_t + \mathrm{div}\,(v_u\, \mu_t) = 0, \quad \mu_0 = \vartheta, \tag{2}$$

$$v_u \doteq (1 - |u|)\, f + g\, u, \quad |u| \leq 1, \tag{3}$$

and an extra constraint on the control of the *"energy"* type:

$$\int_0^T |u(t)|\, \mathrm{d}t = M. \tag{4}$$

Here, $\ell : \mathbb{R}^n \to \mathbb{R}$ is a given cost function; ϑ is a fixed probability measure on the vector space $\mathbb{R}^n$; $f, g : \mathbb{R}^n \to \mathbb{R}^n$ are given vector fields; $M, T > 0$ are constants, and $M < T$ is the "total energy" (resource) of the controller; input functions $u = u(\cdot)$ are (Borel) measurable maps $[0, T] \mapsto [-1, 1]$; $\mathrm{div} \doteq \nabla\cdot$ stands for the divergence operator, while "$\cdot$" denotes the scalar product, and $\nabla \doteq \frac{\partial}{\partial x}$. Equation (2) is understood in the distributional sense [1].

Systems of type (2) are familiar in mathematical modeling of multi-agent dynamic systems. They typically arise as a way of macroscopic representation ("mean field approximation") of infinite dynamic ensembles of indistinguishable microscopic objects (called agents, or particles). Such a lifting to the macroscopic level is achieved by considering, instead of the set of individual states $x_i(t)$, their distribution over $\mathbb{R}^n$ at time t, described by a measure μ_t from the space of probabilities $\mathcal{P}(\mathbb{R}^n)$. The bibliography on such systems and associated control problems is rich enough, including different aspects of modeling, functional properties of measure-valued solutions, and control-theoretical issues like necessary optimality conditions, dynamic programming, and viability. To mention a few, [2–4,10–12].

Let us stress that control signals u in problem (P) are functions of time variable only, and do not depend on the spatial position $x \in \mathbb{R}^n$. On the language of multi-agent systems, this is to say that (P) is a problem of *ensemble* control. Such problems appear in several applications to mathematical biology, physics, and control engineering, see, e.g. [12]. As a paradigmatic example, one can recall the problem of focusing a beam of charged particles in a common magnetic field regarded as control.

Notice that (2), (4) can be equivalently represented in the form of a coupled system, where PDE (2) is paired with the simplest scalar ODE subject to a *terminal constraint*:

$$\begin{cases} \partial_t \mu_t + \mathrm{div}\,(v_u\, \mu_t) = 0, \quad \mu_0 = \vartheta, \\ \dot{y} = |u|, \quad y(0) = 0, \quad y(T) = M. \end{cases} \tag{5}$$

The presence of such a terminal condition (constraint (4)) brings a principal feature of the addressed model by establishing its connection with impulsive continuity equations of the sort [14, 15, 19–21].

The result, we are going to obtain, is trivially extended to systems (2) driven by vector fields of a more general structure:

$$v_u = (1 - |u|) f_0 + \sum_{k=1}^{m} f_k\, u_k, \qquad (6)$$

where

$$u = (u_1, \ldots, u_m) \in \mathbb{R}^m,$$

is subject to the same constraint (4) ($|\cdot|$ is already a norm in $\mathbb{R}^m$), and

$$f_k : \mathbb{R}^n \to \mathbb{R}^n,\ k = \overline{0, m},$$

are given functions of sufficient regularity. The restriction to the case of scalar u is not crucial, but it essentially simplifies the presentation.

As a final introductory remark, we recall that systems of type (2), (4), (6) are derived from linear impulsive continuity equations via the so-called discontinuous time reparameterization. We skip the details and refer to [19], where the mentioned relationship with the impulsive control framework is accurately discussed.

1.1   Notations

Given a metric space X, we denote by $C([0, T]; X)$ the space of continuous mappings $[0, T] \mapsto X$, and endow this space with the usual sup-norm.

By $L_1([0, T]; \mathbb{R}^n)$ and $L_\infty([0, T]; \mathbb{R}^n)$ we mean the Lebesgue quotient spaces of Lebesgue integrable and bounded measurable functions $[0, T] \mapsto \mathbb{R}^n$, respectively.

$\mathcal{P} = \mathcal{P}(\mathbb{R}^n)$ stands for the set of probability measures on $\mathbb{R}^n$, and $\mathcal{P}_1 = \mathcal{P}_1(\mathbb{R}^n)$ denotes the subset of $\mathcal{P}$ composed by measures having finite first moment, i.e., such that

$$\mathrm{m}_1(\mu) \doteq \int_{\mathbb{R}^n} |\eta|\, d\mu(\eta) < \infty.$$

Recall that $\mathcal{P}_1$ turns into a complete separable metric space as soon as it is endowed with the 1-Kantorovich (Wasserstein) distance

$$W_1(\mu, \nu) \doteq \sup \left\{ \int_{\mathbb{R}^n} \varphi\, d(\nu - \mu) \ \middle|\ \begin{array}{l} \varphi \in C(\mathbb{R}^n; \mathbb{R}), \\ \mathrm{Lip}(\varphi) \leq 1 \end{array} \right\},$$

(here, Lip is the minimal Lipschitz constant of a function).

Given $\mu \in \mathcal{P}$ and a Borel measurable map $F : \mathbb{R}^n \mapsto \mathbb{R}^n$, we use the standard notation $F_\sharp \mu$ for the push-forward $\mu \circ F^{-1}$ of μ through F.

$\mathcal{L}^n$ designates the n-dimensional Lebesgue measure.

In what follows, we agree to abbreviate $\int_{\mathbb{R}^n} = \int$ and drop the arguments of integrands when it is possible, to shorten the notation.

1.2 Regularity Hypotheses

We accept the standard regularity assumption

(H_1): There exists $L > 0$ such that, for all $x, y \in \mathbb{R}^n$, it holds

$$|f(x) - f(y)| + |g(x) - g(y)| \le L\,|x - y|.$$

The space $L_\infty([0, T]; [-1, 1])$ is assumed to be endowed with the weak* topology $\sigma(L_\infty, L_1)$.

2 Pontryagin's Maximum Principle

Recall the formulation of Pontryagin's Maximum Principle (PMP) for problem (P), proved in [14]:

Proposition 1. *Assume that hypothesis (H_1) holds together with*

(H_2): *f and g are continuously differentiable in x.*

Let a control process $\bar{\sigma} = (\bar{\mu}, \bar{u})$ be optimal for (P). Then there exists $\lambda \in \mathbb{R}$ such that the following maximum condition holds for $\mathcal{L}^1$-a.a. $t \in [0, T]$:

$$(1 - |\bar{u}(t)|)\,\big(\mathfrak{f}(t) + \lambda\big) + \mathfrak{g}(t)\,\bar{u}(t) = \max\big\{\mathfrak{f}(t) + \lambda, |\mathfrak{g}(t)|\,\big\}, \tag{7}$$

where

$$\mathfrak{f}(t) \doteq \int \nabla \bar{p}_t \cdot f \, \mathrm{d}\bar{\mu}_t, \quad \mathfrak{g}(t) \doteq \int \nabla \bar{p}_t \cdot g \, \mathrm{d}\bar{\mu}_t, \tag{8}$$

and $\bar{p} = \bar{p}_t(x)$ is a solution of the dual *transport equation*

$$\partial_t p_t + v_{\bar{u}} \cdot \nabla p_t = 0, \quad p_T = -\ell. \tag{9}$$

From (7) we see that an optimal control provides an interplay of the two alternatives, $\mathfrak{f} + \lambda$ and $|\mathfrak{g}|$. In terms of impulsive control, the domination of the first option implies that the regular dynamics under the vector field f should be switched on, while another option says that it is time to jump along the vector field g.

Note that PMP is not a sufficient optimality condition for problem (P) even if constraint (4) is dropped, since the problem is non-linear (in fact, it is a sort of "bi-linear" problem: the cost functional is linear in μ, and the dynamics contains a product $u\,\mu$).

3 Feedback Maximum Principle

Feedback necessary optimality conditions, developed for continuous and impulsive ODEs in a series of papers [5–7,9,16,18], bring a sort of compromise between the constructive feature of PMP and the global nature of the Dynamic Programming.

The idea is roughly to recruit the maximum condition (7) for the identification of feedback controls with an "extremal property", which should be related with a given reference process (use the local information), and, at the same time, produce relatively strong, "nonlocal" variation of the reference control.

Though, in general, this idea relies on rather subtle arguments involving weakly monotone functions—solutions of a Hamilton-Jacobi inequality,—in our particular case, the intuition is pretty simple.

3.1   "Exact" Increment Formula of the Cost Functional: An Informal Discussion

In this section, we propose a hint as why the above mentioned idea "to use feedback controls of the PMP extremal structure" would work for problem (P).

To simplify the arguments, we assume that $\vartheta = \rho_0 \mathcal{L}^n$, where $\rho_0 : \mathbb{R}^n \to \mathbb{R}$ is a *compactly supported* density function. Hence, for any t and u, the respective solution $t \mapsto \mu_t[u]$ of (2) takes the form $\mu_t = \rho_t \mathcal{L}^n$ with certain $\rho_t : \mathbb{R}^n \to \mathbb{R}$, where all ρ_t, $t \in [0, T]$, are supported on a common compact subset of $\mathbb{R}^n$ [15, Lemma A.2]. Note that for the case of general measure the desired increment formula is obtained in [14].

Consider problem (P) in its equivalent form (1), (3), (5). As it is standard for optimal control problems with terminal constraints, we introduce the respective Lagrangian

$$L \doteq I + \lambda(y(T) - M).$$

Given a couple of admissible (satisfying all the conditions (3), (5)) triples $\bar{\sigma} = (\bar{\mu}, \bar{y}, \bar{u})$ and $\sigma = (\mu, y, u)$, define by $\bar{p}$ the solution of the dual PDE (9) under the control $\bar{u}$, and consider the increment

$$\Delta_u L \doteq L[u] - L[\bar{u}] = \Delta_u I + \lambda \Delta_u y(T).$$

Clearly,

$$\Delta_u L = I[u] - I[\bar{u}] \doteq \Delta_u I,$$

as soon as $\bar{u}$ and u satisfy (4).

Now, we shall derive an exact representation of $\Delta_u L$ (and therefore, for $\Delta_u I$) by the standard trick. Rewrite

$$\Delta_u L \doteq \int \ell \, \mathrm{d}\Delta_u \mu_T + \lambda \Delta_u y(T) = -\int \bar{p}_T \, \Delta_u \rho_T \, \mathrm{d}x - \lambda \int_0^T \Delta|u| \, \mathrm{d}t.$$

Adding the term $0 \equiv \int \bar{p}_0 \, \Delta_u \rho_0 \, \mathrm{d}x$, we can formally write (believing that the derivatives below make sense):

$$\Delta_u L + \lambda \int_0^T \Delta|u| \, \mathrm{d}t = -\int \mathrm{d}x \int_0^T \partial_t \left(\bar{p}_t \, \Delta_u \rho_t \right) \, \mathrm{d}t$$

$$= - \int \mathrm{d}x \int_0^T (\partial_t \bar{p}_t \, \varDelta_u \rho_t + \bar{p}_t \, \partial_t \varDelta_u \rho_t) \, \mathrm{d}t$$

$$= \int \mathrm{d}x \int_0^T \left(\nabla \bar{p}_t \cdot v_{\bar{u}} \, \varDelta_u \rho_t + \bar{p}_t \, \mathrm{div} \, \varDelta_u(v\rho_t) \right) \mathrm{d}t.$$

Note that

$$\int \mathrm{d}x \int_0^T \bar{p}_t \, \mathrm{div} \, \varDelta_u(v\rho_t) \, \mathrm{d}t = - \int_0^T \mathrm{d}t \int \nabla \bar{p}_t \cdot \varDelta_u(v\rho_t) \, \mathrm{d}x,$$

and we come to

$$\varDelta_u L = \int_0^T \left(\lambda \varDelta(1 - |u(t)|) + \int \nabla \bar{p}_t \cdot \left(v_{u(t)} - v_{\bar{u}(t)} \right) \mathrm{d}\mu_t \right) \mathrm{d}t.$$

In view of the structure of the vector field v_u, the latter expression rewrites:

$$\varDelta_u L = - \int_0^T \varDelta_u H^\lambda \left(\mu_t, \bar{p}_t, u(t) \right) \mathrm{d}t, \tag{10}$$

where we denote

$$H^\lambda(\mu, p, u) = (1 - |u|) \left(\mathfrak{f}[\mu, p] + \lambda \right) + u \, \mathfrak{g}[\mu, p],$$

and

$$\mathfrak{f}[\mu, p] \doteq \int \nabla p \cdot f \, \mathrm{d}\mu, \quad \mathfrak{g}[\mu, p] \doteq \int \nabla p \cdot g \, \mathrm{d}\mu, \tag{11}$$

Based on the increment formula (10) and aimed at the "improvement" of the reference control $\bar{u}$,

$$\varDelta_u I < 0,$$

we shall decide to take the "new" control u in the ensemble-feedback form

$$u_t[\mu] \in \mathbf{U}_t^\lambda(\mu) \doteq W^\lambda(\mu, \bar{p}_t), \tag{12}$$

defined via the contraction of the extremal multivalued map

$$W^\lambda(\mu, p) \doteq \arg \max_{|u| \le 1} H^\lambda(\mu, p, u) \tag{13}$$

to the reference dual trajectory $\bar{p}$. The structure of our control system allows us to find $W^\lambda(\mu, p)$ explicitly:

$$W^\lambda(\mu, p) = \begin{cases} \{0\}, & \text{if } |\mathfrak{g}[\mu, p]| - \mathfrak{f}[\mu, p] < \lambda, \\ \mathrm{sign} \, \mathfrak{g}[\mu], & \text{if } |\mathfrak{g}[\mu, p]| - \mathfrak{f}[\mu, p] > \lambda, \\ [0, 1], & \text{if } |\mathfrak{g}[\mu, p]| - \mathfrak{f}[\mu, p] = \lambda > 0, \\ [-1, 0], & \text{if } |\mathfrak{g}[\mu, p]| + \mathfrak{f}[\mu, p] = -\lambda > 0, \\ [-1, 1], & \text{otherwise.} \end{cases}$$

If a realization of a feedback control $u_t[\mu] \in \mathbf{U}_t^\lambda(\mu)$ led to a "usual" solution $t \mapsto \mu_t$ such that the composition $t \mapsto u_t[\mu_t]$ were an admissible open-loop control (to meet the integral constraint (4) we find an appropriate λ), then we could have a reason to expect the desired improvement. However, operating with (12) requires an accurate approach: the situation here is the same as in the case of discontinuous ODEs.

3.2   Ensemble-Feedback Controls and Sampling Solutions

The notion of feedback control we use is adopted from the finite-dimensional control theory. By a *feedback control* we mean an *arbitrary* single-valued map

$$\mathbf{u} : [0, T] \times \mathcal{P} \to [-1, 1].$$

Its realization as a control signal of system (2) is given by the usual Krasovskii-Subbotin sampling scheme (Euler polygons), based on a step-by-step integration of the continuity equation with a piecewise constant control, computed over the data from the previous step.

Sampling Scheme: Given $\mathbf{u}$, and a partition $\pi = \{t_k\}_{k=0}^K$ of the interval $[0, T]$,

$$t_0 = 0, \ t_{k-1} < t_k, \ k = \overline{1, K}, \ t_K = T,$$

the polygonal arc

$$t \mapsto \mu_t^\pi[\mathbf{u}] \in \mathcal{P}$$

is calculated via step-by-step integration of (2) on intervals $[t_{k-1}, t_k]$, as $k = \overline{1, K}$:

$$\mu_t^\pi[\mathbf{u}] = \mu_t^k, \quad t \in [t_{k-1}, t_k), \tag{14}$$

where

$$\mu^0 \equiv \vartheta, \quad \mu^k \doteq \mu\big[\mathbf{u}_{t_{k-1}}[\mu_{t_k}^{k-1}]\big](t_{k-1}, \mu_{t_k}^{k-1}), \tag{15}$$

and $\mu[u](\tau, \vartheta)$ denotes the distributional solution of the continuity equation under control u, starting from the position $\mu_\tau = \vartheta$.

Together with the polygonal arc, we define the piecewise constant control

$$u^\pi \doteq \mathbf{u}_{t_{k-1}}[\mu^{k-1}] \text{ on } [t_{k-1}, t_k), \ k = \overline{1, K}. \tag{16}$$

Finally, by a sampling solution (Krasovskii-Subbotin constructive motion), we mean any *partial* limit in $C([0, T]; \mathcal{P}_1)$ of a sequence of the above polygons as

$$\mathrm{diam}(\pi) \doteq \max_{k=1}^{K}(t_k - t_{k-1}) \to 0.$$

The set of all sampling solutions produced by feedback $\mathbf{u}$ is denoted by $\mathfrak{S}^{KS}(\mathbf{u})$.

Proposition 2. *Assume that (H_1) hold. Then $\mathfrak{S}^{KS}(\mathbf{u}) \neq \emptyset$, for any $\mathbf{u} \in \mathbf{U}$.*

Proof of this assertion is a simple consequence of the Arzela-Ascoli theorem. Indeed, by hypothesis (H_1), polygonal arcs $\mu_t^\pi \in C([0, T]; \mathcal{P}_1)$ are uniformly Lipschitz continuous with a constant depending only on C, T, and $\mathrm{m}_1(\vartheta)$ [13, Lemma 3], and therefore the family of polygons is equicontinuous and uniformly bounded by standard arguments based on the Gronwall's inequality.

Along with sampling solutions, we shall consider classical feedback solutions such that the following coincidence holds for some open-loop control $u \in L_\infty([0, T]; [-1, 1])$ satisfying (3), (4):

$$u(t) = v(t) \quad \mathcal{L}^1\text{-a.e. on } [0, T],$$

where
$$v(t) = \mathbf{u}_t[\mu[u]]\ \mathcal{L}^1\text{-a.e. on } [0, T],$$

and $\mu[u]$ denotes the solution of (2) under control input u.

In general, for an arbitrary $\mathbf{u}$, the existence of a classical feedback solution is not guaranteed (not to mention that its definition is not constructive). Such solutions come on scene out of the PMP extremality. We denote the (possibly empty) set of classical $\mathbf{u}$-solutions by $\mathfrak{S}^C(\mathbf{u})$ and abbreviate

$$\mathfrak{S}(\mathbf{u}) \doteq \mathfrak{S}^{KS}(\mathbf{u}) \cup \mathfrak{S}^C(\mathbf{u}).$$

3.3   Integral Constraint on Control

Now we shall discuss, how to take care of the energy bound (4) during the sampling scheme. Again, we shall regard system (2)–(4) as a coupled system (5). As it is clear, sampling solutions, produced by feedback controls $w \in \widetilde{\mathbf{U}}_t^\lambda$ generically violate the terminal condition $y(T) = M$. Here, we can adapt the idea from [16] and provide a "correction" of extremal multifunction $\widetilde{\mathbf{U}}_t^\lambda$ using a simple description of the controllability set of the state y to the point (T, M):

$$\widetilde{\mathbf{U}}_t^\lambda(\mu, y) = \begin{cases} \text{Sign } \mathfrak{g}(x, \bar{p}_t), & y \leq t - T + M, \\ \{0\}, & y \geq M, \\ U_t^\lambda(\mu), & \text{otherwise.} \end{cases} \tag{17}$$

As one can easily check, realization of feedback controls $w \in \widetilde{\mathbf{U}}_t^\lambda(\mu, y)$ via the sampling scheme results in open loop controls u^π with the property:

$$\left| \int_0^T |u^\pi|\, dt - M \right| \to 0 \text{ as diam}(\pi) \to 0.$$

Note that the use of corrected extremal multifunction is possible thanks to the simplest form of the ODE in (5). In general, tackling terminal constraints in FMP turns to be a much more delicate problem, see e.g. [8].

3.4   Formulation of the Feedback Maximum Principle

Given a reference process
$$\bar{\sigma} = (\bar{\mu}, \bar{y}, \bar{u}),$$

introduce the following *accessory problem* $(AP_{\bar{\sigma}})$:

$$\int \ell \, d\xi_T \to \inf, \quad \xi \in \bigcup_{\lambda \in \mathbb{R}} \bigcup_{w \in \widetilde{\mathbf{U}}_t^\lambda} \mathfrak{S}(w).$$

Note that the accessory problem is associated with a reference process, whose optimality is to be checked.

The *feedback maximum principle* (FMP) is formulated as follows.

Theorem 1. *Assume that* (H_1) *holds, and let* $\bar{\sigma} = (\bar{\mu}, \bar{y}, \bar{u})$ *be optimal for* (P). *Then* $\bar{\sigma}$ *is also optimal for* $(AP_{\bar{\sigma}})$.

Proof of this assertion is simple and almost literally repeats [16, Theorem 3.1]; it is based on the two obvious facts: 1) optimality (and therefore, PMP extremality) of σ for (P) implies its admissibility in $(AP_{\bar{\sigma}})$, and 2) any sampling solution is uniformly approximated by usual solutions corresponding to piecewise constant controls, and therefore, to find $w \in \widetilde{\mathbf{U}}_t^\lambda$ and $\xi \in \mathfrak{S}^{KS}(w)$ with

$$\int \ell \, d\bar{\mu}_T > \int \ell \, d\xi_T$$

would imply

$$\int \ell \, d\bar{\mu}_T > \int \ell \, d\xi_T^\pi$$

for certain polygonal approximation ξ^π of ξ, which should contradict the optimality of $\bar{\sigma}$.

As in the case of ODEs, the constructive property of FMP comes out of its counter-positive version, i.e., its interpretation as a *sufficient condition for non-optimality*. In other words, one can use Theorem 1 to discard non-optimal processes (in particular, local PMP extremals) via the qualification condition:

$$\int \ell \, d\bar{\mu}_T \le \int \ell \, d\xi_T \quad \forall \xi \in \mathfrak{S}(w) \quad \forall w \in \widetilde{\mathbf{U}}_t^\lambda, \quad \lambda \in \mathbb{R}. \tag{18}$$

For practical implementation of this condition, the interval of variation of the parameter λ can be specified, based on a priori estimates of the initial data, similarly to [16, §4].

In the following simple example, PMP does not distinguish the worst process (giving the absolute maximum) from the best one (corresponding to the absolute minimum), while FMP does.

3.5 Example: Discarding a PMP Extremal by FMP

Let $n = 2$, $\ell(a, b) = b$,

$$f(a, b) \equiv 0, \quad g(a, b) = \begin{pmatrix} 1 \\ -a \end{pmatrix},$$

and

$$T = 3, \quad \vartheta(a, b) = \delta_0(a) \otimes \eta(b),$$

where $\eta \in \mathcal{P}(\mathbb{R})$ is an arbitrary measure. Take

$$\bar{u}(t) = \begin{cases} 0, \, t \in [0, 1), \\ 1, \, t \in [1, 2), \\ -1, \, t \in [2, 3]. \end{cases}$$

The respective curve $\bar{\mu}$ and dual state $\bar{p}$ can be constructed via the flow $t \mapsto \Phi_t(a, b)$ of the characteristic ODE

$$\dot{x} = g\bar{u}.$$

The reference state is defined as

$$\bar{\mu}_t = (\Phi_t)_\sharp \vartheta, \quad t \in [0, T],$$

where

$$\Phi_t(a, b) = \begin{cases} (a, b), & t \in [0, 1), \\ \left(a - 1 + t, \ b + a - \frac{1}{2} + (1 - a)t - \frac{t^2}{2}\right), & t \in [1, 2), \\ \left(a + 3 - t, \ b - 3a - \frac{9}{2} + (a + 3)t - \frac{t^2}{2}\right), & t \in [2, 3], \end{cases}$$

and the dual one takes the form

$$-\bar{p}_t(a, b) = -\ell\left(\Phi_t^{-1}(a, b)\right) = \begin{cases} b, & t \in [0, 1), \\ b + (t - 1)a - \frac{1}{2} + t - \frac{t^2}{2}, & t \in [1, 2), \\ b + (3 - t)a - \frac{9}{2} + 3t - \frac{t^2}{2}, & t \in [2, 3], \end{cases}$$

in which construction we use the inverse of Φ_t:

$$\Phi_t^{-1}(a, b) = \begin{cases} (a, b), & t \in [0, 1), \\ \left(a + 1 - t, \ b + (t - 1)a - \frac{1}{2} + t - \frac{t^2}{2}\right), & t \in [1, 2), \\ \left(a - 3 + t, \ b + (3 - t)a - \frac{9}{2} + 3t - \frac{t^2}{2}\right), & t \subset [2, 3]. \end{cases}$$

The Hamiltonian boils down to

$$H^\lambda(\mu, p, u) = (1 - |u|)\lambda + u \int \nabla p \cdot g \, d\mu,$$

and, observed that

$$\nabla \bar{p}_t(a, b) = \begin{cases} (0, -1), & t \in [0, 1), \\ (1 - t, -1), & t \in [1, 2), \\ (t - 3, -1), & t \in [2, 3], \end{cases}$$

we have

$$H^\lambda(\mu, \bar{p}_t, u) = (1 - |u|)\lambda + u\, h_t[\mu],$$

where

$$h_t[\mu] = \begin{cases} \xi[\mu], & t \in [0, 1), \\ 1 - t + \xi[\mu], & t \in [1, 2), \\ t - 3 + \xi[\mu], & t \in [2, 3], \end{cases}$$

and

$$\xi[\mu] \doteq \int a \, d\mu(a, b).$$

Finally, the multivalued map $\mathbf{U}_t^\lambda(\mu) \doteq W^\lambda(\mu, \bar{p}_t)$ is specified as

$$
\mathbf{U}_t^\lambda(\mu) = \begin{cases}
\{0\}, & \text{if } |h_t[\mu]| < \lambda, \\
\operatorname{sign} h_t[\mu], & \text{if } |h_t[\mu]| > \lambda, \\
[0,1], & \text{if } h_t[\mu] = \lambda > 0, \\
[-1,0], & \text{if } h_t[\mu] = \lambda < 0, \\
[-1,1], & \text{otherwise.}
\end{cases}
$$

Now, noted that

$$
\xi[\vartheta] \doteq \int a \, \mathrm{d}\vartheta = \int a \, \mathrm{d}\delta_0(a) = 0,
$$

and

$$
\xi[\bar{\mu}_t] \doteq \int a \, \mathrm{d}(\varPhi_t)_\sharp \vartheta = \int \varPhi_t^1 \, \mathrm{d}\vartheta = \begin{cases}
0, & t \in [0,1), \\
t - 1, \, t \in [1,2), \\
3 - t, \, t \in [2,3]
\end{cases}
$$

($\varPhi_t^1$ denotes the first component of $\varPhi_t$), it is easy to check the inclusion

$$
\bar{u}(t) \in \mathbf{U}_t^{\bar{\lambda}}(\bar{\mu}) = [-1,1] \text{ with } \bar{\lambda} = 0,
$$

which means that $\bar{\sigma} = (\bar{\mu}, \bar{u})$ is a PMP extremal.

Now we are going to apply Theorem 1. Let $\lambda = \bar{\lambda} = 0$. The structure of the extremal map $\mathbf{U}_t^\lambda(\mu)$ leaves us an option to choose any signal $u \in [-1,1]$ as a matter of the initial guess. To our preference, take $u = \bar{u} = 0$.

Sampling along $[0,1)$ leaves this strategy intact, and the respective solution stays in rest:

$$
\mu_t \equiv \vartheta, \quad \mathbf{U}_t^\lambda(\mu_t) = [-1,1], \quad t \in [0,1).
$$

At $t = 1$, we switch the control signal to another admissible option $u = 1$. The resulted vector field starts to shift the mass distributed on the axis $a = 0$ to the right half plane ($a > 0$), and sampling over the interval $[1,3]$ gives:

$$
\mu_t = (\varPsi_t)_\sharp \vartheta, \quad \varPsi_t(a,b) = \left(a - 1 + t, b + a - \frac{1}{2} + (1-a)t - t^2/2\right), \quad t \in [1,3],
$$

$$
\xi[\mu_t] = \begin{cases} 0, & t \in [0,1), \\ t - 1, \, t \in [1,3], \end{cases} \Rightarrow h_t[\mu_t] = \begin{cases} 0, & t \in [0,2), \\ 2t - 4 \geq 0, \, t \in [2,3], \end{cases}
$$

and the inclusion

$$
u(t) = u[\mu_t] \in \mathbf{U}_t^\lambda(\mu_t) = \begin{cases} [-1,1], \, t \in [0,2), \\ \{1\}, & t \in [2,3], \end{cases}
$$

does hold, indeed.

It remains to notice that

$$
\int b \, \mathrm{d}\eta(b) - 2 = \int b \, \mathrm{d}(\varPsi_3)_\sharp \vartheta(a,b) \doteq I[u] < I[\bar{u}] = \int b \, \mathrm{d}\eta(b).
$$

We thus conclude that the reference extremal $(\bar{u}, \bar{\mu})$ is improved, and therefore, it is non-optimal. One can check that the obtained process is, in fact, a global solution (and the choice $u = -1$ above would lead to the same result).

Finally, note that, when designing u, we used the "pure", uncorrected extremal multifunction $\mathbf{U}_t^\lambda$.

3.6 Concluding Remark: FMP as Numerical Method

As a concluding note, we shall comment on the algorithmic feature of FMP. As it immediately comes to mind, by applying Theorem 1 iteratively, we can generate a sequence of control processes, which is monotone with respect to the cost functional. In other words, FMP can be realized as a so-called control "improvement" algorithm for optimal control. In the case of ODEs, such an algorithm demonstrates its efficiency as a global search numerical method [17]. Our present case is, however, much more delicate than the one, addressed by [17], since now we shall deal with the numerical solution of hyperbolic PDEs. For example, the numerical integration of the dual transport equation (9) turns to be a surprisingly complicated task from the computational viewpoint, even for the 1D case. At the same time, we do not see alternative *continuous* algorithms for a such type of optimal control problems in the available literature. Practical elaboration of the numeric algorithm, based on FMP thus remains a challenging direction of our future study.

References

1. Ambrosio, L., Savaré, G.: Gradient flows of probability measures. In: Handbook of Differential Equations: Evolutionary Equations. Vol. III, pp. 1–136. Handbook of Differential Eqution. Elsevier/North-Holland, Amsterdam (2007)
2. Carrillo, J.A., Fornasier, M., Toscani, G., Vecil, F.: Particle, Kinetic, and Hydrodynamic Models of Swarming, pp. 297–336. Birkhäuser, Boston (2010)
3. Colombo, R.M., Garavello, M., Lécureux-Mercier, M., Pogodaev, N.: Conservation laws in the modeling of moving crowds. In: Hyperbolic Problems: Theory, Numerics, Applications, AIMS Ser. Appl. Math., vol. 8, pp. 467–474. American Institute of Mathematical Science (AIMS), Springfield (2014)
4. Cucker, F., Smale, S.: Emergent behavior in flocks. IEEE Trans. Autom. Control **52**(5), 852–862 (2007)
5. Dykhta, V.A.: Nonstandard duality and nonlocal necessary optimality conditions in nonconvex optimal control problems. Autom. Remote Control **75**(11), 1906–1921 (2014). https://doi.org/10.1134/S0005117914110022
6. Dykhta, V.A.: Weakly monotone solutions of the hamilton-jacobi inequality and optimality conditions with positional controls. Autom. Remote Control **75**(5), 829–844 (2014). https://doi.org/10.1134/S0005117914050038
7. Dykhta, V.A.: Positional strengthenings of the maximum principle and sufficient optimality conditions. Proc. Steklov Inst. Math. **293**(1), 43–57 (2016). https://doi.org/10.1134/S0081543816050059
8. Dykhta, V.A.: Feedback minimum principle for quasi-optimal processes of terminally-constrained control problems. Bull. Irkutsk State Univ. Ser. Math. **19**, 113–128 (2017). https://doi.org/10.26516/1997-7670.2017.19.113

9. Dykhta, V.A., Samsonyuk, O.N.: Feedback minimum principle for impulsive processes. Bull. Irkutsk State Univ. Ser. Math. **25**, 46–62 (2018). https://doi.org/10.26516/1997-7670.2018.25.46

10. Marigonda, A., Quincampoix, M.: Mayer control problem with probabilistic uncertainty on initial positions. J. Diff. Eq. **264**(5), 3212–3252 (2018). https://doi.org/10.1016/j.jde.2017.11.014

11. Mogilner, A., Edelstein-Keshet, L.: A non-local model for a swarm. J. Math. Biol. **38**(6), 534–570 (1999). https://doi.org/10.1007/s002850050158

12. Pogodaev, N.: Optimal control of continuity equations. NoDEA Nonlinear Differ. Equ. Appl. **23**(2) (2016). Art. 21, 24

13. Pogodaev, N.: Program strategies for a dynamic game in the space of measures. Opt. Lett. **13**(8), 1913–1925 (2019). https://doi.org/10.1007/s11590-018-1318-y

14. Pogodaev, N., Staritsyn, M.: On a class of problems of optimal impulse control for a continuity equation. Trudy Instituta Matematiki i Mekhaniki UrO RAN **25**(1), 229–244 (2019)

15. Pogodaev, N., Staritsyn, M.: Impulsive control of nonlocal transport equations. J. Differ. Equ. **269**(4), 3585–3623 (2020). https://www.sciencedirect.com/science/article/pii/S002203962030108X

16. Sorokin, S., Staritsyn, M.: Feedback necessary optimality conditions for a class of terminally constrained state-linear variational problems inspired by impulsive control. Numer. Algebra Control Opt. **7**(2), 201–210 (2017)

17. Sorokin, S., Staritsyn, M.: Numeric algorithm for optimal impulsive control based on feedback maximum principle. Opt. Lett. **13**(6), 1953–1967 (2019). https://doi.org/10.1007/s11590-018-1344-9

18. Staritsyn, M., Sorokin, S.: On feedback strengthening of the maximum principle for measure differential equations. J. Global Optim. **76**, 587–612 (2020). https://doi.org/10.1007/s10898-018-00732-3

19. Staritsyn, M.: On "discontinuous" continuity equation and impulsive ensemble control. Syst. Control Lett. **118**, 77–83 (2018)

20. Staritsyn, M., Pogodaev, N.: Impulsive relaxation of continuity equations and modeling of colliding ensembles. In: Evtushenko, Y., Jaćimović, M., Khachay, M., Kochetov, Y., Malkova, V., Posypkin, M. (eds.) Optimization and Applications, pp. 367–381. Springer, Cham (2019). https://doi.org/10.1007/978-3-030-10934-9

21. Staritsyn, M.V., Pogodaev, N.I.: On a class of impulsive control problems for continuity equations. IFAC-PapersOnLine **51**(32), 468–473 (2018). https://doi.org/10.1016/j.ifacol.2018.11.429.   http://www.sciencedirect.com/science/article/pii/S2405896318331264. 17th IFAC Workshop on Control Applications of Optimization CAO 2018

Nonlocal Optimization Methods for Quadratic Control Systems with Terminal Constraints

Dmitry Trunin[✉][ID]

Buryat State University, Ulan-Ude, Russia
tdobsu@yandex.ru

Abstract. A new approach to optimization of state-quadratic optimal control problems with terminal constraints based on the sequential solution of control improvement problems in the form of special boundary value problems is considered. The developed approach for improving the admissible controls is based on the formulas for the functional increment without the remainder of the expansions. Such formulas make it possible to avoid the laborious operation of parametric variation to improve control, which ultimately leads to increased efficiency of the developed optimization procedures. The nonlocality of improving control is achieved by solving a special boundary value problem, which is much simpler than the boundary value problem of the maximum principle. To solve the boundary-value improvement problem, an iterative algorithm is constructed with the fulfillment of all terminal constraints at each iteration, based on the known perturbation principle. The proposed approach allows the formulation of new necessary optimality conditions that strengthen the known maximum principle in the class of problems under consideration and makes it possible to strictly improve non-optimal controls that satisfy the maximum principle. The comparative efficiency of the considered nonlocal methods with the known methods is illustrated by numerical calculations of model examples.

Keywords: Quadratic control system · Terminal constraints · Control improvement problem · Optimality conditions · Iterative algorithm

Introduction

In [1], methods of non-local improvement of controls in the class of state-linear optimal control problems with a free right endpoint with a linear and quadratic state-oriented objective functional are proposed. These methods are based on special formulas for the increment of the objective functional without the remainder of the expansions and do not contain the laborious operation of parametric variation of the control in the vicinity of the current approximation. Improving

This work supported by RFBR, project 18-41-030005, and Buryat State University, the project of the 2021 year.

© Springer Nature Switzerland AG 2021
P. Pardalos et al. (Eds.): MOTOR 2021, LNCS 12755, pp. 369–378, 2021.
https://doi.org/10.1007/978-3-030-77876-7_25

control is achieved at the cost of solving two special Cauchy problems. The indicated features of the methods are essential factors for increasing the efficiency of solving problems of the class under consideration.

In [2], methods were developed for the non-local improvement of control in the class of state-polynomial optimal control problems with a free right endpoint, generalizing methods [1]. These methods are based on the formulas for the increment of the objective functional without the remainder of the increments, which were obtained using modifications of the adjoint system. In this case, to improve control, it is required to solve a special boundary-value improvement problem. To solve the indicated boundary value problem, the perturbation approach known in mathematics is used.

In papers [3,4], methods of nonlocal improvement [1] are generalized to the class of quadratic in state and linear in control optimal control problems with a partially fixed right endpoint. To improve the admissible control while preserving all terminal constraints, it is required to solve a special boundary value problem.

In this article, to solve this boundary value problem, it is proposed to use an iterative method for solving a system of functional equations in the control space, which is equivalent to the boundary value problem.

1   Improving Control Problem

We consider a class of optimal control problems with terminal constraints, which reduce to a quadratic in state and linear in control optimal control problem with one terminal constraint

$$\dot{x} = A(x,t)u + b(x,t), t \in T = [t_0, \ t_1], \tag{1}$$

$$x(t_0) = x^0, u(t) \in U, \tag{2}$$

$$\Phi(u) = \langle c, x(t_1) \rangle \to \min, \tag{3}$$

$$x_1(t_1) = x_1^1. \tag{4}$$

The functions $A(x,t)$ and $b(x,t)$ are quadratic in x and continuous in t on the set $R^n \times T$; $c \in R^n$ is a given vector, and $c_1 = 0$; time interval T and end state x_1^1 are set.

By accessible controls in problem $(1), (2)$, (3) and (4) we mean functions that are piecewise continuous on an interval and have values in a compact and convex set $U \subset R^r$:

$$V = \{u \in PC^r(T) : u(t) \in U, t \in T\}.$$

For accessible control $v \in V$, we denote $x(t,v), t \in T$ the solution of the Cauchy problem $(1), (2)$ for $u = v(t), t \in T$.

By admissible controls W we mean accessible controls if the terminal constraint is satisfied (4):

$$W = \left\{u \in V : x_1(t_1, u) = x_1^1\right\}.$$

In problem (1), (2), (3) and (4), we define the Pontryagin function with the conjugate variable $p \in R^n$:

$$H(p, x, u, t) = H_0(p, x, t) + \langle H_1(p, x, t), u \rangle,$$

where $H_0(p, x, t) = \langle p, b(x, t) \rangle$, $H_1(p, x, t) = A(x, t)^T p$.

Consider the regular Lagrange functional:

$$L(u, \lambda) = \langle c, x(t_1) \rangle + \lambda(x_1(t_1) - x_1^1), \lambda \in R.$$

Following [2], the formula for the increment of the Lagrange functional, which does not contain the remainder of the expansion, takes the form:

$$\Delta_v L(u^0, \lambda) = -\int_T \langle H_1(p(t, u^0, v, \lambda), x(t, v), t), v(t) - u^0(t) \rangle dt,$$

where (u^0, v) are accessible controls; $p(t, u^0, v, \lambda), t \in T$ is a solution of the modified conjugate system

$$\dot{p} = -H_x(p, x, u, t) - \frac{1}{2} H_{xx}(p, x, u, t) y,$$

$$p_1(t_1) = -\lambda,$$

$$p_i(t_1) = -c_i, \ i = \overline{2, n},$$

for $u = u^0(t)$, $x = x(t, u^0)$, $y = x(t, v) - x(t, u^0)$.

For accessible control $u^0 \in V$ and a fixed projection parameter $\alpha > 0$ similarly to [2], we form the vector function

$$u^\alpha(p, x, t) = P_U \left(u^0(t) + \alpha H_1(p, x, t) \right), \ p \in R^n, \ x \in R^n, \ t \in T, \ \alpha > 0,$$

where P_U is an operator of projection onto a set U in the Euclidean norm.

It was shown in [5] that for nonlocal improvement of an admissible control $u^0 \in W$ it suffices to solve the following boundary value problem:

$$\dot{x} = A(x, t) u^\alpha(p, x, t) + b(x, t), \ t \in T,$$
$$\dot{p} = -H_x(p, x(t, u^0), u^0(t), t) - \frac{1}{2} H_{xx}(p, x(t, u^0), u^0(t), t)(x - x(t, u^0)), \quad (5)$$
$$x(t_0) = x^0, \ x_1(t_1) = x_1^1, \ p_i(t_1) = -c_i, \ i = \overline{2, n}.$$

Let the pair $(x(t), p(t))$, $t \in T$ is a solution to the boundary value problem (5). Let's form the output control $v(t) = u^\alpha(p(t), x(t), t)$, $t \in T$. Improvement is assessed:

$$\Delta_v \Phi(u^0) \leq -\frac{1}{\alpha} \int_T ||v(t) - u^0(t)||^2 dt.$$

It follows from the estimate that if control v differs from control u^0, then a strict improvement of the target functional is provided.

Based on the estimate in [5], it is also shown that the non-uniqueness of the solution to the boundary value improvement problem (5) allows us to strictly improve the admissible control $u^0 \in V$, that satisfies the maximum principle in the regular problem (1)–(4). In this case, the maximum principle can be formulated as follows.

Theorem 1. *Let the control $u^0 \in V$ is optimal in the regular problem (1)–(4). Then $u^0 \in V$ is the output control of the boundary value problem (5) for some $\alpha > 0$.*

This estimate allows us to formulate a strengthened necessary optimality condition in the regular problem (1)–(4).

Theorem 2. *Let the control $u^0 \in V$ be optimal in the regular problem (1)–(4). Then for all $\alpha > 0$, control $u^0 \in V$ is the only output control of the boundary value problem (5).*

Indeed, in the case of the existence of some $\alpha > 0$ output control $v \neq u$, by the estimate, we obtain a strict improvement $\Delta_v \Phi(u^0) < 0$, which contradicts the optimality of the control $u^0 \in V$.

The nonlocal optimization methods proposed in this paper are based on the following statement.

Theorem 3. *Boundary value problem (5) is equivalent to a system of functional equations in the control space with some $\lambda \in R$:*

$$v(t) = u^\alpha(p(t, u^0, v, \lambda), x(t, v), t), \alpha > 0, t \in T,$$
$$x_1(t_1, v) = x_1^1. \tag{6}$$

Indeed, let the pair $(x(t), p(t))$, $t \in T$ is a solution to the boundary value problem (5). Let's build the output control $v(t) = u^\alpha(p(t), x(t), t), t \in T$. Then $x(t) = x(t, v)$, $p(t) = p(t, u^0, v, \lambda)$, $t \in T$ for $\lambda = -p_1(t_1)$. Consequently, the control $v(t), t \in T$, satisfies the system of equations (6) with the indicated $\lambda \in R$.

Conversely, let the control $v(t), t \in T$ is a solution to system (6) for some $\lambda \in R$. Then the pair $(x(t, v), p(t, u^0, v, \lambda))$, $t \in T$ obviously satisfies the boundary value problem (5).

The system of equations (6) is considered as a fixed point problem in the control space with an additional algebraic equation. This allows us to apply and modify the well-known iterative fixed-point methods to solve the system (6).

2   Iterative Methods

To solve the system (6), the well-known algorithm of the method of simple iteration [6] is modified in the following implicit form to $k \geq 0$:

$$v^{k+1}(t) = u^\alpha(p(t, u^0, v^k, \lambda^k), x(t, v^{k+1}), t), t \in T,$$
$$x_1(t_1, v^{k+1}) = x_1^1. \tag{7}$$

As an initial approximation of the iterative process (7), the control $v^0 \in V$ is chosen.

To implement the proposed implicit iterative process, at each iteration, an auxiliary boundary value problem is considered:

$$\dot{x}^{k+1} = A(x^{k+1}, t)u^\alpha(p^{k+1}, x^{k+1}, t) + b(x^{k+1}, t), \ t \in T,$$
$$\dot{p}^{k+1} = -H_x(p^{k+1}, x(t, u^0), u^0(t), t)$$
$$- \tfrac{1}{2}H_{xx}(p^{k+1}, x(t, u^0), u^0(t), t)(x^k(t) - x(t, u^0)), \qquad (8)$$
$$x^{k+1}(t_0) = x^0, \ x_1^{k+1}(t_1) = x_1^1, \ p_i^{k+1}(t_1) = -c_i, \ i = \overline{2, n}.$$

In problem (8), the equation for conjugate variables does not depend on x. Thus, to the solution of the problem (8), we can apply, in analogy with [3], a modification of the well-known [6] shooting method.

We put $p_1(t_1) = \mu$, where $\mu \in R$ is the unknown parameter to be determined. Let us denote $p^\mu(t)$, $t \in T$ the solution of the Cauchy problem:

$$\dot{p} = -H_x(p, x(t, u^0), u^0(t), t)$$
$$- \tfrac{1}{2}H_{xx}(p, x(t, u^0), u^0(t), t)(x^k(t) - x(t, u^0)), \qquad (9)$$
$$p_1(t_1) = \mu, \ p_i^{k+1}(t_1) = -c_i, \ i = \overline{2, n}.$$

Let $x^\mu(t)$, $t \in T$ the solution to the special Cauchy problem:

$$\dot{x} = A(x, t)u^\alpha(p^\mu(t), x, t) + b(x, t), \ t \in T, x(t_0) = x^0.$$

Then the solution to a problem (8) is reduced to finding a solution to the equation concerning the parameter μ:

$$x_1^\mu(t) = x_1^1. \qquad (10)$$

Consider $(x^{k+1}(t), p^{k+1}(t))$, $t \in T$ is a solution of the auxiliary boundary-value problem (8) with a parameter μ, that is a solution to equation (10). Then

$$p^{k+1}(t) - p(t, u^0, v^k, \lambda^k),$$

where $\lambda^k = -p_1^{k+1}(t_1)$.

We form the next control approximation according to the rule:

$$v^{k+1}(t) = u^\alpha(p^{k+1}(t), x^{k+1}(t), t), t \in T.$$

It is clear that $x^{k+1}(t) = x(t, v^{k+1}), t \in T$.

Thus, the implementation of the implicit process (7) at each iteration is reduced to solving the algebraic equation (10).

Another modification of the algorithm of the simple iteration method for solving system (6) has a more familiar standard explicit form for $k \geq 0$:

$$v^{k+1}(t) = u^\alpha(p(t, u^0, v^k, \lambda^k), x(t, v^k), t), t \in T,$$
$$x_1(t_1, v^{k+1}) = x_1^1. \qquad (11)$$

For this modification, at each iteration of the process (11), after calculating the solution $p^\mu(t)$, $t \in T$ to the conjugate Cauchy problem (11) with the parameter $\mu \in R$ the control is formed:

$$v^\mu(t) = u^\alpha(p^\mu(t), x^k(t), t), t \in T.$$

For the obtained control, a solution $x(t, v^\mu)$, $t \in T$ to the standard Cauchy problem is found

$$\dot{x} = A(x,t)v^\mu(t) + b(x,t), \ t \in T, x(t_0) = x^0.$$

The parameter $\mu \in R$ at each iteration of the process (11) is selected from the condition of the terminal constraint:

$$x_1(t_1, v^\mu) = x_1^1. \tag{12}$$

For the obtained solution $\mu \in R$ of Eq. (12), the following control approximation is determined $v^{k+1}(t) = v^\mu(t)$, $t \in T$.

A feature of the proposed iterative algorithms for solving the fixed point problem (6) is the execution of the terminal constraint at each iteration of the process of successive control approximations. In this case, the initial approximation of iterative processes may not satisfy the terminal constraint, which is important for the practical implementation of algorithms.

The convergence of iterative processes can be substantiated using the perturbation principle similarly [2].

Iterative processes are applied until the first improvement in control u^0. Next, a new improvement task is constructed for the obtained control, and the process is repeated. The criterion for stopping the control improvement iterations is the absence of control improvement in terms of the target functional.

Thus, iterative methods for constructing relaxation sequences of admissible controls are formed, i.e. satisfying the terminal constraint.

3   Example

A state-quadratic problem of optimal control of the immune process is considered. In dimensionless form, the controlled model has the form [7]

$$\begin{aligned}
\dot{x}_1 &= h_1 x_1 - h_2 x_1 x_2 - u x_1, t \in T = [0, t_1], \\
\dot{x}_2 &= h_4(x_3 - x_2) - h_8 x_1 x_2, u(t) \in [0, u_{max}], t \in T, \\
\dot{x}_3 &= h_3 x_1 x_2 - h_5(x_3 - 1), \\
\dot{x}_4 &= h_6 x_1 - h_7 x_4, \\
x_1(0) &= x_1^0 > 0, \ x_2(0) = 1, \ x_3(0) = 1, \ x_4(0) = 0, \\
&\Phi_0(u) = x_1(t_1) \to \min,
\end{aligned} \tag{13}$$

$$\Phi_1(u) = \int_T x_4(t)dt \le m, m > 0. \tag{14}$$

Here $x_1 = x_1(t)$ characterizes an infectious pathogen (virus), variables $x_2 = x_2(t)$, $x_3 = x_3(t)$ characterize the body's defenses (antibodies and plasma cells), $x_4 = x_4(t)$ is a degree of damage to the body, $h_i > 0$, $i = \overline{1,8}$ are given constant coefficients. Initial conditions simulate the situation of infection of the organism with a small initial dose of the virus at the initial moment $t = 0$. The control effect $u(t), t \in T$ characterizes the intensity of the administration of immunoglobulins that neutralize the virus. Control $u(t) \equiv 0, t \in T$ corresponds to the case of no treatment. In this case, the model describes the acute course of the disease with recovery.

The values of the coefficients in the case under consideration:

$$h_1 = 2, \quad h_2 = 0.8, \quad h_3 = 10^4, \quad h_4 = 0.17, \quad h_5 = 0.5,$$

$$h_6 = 10, \quad h_7 = 0.12, \quad h_8 = 8, \quad m = 0.1.$$

The initial value x_1^0 was set equal to 10^{-6}.

In the model under consideration, a unit of time corresponds to one day. The maximum value of the control action was set equal to $u_{max} = 0.5$. The time interval T was set equal to 20 days: $t_1 = 20$.

The purpose of the control is to minimize the concentration of the virus by the end of treatment at a given time interval while limiting the damage to the body by introducing immunoglobulins that neutralize the virus.

The limitation (14) is significant when modeling an acute form of a viral disease when the consequences of damage to the body cannot be neglected and one of the goals of treatment is to limit the total load of damage to the body.

Integral constraint (14) by introducing an additional variable according to the rule

$$\dot{x}_5 = x_4, \quad x_5(0) = 0 \tag{15}$$

can be reduced to a terminal constraint.

As a result, the considered problem (13), (14) is reduced to a state-quadratic problem with a terminal constraint

$$x_5(t_1) \leq m, m > 0.$$

In the course of computational experiments, the activity of the indicated functional inequality constraint (15) was established. As a result, the optimal control problem with a partially fixed right end was considered

$$\Phi_1(u) = x_5(t_1) - m = 0, \quad m > 0. \tag{16}$$

To solve a problem (13), (15), (16), we used the method of nonlocal improvement (M2) with exact fulfillment of the terminal constraint with implementation according to rule (7) and the method of penalties (M1), which consists in solving a sequence of optimal control problems with a free right end with penalty target functional

$$\Phi(u) = \Phi_0(u) + \gamma_s \Phi_1^2(u) \rightarrow \min, \tag{17}$$

where the penalty parameter $\gamma_s > 0, s \geq 1$.

Calculation of auxiliary penalty problems (13), (15), (17) was carried out by the conditional gradient method [8]. The practical criterion for stopping the calculation of the penalty problem at a fixed value of the penalty parameter $\gamma_s > 0$ was the condition

$$|\Phi(u^{k+1}) - \Phi(u^k)| < \varepsilon_1 |\Phi(u^k)|, \tag{18}$$

where $k > 0$ is an inner iteration number of the conditional gradient method, $\varepsilon_1 = 10^{-5}$.

Under condition (18), if the specified accuracy of the terminal constraint was not achieved

$$|x_5(t_1, u^{k+1}) - m| < \varepsilon_2, \tag{19}$$

where $\varepsilon_2 = 10^{-4}$, then the penalty parameter $\gamma_s > 0$ was recalculated according to the rule

$$\gamma_{s+1} = \beta\gamma_s.$$

The multiplier value $\beta > 1$ was set equal to 10. The initial value of the penalty parameter γ_0 was set equal to 10^{-10}.

As an initial approximation for the conditional gradient method when calculating a new penalty problem, the obtained computational control u^{k+1} for the previous penalty problem was chosen. The final criterion for stopping the calculation by the M1 method was the simultaneous achievement of conditions (18) and (19).

The numerical solution of the phase and conjugate Cauchy problems was carried out by the Runge-Kutta-Werner method of variable (5–6) order of accuracy using the $DIVPRK$ program of the IMSL Fortran PowerStation 4.0 library [9]. The values of the controlled, phase and conjugate variables were stored in the nodes of a fixed uniform grid T_h with a sampling step h on an interval T. In the intervals between adjacent grid nodes T_h, the control value was taken constant and equal to the value at the left node.

In the M2 method, the solution to equation (10) was calculated using the standard procedure of the Fortran $DUMPOL$ software package [9], which implements the deformable polyhedron method, with criterion (19) to achieve the specified accuracy of the terminal constraint.

A practical criterion for stopping the calculation of sequential improvement problems in the M2 method was the condition

$$|\Phi_0(u^{k+1}) - \Phi_0(u^k)| < \varepsilon_3|\Phi_0(u^k)|,$$

where $k > 0$ is an iteration number, $\varepsilon_3 = 10^{-5}$.

In both methods, the control was chosen as the initial approximation $u(t) \equiv 0, t \in T$.

Comparative calculation results are shown in Table 1.

Table 1. Comparative calculation results

Method	Φ_0	$|\Phi_1|$	N	Note
M1	2.686698×10^{-19}	1.854861×10^{-5}	464	10^{-6}
M2	1.172261×10^{-20}	1.534792×10^{-5}	88	10^3

In Table 1 Φ_0 is a calculated value of the objective functional of the problem, $|\Phi_1|$ is a module of the calculated value of the functional-constraint (16), N is a total number of solved Cauchy problems. In the note for method M1, the value of the penalty parameter is given, at which the specified accuracy (19) of the terminal constraint execution is provided; for the proposed method M2 is a value of the projection parameter α providing convergence.

The computational control in the M1 and M2 methods, with an accuracy of a day, is a piecewise constant function with a switching point at the moment $t = 5$ from the maximum value to the minimum and reverse switching at the moment $t = 14$.

Within the framework of the example, the proposed approach makes it possible to achieve a significant reduction in the computational complexity, estimated by the total number of calculation Cauchy problems, in comparison with the standard penalty method.

Conclusion

In the class of state-quadratic and control-linear optimal control problems with constraints for the regular case, the following results are obtained.

1. Based on the well-known condition for a nonlocal condition for improving control in the form of a special boundary value problem with a projection operator, the necessary conditions for optimality of control in terms of the boundary value problem are obtained.
2. A new condition for nonlocal improvement of control is constructed in the form of a fixed point problem in the control space, which is equivalent to a special boundary value problem.
3. Based on the constructed condition for improving control in the form of a fixed point problem, nonlocal optimization methods are constructed in the class of problems under consideration.
4. On a model problem, a comparative analysis of the computational efficiency of one of the designed optimization methods with the well-known gradient method based on penalties is carried out.

The proposed methods for the nonlocal improvement of admissible controls in the considered class of quadratic problems with constraints are characterized by the following properties:

1. the absence of a procedure for varying the control in a small neighborhood of the improved control, which is typical for gradient methods;
2. the exact fulfillment of terminal constraints at each iteration of control improvement;
3. the possibility of strict improvement of management that satisfies the principle of maximum. Gradient methods do not have this capability.

The indicated properties of the methods are important factors for increasing the efficiency of solving optimal control problems with functional constraints.

References

1. Srochko, V.: Iterative Methods for Solving Optimal Control Problems. Fizmatlit, Moscow (2000)
2. Buldaev, A.: Perturbation Methods in Problem of the Improvement and Optimization of the Controlled Systems. Buryat State University, Ulan-Ude (2008)
3. Trunin, D.: About one approach to nonlocal control improvement in quadratic-in-the-state systems with terminal constraints. Bull. Buryat State Univ. Math. Inform. **2**, 40–45 (2017). https://doi.org/10.18101/2304-5728-2017-2-40-45
4. Trunin, D.: On a certain procedure of non-local improvement of controls in quadric-in-state systems with terminal restrictions. Bull. Buryat State Univ. Math. Inform. **2**, 42–49 (2018). https://doi.org/10.18101/2304-5728-2018-2-42-49
5. Trunin, D.: Projecting procedure of nonlocal improving control in polynomial optimal control problems with terminal constraints. Bull. Buryat State Univ. **9**, 52–57 (2009)
6. Samarskii, A., Gulin, A.: Numerical Methods. Nauka, Moscow (1989)
7. Marchuk, G.: Mathematical Models of Immune Response in Infectious Diseases. Kluwer Press, Dordrecht (1997)
8. Vasiliev, O.: Optimization Methods. World Federation Publishers Company INC, Atlanta (1996)
9. Bartenev, O.: Fortran for Professionals. IMSL Math Library. Part 2. Dialog-MIFI, Moscow (2001)

Identification of the Thermal Conductivity Coefficient of a Substance from a Temperature Field in a Three-Dimensional Domain

Vladimir Zubov$^{(\boxtimes)}$(ID) and Alla Albu(ID)

Dorodnicyn Computing Centre, Federal Research Center "Computer Science and Control" of Russian Academy of Sciences, Moscow, Russia

Abstract. The problem of determining the temperature-dependent thermal conductivity coefficient of a substance in a parallelepiped is considered and investigated. The consideration is carried out on the basis of the first boundary value problem for a three-dimensional non-stationary heat conduction equation. The inverse coefficient problem is reduced to a variational problem and is solved numerically using gradient methods for minimizing the cost functional. The mean-root-square deviations of the temperature field from the experimental data is used as the cost functional. It is well known that it is very important for the gradient methods to determine accurate values of the gradients. For this reason, in this paper we used the efficient Fast Automatic Differentiation technique, which gives the exact functional gradient for the discrete optimal control problem. In this work special attention is paid to the practically important cases when the experimental field is specified only in the subdomain of the object under consideration. The working capacity and effectiveness of the proposed approach are demonstrated by solving a number of nonlinear inverse problems.

Keywords: Heat conduction · Inverse coefficient problems · Gradient · Fast automatic differentiation · Numerical algorithm

1 Introduction

In [1–7] the inverse coefficient problem of identification of the temperature-dependent thermal conductivity coefficient of a substance was studied in one- and two-dimensional variants. However, in practice experimental data is collected for three-dimensional objects. Therefore, it is desirable to solve the problem of identifying the thermal conductivity coefficient in the three-dimensional formulation.

This work was supported by the Russian Foundation for Basic Research (project no. 19-01-00666 A).

© Springer Nature Switzerland AG 2021
P. Pardalos et al. (Eds.): MOTOR 2021, LNCS 12755, pp. 379–393, 2021.
https://doi.org/10.1007/978-3-030-77876-7_26

In this paper the consideration of the inverse problem for a three-dimensional object is carried out on the basis of the first boundary value problem for a three-dimensional non-stationary heat conduction equation. The inverse coefficient problem is reduced to a variational problem. The mean-root-square deviation of the temperature field from the experimental data is used as the cost functional. The optimization problem is solved numerically using gradient methods for minimizing the cost functional, and to calculate the gradient an effective methodology of Fast Automatic Differentiation (FAD-methodology) is used (see [8,9]).

One of the difficulties that one has to face when solving a problem in a three-dimensional formulation is associated with solving a direct problem. In [10], using the examples of a number of nonlinear problems for a three-dimensional heat equation whose coefficients depend on temperature, a comparative analysis of several schemes of alternating directions was performed. The following schemes were examined: a locally one-dimensional scheme [11], a Douglas-Reckford scheme [12], and a Pisman-Reckford scheme [13]. When comparing methods, the accuracy of the obtained solution and the computer time to achieve the required accuracy were taken into account.

The conjugate equations and the formula for calculating the gradient of the cost functional, used in this work, were obtained on the basis of a locally one-dimensional scheme, since the results of numerical experiments showed that it was the least "capricious".

The problem of identifying the thermal conductivity coefficient of a substance is related to the study of the characteristics of newly created materials. When experimental studies are being carrying out, as a rule, samples of material of a simple form are used (usually this is a parallelepiped). It is reasonable to consider the inverse coefficient problem arising in the three-dimensional case also for a parallelepiped object.

In this work, special attention is paid to the practically important cases when the experimental field is specified only in the subdomain of the object under consideration. The working capacity and effectiveness of the proposed approach are demonstrated by solving a number of nonlinear inverse problems.

2   Formulation of the Problem

Suppose that the specimen under study is a parallelepiped of length R, width E and height H. The initial temperature T of the parallelepiped is known. The law of temperature variation on the surfaces of the parallelepiped is also known. For a mathematical description of the heat conduction process in a parallelepiped we use the Cartesian coordinates x, y and z. The points $s = (x, y, z)$ of the parallelepiped create a domain $Q = \{[0, R] \times [0, E] \times [0, H]\}$ with a boundary $\Gamma = \partial Q$. The temperature field at each time is described by the initial-boundary value (mixed) problem:

$$C(s)\frac{\partial T(s,t)}{\partial t} = div_s(K(T(s,t))\nabla_s T(s,t)), \qquad (s,t)\{\in Q \times (0,\Theta]\}, \qquad (1)$$

$$T(s,0) = w_0(s), \qquad\qquad s \in \overline{Q}, \qquad (2)$$

$$T(s,t) = w_\Gamma(s,t), \qquad\qquad s \in \Gamma, \quad 0 \le t \le \Theta. \qquad (3)$$

Here t is time; $T(s,t) \equiv T(x,y,z,t)$ is the temperature of the material at the point s with the coordinates (x,y,z) at time t; $C(s)$ is the volumetric heat capacity of the material; $K(T)$ is the thermal conductivity coefficient; $w_0(s)$ is the given temperature at the initial time $t = 0$; $w_\Gamma(s,t)$ is the given temperature on the boundary of the object. The volumetric heat capacity of a substance $C(s)$ is considered as known function of the coordinates. If the dependence of the thermal conductivity coefficient $K(T)$ on the temperature T is known, then we can solve the mixed problem (1)–(3) to find the temperature distribution $T(s,t)$ in $Q \times [0,\Theta]$.

The inverse coefficient problem is reduced to the following variational problem: it is required to find such a dependence of the thermal conductivity coefficient of a substance on temperature at which the temperature field $T(s,t)$ obtained as a result of solving the direct problem (1)–(3) differs little from the temperature field $Y(s,t)$ obtained experimentally. The quantity:

$$\Phi(K(T)) = \int\limits_0^\Theta \int\limits_Q [T(s,t) - Y(s,t)]^2 \cdot \mu(s,t)ds dt \qquad (4)$$

can be used as the measure of difference between these functions. Here, $\mu(s,t) \ge 0$ is given weight function. Thus, the optimal control problem is to find the optimal control $K(T)$ and the corresponding solution $T(s,t)$ of problem (1)–(3) that minimize functional (4).

3   Numerical Solution of the Inverse Coefficient Problem

The optimal control problem formulated above was solved numerically. Spatial and time grids (generally nonuniform) have been introduced to solve the problem numerically.

The time grid was constructed by a set of nodal values $\{t^j\}_{j=0}^J$, $t^0 = 0$, $t^J = \Theta$. The steps τ^j of this grid were determined by the relations $\tau^j = t^{j+1} - t^j$, $j = \overline{0, J-1}$.

A spatial grid is based by a set of points $\{x_n, y_i, z_l\}$, where $n = \overline{0, N}$, $i = \overline{0, I}$, $l = \overline{0, L}$, h_n^x is the distance between x_n and x_{n+1}, i.e. $h_n^x = x_{n+1} - x_n$, $n = \overline{0, N-1}$. Similarly: $h_i^y = y_{i+1} - y_i$, $i = \overline{0, I-1}$ and $h_l^z = z_{l+1} - z_l$, $l = \overline{0, L-1}$. At each node of the computational domain $\overline{Q} \times [0, \Theta]$ all the functions are determined by their point values.

The temperature interval $[a, b]$ (the interval of interest) on which the function $K(T)$ will be restored is defined as the set of values of the given functions $w_0(s)$

and $w_\Gamma(s,t)$. This interval is partitioned by the points $\widetilde{T}_0 = a, \widetilde{T}_1, \widetilde{T}_2, \ldots, \widetilde{T}_M = b$ into M parts (they can be equal or of different lengths). Each point $\widetilde{T}_m$ ($m = 0, \ldots, M$) is connected with a number $k_m = K(\widetilde{T}_m)$. The function $K(T)$ to be found is approximated by a continuous piecewise linear function with the nodes at the points $\left\{(\widetilde{T}_m, k_m)\right\}_{m=0}^M$ so that $K(T) = k_{m-1} + \frac{k_m - k_{m-1}}{\widetilde{T}_m - \widetilde{T}_{m-1}}(T - \widetilde{T}_{m-1})$ for $\widetilde{T}_{m-1} \leq T \leq \widetilde{T}_m$, ($m = 1, \ldots, M$). If the temperature at the point fell outside the boundaries of the interval $[a; b]$, then the linear extrapolation was used to determine the function $K(T)$.

One of the main elements of the proposed numerical method for solving inverse coefficient problem is the solution of the mixed problem (1)–(3). To approximate the heat equation was used the locally one-dimensional scheme that is a scheme of alternating directions. The cost functional (4) was approximated by a function $F(k_0, k_1, \ldots, k_M)$ of the finite number of variables using the method of rectangles. Minimization of the function $F(k_0, k_1, \ldots, k_M)$ was carried out numerically using the gradient method.

It is well known that it is very important for the gradient methods to determine accurate values of the gradients. For this reason, we used the efficient approach of Fast Automatic Differentiation to calculate the components of gradient. The effectiveness of this methodology is ensured by using the solution of the conjugate problem for calculating the gradient of the function. The FAD-methodology allows us to formulate such adjoint problem that is coordinated with the chosen approximation of the direct problem. As a result, the FAD-methodology delivers canonical formulas by means of which the calculated value of the gradient of the cost functional is precise for the chosen approximation of the optimal control problem.

To verify the performance of the proposed algorithm a huge number of numerical experiments were carried out. The most interesting of them are given in this section in the form of three series of calculations.

The difficulties that one has to face when solving the problem of identifying the thermal conductivity coefficient in a three-dimensional formulation are associated not only with the choice of the discretization scheme for the direct problem, but also with the choice of the computational grid. The analysis of the obtained numerical results showed that if the number of nodes of the spatial grid is not less than 20 in each direction, then the thermal conductivity coefficient is restored with a sufficiently high accuracy, provided that the time step is correctly selected.

The variational problem, associated with the problem of identifying the thermal conductivity coefficient is reduced, was solved for all examples on several computational grids. For each grid, calculations were performed using different initial approximations.

The locally one-dimensional scheme approximating the heat equation chosen in this paper is stable and allows working with a sufficiently large time step. Nevertheless, it is necessary to carry out studies concerning the selection of the time grid, not only for each spatial grid used, but also for each division of the temperature interval on which the thermal conductivity coefficient is restored. Studies have shown that the greater the number of partitions of the temperature interval, the smaller the time step should be.

Two criteria were used to evaluate the accuracy of the obtained numerical solutions of the inverse problem:

$$\varepsilon_1 = \max_{0 \leq m \leq M} \frac{|K_{opt}(\widetilde{T}_m) - K(\widetilde{T}_m)|}{K^*} \tag{5}$$

and

$$\varepsilon_2 = \frac{1}{K^*} \sqrt{\sum_{m=0}^{M} \frac{(K_{opt}(\widetilde{T}_m) - K(\widetilde{T}_m))^2}{M+1}}, \tag{6}$$

where $K(\widetilde{T}_m)$ are the values of the analytical thermal conductivity coefficient calculated at the reference points of the temperature interval, $K_{opt}(\widetilde{T}_m)$ are the values of the thermal conductivity coefficient obtained as a result of solving the optimization problem, and K^* is some characteristic value of the function $K(T)$. In the paper $K^* = \dfrac{\sum_{m=0}^{M} K(\widetilde{T}_m)}{M+1}$ was used.

The numerical experiments carried out have shown that the quality of the recovery of the thermal conductivity coefficient strongly depends on the distribution of the "experimental" temperature field. There are cases when in some sections of the segment $[a, b]$ there is too little data necessary to identify the thermal conductivity coefficient. Therefore, in each specific case, it is necessary to analyze the distribution of experimental data over the intervals of the temperature segment of interest to us.

3.1   The First Series of Calculations

In the first series of calculations, the problem of finding the thermal conductivity coefficient of a substance was considered for the following input data: the traces of the function

$$\Lambda(x, y, z, t) = \frac{1}{2} \sqrt{\frac{x^2 + y^2 + z^2}{3 - 2t}} \tag{7}$$

were chosen as the initial function $w_0(x, y, z)$ and as the boundary function $w_\Gamma(x, y, z, t)$ on the parabolic boundary of the domain $Q \times (0, \Theta) = (0, 1) \times (0, 1) \times (0, 1) \times (0, 1)$. In this case, function (7) is a solution of the mixed problem (1)–(3) for $C(x) = 1$ and $K(T) = T^2$. The temperature at the parabolic boundary of the considered domain varies from $a = 0.0$ to $b = 0.866$. In this series of calculations, it was assumed that in the cost functional (4) the weight function $\mu(x, y, z, t) \equiv 1$, $(x, y, z) \in Q$.

To analyze the distribution of experimental data over the intervals of the temperature segment of interest we introduce the function $W_*(T)$ of the relative measure of that subdomain of the domain $Q \times (0, \Theta)$ in which the function

$\Lambda(x, y, z, t)$ satisfies the condition $\Lambda(x, y, z, t) < T$, and denote by $W(T)$ the derivative of the function $W_*(T)$ with respect to the variable T.

Figure 1 shows the distribution of the function $W(T)$ along the segment $[0.0, 0.866]$. Analysis of this distribution shows that too few "experimental" data correspond to the right end of the segment $[0.0, 0.866]$. Therefore, it can be assumed that there will be difficulties in restoring the thermal conductivity coefficient for $T > 0.8$. The calculations carried out confirmed this assumption: when the temperature segment is evenly divided into $M = 80$ intervals, the last two components of the control vector do not change (the gradient components are zero with machine accuracy).

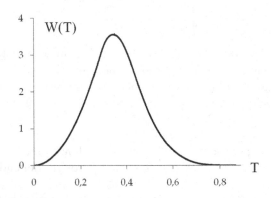

Fig. 1. Distribution of the function $W(T)$.

All calculations in this series can be divided into two groups.

In the calculations of the first group the "analytical" field was used as an "experimental" temperature field:

$$Y_{nil}^j = \Lambda(x_n, y_i, z_l, t^j) = \frac{1}{2}\sqrt{\frac{x_n^2 + y_i^2 + z_l^2}{3 - 2t^j}}$$

$$(n = \overline{1, N-1}, i = \overline{1, I-1}, l = \overline{1, L-1}, j = \overline{1, J}).$$

The calculations of this group revealed the dependence of the numerical solution of the inverse problem on the accuracy of the solution of the direct problem. All the results of calculations of the first group presented below were obtained using a uniform spatial grid with parameters $N = I = L = 25$. The number J of intervals for dividing the time segment $[0, \Theta]$ varied.

a) A Uniform Time Grid with a Number $J = 25$ of Intervals was Used.

To accelerate the iterative process of descent along the gradient, in all the examples presented in this work, the approach proposed in [4] was used. It is based on a sequential increase in the number M of partitions of a segment $[a, b]$. It is advisable to start the process with $M = 1$. After obtaining the optimal solution, it should be used as an initial approximation for variant with $M = 2$. Having obtained the optimal solution for $M = 2$, use it as an initial approximation for $M = 4$, etc.

Figure 2 shows the functions at different stages of the iterative process in the case when the function $K_{ini}(T) = 2.5$ was chosen as the initial approximation. There the function $K(T) = T^2$ and optimal controls obtained for $M = 1$, $M = 2$, and $M = 4$ are shown. It is seen that the support points of the piecewise-linear optimal control obtained at $M = 4$ almost lie on the line $K(T) = T^2$. As for the optimal controls obtained for $M = 8$ and $M = 16$, they practically coincide with the function $K(T) = T^2$. For $M = 8$ the deviations of the obtained thermal conductivity coefficient $K_{opt}(T)$ from its analytical value $K(T) = T^2$ calculated by formulas (5) and (6) are $\varepsilon_1 = 2.1267 \cdot 10^{-2}$ and $\varepsilon_2 = 1.3190 \cdot 10^{-2}$, respectively. For $M = 16$ $\varepsilon_1 = 1.8367 \cdot 10^{-2}$ and $\varepsilon_2 = 1.0468 \cdot 10^{-2}$.

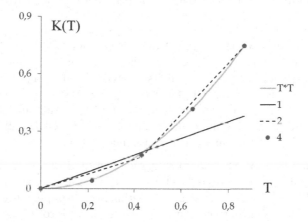

Fig. 2. Control functions at different stages of the iterative process.

However, further dividing the temperature segment into smaller intervals does not lead to the desired results on the selected space-time grid. Figure 3 shows the function $K(T) = T^2$ and optimal control obtained for $M = 80$. It is seen that the obtained solution contains oscillations. Apparently, this is due to the fact that the solution of the direct problem is determined insufficiently accurately on the computational grid used. The use of a small number of partitions (for example, $M = 8$ or $M = 16$) of the temperature segment in this case acts as a smoothing factor in solving the problem.

b) A Uniform Time Grid with a Number $J = 100$ of Intervals was Used.

As a result of solving the inverse problem at $M = 80$, a smooth, oscillation-free solution of the inverse problem was obtained, the graph of which coincides with the graph of the function $K(T) = T^2$. In the process of solving the optimization problem, the cost functional varies from $3.9708 \cdot 10^{-4}$ under the initial control ($K_{ini}(T) = 2.5$) to $7.5517 \cdot 10^{-9}$ under the optimal control. At the same time,

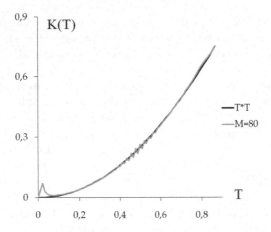

Fig. 3. Control functions at $M = 80$ and analytical thermal conductivity coefficient.

the maximum modulus of the gradient decreased from $1.2422 \cdot 10^{-5}$ to $1.4525 \cdot 10^{-10}$. Calculated by formulas (5) and (6), the deviation of the obtained thermal conductivity coefficient $K_{opt}(T)$ from its analytical value $K(T) = T^2$ on the segment $[0.0, 0.844]$ is $\varepsilon_1 = 1.2764 \cdot 10^{-2}$ and $\varepsilon_2 = 3.8048 \cdot 10^{-3}$, respectively.

It should also be noted that for the example under consideration, the solution to the inverse problem is unique. Calculations performed for different initial approximations $K_{ini}(T)$ always led to the same solution.

In the calculations of the second group the field obtained as a result of the numerical solution of problem (1)–(3) with a known thermal conductivity coefficient $K(T) = T^2$ was used as an "experimental" temperature field.

In this case, for dividing the temperature segment into $M = 80$ parts, the thermal conductivity coefficient was identified with high accuracy even when using a uniform spatial grid with parameters $N = I = L = 25$ and the number $J = 25$ of time intervals. The calculation results obtained on this grid practically do not differ from those obtained as a result of solving the optimization problem on a grid with a time step 4 times less ($J = 100$). In both these cases, the cost functional dropped to values less than 10^{-20}, the maximum of the gradient modulus decreased to $3 \cdot 10^{-14}$, $\varepsilon_1 = 4.2 \cdot 10^{-6}$ and $\varepsilon_2 < 10^{-6}$.

3.2   The Second Series of Calculations

In the second series of calculations, the problem of finding the thermal conductivity coefficient of a substance was considered for the following input data: the traces of the function

$$\Lambda(x, y, z, t) = \sqrt{\frac{9(x + 1)^2 + 20y^2 + 25z^2}{9 - 8t}} \tag{8}$$

were chosen as the initial function $w_0(x, y, z)$ and as the boundary function $w_\Gamma(x, y, z, t)$ on the parabolic boundary of the domain $Q \times (0, \Theta) = (0, 1) \times (0, 1) \times (0, 1) \times (0, 1)$.

In this series of calculations, it was assumed that in the cost functional (4) the weight function $\mu(x, y, z, t) \equiv 1$, $(x, y, z) \in Q$. As the experimental field $Y(x, y, z, t)$ was chosen the temperature field, which was obtained as a result of solving the direct problem (1)–(3) at $C(s) = 1$ and with the thermal conductivity coefficient $K(T) = k(T)$, where the function $k(T)$ is determined by the following equality:

$$k(T) = \begin{cases} 0.1(T-3)(T-6)(T-7) + 3.4, & T \geq 3, \\ 1.2(T-3) + 3.4, & T < 3. \end{cases}$$

The temperature at the parabolic boundary of the considered domain varies from $a = 1.0$ to $b = 9.0$. The segment $[1.0, 9.0]$ was divided into 32 intervals, i.e. $M = 32$. Figure 4 shows the distribution of the function $W(T)$ along the segment $[1.0, 9.0]$. Analysis of this distribution shows that too few "experimental" data correspond to the right end of the segment $[1.0, 9.0]$ (for $T > 8$). Therefore, it can be assumed that there will be difficulties in restoring the thermal conductivity coefficient for $T > 8$.

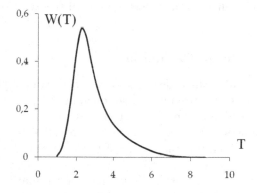

Fig. 4. Distribution of the function $W(T)$.

Numerical calculations were carried out on a uniform spatial grid with parameters $N = I = L = 25$ and the number $J = 50$ of time intervals. As in the first series of calculations, to speed up the iterative process we begin with $M = 1$. Figure 5 shows the functions at different stages of the iterative process in the case when the function $K_{ini}(T) = 2.5$ was chosen as the initial approximation. There the function $K(T) = k(T)$ and optimal controls obtained for $M = 1$, $M = 2$, $M = 4$ and $M = 8$ are given. It can be seen that the support points of the piecewise linear optimal control obtained at $M = 8$ almost lie on the line $K(T) = k(T)$. As for the optimal controls obtained for $M = 16$ and $M = 32$, they practically coincide with the function $K(T) = k(T)$. A small deviation is noted only at the right end of the temperature range, i.e. where there is no experimental data. For dividing the temperature interval into parts $M = 32$ the deviations of the obtained thermal conductivity coefficient $K_{opt}(T)$ from its analytical value $K(T) = k(T)$ on the segment $[1.0, 8.0]$ are $\varepsilon_1 = 7.2876 \cdot 10^{-6}$ and $\varepsilon_2 = 1.5006 \cdot 10^{-6}$, respectively.

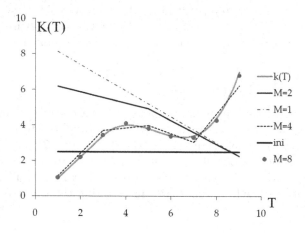

Fig. 5. Control functions at different stages of the iterative process.

3.3   The Third Series of Calculations

The third series of calculations is devoted to the practically important case when the experimental field is given only in a subdomain of the object under consideration. Here we considered the problem of finding the thermal conductivity coefficient with the same input data as in the second series. The third series differs from the second only in the choice of the weighting function in the cost functional.

Numerical calculations were carried out on a uniform spatial grid with parameters $N = I = L = 25$ and the number $J = 50$ of time intervals.

In the first example of this series, the weighting function had a δ-shaped character and in the numerical algorithm was determined by the formula (216 "control" points in the functional):

$$\mu_{nil}^{j} = \begin{cases} 1, & n, i, l = 3, 7, 11, 15, 19, 23, \\ 0, & in \ other \ cases. \end{cases}$$

The function $K_{ini}(T) = 2.5$ was chosen as an initial approximation. The distribution of experimental data along the temperature segment $[1.0, 9.0]$ at the above weighting function is qualitatively similar to that shown in Fig. 4, but quantitatively there are much less data here than in the second series of calculations. However, despite this and the "discrete" nature of the weight function in this case, no qualitative differences from the previous example were noticed when solving the optimization problem. For dividing the temperature interval into $M = 32$ parts the deviations of the obtained thermal conductivity coefficient $K_{opt}(T)$ from its analytical value $K(T) = k(T)$ on the segment $[1.0, 8.0]$ are $\varepsilon_1 = 8.8594 \cdot 10^{-6}$ and $\varepsilon_2 = 1.6499 \cdot 10^{-6}$, respectively.

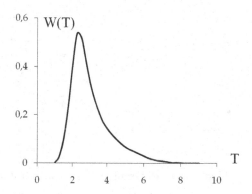

Fig. 6. Distribution of the function $W(T)$.

In the second example of this series, the weight function in the cost functional has the form

$$\mu_{nil}^j = \begin{cases} 1, & i \leq 3, \\ 0, & in \ other \ cases. \end{cases}$$

Figure 6 shows the distribution of experimental data along the temperature segment $[1.0, 9.0]$. Analysis of this distribution shows that there is very little "experimental" data at the right end of the temperature interval at $T > 8.0$.

In this example, a function $K_{ini}(T) = 7.0$ was chosen as the initial approximation. Figure 7 presents solutions to this inverse problem for $M = 2$, $M = 4$, $M = 8$ and function $k(T)$. Figure 8 presents the obtained thermal conductivity coefficient $K_{opt}(T)$ for $M = 32$ and function $k(T)$. Despite the fact that there are fewer experimental data, and that another function was chosen as the initial approximation, nevertheless, the iterative process proceeded in a similar way as in the previous example. In this case, there is also a convergence of optimal solutions to the function "opt" with increasing number M, but this function differs from $k(T)$ when $T > 8$. As for the restoration of the thermal conductivity coefficient for a given experimental field on the interval $[1.0, 8.0]$, here the graphs of the functions $K_{opt}(T)$ and $k(T)$ completely coincide, and the deviations of the obtained thermal conductivity coefficient from its analytical value $K(T) = k(T)$ are $\varepsilon_1 = 3.6477 \cdot 10^{-2}$ and $\varepsilon_2 = 6.7738 \cdot 10^{-3}$, respectively.

In the third example of this series, the weight function in the cost functional has the form

$$\mu_{nil}^j = \begin{cases} 1, & x_n + y_i + z_l + t^j \geq 3.5, \\ 0, & in \ other \ cases. \end{cases}$$

Figure 9 shows the distribution of experimental data along the temperature segment $[1.0, 9.0]$ at the above weighting function. The analysis of this distribution

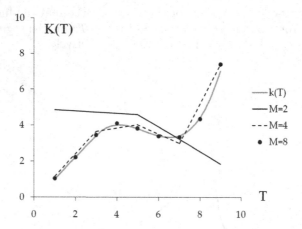

Fig. 7. Control functions at different stages of the iterative process.

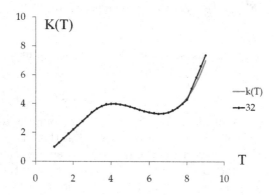

Fig. 8. The thermal conductivity coefficient $K_{opt}(T)$ for $M = 32$ and function $k(T)$.

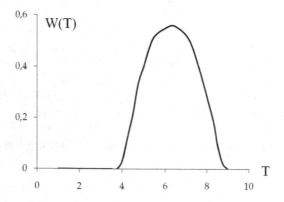

Fig. 9. Distribution of the function $W(T)$.

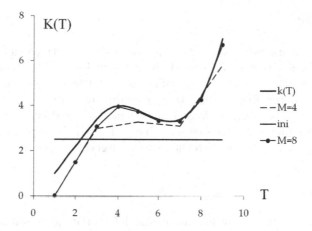

Fig. 10. Control functions at different stages of the iterative process.

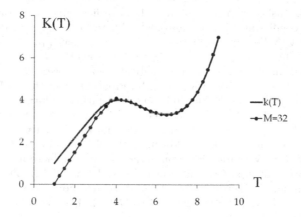

Fig. 11. The thermal conductivity coefficient $K_{opt}(T)$ for $M = 32$ and function $k(T)$.

shows that here most of the "experimental" data is concentrated on the interval $[4.0, 9.0]$. However, there are practically no "experimental" data at $T < 4.0$. In this example, a function $K_{ini}(T) = 2.5$ was chosen as the initial approximation. Figure 10 presents solutions to this inverse problem for $M = 4$, $M = 8$ and function $k(T)$. Figure 11 presents the obtained thermal conductivity coefficient $K_{opt}(T)$ for $M = 32$ and function $k(T)$. In this case, there is a convergence of optimal solutions to the function "opt" with increasing number M, but this function differs from $k(T)$ when $T < 4$. As for the restoration of the thermal conductivity coefficient for a given experimental field on the interval $[4.0, 9.0]$, here the graphs of the functions $K_{opt}(T)$ and $k(T)$ completely coincide, and the deviations of the obtained thermal conductivity coefficient from its analytical value $K(T) = k(T)$ are $\varepsilon_1 = 3.6477 \cdot 10^{-2}$ and $\varepsilon_2 = 6.7738 \cdot 10^{-3}$, respectively.

4   Conclusion

An algorithm is proposed to determine the thermal conductivity of the substance with the help of a given experimental three-dimensional temperature field. On the basis of the calculations, it is concluded that the thermal conductivity is more effectively determined in the temperature range, which corresponds to a larger number of experimental data. If it is important to determine the thermal conductivity in a particular temperature interval, then the experimental data in which the temperature belongs to the desired interval should be used.

References

1. Zubov, V.I.: Application of fast automatic differentiation for solving the inverse coefficient problem for the heat equation. Comput. Math. Math. Phys. **56**(10), 1743–1757 (2016). https://doi.org/10.1134/S0965542516100146
2. Albu, A.F., Evtushenko, Y.G., Zubov, V.I.: Identification of discontinuous thermal conductivity coefficient using fast automatic differentiation. In: Battiti, R., Kvasov, D.E., Sergeyev, Y.D. (eds.) LION 2017. LNCS, vol. 10556, pp. 295–300. Springer, Cham (2017). https://doi.org/10.1007/978-3-319-69404-7_21
3. Zubov, V.I., Albu, A.F.: The FAD-methodology and recovery the thermal conductivity coefficient in two dimension case. In: Proceedings of the VIII International Conference on Optimization Methods and Applications "Optimization and applications", pp. 39–44 (2017). https://doi.org/10.1007/s11590-018-1304-4
4. Albu, A.F., Zubov, V.I.: Identification of thermal conductivity coefficient using a given temperature field. Comput. Math. Math. Phys. **58**(10), 1585–1599 (2018)
5. Albu, A.F., Zubov, V.I.: Identification of the thermal conductivity coefficient using a given surface heat flux. Comput. Math. Math. Phys. **58**(12), 2031–2042 (2018)
6. Albu, A., Zubov, V.: Identification of the thermal conductivity coefficient in two dimension case. Optim. Lett. **13**(8), 1727–1743 (2018). https://doi.org/10.1007/s11590-018-1304-4
7. Albu, A., Zubov, V.: On the stability of the algorithm of identification of the thermal conductivity coefficient. In: Evtushenko, Y., Jaćimović, M., Khachay, M., Kochetov, Y., Malkova, V., Posypkin, M. (eds.) OPTIMA 2018. CCIS, vol. 974, pp. 247–263. Springer, Cham (2019). https://doi.org/10.1007/978-3-030-10934-9_18
8. Evtushenko, Y.G.: Computation of exact gradients in distributed dynamic systems. Optim. Meth. Softw. **9**, 45–75 (1998)
9. Evtushenko, Y.G., Zubov, V.I.: Generalized fast automatic differentiation technique. Comput. Math. Math. Phys. **56**(11), 1819–1833 (2016). https://doi.org/10.1134/S0965542516110075
10. Albu, A.F., Evtushenko, Y.G., Zubov, V.I.: Choice of finite-difference schemes in solving coefficient inverse problems. Comput. Math. Math. Phys. **60**(10), 1589–1600 (2020). https://doi.org/10.1134/S0965542520100048
11. Samarskii, A.A., Vabishchevich, P.N.: Computational Heat Transfer Editorial URSS, Moscow (2003). (in Russian)

12. Douglas, J., Rachford, H.H.: On the numerical solution of heat conduction problems in two and three space variables. Trans. Amer. Math. Soc. **8**, 421–439 (1956)
13. Peaceman, D.W., Rachford, H.H.: The numerical solution of parabolic and elliptic differential equations. J. Soc. Ind. Appl. Math. **3**(1), 28–41 (1955)
14. Albu, A.F., Evtushenko, Y.G., Zubov, V.I.: Application of the fast automatic differentiation technique for solving inverse coefficient problems. Comput. Math. Math. Phys. **60**(1), 15–25 (2020). https://doi.org/10.1134/S0965542520010042

Operational Research and Mathematical Economics

Dixit-Stiglitz-Krugman Model with Investments in R&D

Igor Bykadorov[(✉)]

Sobolev Institute of Mathematics SB RAS, Novosibirsk, Russia

Abstract. We study a monopolistic competition model in the open economy case. The utility of consumers is additive separable. The producers can choose the technology (R&D) endogenously. We examine the local comparative statics of market equilibrium with respect to transport cots (of iceberg type). Early, we find the following preliminary result: increasing transport cost has opposite impacts on the mass of firms and productivity. In the present paper, we study, mainly, the local comparative statics w.r.t. transport costs for the case of autarky (the very big transport costs). For the case of linear production costs, the known (and counter-intuitive!) result is that the social welfare increases near autarky. We generalize this result for the model with investments in R&D, this is the main result of the paper.

Keywords: Dixit-Stiglitz-Krugman model · Market equilibrium · Investments in R&D · Comparative statics · Social welfare · Autarky

1 Introduction

The monopolistic competition theory [16] began to develop rapidly after the famous works of Dixit and Stiglitz [18] and Krugman [20]. Usually, the study is in the framework of linear production costs. The more economically adequate case, when marginal costs decrease when fixed costs ("investments in R&D") increase, is studied not enough.[1]

In [14] we get the results for the model of a closed economy. In [11] we expand this analysis to the trade model. Usually, in monopolistic competition trade models, the study focuses in the comparative statics (with respect to transport costs) of equilibrium variables – individual consumption, the mass of firm, size of the firm, price, etc. The social welfare studies do not a lot.

In 2012, Arkolakis et al. (see [2]), by studying international trade in the USA, concluded: How large are the welfare gains from trade? A crude summary of our results is: "So far, not much."

Because of this famous work, there was a great interest in the theoretical study of the consequences of the "disappearance" of international trade.

---

[1] Some stylized facts for theory are, e.g., in [1, 15, 19, 25, 26].

© Springer Nature Switzerland AG 2021
P. Pardalos et al. (Eds.): MOTOR 2021, LNCS 12755, pp. 397–409, 2021.
https://doi.org/10.1007/978-3-030-77876-7_27

Some results about social welfare can be found in [3, 22, 23], where the linear productions costs are considered. As to non-linear production costs, the results are not strong (see, e.g., [6, 7]).

In [12] (and later in [24]), for the case of linear production costs, the following (counter-intuitive!) result has got: near autarky, the social welfare increases. The question arises: can investments in R&D help avoid this effect?

In this paper, we show that the answer is "no": even if the producers can choose the investments in R&D to decrease the marginal costs, near autarky, the social welfare increases.[2]

The paper is organized as follows. In Sect. 2, we set the model, the main notations, describe the equilibrium, and discuss the main equilibrium equations. In Sect. 3, we study the comparative statics with respect to transport costs in the situation of autarky. Here we formulate the main result (see Proposition 1): near autarky, the Social welfare increases. Besides, we study individual consumptions, size of the firms, mass of the firms, and prices (inverse demand), see Proposition 2. In Sect. 4 the reader can find proofs of Proposition 1 and Proposition 2. Section 5 concludes.

2   Model

The setting of the model is as in [11], but taking into account the symmetry of the countries.

The model of monopolistic competition [4, 5, 8–10, 13, 17, 18, 20, 21, 29] is based on the following assumptions:

– the manufacturers produce goods of the same nature, but not completely interchangeable (product diversity);
– each firm produces one type of product diversity and sets its price;
– the number (mass) of firms is large enough;
– the firms enter the market as long as their profits are positive.

In this paper, for simplicity, we consider the very stylized model of monopolistic competition with the trade of two symmetric (on population) countries. There are one industry and one production factor, interpreted as labor. We introduce the basic concepts and notation. Let

– L be the number of consumers in a country,
– N be the mass of firms in a country;
– $X_i = X(i)$ be the individual domestic consumption of the goods produced by the firm $i \in [0, N]$;
– Z_i be the individual foreign consumption of the goods produced by the firm $i \in [0, N]$;
– Q_i be the output (size) of the firm $i \in [0, N]$;
– ω be the wage in a country, normalized to 1.

---

[2] For simplicity, we study the case of two symmetric (on population) countries. Of course, the result can be developed for non-symmetric countries.

2.1   Consumers

We assume that each consumer share the same twice differentiable sub-utility function, such that

$$u(0) = 0, \qquad u'(\xi) > 0, \qquad u''(\xi) < 0. \tag{1}$$

Thus, function u is increasing and strictly concave.

The problem of a representative consumer in each country is

$$\int_0^N u(X_i)\, di + \int_0^N u(Z_i)\, di \rightarrow \max_{X_i \geq 0, i \in [0,N], Z_i \geq 0, i \in [0,N]}$$

subject to

$$\int_0^N p_i^X(X_i)\, di + \int_0^N p_i^Z Z_i\, di \leq w \equiv 1.$$

From the consumer's First Order Conditions (FOC), we get the inverse demand functions

$$p_i^X(X_i, \Lambda) = \frac{u'(X_i)}{\Lambda}, \qquad p_i^Z(Z_i, \Lambda) = \frac{u'(Z_i)}{\Lambda}, \qquad i \in [0,N]. \tag{2}$$

where Λ is the Lagrange multiplier of the problem of a representative consumer. Note that

$$\Lambda > 0 \tag{3}$$

due to (1), (3).

2.2   Producers

Let F be fixed costs (chosen endogenously); $c(F)$ be the corresponding marginal costs. We assume that $c'(\cdot) < 0$. Besides, let us assume, standardly, that the trade incurs some transport costs of "iceberg type"[3]. Then the sizes of the firms are

$$Q_i = LX_i + \tau LZ_i, \quad i \in [0,N]. \tag{4}$$

while the production costs are

$$V(Q_i, F) = c(F)Q_i + F, \quad i \in [0,N]. \tag{5}$$

Let

$$R(\xi) = u'(\xi) \cdot \xi \tag{6}$$

---

[3] To sell a unit, the firm produces $\tau \cdot 1$.

be the "normalized" revenue. Note that, due to (2), the normalized revenue equals the costs that one consumer spends on the purchase of products of one company (divided by the Lagrange multiplier). Moreover, $R'(\xi) = u'(\xi)(1 - r_u(\xi))$, where

$$r_g(\xi) = -\frac{g''(\xi)\xi}{g'(\xi)} = -\varepsilon_{g'}(\xi) \tag{7}$$

is Arrow-Pratt measure of function g while ε_h is the elasticity of function h:

$$\varepsilon_h(\xi) = \frac{h'(\xi)\xi}{h(\xi)}. \tag{8}$$

Using the inverse demand functions (2) and "normalized" revenue (6), the profit of firm i in each country can be written as

$$\Pi_i = L \cdot \frac{R(X_i)}{\Lambda} + L \cdot \frac{R(Z_i)}{\Lambda} - wV(Q_i, F), \quad i \in [0, N]. \tag{9}$$

Labor Balance. In each country, the labor balance ("total production costs equal total labor") is

$$\int_0^N V(Q_i, F)\, di = L. \tag{10}$$

2.3   Symmetric Case

In each country, all consumers are assumed identical. So we will consider the symmetric case, omitting index i. This way, the individual consumptions are

$$X_i = X, \quad Z_i = Z, \quad i \in [0, N].$$

Therefore, we rewrite the inverse demand functions (2) as

$$p(X, \Lambda) = \frac{u'(X)}{\Lambda}, \qquad p(Z, \Lambda) = \frac{u'(Z)}{\Lambda}, \tag{11}$$

sizes of the firms (4) as
$$Q = LX + \tau LZ, \tag{12}$$

production costs (5) as
$$V(Q, F) = c(F)Q + F, \tag{13}$$

profits (9) as
$$\Pi = L \cdot \frac{R(X)}{\Lambda} + L \cdot \frac{R(Z)}{\Lambda} - V(Q, F), \tag{14}$$

labor balance (10) as
$$N \cdot V(Q, F) = L. \tag{15}$$

2.4   Symmetric Equilibrium

Here, size of the firms are (12), the production costs are (13), the profits are (14).

As it is usual in monopolistic competition, we assume that firms enter the market while their profit remains positive, which implies zero-profit (free-entry) conditions. In the symmetric case, we get

$$\Pi = 0. \tag{16}$$

Each firm maximizes its profits:

$$\Pi \to \max_{X,Z,F \geq 0}.$$

Thus, the Producer's First order conditions (FOC) in the symmetric case are

$$\frac{\partial \Pi}{\partial X} = 0, \qquad \frac{\partial \Pi}{\partial Z} = 0, \qquad \frac{\partial \Pi}{\partial F} = 0. \tag{17}$$

Second order conditions (SOC) are that the matrix

$$\Pi'' = \begin{pmatrix} \dfrac{\partial^2 \Pi}{\partial X^2} & 0 & \dfrac{\partial^2 \Pi}{\partial X \partial F} \\[2mm] 0 & \dfrac{\partial^2 \Pi}{\partial Z^2} & \dfrac{\partial^2 \Pi}{\partial Z \partial F} \\[2mm] \dfrac{\partial^2 \Pi}{\partial X \partial F} & \dfrac{\partial^2 \Pi}{\partial Z \partial F} & \dfrac{\partial^2 \Pi}{\partial F^2} \end{pmatrix}.$$

is negatively defined. In particular,

$$\frac{\partial^2 \Pi}{\partial X^2} < 0, \qquad \frac{\partial^2 \Pi}{\partial Z^2} < 0, \qquad \frac{\partial^2 \Pi}{\partial F^2} < 0 \tag{18}$$

and

$$\frac{\partial^2 \Pi}{\partial X^2} \cdot \frac{\partial^2 \Pi}{\partial Z^2} \cdot \frac{\partial^2 \Pi}{\partial F^2} - \frac{\partial^2 \Pi}{\partial X^2} \cdot \left(\frac{\partial^2 \Pi}{\partial Z \partial F}\right)^2 - \frac{\partial^2 \Pi}{\partial Z^2} \cdot \left(\frac{\partial^2 \Pi}{\partial X \partial F}\right)^2 < 0. \tag{19}$$

Symmetric equilibrium is a bundle

$$(X^*, Z^*, F^*, N^*, \Lambda^*,)$$

satisfying the following:

- profit maximization (17) – (19);
- free entry conditions (16);
- labor balance (15).

Note that the equilibrium prices (inverse demand functions)

$$p(X^*, \Lambda^*), \quad p(Z^*, \Lambda^*)$$

can be obtained from (11). Moreover, First Order Conditions (17) are

$$\frac{R'(X)}{\Lambda} = c(F), \qquad \frac{R'(Z)}{\Lambda} = \tau c(F), \qquad c'(F) q^H = -1. \qquad (20)$$

Due to (3), Second Order Conditions (18) are

$$R''(X) \equiv u''(X)(2 - r_{u'}(X)) < 0, R''(Z) \equiv u''(Z)(2 - r_{u'}(Z)) < 0, c''(F) > 0, \qquad (21)$$

while Second Order Condition (19) is

$$\frac{R''(X)}{R'(X)} \cdot \tau + \frac{R''(Z)}{R'(Z)} - \frac{R''(X)}{R'(X)} \cdot \frac{R''(Z)}{R'(Z)} \cdot \frac{r_c(F)}{\mathcal{E}_c(F)} \cdot \frac{Q}{L} > 0. \qquad (22)$$

3   The Local Comparative Statics w.r.t. τ

The system of equilibrium equations is

$$\frac{\partial \Pi}{\partial X} = 0$$

$$\frac{\partial \Pi}{\partial Z} = 0$$

$$\frac{\partial \Pi}{\partial F} = 0$$

$$\Pi = 0$$

Thus, the system of the local comparative statics w.r.t. τ is

$$\frac{d}{d\tau}\left(\frac{\partial \Pi}{\partial X}\right) = 0$$

$$\frac{d}{d\tau}\left(\frac{\partial \Pi}{\partial Z}\right) = 0$$

$$\frac{d}{d\tau}\left(\frac{\partial \Pi}{\partial F}\right) = 0$$

$$\frac{d\Pi}{d\tau} = 0$$

i.e., due to (20),

$$\frac{R''(X)}{R'(X)} \cdot \frac{dX}{d\tau} - \frac{c'(F)}{c(F)} \cdot \frac{dF}{d\tau} - \frac{1}{\Lambda} \cdot \frac{d\Lambda}{d\tau} = 0$$

$$\frac{R''(Z)}{R'(Z)} \cdot \tau \cdot \frac{dZ}{d\tau} - \frac{c'(F)}{c(F)} \cdot \tau \cdot \frac{dF}{d\tau} - \frac{\tau}{\Lambda} \cdot \frac{d\Lambda}{d\tau} = 1$$

$$-\frac{dX}{d\tau} - \tau \cdot \frac{dZ}{d\tau} - \frac{c''(F)}{c'(F)} \cdot \frac{Q}{L} \cdot \frac{dF}{d\tau} = Z$$

$$\frac{1}{\Lambda} \cdot \frac{d\Lambda}{d\tau} = -\frac{c(F)}{c(F)Q + F} \cdot L \cdot Z$$

In what follows, we will consider not only elasticity of function with respect to a variable (see (7) and (8)), but also elasticity of variable ξ with respect to parameter τ, i.e.,

$$E_\xi = E_{\xi/\tau} = \frac{d\xi}{d\tau} \cdot \frac{\tau}{\xi}.$$

Thus, the system of the local comparative statics w.r.t. τ is

$$\frac{dX}{d\tau} = \frac{1}{\tau} \cdot \frac{R'(X)}{R''(X)} \cdot (\mathcal{E}_c(F) \cdot E_F + E_\Lambda) \tag{23}$$

$$\tau \cdot \frac{dZ}{d\tau} = \frac{R'(Z)}{R''(Z)} \cdot (\mathcal{E}_c(F) \cdot E_F + E_\Lambda + 1) \tag{24}$$

$$\frac{dX}{d\tau} + \tau \cdot \frac{dZ}{d\tau} = r_c(F) \cdot \frac{Q}{L} \cdot \frac{1}{\tau} \cdot E_F - Z \tag{25}$$

$$E_\Lambda = -\frac{c(F)}{c(F)Q + F} \cdot L \cdot \tau \cdot Z \tag{26}$$

As to Social Welfare

$$W = N \cdot (u(X) + u(Z)),$$

we get due to Labor Balance (15)

$$W = L \cdot \frac{(u(X) + u(Z))}{c(F) \cdot Q + F}.$$

Hence

$$\frac{dW}{d\tau} = -N \cdot \frac{W}{L} \cdot \left(c'(F) \cdot Q \cdot \frac{dF}{d\tau} + c(F) \cdot L \cdot \left(\frac{dX}{d\tau} + \tau \cdot \frac{dZ}{d\tau} \right) + \frac{dF}{d\tau} \right)$$

$$+ N \cdot \left(u'(X) \cdot \frac{dX}{d\tau} + u'(Z) \cdot \frac{dZ}{d\tau} \right) =$$

(since $c'(F) \cdot Q + 1 = 0$ due to (20))

$$= -N \cdot \left(W \cdot c(F) \cdot \left(\frac{dX}{d\tau} + \tau \cdot \frac{dZ}{d\tau} \right) + u'(X) \cdot \frac{dX}{d\tau} + u'(Z) \cdot \frac{dZ}{d\tau} \right).$$

Thus, due to (25),

$$\frac{dW}{d\tau} = -N \cdot \left(W \cdot c(F) \cdot \left(r_c(F) \cdot \frac{Q}{L} \cdot \frac{E_F}{\tau} - Z \right) + u'(X) \cdot \frac{dX}{d\tau} + u'(Z) \cdot \frac{dZ}{d\tau} \right). \tag{27}$$

3.1   Autarky

In autarky, τ is such that $Z^* = 0$.

Note that the questions of the existence and uniqueness of equilibrium and optimality are separate problems (often not quite simple), which is not the subject of this study. Moreover, we do not study the existence of autarky. Note that if $u(\cdot)$ is so-called CES-function

$$u(\xi) = \xi^\rho, \rho \in (0,1),$$

the autarky is not possible. So we consider non-CES sub-utility. More precisely, we consider so-called "pro-competitive" sub-utilities, i.e., such that Arrow-Pratt measure r_u is increasing function. Moreover, in addition to (1), in what follows we will assume that

$$u'(0) < \infty, \qquad u''(0) > -\infty, \qquad -\infty < u'''(0) < \infty. \tag{28}$$

Obviously, conditions (28) not valid for CES-function.

Thus, $Q = L \cdot X$, moreover, (23)–(26) are

$$\frac{dX}{d\tau} = \frac{1}{\tau} \cdot \frac{R'(X)}{R''(X)} \cdot \mathcal{E}_c(F) \cdot E_F \tag{29}$$

$$\tau \cdot \frac{dZ}{d\tau} = \frac{R'(0)}{R''(0)} \cdot (\mathcal{E}_c(F) \cdot E_F + 1) \tag{30}$$

$$\left(\left(\frac{1}{\tau} \cdot \frac{R'(X)}{R''(X)} + \frac{R'(0)}{R''(0)} \right) \cdot \mathcal{E}_c(F) - r_c(F) \cdot \frac{X}{\tau} \right) \cdot E_F = -\frac{R'(0)}{R''(0)} \tag{31}$$

$$E_\Lambda = 0 \tag{32}$$

Note that, due to (28),

$$R'(0) = u'(0) > 0$$

and

$$R''(0) = 2u''(0) < 0.$$

Moreover, (22) is

$$\frac{R''(X)}{R'(X)} \cdot \tau + \frac{R''(0)}{R'(0)} - \frac{R''(X)}{R'(X)} \cdot \frac{R''(0)}{R'(0)} \cdot \frac{r_c(F)}{\mathcal{E}_c(F)} \cdot X > 0,$$

i.e. (let us recall that $\mathcal{E}_c(F) < 0$),

$$\left(\frac{1}{\tau} \cdot \frac{R'(X)}{R''(X)} + \frac{R'(0)}{R''(0)} \right) \cdot \mathcal{E}_c(F) - r_c(F) \cdot \frac{X}{\tau} < 0.$$

Hence

$$E_F = -\frac{R'(0)}{R''(0)} \cdot \frac{1}{\left(\frac{1}{\tau} \cdot \frac{R'(X)}{R''(X)} + \frac{R'(0)}{R''(0)} \right) \cdot \mathcal{E}_c(F) - r_c(F) \cdot \frac{X}{\tau}} < 0. \tag{33}$$

Now, let us formulate the main result of the paper.

Proposition 1. *Near autarky, the Social welfare increases.*

Moreover, (33) admits us to understand the behavior of the individual consumptions, the size of the firms, the mass of the firms, and prices (inverse demand).

Proposition 2. *Near autarky,*

$$\frac{dX}{d\tau} < 0, \qquad \frac{dZ}{d\tau} < 0, \qquad \frac{dQ}{d\tau} < 0, \qquad \frac{dN}{d\tau} > 0,$$

$$\frac{d}{d\tau}\left(p\left(X, \Lambda\right)\right) > 0, \qquad \frac{d}{d\tau}\left(p\left(Z, \Lambda\right)\right) > 0.$$

4 Proofs

4.1 Proof of Proposition 1

Due to (27),

$$\frac{dW}{d\tau} = -N \cdot \left(W \cdot c\left(F\right) \cdot r_c\left(F\right) \cdot \frac{X}{\tau} \cdot E_F + u'\left(X\right) \cdot \frac{dX}{d\tau} + u'\left(0\right) \cdot \frac{dZ}{d\tau}\right). \tag{34}$$

Note that

$$u'\left(X\right) \cdot \frac{dX}{d\tau} + u'\left(0\right) \cdot \frac{dZ}{d\tau}$$

$$= \frac{u'\left(X\right)}{\tau} \cdot \frac{R'\left(X\right)}{R''\left(X\right)} \cdot \mathcal{E}_c\left(F\right) \cdot E_F + \frac{u'\left(0\right)}{\tau} \cdot \frac{R'\left(0\right)}{R''\left(0\right)} \cdot \left(\mathcal{E}_c\left(F\right) \cdot E_F + 1\right)$$

$$= \left(\frac{u'\left(X\right)}{\tau} \cdot \frac{R'\left(X\right)}{R''\left(X\right)} + \frac{u'\left(0\right)}{\tau} \cdot \frac{R'\left(0\right)}{R''\left(0\right)}\right) \cdot \mathcal{E}_c\left(F\right) \cdot E_F + \frac{u'\left(0\right)}{\tau} \cdot \frac{R'\left(0\right)}{R''\left(0\right)}$$

(due to (31))

$$= \left(\frac{u'\left(X\right)}{\tau} \cdot \frac{R'\left(X\right)}{R''\left(X\right)} + \frac{u'\left(0\right)}{\tau} \cdot \frac{R'\left(0\right)}{R''\left(0\right)}\right) \cdot \mathcal{E}_c\left(F\right) \cdot E_F$$

$$- \frac{u'\left(0\right)}{\tau} \cdot \left(\left(\frac{1}{\tau} \cdot \frac{R'\left(X\right)}{R''\left(X\right)} + \frac{R'\left(0\right)}{R''\left(0\right)}\right) \cdot \mathcal{E}_c\left(F\right) - r_c\left(F\right) \cdot \frac{X}{\tau}\right) \cdot E_F$$

$$= \left(\left(\frac{u'\left(X\right)}{\tau} \cdot \frac{R'\left(X\right)}{R''\left(X\right)} - \frac{u'\left(0\right)}{\tau^2} \cdot \frac{R'\left(X\right)}{R''\left(X\right)}\right) \cdot \mathcal{E}_c\left(F\right) + \frac{u'\left(0\right)}{\tau} \cdot r_c\left(F\right) \cdot \frac{X}{\tau}\right) \cdot E_F$$

$$= \frac{1}{\tau} \cdot \left(\frac{R'\left(X\right)}{R''\left(X\right)} \cdot \left(u'\left(X\right) - \frac{R'\left(0\right)}{\tau}\right) \cdot \mathcal{E}_c\left(F\right) + \frac{R'\left(0\right)}{\tau} \cdot r_c\left(F\right) \cdot X\right) \cdot E_F.$$

Thus, since

$$R'\left(0\right) = R'\left(X\right) \cdot \tau$$

due to (20),

$$u'\left(X\right) \cdot \frac{dX}{d\tau} + u'\left(0\right) \cdot \frac{dZ}{d\tau}$$

$$= \frac{1}{\tau} \cdot \left(\frac{R'(X)}{R''(X)} \cdot (u'(X) - R'(X)) \cdot \mathcal{E}_c(F) + R'(X) \cdot r_c(F) \cdot X \right) \cdot E_F.$$

Hence

$$\frac{dW}{d\tau} = -N \cdot \left(W \cdot c(F) \cdot r_c(F) \cdot \frac{X}{\tau} \cdot E_F + u'(X) \cdot \frac{dX}{d\tau} + u'(0) \cdot \frac{dZ}{d\tau} \right)$$

$$= -N \cdot \left(W \cdot c(F) \cdot r_c(F) \cdot \frac{X}{\tau} \right) \cdot E_F$$

$$- N \cdot \frac{1}{\tau} \cdot \left(\frac{R'(X)}{R''(X)} \cdot (u'(X) - R'(X)) \cdot \mathcal{E}_c(F) + R'(X) \cdot r_c(F) \cdot X \right) \cdot E_F$$

$$= -\frac{N}{\tau} \cdot \left(\frac{R'(X)}{R''(X)} \cdot (u'(X) - R'(X)) \cdot \mathcal{E}_c(F) \right) \cdot E_F$$

$$- \frac{N}{\tau} \cdot ((W \cdot c(F) \cdot X + R'(X)) \cdot r_c(F) \cdot X) \cdot E_F$$

$$= -\frac{N}{\tau} \cdot \left(-R(X) \cdot \mathcal{E}_c(F) + \left(\frac{c(F) \cdot L \cdot X}{V(Q,F)} \cdot u(X) + R'(X) \right) \cdot r_c(F) \cdot X \right) \cdot E_F.$$

Note that

$$\frac{c(F) \cdot L \cdot X}{V(Q,F)} = \frac{R'(X) \cdot X}{R(X)} = \mathcal{E}_R(X)$$

due to (16) and (20). Hence

$$\frac{dW}{d\tau} = -\frac{N}{\tau} \cdot (-R(X) \cdot \mathcal{E}_c(F) + (\mathcal{E}_R(X) \cdot u(X) + R'(X)) \cdot r_c(F) \cdot X) \cdot E_F$$

$$= -\frac{N}{\tau} \cdot (-R(X) \cdot \mathcal{E}_c(F) + \mathcal{E}_R(X) \cdot (u(X) + u'(X)) \cdot r_c(F) \cdot X) \cdot E_F > 0.$$

4.2   Proof of Proposition 2

Due to (29), (30), (21), and (33),

$$\frac{dX}{d\tau} < 0, \qquad \frac{dZ}{d\tau} < 0.$$

Further,

$$\frac{dQ}{d\tau} = L \cdot \left(\frac{dX}{d\tau} + \tau \cdot \frac{dZ}{d\tau} + Z \right) = L \cdot \left(\frac{dX}{d\tau} + \tau \cdot \frac{dZ}{d\tau} \right) < 0.$$

Further,

$$\frac{dN}{d\tau} = \frac{d}{d\tau} \left(\frac{L}{V(Q,F)} \right)$$

$$= -\frac{L}{(V(Q,F))^2} \cdot \left(\frac{\partial V(Q,F)}{\partial Q} \cdot \frac{dQ}{d\tau} + \frac{\partial V(Q,F)}{\partial F} \cdot \frac{dF}{d\tau} \right)$$

$$= -\frac{L}{\left(V\left(Q,F\right)\right)^2} \cdot \left(c\left(F\right) \cdot \frac{dQ}{d\tau} + \left(c'\left(F\right) \cdot Q + 1\right) \cdot \frac{dF}{d\tau}\right)$$

$$= -\frac{L \cdot c\left(F\right)}{\left(V\left(Q,F\right)\right)^2} \cdot \frac{dQ}{d\tau} > 0.$$

Further,

$$E_{p(X,\Lambda)} = E_{\frac{u'(X)}{\Lambda}} = E_{u'(X)} - E_\Lambda = \mathcal{E}_{u'}\left(X\right) \cdot E_X - E_\Lambda = -r_u\left(X\right) \cdot E_X > 0.$$

Finally,

$$\frac{d}{d\tau}\left(p\left(Z,\Lambda\right)\right) = \frac{d}{d\tau}\left(\frac{u'\left(Z\right)}{\Lambda}\right) = -\frac{u'\left(Z\right) \cdot \Lambda \cdot \dfrac{dZ}{d\tau} - u'\left(Z\right) \cdot \dfrac{d\Lambda}{d\tau}}{\Lambda^2}$$

$$= -\frac{u'\left(Z\right)}{\Lambda} \cdot \frac{dZ}{d\tau} = -p\left(Z,\Lambda\right) \cdot \frac{dZ}{d\tau} > 0.$$

5   Conclusion

In this paper, we continue to examine the impact of R&D investment on market equilibrium in the Dixit–Stiglitz–Krugman model.

In contrast to previous work, here we focus on the behavior of social welfare near autarky when transport costs are too high. We show that, near autarky, social welfare increases with respect to transport costs. This result seems counter-intuitive, it was previously known only for the linear production cost case.

For simplicity, we study the case of two symmetric (on population) countries. Note that the result can be developed for non-symmetric countries But in this case, it is necessary to use the addition condition – so-called "trade balance" (cf. [12]). We plan to study the "asymmetric" case in the near future.

Moreover, it seems interesting to generalize the result for the trade with retailing (cf. [27,28]).

For policy-making, our topic may be interesting because of a new understanding of gains from trade: technological changes in response to trade liberalization. Furthermore, for modernization and active industrial policy practiced in some countries it can be interesting, which equilibrium outcome in various sectors may follow from some stimulating measures like tax reductions conditional on R&D.

Acknowledgments. The study was carried out within the framework of the state contract of the Sobolev Institute of Mathematics (project no. 0314-2019-0018). The work was supported in part by the Russian Foundation for Basic Research (project no. 19-010-00910).

References

1. Aw, B.Y., Roberts, M.J., Xu, D.Y.: R&D investments, exporting, and the evolution of firm productivity. Am. Econ. Rev. Papers Proc. **98**(2), 451–456 (2008)
2. Arkolakis, C., Costinot, A., Rodríguez-Clare, A.: New trade models, same old gains? Am. Econ. Rev. **102**(1), 94–130 (2012)
3. Arkolakis, C., Costinot, A., Donaldson, D., Rodríguez-Clare, A.: The elusive pro-competitive effects of trade. Rev. Econ. Stud. **86**(1), 46–80 (2019)
4. Baldwin, R.E., Forslid, R.: Trade liberalization with heterogeneous firms. Rev. Dev. Econ. **14**(2), 161–176 (2010)
5. Behrens, K., Murata, Y.: General equilibrium models of monopolistic competition: a new approach. J. Econ. Theor **136**(1), 776–787 (2007)
6. Belyaev, I., Bykadorov, I.: International trade models in monopolistic competition: the case of non-linear costs. In: IEEE Xplore, pp. 12–16 (2019)
7. Belyaev, I., Bykadorov, I.: Dixit-Stiglitz-Krugman Model with Nonlinear Costs. Lect. Notes Comput. Sci. **12095**, 157–169 (2020)
8. Bykadorov, I.: Monopolistic competition model with different technological innovation and consumer utility levels. CEUR Workshop Proc. **1987**, 108–114 (2017)
9. Bykadorov, I.: Monopolistic competition with investments in productivity. Optim. Lett. **13**(8), 1803–1817 (2018). https://doi.org/10.1007/s11590-018-1336-9
10. Bykadorov, I.: Social optimality in international trade under monopolistic competition. Commun. Comput. Inf. Sci. **1090**, 163–177 (2019)
11. Bykadorov, I.: Investments in R&D in monopolistic competitive trade model. Lecture Notes Comput. Sci. **12095**, 170–183 (2020)
12. Bykadorov, I., Ellero, A., Funari, S., Kokovin, S., Molchanov, P.: Painful Birth of Trade under Classical Monopolistic Competition, National Research University Higher School of Economics, Basic Research Program Working Papers, Series: Economics, WP BRP 132/EC/2016. https://doi.org/10.2139/ssrn.2759872
13. Bykadorov, I., Gorn, A., Kokovin, S., Zhelobodko, E.: Why are losses from trade unlikely? Econ. Lett. **129**, 35–38 (2015)
14. Bykadorov, I., Kokovin, S.: Can a larger market foster R&D under monopolistic competition with variable mark-ups? Res. Econ. **71**(4), 663–674 (2017)
15. Campbell, J.R., Hopenhayn, H.A.: Market size matters. J. Ind. Econ. **53**(1), 1–25 (2005)
16. Chamberlin, E.H.: The Theory of Monopolistic Competition: A Re-Orientation of the Theory of Value. Harvard University Press, Cambridge (1933)
17. Dhingra, S., Morrow, J.: Monopolistic competition and optimum product diversity under firm heterogeneity. J. Polit. Econ. **127**(1), 196–232 (2019)
18. Dixit, A., Stiglitz, J.: Monopolistic competition and optimum product diversity. Am. Econ. Rev. **67**(3), 297–308 (1977)
19. Hummels, D., Klenow, P.T.: The variety and quality of a nation's exports. Am. Econ. Rev. **95**(3), 704–723 (2005)
20. Krugman, P.R.: Increasing returns, monopolistic competition, and international trade. J. Int. Econ. **9**(4), 469–479 (1979)
21. Melitz, M.J.: The impact of trade on intra-industry reallocations and aggregate industry productivity. Econometrica **71**(6), 1695–1725 (2003)
22. Melitz, M.J., Redding, S.J.: Missing gains from trade? Am. Econ. Rev. **104**(5), 317–321 (2014)
23. Melitz, M.J., Redding, S.J.: New trade models. New welfare implications. Am. Econ. Rev. **105**(3), 1105–1146 (2015)

24. Morgan, J., Tumlinson, J., Vardy, F.: Bad Trade: The Loss of Variety. SSRN WP (2020). https://doi.org/10.2139/ssrn3529246
25. Redding, S.: Theories of heterogeneous firms and trade. Ann. Rev. Econ. **3**, 1–24 (2011)
26. Syverson, C.: Prices, spatial competition, and heterogeneous producers: an empirical test. J. Ind. Econ. **55**(2), 197–222 (2007)
27. Tilzo, O., Bykadorov, I.: Retailing under monopolistic competition: a comparative analysis. In: IEEE Xplore, pp. 156–161 (2019)
28. Tilzo, O., Bykadorov, I.: Monopolistic competition model with retailing. Commun. Comput. Inf. Sci. **1275**, 287–301 (2020)
29. Zhelobodko, E., Kokovin, S., Parenti, M., Thisse, J.-F.: Monopolistic competition in general equilibrium: beyond the constant elasticity of substitution. Econometrica **80**(6), 2765–2784 (2012)

On Contractual Approach for Non-convex Production Economies

Valeriy Marakulin[✉] [iD]

Sobolev Institute of Mathematics, Russian Academy of Sciences,
4 Acad. Koptyug Avenue, Novosibirsk 630090, Russia
http://www.math.nsc.ru/~mathecon/marakENG.html

Abstract. The paper investigates a contractual approach and studies economies with non-convex production. The contractual theory developed in [15,16] for exchange and in [19] for production economies is modified and adapted to the models with non-convex and non-smooth production sector. We clarify an appropriate notion of the web of contracts, their dominance by coalitions, the partial break of contracts, etc. A generalized notion of marginal contractual allocation (called K-marginal fuzzy contractual) is introduced and used in equilibrium analysis together with marginal cost pricing (MCP-equilibrium) that is applied in literature instead of Walrasian equilibrium for production economies with increasing returns to scale. We analyze marginal pricing rules that can be specified as Clark's derivative and similar ones. The equivalence between MCP-equilibria and K-marginal fuzzy contractual allocations (K is a convex cone) presents the main result. It can be viewed as a theoretical substantiation of the concept of MCP-equilibrium in non-convex economies. The work develops a contractual approach as a universal way for perfect competition modeling.

Keywords: Contractual production economies · Non-convex technologies · Marginal cost pricing · MCP-equilibrium · Clark's derivative

1 Introduction

One of the major objectives of the economic theory and its basic part—General Equilibrium Theory—consists of the description of resource allocation implemented via the system of the markets. In the classical Arrow–Debreu model resulting allocation arrives as Walrasian (competitive) equilibrium that is the basic object of the theoretical analysis, see [1,2,14,20], etc. Arrow–Debreu economies were developed and generalized in different directions, one of them being the study of models with non-convex technological sets. Structurally they are Arrow–Debreu models for which increasing returns from the scale in production is possible; the

The study was carried out within the framework of the state contract of the Sobolev Institute of Mathematics (project no. 0314-2019-0018).

© Springer Nature Switzerland AG 2021
P. Pardalos et al. (Eds.): MOTOR 2021, LNCS 12755, pp. 410–429, 2021.
https://doi.org/10.1007/978-3-030-77876-7_28

convexity of technological sets (and sets of the preferred consumption bundles) is a very important assumption, otherwise, equilibria may not exist. However, non-convexity in technologies is a characteristic property for many modern industrial spheres (for example, for private municipal enterprises). Therefore, the case of non-convex technological sets is a very important theoretical problem.

The non-convexity in technologies leads to the known concept of equilibrium with pricing by marginal costs, so-called MCP-equilibrium. For the first time, the existence of MCP-equilibrium for a monopolistic economy with one firm has been established in [13]. Further, in [3] the existence of an equilibrium with marginal costs pricing has been proved for several firms with non-convex technologies, most general results have been obtained in [4,5]; a survey of literature one can find in [6]. Notice that equilibrium with marginal costs pricing implements only necessary conditions for Pareto optimality of current (equilibrium) production allocation. In general, MCP-equilibrium may not be Pareto optimal; however, these conditions are sufficient in a convex case and correspond with the profit maximization of producers.

Our analysis is based on a contractual approach extended to the production model in an appropriate way. In the consumption sector, we consider barter contracts delivering commodities for exchange, but for production contracts, agents have taken material expenses related to the production of goods. The collections of contracts can be transformed via concluding new mutually beneficial barter contracts and breaking (possible partial) existing ones, production plans can now also be changed. For non-convex technologies, a transformation of the production program is possible only within the limits of the specific marginal convex cone. Stable collections (webs) of contracts are the subject of the study, these webs allow to characterize Walrasian equilibrium for the convex economy and with MCP-pricing (marginal cost pricing) equilibrium for non-convex one. A specific property of the contractual approach is that all processes of production and exchange are going without any kind of value parameters.

In the next section, we describe production economies, study the Pareto frontier in non-convex and non-differential case, and present crucial necessary conditions in terms of tangent cones. In the third section, we introduce specific generalized MCP-equilibrium and study contractual approach for non-convex case: the model is presented here and the equivalence theorem is proved.

2   Non-convex Technologies in Arrow–Debreu Economies

2.1   Production Economies

The formal classical Arrow–Debreu economy in its shortest form is presented by the following bundle of parameters:

$$\mathcal{E}^{AD} = \langle \mathcal{I}, \mathcal{J}, \mathbb{R}^l, \{X_i, \mathcal{P}_i(\cdot), \mathbf{e}_i, \{\theta_i\}\}_{i \in \mathcal{I}}, \{Y_j\}_{j \in \mathcal{J}} \rangle. \tag{1}$$

Here $\mathcal{I} = \{1, \ldots, n\}$ is the set of consumers, $\mathcal{J} = \{1, \ldots, m\}$ is the set of producers (firms), l is a number of commodities and $\mathbb{R}^l = L$ is the commodity space.

Consumption sets are denoted as $X_i \subset \mathbb{R}^l$ and $X = \prod_{i \in \mathcal{I}} X_i$; agents' preferences are presented by point-to-set mappings $\mathcal{P}_i : X_i \Rightarrow X_i$, $i \in \mathcal{I}$ where $\mathcal{P}_i(x_i) = \{y_i \in X_i \mid y_i \succ_i x_i\}$ is a set of all consumption bundles strictly preferred by the i-th agent to the bundle x_i. It is also applied the notation $y_i \succ_i x_i$ which is equivalent to $y_i \in \mathcal{P}_i(x_i)$. Consumers have also initial endowments $\mathbf{e}_i \in X_i$, $i \in \mathcal{I}$. Determine $\mathbf{e} = (\mathbf{e}_1, \dots, \mathbf{e}_n)$. A producer $j \in \mathcal{J}$ is described by a technological set $Y_j \subset L$, $Y = \prod_{j \in \mathcal{J}} Y_j$, defined in terms of material flows, i.e., a non-negative component of $y_j \in Y_j$ is an output but if it is negative then it is an input of commodity in the units of counting. Now for a vector of prices $p = (p_1, \dots, p_l) \in L' = \mathbb{R}^l$ profit $\pi_j(p, y_j)$ for a plan $y_j \in Y_j$ can be calculated in the form of inner product $\pi_j(p, y_j) = \langle p, y_j \rangle = p \cdot y_j$. There are also nm scalar values $\theta_i^j \geq 0$ they being the components of vectors $\theta_i = (\theta_i^1, \dots, \theta_i^m)$ present the shares of i in the profits π_j of producers $j \in \mathcal{J}$; by the definition $\sum_{i=1}^n \theta_i = (1, \dots, 1)$. Further, I recall the definition of **competitive (Walrasian)** equilibrium.

Definition 1. *A triplet (x, y, p), where $x = (x_i)_{i \in \mathcal{I}} \in X$ is a family of consumption plans, $y = (y_j)_{j \in \mathcal{J}} \in Y$ are production plans and $p = (p_1, \dots, p_l) \neq 0$, $p \in L'$ is a price vector is said to be **quasi-equilibrium**, if* [1]:

$$p \cdot y_j \geq \langle p, Y_j \rangle \; \forall j \in \mathcal{J}, \tag{2}$$

$$\langle p, \mathcal{P}_i(x_i) \rangle \geq p \cdot \mathbf{e}_i + \sum_{j=1}^m \theta_i^j p \cdot y_j = p \cdot x_i \; \forall i \in \mathcal{I}, \tag{3}$$

$$\sum_{i=1}^n x_i = \sum_{i=1}^n \mathbf{e}_i + \sum_{j=1}^m y_j. \tag{4}$$

*If all inequalities in (3) have strict sign then the triplet (x, y, p) is called **competitive (Walrasian)** equilibrium.*

Requirements (2)–(4) have a familiar economic sense. If inequality (3) has strict form it means that consumption plan x_i is an optimal choice (demand) for individual i under his/her budget constraint $p \cdot z_i \leq p \cdot \mathbf{e}_i + \sum_{j=1}^m \theta_i^j p \cdot y_j = r_i(p, y)$, $z_i \in X_i$, where the right-hand side presents the total agent's income from all channels under prices $p = (p_1, \dots, p_l) \in L'$. Condition (2) says that producers maximize profit and (4) is a material balance condition, that usually is presented as the equality of demand and supply.

Conditions guaranteeing the existence of equilibria in Arrow–Debreu model are well known in the literature, e.g. see [1,17,20]. In the consumption sector, they are the continuity (different versions are applied), open-convex values, irreflexivity, and local non-satiation of agents' preferences $\mathcal{P}_i : X_i \Rightarrow X_i \; \forall i \in \mathcal{I}$. Consumption sets have to be *convex* and closed; moreover, they have to provide a bounded (compact) set of all feasible allocations. These requirements are sufficient for quasi-equilibria or more refined notion of equilibria with non-standard prices do exist in exchange economy, see [11,17]. For strict equilibria to exist one

---

[1] $\langle A, B \rangle = \{\langle a, b \rangle = a \cdot b \mid a \in A, \; b \in B\}$ for all $A, B \subset L$; $A \geq b \iff a \geq b \; \forall a \in A$.

needs to require additional survival assumptions: it may be resource relatedness, irreducibility, or something like this one. Further, let us turn to the production sector and consider it in more detail.

For the production sector, it is usually assumed that for all $j \in \mathcal{J}$ technological sets Y_j have the following properties:

- Y_j—*convex*, closed sets (i.e., limit and mixed technological processes are permissible),
- $Y_j - \mathbb{R}_+^l \subset Y_j$—free disposal condition,
- $Y_j \cap \mathbb{R}_+^l = \{0\}$—no free lunch, where $\mathbb{R}_+^l$ is a positive orthant of commodity space,
- $Y \cap (-Y) = \{0\}$—the irreversibility of production processes.

In spite of the latter three requirements have an own economic sense, they are really needed to provide, together with consumption sets properties (boundedness from below), that the set of all feasible (balanced) allocations is bounded one. Nowadays one can often meet a direct requirement for the feasible allocation set to be bounded. The first assumption is for us now the most of interest and as a part of it the convexity of production sets. Without this requirement, equilibria may not exist.

2.2   Pareto Frontier and Tangent Cones

We start our analysis from the characterization of Pareto optimality. Everywhere below we will suppose

(A) *For each $i \in \mathcal{I}$, X_i is a convex solid[2] closed set, $\mathbf{e}_i \in X_i$, and for every $x_i \in X_i$ there exists an open convex $G_i \subset L$ such that $\mathcal{P}_i(x_i) = G_i \cap X_i$ and if $\mathcal{P}_i(x_i) \neq \emptyset$ (non-satiated preferences) then $x_i \in \mathrm{cl}\,\mathcal{P}_i(x_i) \backslash \mathcal{P}_i(x_i)$.*[3]

Let us specify the set of all feasible allocations in model $\mathcal{E}$

$$\mathcal{A}(\mathcal{E}) = \{z = (x, y) \in X \times Y \mid \sum_{i \in \mathcal{I}} x_i = \sum_{i \in \mathcal{I}} \mathbf{e}_i + \sum_{j \in \mathcal{J}} y_j\}.$$

Now I recall a standard definition of Pareto optimality.

- *A feasible allocation $(x, y) \in \mathcal{A}(\mathcal{E})$ is said to be (weakly) Pareto optimal if there is no a family $((x'_i)_{i' \in \mathcal{I}}, (y'_j)_{j \in \mathcal{J}}) \in \mathcal{A}(\mathcal{E})$ such that $x'_i \succ_i x_i$ for all $i \in \mathcal{I}$.*

In our analysis we shall apply several notions of tangent cones, see [7, 8, 21]. Now first I recall some Minkowski algebraic operations. The product by the number, the sum and the difference of Minkowski are determined by the formulas: if $A, B \subset L$, $\lambda \in \mathbb{R}$ then

$$\lambda A = \{\lambda a \mid a \in A\}, \quad A + b = b + A = \{a + b \mid a \in A\}, \ b \in L,$$

$$A + B = \{a + b \mid a \in A, \ b \in B\}, \quad A \overset{*}{-} B = \{a \in L \mid a + B \subseteq A\},$$

$$\varrho(b, A) = \inf\{\|b - a\| \mid a \in A\} - \text{ is distance from a point } b \in L \text{ to the set } A.$$

---

[2] Here "solid" is equivalent to "having a nonempty interior.".

[3] The symbol $\mathrm{cl}\,A$ denotes the closure of A and $\backslash$ is set for the set-theoretical difference. The requirement means the preferences are locally non-satiated.

Let $\Omega \subset \mathbb{R}^l$ be a set and $a \in \mathrm{cl}(\Omega)$. A vector g is called an *admissible* direction of the set Ω at a point a if there is a number $\alpha_g > 0$ such that

$$a + \alpha g \in \Omega, \quad \alpha \in (0, \alpha_g).$$

The set $T_A(\Omega, a)$ of all *admissible* directions is called *adjacent* tangent cone to $\Omega \subset \mathbb{R}^l$ at $a \in \mathrm{cl}(\Omega)$. It can be shortly presented as

$$T_A(\Omega, a) = \lim_{\alpha \to +0} \frac{\Omega - a}{\alpha} = \bigcup_{\alpha_0 > 0} \bigcap_{\alpha \in (0, \alpha_0)} \frac{\Omega - a}{\alpha}.$$

The vector g is called the *tangent* direction at the point a to the set Ω, if for $\alpha \to +0$ there is $o(\alpha) : (0, \alpha_0) \to \mathbb{R}^l$ such that

$$a + \alpha g + o(\alpha) \in \Omega,$$

where $o(\alpha)$ is so that $\frac{o(\alpha)}{\alpha} \to 0$ for $\alpha \to +0$. The set $T_L(\Omega, a)$ of all tangent directions is called (lower) *tangent* cone to $\Omega \subset \mathbb{R}^l$ at $a \in \mathrm{cl}(\Omega)$ and can also be presented in the form

$$T_L(\Omega, a) = \liminf_{\alpha \downarrow 0} \frac{\Omega - a}{\alpha} = \{ g \in \Omega \mid \lim_{\alpha \downarrow 0} \varrho(g, \alpha^{-1}(\Omega - a)) = 0 \}.$$

An element $g \in \mathbb{R}^l$ is called the *possible* tangent direction to Ω at a, if there is a sequence $g_k \in \mathbb{R}^l$, $k \in \mathbb{N}$ and a sequence of positive real numbers $\alpha_k > 0$, $k \in \mathbb{N}$ so that

$$g_k \to g, \quad \alpha_k \downarrow 0, \quad a + \alpha_k g_k \in \Omega.$$

The set $T_B(\Omega, a)$ of all tangent vectors is called the *contingent* cone (or the *Bouligand tangent* cone) to $\Omega \subset \mathbb{R}^l$ at $a \in \mathrm{cl}(\Omega)$. In a short form, it is

$$T_B(\Omega, a) = \limsup_{\alpha \downarrow 0} \frac{\Omega - a}{\alpha} = \{ g \in \mathbb{R}^l \mid \liminf_{\alpha \downarrow 0} \varrho(g, \alpha^{-1}(\Omega - a)) = 0 \}.$$

An element $g \in \mathbb{R}^l$ belongs to *Clarke cone* if for any sequences $x_k \in \Omega$ and positive real numbers $\alpha_k > 0$, $k \in \mathbb{N}$ such that $x_k \to a$, $\alpha_k \downarrow 0$ there exists a sequence $g_k \in \mathbb{R}^l$ for which $x_k + \alpha_k g_k \in \Omega$.[4] So, the Clarke tangent cone to a set Ω at a point a can be described as

$$T_C(\Omega, a) = \liminf_{\alpha \downarrow 0, x \to a} \frac{\Omega - x}{\alpha} = \{ g \in \mathbb{R}^l \mid \lim_{\alpha \downarrow 0, x \to a} \varrho(g, \alpha^{-1}(\Omega - a)) = 0 \}.$$

where the convergence $x \to a$ occurs over the set Ω.

The inclusions are obvious

$$T_C(\Omega, a) \subseteq T_L(\Omega, a) \subseteq T_B(\Omega, a).$$

If the set Ω is convex (or locally convex), then all the three indicated cones are equal. Clearly also that $T_A(\Omega, a) \subseteq T_L(\Omega, a)$.

---

[4] This is not Clarke's original definition from [7], but it is equivalent to it.

Lemma 1. *Let an economy* $\mathcal{E}$ *obey* **(A)** *and be non-satiated at a point* $((\bar{x}_i)_{i\in\mathcal{I}}, (\bar{y}_j)_{j\in\mathcal{J}}) \in \mathcal{A}(\mathcal{E})$. *Assume that for each* $j \in \mathcal{J}$ *there is a* **convex** *cone* $K_j(\bar{y}_j) \subset \mathbb{R}^l$ *such that* $K_j(\bar{y}_j) \subset T_B(Y_j, \bar{y}_j)$. *Now if the allocation* $(\bar{x}, \bar{y}) \in \mathcal{A}(\mathcal{E})$ *is Pareto optimal then* $\bar{y}_j \in \partial Y_j$ $\forall j \in \mathcal{J}$ *and there is* $p \in \mathbb{R}^l$, $p \neq 0$ *such that*

$$\langle \mathcal{P}_i(\bar{x}_i), p \rangle \geq \langle \bar{x}_i, p \rangle, \quad i \in \mathcal{I}, \tag{5}$$

$$\langle K_j(\bar{y}_j), p \rangle \leq 0, \quad j \in \mathcal{J}. \tag{6}$$

Proof. The Pareto optimality of an allocation can be rewritten in the following form: $((\bar{x}_i)_{i\in\mathcal{I}}, (\bar{y}_j)_{j\in\mathcal{J}}) \in \mathcal{A}(\mathcal{E})$,

$$\left[\sum_{i\in\mathcal{I}} \mathcal{P}_i(\bar{x}_i) - \sum_{i\in\mathcal{I}} \mathbf{e}_i\right] \bigcap \sum_{j\in\mathcal{J}} Y_j = \emptyset. \tag{7}$$

Assume $\bar{y}_{j'} \in \mathrm{int}\, Y_{j'}$ for some $j' \in \mathcal{J}$. Now via **(A)** one can find $x_i' \in \mathcal{P}_i(\bar{x}_i)$ enough close to $\bar{x}_i$, $i \in \mathcal{I}$ and such that

$$\sum_{i\in\mathcal{I}}(x_i' - \bar{x}_i) + \bar{y}_{j'} = y_{j'}' \in \mathrm{int}\, Y_{j'} \quad \Rightarrow \quad \sum_{i\in\mathcal{I}} x_i' = \sum_{j\neq j', j\in\mathcal{J}} \bar{y}_j + y_{j'}' + \sum_{i\in\mathcal{I}} \mathbf{e}_i.$$

So we find an allocation $(x', (y_j')_{j\in\mathcal{J}})$, where $y_j' = \bar{y}_j$ for $j \neq j'$, which dominates $(\bar{x}, \bar{y})$, that is impossible for Pareto optimal allocation.

Further, due to $\sum_{i\in\mathcal{I}} \bar{x}_i - \sum_{i\in\mathcal{I}} \mathbf{e}_i - \sum_{j\in\mathcal{J}} \bar{y}_j$, (7) and Bouligand tangent cone definition one concludes

$$\left[\sum_{i\in\mathcal{I}} \mathrm{int}\, \mathcal{P}_i(\bar{x}_i) - \sum_{i\in\mathcal{I}} \mathbf{e}_i\right] \bigcap \left(\sum_{j\in\mathcal{J}} \bar{y}_j + T_B\left(\sum_{j\in\mathcal{J}} Y_j, \sum_{j\in\mathcal{J}} \bar{y}_j\right)\right) = \emptyset.$$

Now since $T_B(\sum_{j\in\mathcal{J}} Y_j, \sum_{j\in\mathcal{J}} \bar{y}_j) \supseteq \sum_{j\in\mathcal{J}} T_B(Y_j, \bar{y}_j)$ we find

$$\left[\sum_{i\in\mathcal{I}} \mathrm{int}\, \mathcal{P}_i(\bar{x}_i) - \sum_{i\in\mathcal{I}} \mathbf{e}_i\right] \bigcap \sum_{j\in\mathcal{J}} [\bar{y}_j + T_B(Y_j, \bar{y}_j)] = \emptyset \quad \Rightarrow$$
$$\left[\sum_{i\in\mathcal{I}} \mathrm{int}\, \mathcal{P}_i(\bar{x}_i) - \sum_{i\in\mathcal{I}} \mathbf{e}_i\right] \bigcap \sum_{j\in\mathcal{J}} [\bar{y}_j + K_j(\bar{y}_j)] = \emptyset.$$

Here two convex sets have empty intersection, one of which has a nonempty interior. We can apply separation theorem and find $p \in \mathbb{R}^l$, $p \neq 0$ such that

$$\langle p, \sum_{i\in\mathcal{I}} \mathrm{int}\, \mathcal{P}_i(\bar{x}_i) - \sum_{i\in\mathcal{I}} \mathbf{e}_i \rangle \geq \langle p, \sum_{j\in\mathcal{J}} [\bar{y}_j + K_j(\bar{y}_j)] \rangle.$$

The set of values on the right-hand side of this inequality is upper-bounded, therefore because every $K_j(\bar{y}_j)$ is a conic set with a vertex at zero, we conclude

$$\langle p, K_j(\bar{y}_j) \rangle \leq 0 \quad \forall j \in \mathcal{J} \quad \& \quad \langle p, \sum_{i\in\mathcal{I}} \mathrm{int}\, \mathcal{P}_i(\bar{x}_i) - \sum_{i\in\mathcal{I}} \mathbf{e}_i \rangle \geq \langle p, \sum_{j\in\mathcal{J}} \bar{y}_j \rangle.$$

The left inequality states (6). Now applying $\bar{x}_i \in \operatorname{cl} \mathcal{P}_i(\bar{x}_i)$, $i \in \mathcal{I}$, taking into account $\sum_{i \in \mathcal{I}} \bar{x}_i = \sum_{i \in \mathcal{I}} \mathbf{e}_i + \sum_{j \in \mathcal{J}} \bar{y}_j$, passing to the limits[5] and transforming the right inequality we yield: for each $i' \in \mathcal{I}$

$$\sum_{i \neq i', i \in \mathcal{I}} \langle p, \bar{x}_i \rangle + \langle p, \mathcal{P}_{i'}(\bar{x}_{i'}) \rangle \geq \langle p, \sum_{i \in \mathcal{I}} \mathbf{e}_i \rangle + \langle p, \sum_{j \in \mathcal{J}} \bar{y}_j \rangle = \sum_{i \neq i', i \in \mathcal{I}} \langle p, \bar{x}_i \rangle + \langle p, \bar{x}_{i'} \rangle.$$

Omitting identical terms this proves (5).   □

So, Lemma 1 presents specific necessary conditions for an allocation to be Pareto optimal and this is an analog of the Second Welfare Theorem. Here the family of cones $K_j(\bar{y}_j) \subset T_B(Y_j, \bar{y}_j)$, $j \in \mathcal{J}$ accumulates the form of efficiency in this specification: wider cones provide a better description, and the best among them is a collection of half-spaces. Of course one can also try to apply the asymptotic cones described above, but among them only the Clarke cone is convex. The way to overcome this obstacle is to transform these cones using the method suggested in [21,22] for specifying the derivatives of the point-to-set mapping: I describe this construction below. It is based on the crucial property of Minkowski geometrical difference presented in the following lemma (see Lemma 1.1.4 in [22]).

Lemma 2. *For any cone K, the set $K \overset{*}{-} K$ is its **convex** subcone. For the convex cone K, the equality $K = K \overset{*}{-} K$ holds. If K is closed then $K \overset{*}{-} K$ is closed and*

$$T_C(K, 0) = K \overset{*}{-} K.$$

Proof. For $x \in (K \overset{*}{-} K)$ and real $\lambda > 0$ one has $\lambda(x + K) \subset \lambda K = K \Rightarrow \lambda x + K \subset K$. So, $K \overset{*}{-} K$ is a cone. Moreover, since for $x, y \in (K \overset{*}{-} K)$ one has $x + (y + K) \subseteq x + K \subseteq K$ one concludes $x + y \in (K \overset{*}{-} K)$, i.e., $K \overset{*}{-} K$ is a convex cone. The closeness of $K \overset{*}{-} K$ for closed K is obvious. The proof of relation $T_C(K, 0) = K \overset{*}{-} K$ can be found in [22], Lemma 1.4.5, p. 45.   □

There is another way to associate a cone with the given set.

Definition 2. *Let $\Omega \subset L$. The set*

$$O^+ \Omega = \{ x \in L \mid \forall a \in \Omega, \forall \lambda \in \mathbb{R}, \lambda \geq 0, \ a + \lambda x \in \Omega \}. \tag{8}$$

*is called the **asymptotic** cone of Ω.*

It is easy to see that $O^+ \Omega$ is convex and in general, it can be non-closed. The asymptotic cone is closely related with the cones specified via Minkowski geometrical difference, it can be alternatively defined as

$$O^+ \Omega = \Omega \overset{*}{-} \Omega.$$

---

[5] Here $\forall p \in \mathbb{R}^l$ $\langle p, \operatorname{cl} \mathcal{P}_i(\bar{x}_i) \rangle \geq b \iff \langle p, \mathcal{P}_i(\bar{x}_i) \rangle \geq b \iff \langle p, \operatorname{int} \mathcal{P}_i(\bar{x}_i) \rangle \geq b$, $\forall b \in \mathbb{R}$.

Now applying the geometrical difference to the cones introduced above we are going to the specification of convex approximating cones. They are the following: if $a \in \mathrm{cl}\, \Omega$, $\Omega \subset L$; then **asymptotic lower tangent** cone is

$$T_{AL}(\Omega, a) = T_L(\Omega, a) \overset{*}{-} T_L(\Omega, a),$$

and the **asymptotic upper tangent** cone is

$$T_{AB}(\Omega, a) = \mathrm{cl}\left[(T_B(\Omega, a) \overset{*}{-} T_B(\Omega, a)) + T_{AL}(\Omega, a) \right].$$

The most important properties of these cones are summarized in the following theorem (see Theorem 1.4.1 in [22]).

Theorem 1 (Polovinkin). *The cones $T_{AL}(\Omega, a)$ and $T_{AB}(\Omega, a)$ are convex and closed. Moreover, the equalities and inclusions are valid:*

$$T_{AL}(\Omega, a) = T_C(T_L(\Omega, a); 0), \tag{9}$$

$$T_{AB}(\Omega, a) = \mathrm{cl}\left[T_C(T_L(\Omega, a); 0) + T_C(T_B(\Omega, a); 0)\right] \tag{10}$$

$$T_C(\Omega, a) \subseteq T_{AL}(\Omega, a) \subseteq T_{AB}(\Omega, a) \subseteq T_B(\Omega, a). \tag{11}$$

Moreover, all inclusions in (11) can be strict.

Proof of the theorem can be found in [22]. A useful characterization of the Clarke cone is presented in the following theorem. This specification is based on Theorem 1.2 from [8], p. 84, I omit a detailed proof here.

Theorem 2. *Let $Y \subset \mathbb{R}^l$ be a technological set, satisfying standard assumptions and $\bar{y} \in \partial Y$. Then*

- *For almost all boundary points $\Omega \subset \partial Y$, every point has a tangent hyperplane, i.e. it is the differentiability point of a function f which graph presents*

$$Y = \{y \in \mathbb{R}^l \mid f(y) \le 0\}, \quad \partial Y = \{y \in \mathbb{R}^l \mid f(y) = 0\}.$$

- *Clarke tangent cone at $\bar{y} \in Y$ is a normal cone to*

$$\Delta(\bar{y}) = \left\{ h \in \mathbb{R}^l \mid \exists y^t \in \Omega, t \in \mathbb{N} : \ y^t \underset{t \to \infty}{\to} \bar{y} \ \& \ \frac{\nabla f(y^t)}{\|\nabla f(y^t)\|} \underset{t \to \infty}{\to} h \right\}, \quad i.e.$$
$$T_C(Y, \bar{y}) = \{\kappa \in \mathbb{R}^l \mid \langle h, \kappa \rangle \le 0 \ \forall h \in \Delta(\bar{y})\}.$$

Proof. According to the assumptions for every $z \in \mathbb{R}^l$ one can find the only real $\lambda = \lambda(z)$ such that $[z - \lambda(1, 1, \dots, 1)] \in \partial Y$. Now we have

$$Y = \{z \in \mathbb{R}^l \mid \lambda(z) \le 0\}.$$

Define $f(z) = \lambda(z)$, $z \in \mathbb{R}^l$. Clearly, because of assumptions, $f(\cdot)$ satisfies the Lipschitz condition and, therefore, is an absolutely continuous function. It is known that functions of this type are differentiable almost everywhere. The second statement of the theorem follows from the characterization of Clarke subdifferential for Lipschitz functions presented in Theorem 1.2 from [8].   □

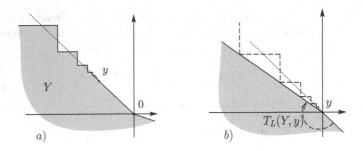

Fig. 1. a) A non-convex set Y; b) Lower tangent cone $T_L(Y, y)$

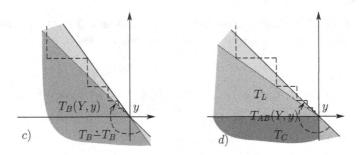

Fig. 2. c) Bouligand tangent cone $T_B(Y, y)$; d) Tangent cones to Y at $y \in Y$

So, as an application one can describe the necessary conditions for an allocation to be Pareto optimal, the best option is to choose an **asymptotic upper tangent** cone $T_{AB}(Y_j, \bar{y}_j)$. In contrast, the worst choice is the Clarke cone $T_C(Y_j, \bar{y}_j)$. Let us call a convex closed cone $K \subseteq L$ **marginal** (at a point $\bar{y}_j \in Y_j$) iff

$$T_C(Y_j, \bar{y}_j) \subseteq K \subseteq T_{AB}(Y_j, \bar{y}_j). \tag{12}$$

A specification of the related equilibrium concepts and related contractual notions is presented in the next section. Below, I give an example showing that the cones presented above can be really different ones.

Example 1. On the two-dimensional plane, Fig. 1 and 2 present possible non-convex technological set Y and mentioned above cones associated with the point $y \in Y$. In Fig. 1b) dark area presents a *lower tangent* cone at y (it is convex now), in Fig. 2c) non-convex *Bouligand* tangent cone is denoted as a light gray area with a solid boundary line; from this, an asymptotic cone $T_B \overset{*}{\cdot} T_B$ is derived, it is shown as a dark gray area. Figure 2d) presents all conic types: negative orthant coincides with $T_C(Y, y)$ since $(1, 0), (0, 1) \in \Delta(y)$ (see Theorem 2), wider middle gray area corresponds to lower tangent cone $T_L(Y, y)$ and the widest one is asymptotic upper cone $T_{AB}(Y, y)$, now it is a half-space, bounder by the diagonal solid line.   □

3 MCP-Equilibrium and Contracts in Economies with Non-convex Technologies

3.1 Marginal Cost Pricing

Below we study an economy $\mathcal{E}$ with non-convex production sets but our approach covers also the convex case as a particular one. Nonconvexity in production may occur, for example, due to increasing returns to scale (firm revenues are increasing per-unit costs). For example, the recording of the CD-ROM is costlier than its replication, overwriting occurs at low cost. However, this possibility (because of technical-mathematical reasons) has not been studied in the classical version of the existence theory. Of course, in order for an equilibrium to exist under non-convex technology, the concept should be appropriately modified. However, this modification should be such that the new concept is resulted in (or at least has chances) the Pareto optimal allocation, as it is in the convex case. Pricing on the basis of average costs does not satisfy this requirement. The key idea of MCP-equilibrium is that profit maximization is replaced by the (necessary) first-order condition (let us say expressed in terms of gradients of the functions that define the production sets), which in the convex case is also sufficient for a plan to be a profit maximizer. Thus the concept directly generalizes the usual competitive equilibrium. As the subject of interpretation, one is usually talking about the social planner, who has the ability to "evaluate" the price obtained according to the principle of marginal cost pricing (MCP), and then force the manufacturers to adhere to the specified production plans. Let's first analyze the simplest version of a non-convex manufacturing sector, described by differentiable functions.

Assume that the production sets Y_j are described via *differentiable functions* φ_j by formula

$$Y_j = \{y \in L \mid \varphi_j(y) \leq 0\}, \ j \in \mathcal{J}. \tag{13}$$

and, moreover, in this case the boundary of production sets can be defined as[6]

$$\partial Y_j = \{y \in L \mid \varphi_j(y) = 0\} \neq Y_j, \ j \in \mathcal{J}. \tag{14}$$

Define $X = \prod_{i \in \mathcal{I}} X_i, Y = \prod_{j \in \mathcal{J}} Y_j$. The following definition of MCP-equilibrium can be found in [6, 13].

Definition 3. MCP–*equilibrium (marginal cost pricing) is a triplet* (x, y, p), *where* $x = (x_i)_{i \in \mathcal{I}} \in X$ *is a family of consumption plans,* $y = (y_j)_{j \in \mathcal{J}} \in Y$ *are production plans and* $p = (p_1, \ldots, p_l) \neq 0$ *is a price vector, that satisfies the following conditions:*

$$y \in \prod_{j \in \mathcal{J}} \partial Y_j \quad \& \quad \exists \lambda_j > 0 : \ p = \lambda_j \nabla \varphi_j(y_j), \ \forall j \in \mathcal{J}, \tag{15}$$

---

[6] Clearly, in the general case, the topological boundary of the set may be narrower than the set described in (14).

$$\langle p, \mathcal{P}_i(x_i)\rangle > p \cdot \mathbf{e}_i + \sum_{j\in\mathcal{J}} \theta_i^j p \cdot y_j = p \cdot x_i, \ \forall i \in \mathcal{I}, \tag{16}$$

$$\sum_{i\in\mathcal{I}} x_i = \sum_{j\in\mathcal{J}} y_j + \sum_{i\in\mathcal{I}} \mathbf{e}_i. \tag{17}$$

The requirement (15) is the above-mentioned first-order conditions which the equilibrium production plans must satisfy instead of the condition of profit maximization (2). Conditions (16), (17) present the optimum of consumer preferences under budget and other constraints and the balance of commodity markets.

In the latter definition, it is implicitly assumed that production can be unprofitable, but total taxes cover the losses of firms with nonconvex production sets. It is important to note that if all firms have convex technologies, the concept of equilibrium with marginal cost pricing is turning to be the Walrasian equilibrium in the classical Arrow–Debreu model. Further well-known in the literature and one of the simplest results (see [6]) on the existence of MCP-equilibria is stated.

Consider a model with one firm and let assumptions (13), (14) hold. In addition, let us assume $0 \in Y$, $Y - \mathbb{R}^l_+ \subseteq Y$, the set $(Y + \sum_{\mathcal{I}} \mathbf{e}_i) \cap \mathbb{R}^l_+$ is bounded and if $y + \sum_{\mathcal{I}} \mathbf{e}_i \in \partial Y \cap \mathbb{R}^l_+$, then $\nabla\varphi(y) \gg 0$.

Let for all $i \in \mathcal{I}$ consumption sets $X_i = \mathbb{R}^l_+$, and preferences are determined via utility functions $u_i : X_i \to \mathbb{R}$, which are continuous, strictly concave, and locally non-satiated ones.

Theorem 3 (Mantel, 1979). *Under presented assumptions an equilibrium with marginal cost pricing does exist.*

Essentially stronger results can be found in [4,5]. Here many firms are considered and they may have a non-smooth boundary, as well as a general rule of pricing is analyzed. In this context, the mapping $\psi : \prod_{\mathcal{J}} \partial Y_j \to \mathbb{R}^l_+$ is considered, which maps a vector of production plans to a set of production prices. Requirements on the map $\psi(\cdot)$ are very general and this approach can present the marginal cost pricing as soon as average cost pricing—ACP, and also other variants, see [4,6]. Nevertheless, the most interesting variants of the price rule are presented in the form of the requirement for prices to satisfy the necessary (first-order) conditions for the Pareto optimality of the equilibrium distribution. In [4] there are several price rules, each can be multi-valued and specified as Clark normal cone:

$$g_j(y_j) = \{p \in \mathbb{R}^l \mid \langle p, T_C(Y_j, y_j)\rangle \le 0\}, \quad y_j \in \partial Y_j, \quad j \in \mathcal{J}.$$

Now condition (15) in Definition 3 is changed to

$$y \in \prod_{j\in\mathcal{J}} \partial Y_j \ \& \ p \in g_j(\bar{y}_j), \ \forall j \in \mathcal{J}.$$

This replacement leads us to MCP-equilibrium by Bonnisseau–Cornet. One can continue the generalization and apply another normal tangent cone, but he/she only needs that presented tangent cone is convex. So, one can apply

$$g_j(y_j) = \{p \in \mathbb{R}^l \mid \langle p, T_{AB}(Y_j, y_j)\rangle \le 0\}, \quad y_j \in \partial Y_j, \quad j \in \mathcal{J},$$

where $T_{AB}(Y_j, y_j)$ was defined in (10).

Further, we are passing on the major purpose of the paper—the analysis of the contractual approach in Arrow–Debreu model with non-convex technologies.

3.2   Contracts in Non-convex Arrow–Debreu Model

First I recall briefly the conceptual apparatus of the theory of barter contracts, see [16, 18, 19], while adapting it to the model with the production sector.

Any vector $v = (v_i)_{i \in \mathcal{I}} \in L^{\mathcal{I}}$ satisfying $\sum_{i \in \mathcal{I}} v_i = 0$ is called a barter (exchange) *contract*. Such barter contracts are used in pure exchange economies, as well as in the consumption sector in the economy with production. In what follows, we assume that any barter agreement is valid. With every finite collection V of (permissible) contracts, it can be associated allocation $x(V) = \mathbf{e} + \sum_{v \in V} v$, where $\mathbf{e} = (\mathbf{e}_1, \dots, \mathbf{e}_n) \in X$ is an initial resource allocation. If $\mathbf{e} + \sum_{v \in U} v \in X$ $\forall U \subseteq V$, i.e., if any part of the contracts is broken one can get anyway a feasible allocation, then we call V a *web* of contracts.

How can contractual concepts be modified and adapted to an economy with production? For the beginning, one can study the model with individualized production, to which the standard convex Arrow–Debreu model can be easily reduced. Indeed, specifying $\bar{Y}_i = \sum_{j \in \mathcal{J}} \theta_i^j Y_j$ one arrives at an equivalent model. In such a case one can introduce the notion of contract as a pair $(v, y) \in L^{\mathcal{I}} \times L^{\mathcal{I}}$, where v is an ordinary barter contract but $y = (y_1, \dots, y_n)$ is a vector that corresponds to production programs y_i for individuals $i \in \mathcal{I}$. Now if $(v, y) \in L^{\mathcal{I}} \times \prod_{i \in \mathcal{I}} \bar{Y}_i$, i.e., if each production program is feasible, $y_i \in \bar{Y}_i$, $i \in \mathcal{I}$, then contract (v, y) is permissible. For a finite collection V of contracts in the model (1) one can put into correspondence (consumption) allocation

$$x(V) = \mathbf{e} + \sum_{(v,y) \in V} y + \sum_{(v,y) \in V} v.$$

This method is well working for convex production economies and was successfully applied in [18, 19]. For a non-convex economy, the presented reduction works incorrectly because leads to an arbitrary change of the model since now relation $\sum_{i \in \mathcal{I}} \bar{Y}_i = \sum_{j \in \mathcal{J}} Y_j$ is violated (it is true only for convex Y_j, $j \in \mathcal{J}$).

The crucial idea for the comparing of production programs is that the difference between a new production plan $y_j' \in Y_j$ and an old one $y_j \in Y_j$ has to belong to a (convex) marginal tangent cone specified at the point y_j for the set Y_j. In other words, one needs

$$y_j' - y_j \in K_j(y_j), \quad j \in \mathcal{J}$$

to be able to compare a bundle $(y_j')_{j \in \mathcal{J}} = y' \in Y$ of production plans with the initial bundle $(y_j)_{j \in \mathcal{J}} = y \in Y$. Cones $K_j(y_j)$ are associated with the contractual interaction and present the first-order conditions specified in Sect. 2.2 and Lemma 1. In particular these cones can be chosen as $T_C(Y_j, y_j)$ either $T_{AL}(Y_j, y_j)$, or $T_{AB}(Y_j, y_j)$ defined in Sect. 2.2. Now we introduce the concept of the web of contracts and their domination in production economies.

Again we specify contract as is a pair $(v, y) \in L^{\mathcal{I}} \times L^{\mathcal{J}}$, where v is an ordinary barter contract $v = (v_1, v_2, \ldots, v_n)$ and $y = (y_1, \ldots, y_m)$ is a bundle of production programs $y_j \in Y_j$, $j \in \mathcal{J}$. Now a reallocation $z = (z_1, \ldots, z_n)$ of production plans can be added to individual consumptions, it can be specified as follows

$$z_i = \sum_{j \in \mathcal{J}} \theta_i^j y_j, \quad i \in \mathcal{I}.$$

So, if we have $(v, y) \in L^{\mathcal{I}} \times Y$ and $\sum_{i \in \mathcal{I}} v_i = 0$, then contract (v, y) is permissible and it implies consumptions $x_i = e_i + v_i + z_i \in L$, $i \in \mathcal{I}$. Now we can define a web of contracts (V, y) associated with the family of production plans $y \in Y$: here V is a finite set of barter contracts such that

$$\sum_{v \in U} v + z + e \in X, \quad \forall U \subseteq V \iff e_i + z_i + \sum_{v \in U} v_i \in X_i, \ \forall i \in \mathcal{I}, \ \forall U \subseteq V,$$

i.e., under the current endowments $e \in X$ and production plans $y \in Y$, individuals enter into contractual relationships so that they can break any contracts. Specify:

$$x(V, y) = e + z(y) + \sum_{v \in V} v \in L^{\mathcal{I}}.$$

I pay your attention that we apply the only family of production plans $y = (y_j)_{\mathcal{J}}$. Now let us pass to the definition of webs domination. This domination assumes that some contracts from a web can be broken (as a whole), some new contracts concluded and new production plans generated.

A web of contracts (W, y') dominates a web (V, y) by coalition $S \subseteq \mathcal{I}$, written as $(W, y') \succ_S (V, y)$, iff

- $\forall v \in V \backslash W$, $S \cap \{i \in \mathcal{I} \mid v_i \neq 0\} \neq \emptyset$—consumers from S can break contracts from $V \backslash W$;
- $\forall v \in W \backslash V$, $S \supseteq \{i \in \mathcal{I} \mid v_i \neq 0\}$—only consumers from S can sign new contracts;
- $(y_j' - y_j) \in K_j(y_j)$, $j \in \mathcal{J}$;
- $x_i(W, y') \succ_i x_i(V, y) \ \forall i \in S$.

An allocation that cannot be dominated is called contractual.

The described contractual interaction also allows us to introduce a concept of contractual core, specified for non-convex production. This weak concept is related with a bundle of marginal conic sets $K_j(y_j) \subseteq L$, $j \in \mathcal{J}$ and is so that MCP-equilibrium with prices from the normal cone, i.e., prices $p \in L'$, $p \neq 0$ satisfying

$$\langle p, K_j \rangle \leq 0, \quad j \in \mathcal{J},$$

belongs to the core.

Definition 4. *Coalition $S \subseteq \mathcal{I}$ dominates an allocation $(x, y) \in \mathcal{A}(\mathcal{E})$ via marginal cones $K = (K_j)_{j \in \mathcal{J}} \subset L^{\mathcal{J}}$ if there is $(x^S, z^S) \in \prod_{i \in S} X_i \times L^S$ such*

*that for some $y'_j \in Y_j$ satisfying $(y'_j - y_j) \in K_j$, $j \in \mathcal{J}$ one has $z_i^S = \sum_{j \in \mathcal{J}} \theta_i^j y'_j$
and $\sum_S x_i^S = \sum_S \mathbf{e}_i + \sum_S z_i^S$ so that $x_i^S \succ_i x_i \ \forall i \in S$.*

*The set $\mathcal{C}^K(\mathcal{E}) \subset \mathcal{A}(\mathcal{E})$ of all allocations that can be dominated by no coalition
is called **weak K-core**.*

This core is really weaker than the standard one as soon as the possibilities of
a coalition to dominate are reduced and the core may contain allocations that
are not Pareto optimal. An advantage of the concept is that generalized MCP-
equilibrium allocation (presented below) always is an element of the core and
it is nonempty. It can be easily proved by arguing by contradiction. Moreover,
we can consider the classical method of modeling perfect competition and apply
the replicas of the economy, having in mind the asymptotic theorem on the
coincidence of the core and equilibrium.

Now we are passing to a discussion of partially broken contracts. It leads us
to the notions of proper and fuzzy contractual allocations.

The simplest contractual model admitting partial break of contracts can be
presented in the following way. Let (V, y) be a web. For real α define

$$\alpha V = \{\alpha \cdot v \mid v \in V\}.$$

So, αV is a web, yielded from V by multiplying contracts on α. For $0 \le \alpha \le 1$
consider web $U = \alpha V \cup (1-\alpha) V$, which obviously implements the same allocation
$x(U, y) = x(V, y)$. The web $U = \alpha V \cup (1 - \alpha) V$ is called α-partition of the web
V. An allocation $x = x(V, y)$ is properly contractual if α-partition of V is stable
for every $\alpha \in [0, 1]$. For the model of an economy with *individualized production*,
I present below a narrative definition in substantial terms.

Definition 5 (Marakulin, 2014). *A pair $(x, y) \in X \times Y$ is called properly
contractual allocation if there is a web V such that the following conditions are
satisfied:*

(i) $x = x(V, y) = \sum_{v \in V} v + \mathbf{e} + y.$

(ii) There is no coalition S, for which it is profitable:

  (α) to partially break barter contracts;

  *(β) to transit from the programs $y = (y_S, y_{\mathcal{I} \setminus S})$ to new production programs
$y' = (y'_S, y_{\mathcal{I} \setminus S})$, where $y'_S \in \prod_{i \in S} Y_i$ and $y_{\mathcal{I} \setminus S} \in \prod_{i \in \mathcal{I} \setminus S} Y_i;$*

  (γ) to sign a new contract.

In [19] it has been proven that for the *smooth convex* Arrow–Debreu
economies every interior properly contractual allocation is an equilibrium, see
also [18]. Moreover, this result has been extended to the non-convex Arrow–
Debreu model with respect to MCP-equilibrium and technological sets having
smooth boundaries. Below we introduce and study the concept of fuzzy contractual allocation and prove the equivalence of it with MCP-equilibrium in a very
general context for non-smooth, non-convex economies and do not assuming an
interior point.

Definition 6. *Let* $K = (K_j)_{j \in \mathcal{J}} \subset L^{\mathcal{J}}$ *be a family of marginal cones. An allocation* $(\bar{x}, \bar{y}) \in \mathcal{A}(\mathcal{E})$ *implemented by a web of barter contracts* $V = \{v\}$*, where*

$$v = \bar{x} - \bar{z} - \mathbf{e}, \quad \bar{z} = (\bar{z}_i)_{i \in \mathcal{I}}, \quad \bar{z}_i = \sum_{j \in \mathcal{J}} \theta_i^j \bar{y}_j$$

*is called **K-marginal fuzzy contractual** if for every* $t = (t_i)_{i \in \mathcal{I}}$, $0 \le t_i \le 1$, $\forall i \in \mathcal{I}$*, there are no other production programs* $y_j^i \in L$, $y_j^i - \bar{y}_j \in K_j$, $j \in \mathcal{J}$ *and a barter contract* $w = (w_1, \ldots, w_n) \in L^{\mathcal{I}}$, $\sum_{i \in \mathcal{I}} w_i = 0$ *such that for*

$$\xi_i = \xi_i(t, v, w, y) = \mathbf{e}_i + t_i v_i + w_i + z_i, \quad z_i = \sum_{j \in \mathcal{J}} \theta_i^j y_j^i, \quad i \in \mathcal{I} \quad (18)$$

one takes place

$$\xi_i \succ_i \bar{x}_i \quad \forall i: \ \xi_i \ne \bar{x}_i. \quad (19)$$

This definition describes the following contractual interaction. First, the individuals are going to partially break barter contracts, each in the measure $(1 - t_i)$ and it is private information: nobody else knows about it. Second, they change production plans for $y_j^i \in L$ in a part θ_i^j that they can control and under the condition $y_j^i \in \bar{y}_j + K_j$, $j \in \mathcal{J}$: this is a form of social production efficiency (notice that $\sum_{i \in \mathcal{I}} \theta_i^j y_j^i = y_j \in \bar{y}_j + K_j$ since K_j is a convex set). Finally, agents conclude a new barter contract $w = (w_1, \ldots, w_n)$. As a result of this interaction, each involved individual has to benefit. Notice that due to (19) contract $w = 0$ is also possible and thereby only partial break can be realized. So, if there is no domination of this kind, then an allocation is called K-marginal fuzzy contractual.

The following lemma characterizes fuzzy contractual allocations in "geometrical" categories. Recall that an allocation $(\bar{x}, \bar{y}) \in \mathcal{A}(\mathcal{E})$ is called *stable relative to asymmetrical partial break (lower stable)* iff $\forall i \in \mathcal{I} \ \forall z_i \in Z_i = \sum_{j \in \mathcal{J}} \theta_i^j(\bar{y}_j + K_j)$ [7]

$$z_i + (1 - \lambda)\bar{x}_i + \lambda(\mathbf{e}_i + \bar{z}_i) \not\succ_i \bar{x}_i \quad \forall 0 \le \lambda \le 1 \iff$$
$$\forall i \in \mathcal{I} \quad (Z_i + [\bar{x}_i, \mathbf{e}_i + \bar{z}_i]) \bigcap \mathcal{P}_i(\bar{x}_i) = \emptyset. \quad (20)$$

The specificity of condition (20) is that if it is not true, then there is an individual who acting separately certainly changes the wcb as soon as it is beneficial for him. Here, there is no contractual interaction.

Lemma 3. *Let* $(\bar{x}, \bar{y}) \in \mathcal{A}(\mathcal{E})$ *be an allocation lower stable relative to a partial break and* $K = (K_j)_{j \in \mathcal{J}} \subset L^{\mathcal{J}}$ *be a family of marginal cones specified for* $(\bar{y}_j)_{j \in \mathcal{J}} = \bar{y}$*. Then* $(\bar{x}, \bar{y})$ *is* K*-fuzzy contractual if and only if*

$$\mathcal{A}(L^{\mathcal{I}}) \bigcap \prod_{i \in \mathcal{I}} \left((\mathcal{P}_i(\bar{x}_i) - Z_i + \mathrm{co}\{(\mathbf{e}_i + \bar{z}_i - \bar{x}_i), 0\}) \bigcup \{\mathbf{e}_i\} \right) = \{\mathbf{e}\}. \quad (21)$$

---

[7] It means that nobody wants to partially break a barter contract without concluding a new one.

Here $Z_i = \sum_{j \in \mathcal{J}} \theta_i^j (\bar{y}_j + K_j)$, $\bar{z}_i = \sum_{j \in \mathcal{J}} \theta_i^j \bar{y}_j$, $i \in \mathcal{I}$, $\mathbf{e} = (\mathbf{e}_1, \mathbf{e}_2, \ldots, \mathbf{e}_n)$ and $\mathcal{A}(L^{\mathcal{I}})$ is a subspace that corresponds to the material balance constraints:

$$\mathcal{A}(L^{\mathcal{I}}) = \{(\zeta_i)_{i \in \mathcal{I}} \in L^{\mathcal{I}} \mid \sum_{i \in \mathcal{I}} \zeta_i = \sum_{i \in \mathcal{I}} \mathbf{e}_i\}.$$

This characterization works very efficiently in applications, but one has to apply (21) and (20) simultaneously. In particular, applying it below we state the equivalence between fuzzy contractual allocations and MCP-equilibria. To do it one has to separate sets from the left-hand part of (21) by linear functional and analyze the result.

Proof. Necessity. Let (21) be false. Therefore in the left-hand side of the intersection (21), there is $\zeta = (\zeta_i)_{i \in \mathcal{I}} \neq \mathbf{e}$. Define

$$S = \{i \in \mathcal{I} \mid \zeta_i \neq \mathbf{e}_i\} \neq \emptyset.$$

Further, find a new contract with this support and the appropriate amounts of contracts breaking. Define $w_i = \zeta_i - \mathbf{e}_i$ that gives $\sum_{i \in \mathcal{I}} w_i = 0$ since $\zeta \in \mathcal{A}(L^{\mathcal{I}})$. For $i \notin S$ one obviously has $w_i = 0$, i.e., $\text{supp}(w) = S$. Also for $i \in S$ one has

$$\zeta_i \in (\mathcal{P}_i(\bar{x}_i) - Z_i + \text{co}\{0, (\mathbf{e}_i + \bar{z}_i - \bar{x}_i)\}),$$

that allows to conclude $\exists 0 \leq t_i \leq 1$ and $\xi_i \succ_i \bar{x}_i$ such that:

$$\mathbf{e}_i + w_i = \zeta_i = \xi_i - z_i + t_i(\mathbf{e}_i + \bar{z}_i - \bar{x}_i) \quad \Rightarrow \quad \xi_i = \mathbf{e}_i + t_i v_i + z_i + w_i,$$

for $v_i = \bar{x}_i - \mathbf{e}_i - z_i$, $i \in S$ (for $i \notin S$ one puts $t_i = 1$ & $z_i = \bar{z}_i \Rightarrow \xi_i = \bar{x}_i$). Here due to $\sum_{\mathcal{I}} \bar{x}_i = \sum_{\mathcal{I}} \mathbf{e}_i + \sum_{\mathcal{I}} \bar{z}_i$ we have $\sum_{\mathcal{I}} v_i = 0$. This contradicts the definition of K-fuzzy contractual allocation.

Sufficiency. Let (21) be true for $(\bar{x}, \bar{y}) \in \mathcal{A}(\mathcal{E})$, and in addition, let the allocation be lower stable relative to the partial breaking. Assume this is not K-fuzzy contractual. Then there are $t = (t_i)_{\mathcal{I}}$, plans $z_i \in Z_i$, $i \in \mathcal{I}$, and a barter contract $w = (w_1, \ldots, w_n) \in \mathbb{R}^{ln}$, $\sum_{\mathcal{I}} w_i = 0$, satisfying all Definition 6 requirements. This for $v_i = \bar{x}_i - \bar{z}_i - \mathbf{e}_i$ due to (18) for the members of a nonempty coalition yields

$$\exists \xi_i \in \mathcal{P}_i(\bar{x}_i): \quad x_i = \xi_i + t_i(\mathbf{e}_i + \bar{z}_i - \bar{x}_i) = \mathbf{e}_i + w_i + z_i. \tag{22}$$

Summing over i by the definition of contract one concludes $\sum_{\mathcal{I}} x_i = \sum_{\mathcal{I}} \mathbf{e}_i + \sum_{\mathcal{I}} z_i$, i.e., for $\zeta_i = x_i - z_i$, $i \in \mathcal{I}$ allocation $\zeta = (\zeta_i)_{\mathcal{I}}$ belongs to the intersection in the left-hand part of (21). If one supposes $\zeta = \mathbf{e}$, then $x_i = \mathbf{e}_i + z_i$ $\forall i \in \mathcal{I}$, that being substituted to the right-hand part of (22) yields $w_i = 0$, $\forall i \in \mathcal{I}$ $\Rightarrow \text{supp}(w) = \emptyset$. Hence, domination is carried out without the exchange and only via a partial break of the gross contract $v = \bar{x} - \bar{z} - \mathbf{e}$. However, this contradicts the lower stability relative to a partial break. Therefore, one finds an allocation $\zeta \neq \mathbf{e}$ which belongs to the intersection of the left-hand side (21); it is a contradiction. $\qquad \square$

Now we formalize generalized MCP-equilibrium and then state equivalence of it and K-marginal fuzzy contractual allocation.

Definition 7. *Let (x, y, p) be a triplet such as $x = (x_i)_{i \in \mathcal{I}} \in X$ is a family of consumption plans, $y = (y_j)_{j \in \mathcal{J}} \in Y$ are production plans and $p = (p_1, ..., p_l) \neq 0$ is a price vector. Given a family $K = (K_j(y_j))_{j \in \mathcal{J}}$ of marginal cones at points $(y_j)_{j \in \mathcal{J}}$, triplet (x, y, p) is called **K-MCP-quasi-equilibrium** if it obeys the following conditions:*

$$y_j \in \partial Y_j \quad \& \quad \langle p, K_j \rangle \leq 0 \quad \forall j \in \mathcal{J}, \tag{23}$$

$$\langle p, \mathcal{P}_i(x_i) \rangle \geq p \cdot \mathbf{e}_i + \sum_{j \in \mathcal{J}} \theta_i^j p \cdot y_j = p \cdot x_i \quad \forall i \in \mathcal{I}, \tag{24}$$

$$\sum_{i \in \mathcal{I}} x_i = \sum_{j \in \mathcal{J}} y_j + \sum_{i \in \mathcal{I}} \mathbf{e}_i. \tag{25}$$

*If all inequalities in (24) have **strict** signs, then the triplet (x, y, p) is called K-MCP-equilibrium.*

Here, in contrast to the convex case, the freedom of the individual in choosing a production plan is limited. It is assumed that mutual cooperation or some authority establishes joint production plans, and the individuals still have the right to decide: do it or not (a specific form of non-binding agreement). However, the individual deviations from a given production plan are only possible within the sets $K_j \cap Y_j$ in a part that an individual $i \in \mathcal{I}$ can control, i.e. $\theta_i^j K_j$. Again, in the case of convex production, this requirement and standard one coincide.

Finally, we note that similarly to the convex Arrow–Debreu economy for non-convex case one can introduce the notion of a fuzzy core that is closely related to fuzzy contractual allocations. The main theoretical value of the fuzzy core is that it can be effectively applied to develop a theory of the existence of competitive equilibria and, in view of results presented here, fuzzy contractual allocations. I omit this presentation, the general methodology of this approach (with regard to the existence of equilibrium in convex economies) one can find in [1], and the most advanced mathematical results in [9,10] and [17], etc.

Remark 1. The existence of MCP-equilibria and the non-emptiness of weak K-core is not studied in this paper, we left this problem for further research. In the literature, it is well known that these problems are closely related and the existence of one can be derived from another one. For Clarke cones $K = K_C(Y_j, y_j)$ we deal with MCP-equilibrium by Bonnisseau–Cornet, which existence was stated in [5] that clearly implies the non-emptiness of our weak core. Notice that under standard assumptions for production set $Y \subset \mathbb{R}^l$ a point-to-set mapping $y \Rightarrow T_C(Y, y)$ has good mathematical properties, it is lower semicontinuous correspondence with non-empty, convex, and closed values. This simplifies the existence problem, for other chosen cones (e.g. $K = T_{AB}(Y, y)$) it is not true and it can be a difficult obstacle. $\square$

Relationship between K-marginal fuzzy contractual allocation and K-MCP-equilibrium is established in the following

Theorem 4. *Let $(\bar{x}, \bar{y}) \in \mathcal{A}(\mathcal{E})$ be an allocation in Arrow–Debreu economy and $K = (K_j(\bar{y}_j))_{j \in \mathcal{J}}$ be a family of closed convex cones such that*

$$T_C(Y_j, \bar{y}_j) \subseteq K_j(\bar{y}_j) \subseteq T_{AB}(Y_j, \bar{y}_j), \quad j \in \mathcal{J}.$$

If $(\bar{x}, \bar{y})$ is K-marginal fuzzy contractual, then it is K-MCP-quasi-equilibrium. On the contrary, assume that there are prices $\bar{p} \in L'$ such that $(\bar{x}, \bar{y}, \bar{p})$ is K-MCP-equilibrium. Then $(\bar{x}, \bar{y})$ is K-marginal fuzzy contractual.

In the theory of the existence of Walrasian equilibrium, the conditions under which each quasi-equilibrium turns into a strict equilibrium are well known: it is necessary to require additionally some survival assumptions, the representative of which can be irreducibility. Here we omit the description of these methods and refer the interrogated reader to [17].

Proof. Necessity. Our analysis is based on Lemma 3 and the application of relation (21). Indeed, $(\bar{x}, \bar{y})$ is K-fuzzy contractual iff (21) and (20) hold. Excluding $\mathbf{e}_i$ from the left-hand side of (21), one concludes

$$\mathcal{A}(L^{\mathcal{I}}) \bigcap \prod_{i \in \mathcal{I}} (\mathcal{P}_i(\bar{x}_i) - Z_i + \operatorname{co}\{(\mathbf{e}_i + \bar{z}_i - \bar{x}_i), 0\}) = \emptyset.$$

Here affine subspace $\mathcal{A}(L^{\mathcal{I}})$ is intersected with a convex solid set. Applying classical separation theorem we can find a nonzero linear functional (vector) $f = (f_1, \ldots, f_n) \in L^{\mathcal{I}}$ separating these sets, that is

$$\langle f, \mathcal{A}(L^{\mathcal{I}}) \rangle \leq \langle f, \prod_{i \in \mathcal{I}} (\mathcal{P}_i(\bar{x}_i) - Z_i + \operatorname{co}\{(\mathbf{e}_i + \bar{z}_i - \bar{x}_i), 0\}) \rangle.$$

The left-hand side of the inequality is upper bounded and, since $\mathcal{A}(L^{\mathcal{I}})$ is a subspace, one concludes

$$\langle (f_1, \ldots, f_n), \{(\zeta_1, \ldots, \zeta_n) \in L^n \mid \sum_{i=1}^{n} \zeta_i = 0\} \rangle = 0 \quad \Rightarrow$$

$$f_i = f_k = p \neq 0 \quad \forall i, k \in \mathcal{I}.$$

Now, substituting this to the latter inequality and transforming it we obtain

$$\sum_{i \in \mathcal{I}} \langle p, \mathcal{P}_i(\bar{x}_i) - Z_i + \operatorname{co}\{(\mathbf{e}_i + \bar{z}_i - \bar{x}_i), 0\} \rangle \geq \sum_{i \in \mathcal{I}} p\mathbf{e}_i. \tag{26}$$

Now we recall that $Z_i = \bar{z}_i + \sum_{j \in \mathcal{J}} \theta_i^j K_j$ where every K_j is a convex conic set, that due to $\sum_{i \in \mathcal{I}} \sum_{j \in \mathcal{J}} \theta_i^j K_j = \sum_{j \in \mathcal{J}} K_j$ and the last inequality allows us to conclude

$$\langle p, -\sum_{j \in \mathcal{J}} K_j \rangle \geq 0 \quad \Rightarrow \quad \langle p, K_j \rangle \leq 0 \quad \forall j \in \mathcal{J},$$

that proves (23). Further, notice that $\bar{x}_i \in \mathrm{cl}\,\mathcal{P}_i(\bar{x}_i)$ $\forall i \in \mathcal{I}$ implies

$$\bar{x}_i - \bar{z}_i + \mathbf{e}_i + \bar{z}_i - \bar{x}_i = \mathbf{e}_i \in \mathrm{cl}[\mathcal{P}_i(\bar{x}_i) - Z_i + \mathrm{co}\{(\mathbf{e}_i + \bar{z}_i - \bar{x}_i), 0\}],$$

that substituting it into left hand side of (26) for $j \neq i$ and then omitting identical terms yields

$$\forall i \in \mathcal{I} \quad \langle p, \mathcal{P}_i(\bar{x}_i) - Z_i + \mathrm{co}\{(\mathbf{e}_i + \bar{z}_i - \bar{x}_i), 0\}\rangle \geq p\mathbf{e}_i \quad \Rightarrow \quad \langle p, \mathcal{P}_i(\bar{x}_i) - \bar{z}_i\rangle \geq p\mathbf{e}_i.$$

Again, taking into account $\bar{x}_i \in \mathrm{cl}\,\mathcal{P}_i(\bar{x}_i)$ and specification of $\bar{z}_i$ the last one implies

$$\langle p, \mathcal{P}_i(\bar{x}_i)\rangle \geq p\mathbf{e}_i + \sum_{j \in \mathcal{J}} \theta_i^j p\bar{y}_j \quad \& \quad p\bar{x}_i \geq p\mathbf{e}_i + \sum_{j \in \mathcal{J}} \theta_i^j p\bar{y}_j \quad i \in \mathcal{I}.$$

Because $(\bar{x}, \bar{y}) \in \mathcal{A}(\mathcal{E})$, one concludes $p\bar{x}_i = p\mathbf{e}_i + \sum_{j \in \mathcal{J}} \theta_i^j p\bar{y}_j$ for all $i \in \mathcal{I}$ that finishes the proving $(\bar{x}, \bar{y}, p)$ is in fact a quasi-equilibrium by Definition 7.

Sufficiency. Let $(\bar{x}, \bar{y}, \bar{p})$ be a K-MCP-equilibrium. It easy to see that now condition (24) in its strict form and (23) for each (fixed) $i \in \mathcal{I}$ imply that

$$\langle \bar{p}, (\mathcal{P}_i(\bar{x}_i) - Z_i + \mathrm{co}\{(\mathbf{e}_i + \bar{z}_i - \bar{x}_i), 0\})\rangle > \bar{p}\mathbf{e}_i.$$

We always have

$$\sum_{i' \in \mathcal{I}, i' \neq i} \langle \bar{p}, (\mathcal{P}_{i'}(\bar{x}_{i'}) - Z_{i'} + \mathrm{co}\{(\mathbf{e}_{i'} + \bar{z}_{i'} - \bar{x}_{i'}), 0\}) \cup \{\mathbf{e}_{i'}\}\rangle \geq \sum_{i' \in \mathcal{I}, i' \neq i} \bar{p}\mathbf{e}_{i'}.$$

Summing this and latter one we get a strict inequality that is true for every $i \in \mathcal{I}$. This can be true only if (21) holds. Condition (20) is also true, it follows from (23), (24) and via Z_i, $i \in \mathcal{I}$ specification. Theorem 4 is proved. ∎

References

1. Aliprantis, C.D., Brown, D.J., Burkinshaw, O.: Existence and Optimality of Competitive Equilibria, 284 p. Springer, Berlin (1989). https://doi.org/10.1007/978-3-662-21893-8
2. Arrow, K., Debreu, G.: Existence of an equilibrium for a competitive economy. Econometrica **22**(3), 265–290 (1954)
3. Beato, P., Mas-Colell, A.: On marginal cost pricing with given tax-subsidy rules. J. Econ. Theory **37**, 356–365 (1985)
4. Bonnisseau, J.M., Cornet, B.: Existence of equilibria when firms follow bonded losses pricing rules. J. Math. Econ. **17**, 103–118 (1988)
5. Bonnisseau, J.M., Cornet, B.: Existence of marginal cost pricing equilibria in economies with several nonconvex firms. Econometrica **58**, 661–682 (1990)
6. Brown, D.J.: Equilibrium analysis with non-convex technologies. In: Hildenbrand, W., Sonneshein, H. (eds.) Handbook of Mathematical Economics IV, pp. 1964–1995 (1991)
7. Clarke, F.H.: Optimization and Nonsmooth Analysis. Wiley, New York (1983)

8. Demianov, V.F., Rubinov, A.M.: Fundamentals of Nonsmooth Analysis and Quasi-Differential Calculus. Science, Moscow (1990). (in Russian)

9. Florenzano, M.: On the non-emptiness of the core of a coalitional production economy without ordered preferences. J. Math. Anal. Appl. **141**, 484–490 (1989)

10. Florenzano, M.: Edgeworth equilibria, fuzzy core and equilibria of a production economy without ordered preferences. J. Math. Anal. Appl. **153**, 18–36 (1990)

11. Konovalov, A., Marakulin, V.: Equilibria without the survival assumption. J. Math. Econ. **42**, 198–215 (2006)

12. Kozyrev, A.N.: The stable systems of contracts in an exchange economy. Optimization **29**(44), 66–78 (1981). (The issue of IM SB AS USSR, in Russian)

13. Mantel, R.: Equilibrio con rendimiento crecientes a escala. Anales de la Asociation Argentine de Economia Politica **1**, 271–283 (1979)

14. McKenzie, L.W.: On equilibrium in Graham's model of world trade and other competitive systems. Econometrica **22**(2), 147–161 (1954)

15. Marakulin, V.M.: Contracts and domination in incomplete markets. Economic Education and Research Consortium. Working Paper Series, 02/04, 104 p. (2003)

16. Marakulin, V.M.: Contracts and domination in competitive economies. J. New Econ. Assoc. **9**, 10–32 (2011). (in Russian)

17. Marakulin, V.M.: Abstract Equilibrium Analysis in Mathematical Economics, 348 p. SB Russian Academy of Science Publisher, Novosibirsk (2012). (in Russian)

18. Marakulin, V.M.: On the Edgeworth conjecture for production economies with public goods: a contract-based approach. J. Math. Econ. **49**(3), 189–200 (2013)

19. Marakulin, V.M.: On contractual approach for Arrow-Debreu-McKenzie economics. Econ. Math. Methods **50**(1), 61–79 (2014). (in Russian)

20. Mas-Colell, A., Whinston, M.D., Green, J.R.: Microeconomic Theory, 981 p. Oxford University Press, New York (1995)

21. Polovinkin, E.S.: On the question of differentiation of multivalued mappings. Some problems of modern mathematics and their applications to problems of math. Physics, pp. 90–97. Moscow Institute of Physics and Technology, Moscow (1985)

22. Polovinkin, E.S., Balashov, M.V.: Elements of Convex and Strongly Convex Analysis, 416 p. Fizmatlit, Moscow (2004)

Optimization of Regulation and Infrastructure of the Electricity Market

Alexander Vasin[ID] and Olesya Grigoryeva[(✉)][ID]

Lomonosov Moscow State University, Moscow 119991, Russia

Abstract. This research examines a model of a wholesale electricity market including consumers and producers located at several nodes, transmitting lines, and energy storages. Tariff regulation aimed at shifting some part of consumption from peak zones of the schedule to off-peak time of the day is considered. Among producers, we distinguish renewable energy sources with stochastic production volumes. They can be used to replace more expensive energy sources, but under adverse conditions should be replaced by reserve capacities with conventional technologies or energy storages. Problems of optimal regulation aimed at maximizing the expected social welfare are studied. We prove that optimal tariffs for consumers at every node and each time interval should correspond to average marginal supply costs for the node and the time. Optimal control strategies for energy storages are determined with account of their total charge volumes and maximum charging rates. We prove that, for every storage, the strategy corresponds to maximization of its profit from energy resale at the competitive market. A problem of the market infrastructure optimization is also studied. Proceeding from Lagrange theorem, we obtain the system of equations for determination of optimal parameters of the storages.

Keywords: Energy storages · Optimal control · Lagrange theorem

1 Introduction

Energy storage is a new tool for increasing the efficiency of wholesale electricity markets. Electric capacity storage devices (storages) permit to redistribute the energy produced within the day, to provide the balance of the supply and the demand at every time and reduce the total production costs. In particular, storages facilitate the efficient use of renewable energy sources (RES). The volume of power they supply is a random variable depending on weather conditions. In a situation where it is necessary to guarantee the supply of energy to all consumers, under adverse conditions it should be replaced by the energy from other sources. Energy storages provide the efficient substitution. Another tool for the same purpose is tariff regulation aimed at transferring part of consumption from the peak zone of the schedule to off-peak time of the day. Aligning

Supported by RFBR № 19-01-00533 A.

the curves of the daily load of consumers reduces the demand for generating capacity, transmission and production costs.

The present paper aims to develop mathematical models for computation of the optimal tariff rates and the optimal control of energy storages for the wholesale electricity market. We study the case where the storage control is based on the reliable forecast of the random factors for the planning interval (for the day ahead). Below we find out the rule that determines the optimal tariff rates under these conditions. We also discuss the problem of computation of optimal parameters of the storages. Our previous research [1, 2] examined some situations with incomplete information on the random factors. On the other hand, the tariff rates are set for a long time, proceeding from the probabilistic distribution of the random factors.

Models of the electricity market, taking into account the mentioned new factors, have been developed in a number of scientific papers. In [3], inelastic demand from consumers is considered, which includes the hourly components of the required volume, as well as the shiftable load, which can be redistributed during the day, taking into account the cost of transferring from the most favorable time at less convenient. Using the theory of contracts, the authors study the problem of optimizing the operation of the energy system by introducing tariffs that encourage consumers to shift the shiftable load at off-peak times. The paper [4] discusses the problem of creating an optimal generation schedule in terms of minimizing costs and emissions. The optimal planning schedule is based on the use of different electricity prices for consumers in the day-ahead market, as well as energy storage systems to achieve optimal volumes of production and consumption. The results show that shiftable loads should be moved at off-peak periods, in particular at midnight. Another conclusion is that the presence of a large number of available blocks with low generating power is preferable to the presence of one block with a large output power. While cost minimization and emission minimization are conflicting goals, a solution can be found that optimizes the vector criterion. Paper [5] considers a similar problem of minimizing costs and emission in framework of a stochastic model and employs the probabilistic concept of confidence intervals in order to evaluate forecasting uncertainty. Particular optimization problems by means of the mentioned tools were also considered in [1, 6–8] for energy systems of various scales.

The main focus of the present research is on maximizing the social welfare for the wholesale electricity market by means of energy storage devices and tariff regulation. Section 2 introduces a model of a network wholesale electricity market with RES and energy storages and establishes the first-order conditions that determine the optimal control of the storages, proceeding from the Welfare Theorem [9]. Section 3 specifies the optimal control and price dynamics for some special cases, where constraints on the storage volume or on the charging rate are not binding. Section 4 studies the problem of computation of optimal parameters of the storage. Section 5 examines the optimal tariff rates. In conclusion, we discuss the main results and some tasks for the future.

2   Spot Market Model and Formulation of Optimization Problem

Consider a model of an electricity market consisting of several local markets and a network transmission system (see [10–12]). Let N be the set of nodes, and $L \subseteq N \times N$ be the set of edges in this network. Every node $i \in N$ corresponds to a local perfectly competitive market. Since the demand for electricity varies significantly during the day, we consider the functioning of the system depending on time $t \in \Theta = \overline{1, T}$, where t is a period of the time with approximately constant needs, T is the length of the planning interval. In particular, an hour of the day may be denoted by t. For any $(i, j) \in L$, the transmission line is characterized by the transmission capacity Q_{ij}^0 and loss coefficient k_{ij}. The energy flow q_{ij}^t from node i to node j at any time t is limited by Q_{ij}^0.

Consider the main groups of agents (producers and consumers) operating in the market. Let $A_1(i)$ denote the set of electricity producers with traditional technologies at node i. Each $a \in A_1(i)$ is characterized by the cost function $C_a(v)$ for providing capacity v at any time. The supply function $S_a(p_i)$ – $Arg \max_{v \geq 0}(vp_i - C_a(v))$ determines the optimal production volume for a given period depending on the price. The total supply function $S_i(p_i) = \sum_{a \in A_1(i)} S_a(p_i)$ characterizes aggregated behavior for this group and determines its total production v_{1i}^t at time t depending on the price p_i^t. Below we consider the aggregated cost function $C_i(v)$ corresponding to this supply function (see [1]). Denote $A_2(i)$ as the set of producers using RES at this node. For them, the total amount of supplied power is a random variable v_{2i}^t depending on the time and weather conditions at node i. The variable costs for solar panels and wind turbines are close to 0, so the capacities of these producers can be used to replace more expensive energy sources. However, under adverse conditions they must be replaced by reserve capacities with conventional technologies.

Consider electricity consumption. Denote B_i as the set of consumers at node i. Each consumer $b \in B_i$ has needs associated with a specific time t (for example, space heating and lighting), as well as several types of needs $l = \overline{1, L}$, which can be satisfied at different times. In the general case, the utility functions of consumers can be described as follows. For every b, his consumption is characterized by vector $\overrightarrow{v}_b = (v_{bl}^t, t \in \Theta, l = \overline{0, L})$, where the volume v_{b0}^t is the consumption associated with the needs for a given hour t. The functions $u_{b0}^t(v_{b0}^t, \psi_b^t), t \in \Theta$, show the utility of such consumption depending on the random factor ψ_b^t characterizing the weather conditions and other random events affecting the need for electricity. We assume below that $u_{b0}^t(v_{b0}^t, \psi_b^t) = u_{b0}^t(v_{b0}^t + \psi_b^t)$. The volume $v_{bl}^t \geq 0$ determines the energy consumption associated with the target l in the period t. For each type l, the utility of consumption depends on the total amount of energy allocated to the corresponding target l, taking into account the costs e_{bl}^t of shifting this consumption to time t. Thus, the total utility of the consumer is as follows:

$$U_b(\overrightarrow{v}_b, \overrightarrow{\psi}_b) = \sum_{t=1}^{T} u_{b0}^t(v_{b0}^t, \psi_b^t) + \sum_{l=1}^{L}(u_{bl}(\sum_{t=1}^{T} v_{bl}^t) - \sum_{t=1}^{T} v_{bl}^t e_{bl}^t).$$

In the whole paper below, the cost and the utility functions meet standard assumptions for micro-economic models (see [13]): they are monotonously increasing, cost functions are convex, utility functions are concave.

Typically consumers cannot plan daily consumption schedules based on current wholesale prices, but determine them proceeding from tariff rates $\overrightarrow{\pi}_b = (\pi_b^t, t \in \Theta)$ set for a long time (about a year), taking into account mean values of the random factors affecting their consumption during the day. Demand function $\overrightarrow{D}_b(\overrightarrow{\pi}_b)$, where $\overrightarrow{D}_b = (D_{bl}^t, t \in \Theta, l = \overline{0, L})$, shows the optimal volume consumer b purchases at each time period t for every target l. The vector $\overrightarrow{D}_b$ is a solution to the following optimization problem:

$$\overrightarrow{v}_b^* \rightarrow \max \Big[\sum_{t=1}^{T} u_{b0}^t(v_{b0}^t) + \sum_{l=1}^{L}(u_{bl}(\sum_{t=1}^{T} v_{bl}^t) - \sum_{t=1}^{T} v_{bl}^t e_{bl}^t)$$
$$- \sum_{t=1}^{T} \overline{p}_b^t(v_{b0}^t + \sum_{l=1}^{L} v_{bl}^t - \mathbb{E}\,\psi_b^t)\Big]. \tag{1}$$

Note that, here v_{b0}^t is a planed consumption volume and the actual consumption volume for each period t is $v_{bo}^t - \psi_b^t$. We bound our study with the case where during all periods the consumption volume of each consumer remains positive, that is, a random factor does not nullify his demand.

The present model takes into account the possibility of using energy storages to optimize the functioning of the system. The characteristics of the storage at node i are the capacity, the rates and the efficiency coefficients of charging and discharging. Following the works [6,13], describe them as follows. Denote as E_i^{min} and E_i^{max} respectively, the minimum and the maximum allowable charge levels of the storage, V_i^{ch} and V_i^{dis}—maximum rates of charging and discharging, η_i^{ch} and η_i^{dis}—as charging and discharging efficiency coefficients, respectively. Denote as v_{iBat}^t the amount of energy the storage charges or discharges during period t, a positive value corresponds to charging; v_{iBat}^0 shows the initial charge. The storage control strategy is specified by vector $\overrightarrow{v}_{iBat} = (v_{iBat}^t, t \in \Theta)$. Feasible controls satisfy the following constraints: $\forall i \in N, \forall t \in \Theta$

$$0 \leq -v_{iBat}^t/\eta_i^{dis} \leq V_i^{dis} \text{ for discharging rates;} \tag{2}$$

$$0 \leq \eta_i^{ch} v_{iBat}^t \leq V_i^{ch} \text{ for charging rates;} \tag{3}$$

$$E_i^{min} \leq \sum_{k=0}^{t} v_{iBat}^k \leq E_i^{max}; \tag{4}$$

$$\sum_{t=1}^{T} v_{iBat}^{t} = 0. \tag{5}$$

In order to simplify formulas below, let $E_i^{min} = 0$, $E_i^{max} = E_i$, $\eta_i^{ch} = \eta_i^{dis} = \eta_i$, $V_i^{dis} = V_i^{ch} = V$, $v_{iBat}^0 = 0$, $\forall i \in N$.

The social welfare in this model is determined as follows. Under given tariff vector $\overrightarrow{p}$, storage control strategies $\overrightarrow{v}_{iBat}, i \in N$, and energy flows $\overrightarrow{q} = (q_{ij}^t, t \in \Theta, (i,j) \in L)$, where $q_{ij}^t = -q_{ji}^t$, the production volumes $\overrightarrow{v}_1 = (v_{1i}^t, t \in \Theta, i \in N)$ should meet the required electricity supply and provide the energy balance at every node i and any time t. That is, the volumes proceed from the equations:

$$\sum_{b \in B_i} D_b^t(\overrightarrow{\pi}) - \psi_i^t = v_{1i}^t + v_{2i}^t - v_{Bat i}^t + \sum_{l:(l,i) \in L} q_{li}^t(1 - k_{li}(q_{li}^t)), \tag{6}$$

$$\text{where} \quad k_{li}(q) = \begin{cases} k_{li}, & \text{if } q > 0, \\ 0, & \text{otherwise.} \end{cases} \tag{7}$$

The total social welfare for the network electricity market under given control strategies is

$$W(\overrightarrow{p}, \overrightarrow{v}_{Bat}, \overrightarrow{q}, \overrightarrow{\psi}) = \sum_{i \in N}(\sum_{b \in B_i} u_b(D_b(\overrightarrow{p})) - \sum_{t \in \Theta} c_i(\sum_{b \in B_i} D_b^t(\overrightarrow{p})$$

$$+ \eta_i^t v_{iBat}^t - \psi_i^t - v_{2i}^t - \sum_{l:(l,i) \in L} q_{li}^t(1 - k_{li}(q_{li}^t)))), \tag{8}$$

$$\text{where} \quad \eta_i^t = \begin{cases} \eta_i^{ch}, & \text{if } v_{iBat}^t > 0; \\ \frac{1}{\eta_i^{dis}}, & \text{if } v_{iBat}^t < 0. \end{cases}$$

This is the total utility of the consumption minus the costs of the production with account of the energy transmission and exchange with the storages. For every node, it includes two stochastic components: the total energy production by the RES and the random change of the planned demand. First, we study the case where the values of the random variables are known till the end of the planning interval, so the control depends on these values. Below we assume that vector $\overrightarrow{\psi}$ shows the sum of the two random variables. The problem of optimization of the mathematical expectation of the social welfare is formulated as

$$\max_{\overrightarrow{\pi}, \overrightarrow{v}_{Bat}(\overrightarrow{\psi}), \overrightarrow{q}(\overrightarrow{\psi})} \mathbb{E} W(\overrightarrow{p}, \overrightarrow{v}_{Bat}, \overrightarrow{q}, \overrightarrow{\psi}) \tag{9}$$

under conditions (2–5) and $q_{ij} \leq Q_{ij}^0$ for any $(i,j) \in L$.

The rest of this section and Sects. 3, 4 examine the corresponding problem of optimization of the social welfare by consumption volumes (instead of tariff rates) under the known random variables:

$$W(\overrightarrow{v}, \overrightarrow{v}_{Bat}, \overrightarrow{q}) = \sum_{i \in N}(\sum_{b \in B_i} u_b(\overrightarrow{v}_b) - \sum_{t \in \Theta} c_i(\sum_{b \in B_i} \sum_l v_{bl}^t + \eta_i^t v_{iBat}^t$$
$$- \psi_i^t - \sum_{l:(l,i) \in L} q_{li}^t(1 - k_{li}(q_{li}^t))))) \to max. \tag{10}$$

With this optimization problem, we can match a competitive network market with several products $g^t, t \in \Theta$, where good g^t corresponds to the energy consumed in period t. Each product is produced by generating companies of the sets A_{1i} and A_{2i} specified above. For every node i, the cost function is determined as $c_i^t(v) = c_i(v - \psi_i^t)$, where $c_i()$ is the minimum cost function determined above for the set of companies A_{1i}. Besides that, product g^t may be produced by energy storages. In this context, we consider the storage at node i as a producing capacity: at every time t it can use the energy produced in the previous periods as the resource for energy production at this time. For storage i, his production strategy is determined by vector $\overrightarrow{v}_{iBat}$, which meets conditions (2–5). Under a given strategy, the profit is $\sum_t \eta_i^t v_{iBat}^t p_i^t$. In order to obtain conditions of the competitive equilibrium for this sector, consider the Lagrange function for every storage i: $L_i(\overrightarrow{v}_{iBat}, \overrightarrow{\lambda}) = \sum_t \eta_i^t v_{iBat}^t p_i^t + \sum_{t \in \Theta} \sum_{r=1}^{5} \lambda_{ir}^t g_r^t(\overrightarrow{v}_{iBat})$, where

$\forall i \in N, t \in \Theta \; g_{i1}^t(\overrightarrow{v}_{iBat}) := v_{iBat}^t/\eta_i^{dis} + V_i^{dis}, g_{i2}^t(\overrightarrow{v}_{iBat}) := V_i^{ch} - \eta_i^{ch} v_{iBat}^t,$

$g_{i3}^t(\overrightarrow{v}_{iBat}) := \sum_{k=0}^{t} v_{iBat}^k, g_{i4}^t(\overrightarrow{v}_{iBat}) := E_i - \sum_{k=0}^{t} v_{iBat}^k, g_{i5}(\overrightarrow{v}_{iBat}) := \sum_{t=1}^{T} v_{iBat}^t.$

Inequalities $g_{ik}^t(...) \geq 0, t \in \Theta, k = \overline{1,4}$, correspond to constraints (2–4) imposed on the storage control, and equality $g_{i5}(\overrightarrow{v}_{iBat}) = 0$—to constraint (5).

Consider also the first-order condition for v_{iBat} to maximize the Lagrange function:

$$\begin{cases} v_{iBat}^t > 0 \Rightarrow \eta p_i^t = \sum_{k=t}^{T}(\lambda_{i3}^k - \lambda_{i4}^k) + \lambda_{i5} - \eta_i \lambda_{i2}^t, \\[2mm] v_{iBat}^t < 0 \Rightarrow p_i^t = \lambda_{i1}^t + \eta_i(\sum_{k=t}^{T}(\lambda_{i3}^k - \lambda_{i4}^k) + \lambda_{i5}), \\[2mm] v_{iBat}^t = 0 \Rightarrow \eta_i(\sum_{k=t}^{T}(\lambda_{i3}^k - \lambda_{i4}^k) + \lambda_{i5}) \geq p_i^t \geq (\sum_{k=t}^{T}(\lambda_{i3}^k - \lambda_{i4}^k) + \lambda_{i5})/\eta_i. \end{cases} \tag{11}$$

The conditions of complementary slackness for the constraints (2, 3, 5) are $\lambda_{i1}^t(V_i^{dis} + \frac{v_{iBat}^t}{\eta_i^{dis}}) = 0, \lambda_{i2}^t(V_i^{ch} - \eta_i^{ch} v_{iBat}^t) = 0, \lambda_{i3}^t(\sum_{k=0}^{t} v_{iBat}^k - E_i^{min}) = 0,$ $\lambda_{i4}^t(E_i^{max} - \sum_{k=0}^{t} v_{iBat}^k) = 0.$

The collection consisting of price vector $\overrightarrow{p} = (p_i^t, t \in \Theta, i \in N)$, consumption volume vector $\overrightarrow{v} = (\overrightarrow{v}_b, b \in B_i)$ and flow vector $\overrightarrow{q} = (q_{ij}^t, t \in \Theta, (i,j) \in L)$, is called a competitive equilibrium of the given market if it satisfies (11) and the following conditions (12–16):

$$(v_{bl}^t > 0) \Rightarrow p_i^t + e_{bl}^t = \arg\min_{\tau \in \Theta}(p_i^\tau + e_{bl}^\tau) = u_{bl}'(\sum_{\tau \in \Theta} v_{bl}^\tau) \tag{12}$$

$$\forall t \in \Theta, b \in B_i, l = \overline{1, L},$$

that is, every consumer b, for any target l, chooses a consumption vector that maximizes his utility, taking into account the cost of purchasing energy and the costs of transferring the consumption to a less convenient time;

$$\forall t \in \Theta, b \in B_i \quad u_{b0}^{t}{}'(v_{b0}^t) = p_i^t, \tag{13}$$

that is, the volume of the current consumption in period t is determined from the equity of the marginal utility of consumption to the energy price;

$$\forall t, i \quad c_i'(\sum_{b \in B_i} \sum_l v_{bl}^t - \psi_i^t + \eta_i^t v_{iBat}{}^t - \sum_{(j,i) \in L} q_{ji}^t(1 - k_{ji}(q_{ji}^t))) = p_i^t, \tag{14}$$

that is, in each period at every node i the marginal cost of production is equal to the price of energy;

$$\forall t, (i,j) \in L \quad (1/(1 - k_{ij}) > p_i^t/p_j^t > 1 - k_{ij}) \Rightarrow q_{ij}^t = 0; \tag{15}$$

$$q_{ij}^t \in (0, Q_{ij}^0) \Rightarrow p_i^t = (1 - k_{ij})p_j^t, p_j^t(1 - k_{ij}) > p_i^t, \Rightarrow q_{ij}^t = Q_{ij}^0. \tag{16}$$

These relations mean that the flow between nodes i and j grows as long as the commodity transportation is profitable for a small dealer (arbitrager).

Theorem 1. *Problem (10) of the social welfare optimization is a convex programming problem. The set of its solutions corresponds to the set of competitive equilibria given by relations (11–16) for the specified market.*

Proof. The market under consideration meets the conditions of the Welfare Theorem, (see [1]). The set of strategies in problem (10) is convex and closed. Given that $c_i(v) \to \infty$ under $v \to \infty$, the area, where the objective function is positive, is bounded. The properties of functions $c_i(v)$, $u_b(v)$ imply concavity of the objective function. Therefore (10) is a convex programming problem. The first-order conditions (11–16) determine its solutions. According to the Welfare Theorem, each solution is a competitive equilibrium.

3   Optimal Control of the Storage at the Local Market

Below we specify the optimal control for some particular cases of problem (10). In order to simplify formulas, consider a local market with one storage. The results are easy to generalize to a network market. First, examine the case where the storage volume is relatively small, so the optimal control does not change the general structure of the price dynamics, and the constraint on the charging rate is not binding.

Consider equilibrium prices $p_0(t)$, $t \in \Theta$, under the unused storage, that is, under $\overrightarrow{v}_{Bat} = 0$. Let p_0^M and p_0^m denote the maximal and the minimal prices. If $p_0^M \leq \eta_i^2 p_0^m$ then the prices and the storage control $\overrightarrow{v}_{Bat} = 0$ correspond to the competitive equilibrium since condition (11) holds. Otherwise, consider essential local extremums $t_1 < \overline{t}_1 < t_2 < \overline{t}_2 < \ldots < t_k < \overline{t}_k$ such that $t_1, \ldots, t_k$ are local minimums, $\overline{t}_1, \ldots, \overline{t}_k$ are local maximums which meet the following conditions:

$$\forall t \in (t_l, \overline{t}_l) \ \ p_0^{\overline{t}_l} \geq p_0^t \geq p_0^{t_l}, \ \ \forall t' \in (t, \overline{t}_l) \ \ \eta^2 p_0^{t'} > p_0^t,$$

$$\forall t \in (\overline{t}_l, t_{l+1}) \ \ p_0^{\overline{t}_l} \geq p_0^t \geq p_0^{t_{l+1}}, \ \ \forall t' \in (t, t_{l+1}) \ \ p_0^t < \eta^2 p_0^t, \tag{17}$$

$$\forall l \ \ p_0^{\overline{t}_l} > \eta^2 p_0^{t_l}, p_0^{\overline{t}_l} > \eta^2 p_0^{t_{l+1}}, \quad here \quad k+1 := 0.$$

These conditions mean that, under fixed prices p_0^t, $t \in \Theta$, for the storage it is profitable to buy the maximal amount of energy at time t_l and sell it at time $\overline{t}_l$ for every $l = \overline{1, k}$, and it is unprofitable to buy or sell energy in any other periods. Figure 1 shows some examples of essential and unessential local extremums.

Determine the equilibrium prices and the storage control that meet conditions (11–16). Let $E(t) = \sum_{k=1}^{t} v_{Bat}^k$. For every $l = \overline{1, k}$ we find time interval $(t_l^{st}, t_l^{fin}) \ni t_l$, storage charging rates v_{Bat}^{t*}, $t \in (t_l^{st}, t_l^{fin})$, and price p_l^m such that

$$E(t_l^{st} - 1) = 0, \ E(t_l^{fin}) = E,$$

$$\forall t \in (t_l^{st}, t_l^{fin}) \ \ v_{Bat}^t \geq 0, \ p^t \geq p_l^m, \ p^t > p_l^m \Rightarrow v_{Bat}^t = 0. \tag{18}$$

These conditions mean that the storage is completely charged up to the end of the interval, and the price stays the same in all periods when $v_{Bat}^{t*} > 0$. In a similar way, we determine time interval $(\overline{t}_l^{st}, \overline{t}_l^{fin}) \ni \overline{t}_l$, storage discharging rates v_{Bat}^{t*}, $t \in (\overline{t}_l^{st}, \overline{t}_l^{fin})$, and price p_l^M such that

$$E(\overline{t}_l^{st} - 1) = E, \ E(\overline{t}_l^{fin}) = 0,$$

$$\forall t \in (\overline{t}_l^{st}, \overline{t}_l^{fin}) \ \ v_{Bat}^t \leq 0, \ p^t \leq p_l^m, \ p^t < p_l^m \Rightarrow v_{Bat}^t = 0. \tag{19}$$

We set

$$v_{Bat}^{t*} = 0 \ \forall t \notin \cup_l (t_l^{st}, t_l^{fin}) \cup (\overline{t}_l^{st}, \overline{t}_l^{fin}) \tag{20}$$

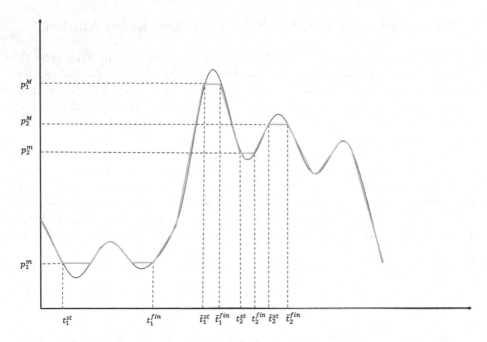

Fig. 1. Initial prices p_0^t and equilibrium prices p^t under binding constraint on the storage volume.

and check conditions

$$\forall t \in (t_l^{fin}, \bar{t}_l^{st}) \ p^t \in (p_l^m, p_l^M), \ \forall t' \in (t, \bar{t}_l) \ \eta^2 p_0^{t'} > p_0^t,$$
$$\forall t \in (\bar{t}_l^{fin}, t_{l+1}^{st}) \ p^t \in (p_{l+1}^m, p_l^M), \ \forall t' \in (t, t_{l+1}) \ p_0^{t'} < \eta^2 p_0^t. \tag{21}$$

Note 1. The equilibrium prices in intervals $(t_l^{fin}, \bar{t}_l^{st})$ and $(\bar{t}_l^{fin}, t_{l+1}^{st})$ do not change ($p^t = p_0^t$) if the shiftable loads stay the same there. Then conditions (21) hold. Figure 1, 2 and 3 represent this case.

Note that, for every l, the intervals monotonously extend, the price p_l^m monotonously and continuously increases, and the price p_l^M monotonously and continuously decreases by the storage volume E. Thus, under sufficiently small E, there exists collection $(\vec{v}, \vec{v}_{Bat}, \vec{p})$ that meets conditions (12–14). In order to be a competitive equilibrium, the collection should also meet relations (11). The necessary and sufficient conditions for this are (21) and

$$\forall l \ \eta^2 p_l^m \le min(p_{l-1}^M, \ p_l^M), \ \text{where} \ p_0^M := p_k^M. \tag{22}$$

Thus, we obtain the following result.

Theorem 2. *Let prices p_l^m and p_l^M, $l = \overline{1,k}$, determined according to conditions (18, 19), meet inequality (22). Consider consumption volume vector $\vec{v}^*$, price vector $\vec{p}^*$ and storage control $\vec{v}^*_{Bat}$, proceeding from conditions (12–14) and (18–20). If the prices meet also (21) then this collection corresponds to the* *competitive equilibrium of the market, and $\overrightarrow{v}^*_{Bat}$ determines the optimal control of the storage for problem (10).*

Consider the other case where the constraint on the charging rate may be binding, while the constraint on the storage volume is never binding. Then condition (11) shows that $p^t = -\lambda^t_2 + \lambda_5/\eta$ in periods with the maximal charging rate, $p^t = \lambda^t_1 + \lambda_5\eta$ in periods with the maximal discharging rate. The results of our previous research ([2], theorems 4 and 5) imply the next theorem for this case (see also Fig. 2).

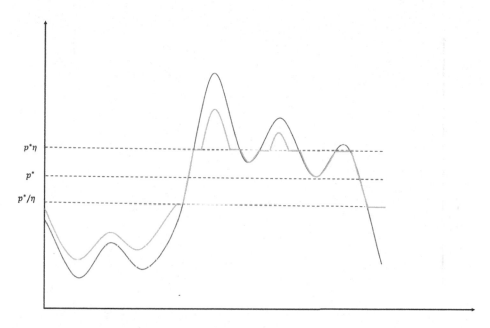

Fig. 2. Initial prices p^t_0 and equilibrium prices p^t under binding constraint on the charging rate.

Theorem 3. *If the constraint on the storage volume is never binding then the optimal collection $(\overrightarrow{v}^*, \overrightarrow{v}_{Bat}, \overrightarrow{p})$ for problem (10) satisfies the following conditions: there is a price p^*, such that:*

- *if price p^t in period t meets condition $p^t > p^*\eta$, then the storage is discharged at the maximal rate: $v^t_{Bat} = -V\eta$;*
- *if $p^t = p^*\eta$, then the storage is discharged;*
- *if $\frac{p^*}{\eta} < p^t < p^*\eta$, then the storage is not charged or discharged, $v^t_{Bat} = 0$;*
- *if $p^t = \frac{p^*}{\eta}$, then the storage is charged;*
- *if $p^t < \frac{p^*}{\eta}$, then the storage is charged at the maximal rate: $v^t_{Bat} = V/\eta$.*

The optimal values v_{Bat}^t, $t \in \Theta$, obtained from the first-order conditions, monotonously increase by p^*, so this price is uniquely determined from equation: $\sum_{t=1}^{T} v_{Bat}^t(p^*) = 0$.

An interesting particular case is where volume E and maximal rate V are so large that none of constraints (2–4) is binding. Then the optimal control almost equalizes the equilibrium prices: $\forall t \; p^*/\eta \leq p^t \leq p^*\eta$; if $p^t \in (p^*/\eta, p^*\eta)$ then $v_{Bat}^* = 0$. Figure 3 shows the price dynamics under the optimal control for the same initial prices.

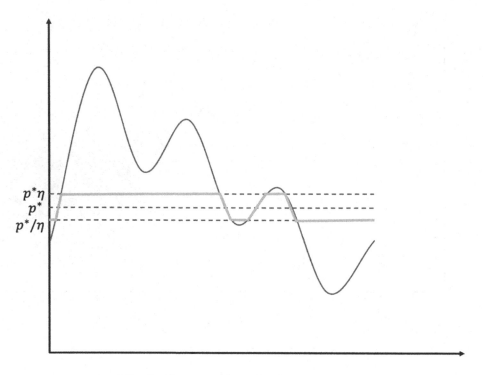

Fig. 3. The case without binding constraints.

Next, consider the opposite case: the storage volume is relatively small, so the optimal control does not change the general structure of the price dynamics, and the constraint on the charging rate is binding in some periods. Then the equilibrium prices and the optimal control are similar to those determined by Theorem 2, with the following modifications. In rule (18) for determination of the time intervals, when the storage accumulates the energy, we change condition $\forall t \in (t_l^{st}, t_l^{fin})$, $l = \overline{1, k}$, $p^t \geq p_l^m$ for $\forall t \in (\overline{t}_l^{st}, \overline{t}_l^{fin})$, $l = \overline{1, k}$, $(p_b^t < p_l^m) \Rightarrow v_{Bat}^{t*} = V/\eta$. Sometimes in these intervals the charging rate V is insufficient to increase the equilibrium price p^t to p_l^m. In a similar way, condition

$\forall t \in (t_l^{st}, t_l^{fin})(p^t > p_l^m) \Rightarrow v_{Bat}^{t*} = -V\eta$ substitutes $\forall t \in (\overline{t}_l^{st}, \overline{t}_l^{fin})$ $p^t \leq p_l^M$ in rule (19) for determination of the time intervals, when the storage returns the energy to the market. With these changes, Theorem 2 holds for this case.

4   Optimization of the Storage Parameters

Assume that the storage volume and the charging rate can be increased, the initial values are E_0 and V_0, and functions $C_1(E)$ and $C_2(V)$ determine the costs of their increasing, reduced to the planning interval Θ (see [14]). Then E and V become the components of control vector $u = (v_{Bat}^t, t \in \Theta, E, V)$ that should meet conditions (2–5) and $E \geq E_0, V \geq V_0$. Consider the corresponding problem of optimization of the social welfare:

$$W(\overrightarrow{v}_{Bat}, u) = \sum_{t=1}^{T} [u^t(v^t) - c(v^t + \eta^t(v_{Bat}^t)v_{Bat}^t)] - C_1(E) - C_2(V) \to \max.$$

(23)

Let marginal costs $C_1'(E)$ and $C_2'(V)$ monotonously increase and tend to ∞ as the arguments tend to ∞. Consider the Lagrange function for the problem of optimization by u: $\overline{L}(\overrightarrow{v}_{Bat}, \overrightarrow{\lambda}, E, V) = L(\overrightarrow{v}_{Bat}, \overrightarrow{\lambda}) - C_1(E) - C_2(V)$. Let $E^* > E_0$, $V^* > V_0$. The first-order conditions include, besides (11), also the following equations:

$$C_1'(E) = \sum_{t=1}^{T} \lambda_4^t; \quad C_2'(V) = \sum_{t=1}^{T} (\lambda_1^t + \lambda_2^t).$$

(24)

Theorem 4. *Problem (23) of the social welfare optimization is a convex programming problem. Any optimal strategy $(\overrightarrow{v}_{Bat}^*, u^*)$ satisfies conditions (11–14, 24).*

Next, consider the case where the optimal storage volume E^* is relatively small, so the optimal control does not change the general structure of the price dynamics, and the constraint on the charging rate is never binding, so $V^* = V_0$. The conditions for this are similar to (18–21). Let $p_l^m(E), p_l^M(E), l = \overline{1, k}$, denote the prices proceeding from this system under a given volume E.

Theorem 5. *Let time intervals (t_l^{st}, t_l^{fin}) and $(\overline{t}_l^{st}, \overline{t}_l^{fin})$, $l = \overline{1, k}$, prices p^t, $t \in \Theta$, storage control vector $\overrightarrow{v}_{Bat}^*$, consumption volume vector $\overrightarrow{v}^*$, and storage volume $E^* \geq E_0$ meet conditions (12–14, 18–22) and*

$$\sum_{l=1}^{k} (p_l^M(E) - p_l^m(E)) = C_1'(E).$$

(25)

Then these values correspond to the competitive equilibrium of the market, and $(\overrightarrow{v}_{Bat}^*, E^*)$ determines the optimal control of the storage for problem (23). If $E^* < E_0$ while the other conditions hold, then E_0 is the optimal volume.

Proof. In this case $\lambda_1^t = \lambda_2^t = 0$ for any $t \in \Theta$; $p_{\bar{t}l}^{t^{st}} = p_{\bar{t}l}^{t^{fin}} = p_l^m(E)$, $\forall t \in [t_l^{st}, t_l^{fin})$ $\lambda_3^t = \lambda_4^t = 0$; $\forall t \in [\bar{t}_l^{st}, \bar{t}_l^{fin})$ $p^t = p_l^M$, $p^{\bar{t}^{st}} = p^{\bar{t}^{fin}} = p_l^M(E)$; $\forall t \in [t_l^{st}, \bar{t}_l^{fin}]$ $\lambda_3^t = 0$, and, proceeding from (11), let $\lambda_4^t = p^{t+1} - p^t$; $\forall t \in [\bar{t}_l^{st}, t_l^{fin}]$ $\lambda_4^t = 0$, and, proceeding from (11), let $\lambda_3^t = p^t - p^{t+1}$. Thus,

$$\sum_{t \in \Theta} \lambda_3^t = \sum_{t \in \Theta} \lambda_4^t = \sum_{l=1}^k (p_l^M(E) - p_l^m(E)) = C_1'(E).$$

Now, consider the case where only constraints (2, 3) on the charging rates may be binding. Then $E^* = E_0$, and for any V, Theorem 3 determines the optimal storage control, the threshold price $p^*(V)$ and the equilibrium prices $p^t(V)$. Proceeding from conditions (11), we obtain for this case: $\lambda_3^t = \lambda_4^t = 0$ for any $t \in \Theta$; $\lambda_2^t = p^*/\eta - p^t \geq 0$ if $v_{Bat}^t = V/\eta$, otherwise $\lambda_2^t = 0$; $\lambda_1^t = p^t - \eta p^* \geq 0$ if $v_{Bat}^t = -V\eta$, otherwise $\lambda_1^t = 0$. Condition (24) implies the next result:

Theorem 6. *If* $\displaystyle\sum_{t:p^t(V_0)<p^*/\eta} (p^*(V_0)/\eta - p^t(V_0)) + \sum_{t:p^t(V_0)>p^*\eta} (p^t(V_0) - \eta p^*(V_0)) <$ $C_2'(V_0)$ *then* $V^* - V_0$, *otherwise the optimal charging rate* V^* *meets equation*

$$\sum_{t:p^t(V)<p^*/\eta} (p^*(V)/\eta - p^t(V)) + \sum_{t:p^t(V)>p^*\eta} (p^t(V) - \eta p^*(V)) = C_2'(V).$$

5 The Optimal Control for the Stochastic Problem

Now, return to problem (9), where, instead of consumption volumes, tariff rates for consumers are the components of the control. Under the known random variables $\vec{\psi}$, the optimal values of social welfare are the same for problems (9) and (10): it is enough to set tariff rates equal to the equilibrium prices specified in Theorem 1, then the demand functions determine the optimal consumption volumes meeting conditions (12–14). However, in reality many consumers cannot plan daily consumption schedules based on current wholesale prices, but determine them, proceeding from random factors and tariff rates set for a long time. Assume that behaviour of consumers meets relation (1). Then, any fixed tariff rates $\vec{\pi}$ determine the corresponding consumption volumes $\vec{v}(\vec{\pi})$. Under the given volumes and random variables $\vec{\psi}$, consider optimization of the welfare in (10) by $\vec{v}_{Bat}$ and $\vec{q}$. The optimal values $(\vec{v}_{Bat}, \vec{q})(\vec{v}, \vec{\psi})$ are determined according to Theorem 1: as well as the equilibrium prices, they proceed from conditions (11, 14–16). Next, consider optimization of the welfare by consumption volumes in problem (10). The optimal volumes $\vec{v}^{**}$ meet the following first-order conditions:

$$(v_{bl}^t > 0) \Rightarrow \pi_b^t + e_{bl}^t = \arg\min_{\tau \in \Theta}(\pi_b^\tau + e_{bl}^\tau) = u_{bl}'\left(\sum_{\tau \in \Theta} v_{bl}^\tau\right) \tag{26}$$

$$\forall t \in \Theta, b \in B_i, l = \overline{1, L};$$

$$\forall t \in \Theta, b \in B_i \quad u_{b0}^{t\,\prime}(v_{b0}^t) = \pi_b^t. \tag{27}$$

For problem (9), we obtain the same value of the welfare if we set:

$$\overrightarrow{\pi} = E(\overrightarrow{p}(\overrightarrow{\psi})). \tag{28}$$

Then the optimal volumes determined by demand functions $D_b(\overrightarrow{p})$ coincide with $\overrightarrow{v}^{**}$ since they meet the same first-order conditions. On the other hand, the optimal value of problem (9) does not exceed the optimal value in problem (10), since determination of consumption volumes by the demand function is a special case of an arbitrary choice of volume vectors in problem (10). Finally we obtain the following result.

Theorem 7. *The optimal tariff rates $\overrightarrow{\pi}^{**}$, storage control and energy flows $(\overrightarrow{v}^{**}_{Bat}, \overrightarrow{q}^{**})(\overrightarrow{\psi})$ for problem (9) proceed from system (11, 14–16, 26–28).*

Thus, the optimal tariff rates are equal to the average values of the equilibrium wholesale prices. Note that storage control and energy flows $(\overrightarrow{v}^{**}_{Bat}, \overrightarrow{q}^{**})(\overrightarrow{\psi})$, as well as the equilibrium wholesale prices $\widetilde{p}(\overrightarrow{\psi})$, do not coincide with the similar values for the model with completely rational and informed consumers studied in the previous sections.

Let $\overline{W}(E, V)$ denote the optimal value of the expected social welfare depending on parameters of the storage for a local market. In practice, these values do not change for a long time including many planning intervals. Thus, the optimal values correspond to the solution of the following optimization problem:

$$\overline{W}(E, V) - C_1(E) - C_2(V) \to \max.$$

The solution may be obtained from the first-order conditions in a similar way as for the model in Sect. 3.

6   Conclusion

The present paper examined several problems of optimal regulation of the electricity market. The regulation aims to maximize the social welfare. We looked at the network wholesale market that uses renewable energy sources, tariff regulation and energy storages. The case, where random variables (energy generation by RES, random components of the energy consumption) are known for the planning interval (the day ahead), was completely examined. We proved the convexity of the problems and obtained the first-order conditions for computation of the optimal energy flows and the optimal control of the energy storages. Also, conditions were obtained to determine the optimal storage volume and charging rate at the local market. For the stochastic problem of choosing the rates of electricity tariffs, it is proved that the optimal rates should correspond to the average marginal costs of energy production.

An important task for future research is the solution of similar problems with incomplete information about the values of random variables in the planning interval.

References

1. Vasin, A., Grigoryeva, O.: On optimizing electricity markets performance. In: Olenev, N., Evtushenko, Y., Khachay, M., Malkova, V. (eds.) OPTIMA 2020. LNCS, vol. 12422, pp. 272–286. Springer, Cham (2020). https://doi.org/10.1007/978-3-030-62867-3_20
2. Vasin, A.A., Grigoryeva, O.M., Shendyapin, A.S.: Izvestiya RAN. Theor. Control Syst. **3** (2021)
3. Aizenberg, N., Stashkevich, T., Voropai, N.: Forming rate options for various types of consumers in the retail electricity market by solving the adverse selection problem. Int. J. Public Adm. **2**(5), 99–110 (2019)
4. Nazari, A.A., Keypour, R.: Participation of responsive electrical consumers in load smoothing and reserve providing to optimize the schedule of a typical microgrid. Energy Syst. **11**(4), 885–908 (2019). https://doi.org/10.1007/s12667-019-00349-9
5. Yaagoubi, N., Mouftah, H.T.: User-aware game theoretic approach for demand management. IEEE Trans. Smart Grid **6**(2), 716–725 (2015). https://doi.org/10.1109/TSG.2014.2363098
6. Gellings, C.W.: The concept of demand-side management for electric utilities. Proc. IEEE **73**(10), 1468–1570 (1985). https://doi.org/10.1109/PROC.1985.13318
7. Conejo, J., Morales, J.M., Baringo, L.: Real time demand response model. IEEE Trans. Smart Grid **1**(3), 236–242 (2010). https://doi.org/10.1109/TSG.2010.2078843
8. Samadi, P., Mohsenian-Rad, H., Schober, R., Wong, V.W.: Advanced demand side management for the future smart grid using mechanism design. IEEE Trans. Smart Grid **3**(3), 1170–1180 (2012). https://doi.org/10.1109/TSG.2012.2203341
9. Arrow, K.J., Debreu, G.: Existence of an equilibrium for a competitive economy. Econometrica **22**, 265–290 (1954)
10. Vasin, A.A., Dolmatova, M.S.: Optimization of transmission capacities for multinodal markets. Procedia Comput. Sci. **91**, 238–244 (2016)
11. Vasin, A.A., Grigoryeva, O.M., Lesik, I.A.: The problem of optimal development of the energy market of the transport system. Paper Presented at the VIII Moscow International Conference on Operation Research (ORM 2018), vol. 2, pp. 247–251 (2018)
12. Vasin, A.A., Grigoryeva, O.M., Tsyganov, N.I.: Optimization of an energy market transportation system. Dokl. Math. **96**(1), 411–414 (2017). https://doi.org/10.1134/S1064562417040202
13. Denzau, A.T.: Microeconomic Analysis: Markets and Dynamics. Richard d Irwin (1992)
14. Stoft, S.: Power System Economics: Designing Markets for Electricity. Wilcy-IEEE Press, New York (2002)

Data Analysis

Weakly Supervised Regression Using Manifold Regularization and Low-Rank Matrix Representation

Vladimir Berikov[1,2]([✉])[iD] and Alexander Litvinenko[3][iD]

[1] Sobolev Institute of Mathematics, Novosibirsk, Russia
[2] Novosibirsk State University, Novosibirsk, Russia
berikov@math.nsc.ru
[3] RWTH Aachen, Aachen, Germany
litvinenko@uq.rwth-aachen.de
http://www.math.nsc.ru
https://www.uq.rwth-aachen.de

Abstract. We solve a weakly supervised regression problem. Under "weakly" we understand that for some training points the labels are known, for some unknown, and for others uncertain due to the presence of random noise or other reasons such as lack of resources. The solution process requires to optimize a certain objective function (the loss function), which combines manifold regularization and low-rank matrix decomposition techniques. These low-rank approximations allow us to speed up all matrix calculations and reduce storage requirements. This is especially crucial for large datasets. Ensemble clustering is used for obtaining the co-association matrix, which we consider as the similarity matrix. The utilization of these techniques allows us to increase the quality and stability of the solution. In the numerical section, we applied the suggested method to artificial and real datasets using Monte-Carlo modeling.

Keywords: Weakly supervised learning · Manifold regularization · Low-rank matrix decomposition · Cluster ensemble · Co-association matrix

1 Introduction

Nowadays, machine learning (ML) theory and methods are rapidly developing and increasingly used in various fields of science and technology. An urgent problem remains a further improvement of ML methodology: the development of methods that allow obtaining accurate and reliable solutions in a reasonable time in conditions of noise distortions, large data size, and lack of training information. In many applications, only a small part of the data can be labeled, i.e., the values of the predicted feature are not provided for all data objects. In the case of a

P. Pardalos et al. (Eds.): MOTOR 2021, LNCS 12755, pp. 447–461, 2021.
https://doi.org/10.1007/978-3-030-77876-7_30

large amount of data and limited resources for its processing, some data objects can be inaccurately labeled.

As a real-world example of such a problem, one can address the task of manual annotation of a large number of computed tomography (CT) digital images. In order to distinguish the brain areas affected by stroke, it is required to engage a highly qualified radiologist, and the process is rather time-consuming. It is possible that some parts of the images will stay without specifying specific regions (for example, it is simply indicated that pathological signs are present in the given CT scan) or are labeled inaccurately. In this case, the assumed region can be outlined with a frame; the closer to the center of the frame, the greater the confidence that the brain tissue is damaged.

Weakly supervised learning is a part of ML research aimed at elaborating models and methods for the analysis of such type of information. In the formulation of a weakly supervised learning problem, it is assumed that some of the sample objects are labeled inaccurately. This inaccuracy can be understood in different ways [29].

In the case of *coarse grained* label information, class labels are provided only for sets of objects. For example, a collection of regulatory SNPs (Single-Nucleotide Polymorphisms) in DNA can be marked as a group linked to a pathology-associated gene. It is required to predict the class (its label) of a new group of objects. This task is also called *multi-instance learning*. This problem is considered, for example, in [26], in which a modification of the SVM method is proposed for the solution.

In another setting, it is assumed that there is an uncertainty in the indication of the exact class label arising from errors or due to the limitations of the observation method itself. Over time various solutions to the problem have been proposed. One of them is based on finding potentially erroneous labels and correcting them [23]. A similar idea (called *censoring* of the sample) was developed in [8]. This approach usually relies on information about the nearest neighbors of points. Therefore it becomes less reliable in high-dimensional feature spaces since the points become approximately equidistant from each other.

Another methodology is used when the labeling is performed by many independent workers (i.e., *crowd-sourcing* technique); among them can be both experienced and inexperienced members (and even deliberately mistaken). To solve the problem, probabilistic or ensemble methods are used [24, 28].

The following approach is based on minimizing theoretical risk estimates taking into account the random labeling error. The authors of [11] propose a method based on the fact that the empirical risk functional can be divided into two parts. The first part does not depend on the noise. Only the second part is affected by noisy labels.

Methods based on the *cluster assumption* and the *manifold assumption* are also used [2, 16]. In [12], the upper bounds of the labels' noise characteristics are obtained, and an algorithm for estimating the degree of noise is proposed using preliminary data partitioning into clusters.

In this paper, we consider a weakly supervised regression problem in the transductive learning setting. It means that the test sample is known, and the values of the predictors can be used as additional information for the target feature prediction.

We propose a novel method using a combination of manifold regularization methodology, cluster ensemble, and low-rank matrix representation. We assume the existence of the dependence between clusters presented in the data and the predicted continuous target feature (cluster assumption). Such dependence can be found, for example, when some hidden structures are present in data, and the belonging of objects to the same structural unit affects the similarity of their target feature values.

In the rest of the paper, we give the formal statement of the problem (Sect. 2), describe the details of the suggested method (Sects. 3 and 4), and present the results of numerical experiments (Sect. 5). Finally, we give some concluding remarks.

2 Problem Description and Notation

Consider a dataset $\mathbf{X} = \{x_1, \ldots, x_n\}$, where $x_i \in \mathbb{R}^d$ is a feature vector, d the dimensionality of feature space, and n the sample size. Suppose that each data point is sampled from an unknown distribution $\mathcal{P}_X(x)$.

Fully supervised learning assumes we are given a set $Y = \{y_1, \ldots, y_n\}$, $y_i \in D_Y$, of target feature labels for each data point. In the regression problem, values from a continuous compact set $D_Y \subset \mathbb{R}$ are understood as target feature labels.

The objective is to find a decision function $y = f(x)$, which should forecast the target feature for new examples from the same distribution. The decision function should optimize a quality metric, e.g., minimize an estimate of the expected loss.

In an unsupervised learning problem, the target feature is not specified. It is necessary to find a meaningful representation of data, i.e., find a partition $P = \{C_1, \ldots, C_K\}$ of $\mathbf{X}$ on a relatively small number K of homogeneous clusters describing the structure of data. The homogeneity criterion depends on the similarity of observations within clusters and the distances between them. Quite often, the optimal number of clusters is unknown and should be determined using a cluster validity index.

The obtained cluster partition can be uncertain due to a lack of knowledge about data structure, vagueness in setting optional parameters of the learning algorithm, or dependence on random initializations. In this case, ensemble clustering is a way of obtaining a robust clustering solution. This methodology aims at finding consensus partition from different partition variants [7]. A properly organized ensemble (even composed of "weak" algorithms) often significantly improves the clustering quality.

In the problem of semi-supervised transductive learning, the target feature labels are known only for a part of the data set $\mathbf{X}_1 \subset \mathbf{X}$ (of comparatively small size as usual). We assume that $\mathbf{X}_1 = \{x_1, \ldots, x_{n_1}\}$, and the unlabeled part is $\mathbf{X}_0 = \{x_{n_1+1}, \ldots, x_n\}$. The set of labels for points from $\mathbf{X}_1$ is denoted by $\mathbf{Y}_1 = \{y_1, \ldots, y_{n_1}\}$. It is required to predict labels $\mathbf{Y}_0 = (y_{n_1+1}, \ldots, y_n)$ in the best way for the unlabeled sub-sample $\mathbf{X}_0$.

This task is essential because in many applied problems only a small part of available data can be labeled due to the considerable cost of target feature registration.

We consider a weakly supervised learning context, i.e., we suppose that for some data points, the labels are known, for some unknown, and for others uncertain due to reasons such as lack of resources for their careful labeling or presence of random distortions arising in the label identification process.

To model the uncertainty in the label identification, we suppose that for each ith data point, $i = 1, \ldots, n_1$, the value y_i of the target feature is a realization of a random variable Y_i with cumulative distribution function (cdf) $F_i(y)$ defined on D_Y. We suppose that $F_i(y)$ belongs to a given distribution family.

In this paper, the regression problem is considered, i.e., the predicted feature is continuous. Further we assume the following normal distribution model for the uncertain target variable:

$$Y_i \sim \mathcal{N}(a_i, \sigma_i), \tag{1}$$

where a_i, σ_i are the mean and the standard deviation respectively. The larger σ_i, the more uncertain is the labelling. It is presumed that parameters $a_i = y_i$ and $\sigma_i = s_i$ are known for each (weakly) labeled observation, $i = 1, \ldots, n_1$. For strictly determined observation y_i, we nevertheless postulate a normal uncertainty model with $a_i = y_i$ and small standard deviation $\sigma_i \approx 0$.

We aim at finding a weak labeling of points from $\mathbf{X}_0$, i.e., determining $F_i(y)$ for $i = n_1 + 1, \ldots, n$ following an objective criterion.

3   Manifold Regularization

Semi-supervised learning and weakly supervised learning assume two basic assumptions: cluster assumption and the assumption that the data with similar labels belong to a low-dimensional manifold.

According to the cluster assumption, one believes that objects from the same cluster often have the same labels or labels close to each other.

The manifold assumption is based on the hypothesis that there is a smooth manifold (for example, a two-dimensional surface in multidimensional space) to which points with similar labels belong. Manifold regularization [2,25] is based on this assumption. In addition to the learning error, the regularizing component is minimized during the model fitting stage. The component characterizes the smoothness of the decision function change. In dense regions, the decision function must change slowly, so its gradient must be small. In other words, data points from $\mathbf{X}$ lie on a low-dimensional non-linear manifold M, and the decision function is smooth on this manifold, i.e., points close to each other possess similar labels.

In semi-supervised learning, the regularization functional to be minimized can be written as following:

$$J(f) = \frac{1}{n_1} \sum_{x_i \in X_1} V(y_i, f_i) + \gamma \, \|f\|_M^2 \,,$$

where $f = (f_1, \ldots, f_n)$ is a vector of predicted labels, V a loss function, $\gamma > 0$ a regularization parameter, and $\|f\|_M^2$ characterizes the smoothness of the function. In dense regions, the decision function should change slowly, i.e., its gradient $\nabla_M f(x)$ should be small. Thus, the manifold regularizer can be chosen in this way:

$$\|f\|_M^2 = \int_{x \in M} \|\nabla_M f(x)\|^2 \, d\mathcal{P}_X(x).$$

Graph Laplacian (GL) [2,27] is a convenient tool for $\|f\|_M^2$ estimation. Let $G = (V, E)$ be a weighted non-oriented complete graph, in which the set of vertices V corresponds to points from $\mathbf{X}$, and the set of edges E corresponds to pairs (x_i, x_j), $i, j = 1, \ldots, n$, $i \neq j$. Each edge (x_i, x_j) is associated with a non-negative weight W_{ij} (the degree of similarity between the points).

The degree of similarity can be calculated by using an appropriate function, for example from the Matérn family [22]. The Matérn function depends only on the distance $h := \|x_i - x_j\|$ and is defined as

$$W(h) = \frac{\sigma^2}{2^{\nu-1}\Gamma(\nu)} \left(\frac{h}{\ell}\right)^\nu K_\nu \left(\frac{h}{\ell}\right)$$

with three parameters ℓ, ν, and σ^2. For instance, $\nu = 1/2$ gives the well-known exponential kernel $W(h) = \sigma^2 \exp(-h/\ell)$, and $\nu \to \infty$ gives the Gaussian kernel $W(h) = \sigma^2 \exp(-h^2/2\ell^2)$. In this paper we use the Gaussian covariance function, also called the radial basis function (RBF kernel), with $\sigma = 1$:

$$W_{ij} = \exp\left(-\frac{\|x_i - x_j\|^2}{2\ell^2}\right). \tag{2}$$

By $L := D - W$ we denote the standard GL, where D is a diagonal matrix with elements $D_{ii} = \sum_j W_{ij}$. There are also normalized GL: $L_{norm} = D^{-1/2} W D^{-1/2}$ and the random walk GL: $L_{rw} = D^{-1} W$.

One can show that in semi-supervised regression, the regularization term can be expressed as following:

$$\|f\|_M^2 \approx \frac{1}{n^2} f^\top L f = \frac{1}{n^2} \sum_{x_i, x_j \in \mathbf{X}} W_{ij}(f_i - f_j)^2.$$

4   Proposed Method

Consider a modification of the manifold regularization scheme for a considered weakly supervised transductive learning problem.

Let $F = \{F_1, \ldots, F_n\}$ denote the set of cdfs for data points; each cdf F_i is represented by a pair of parameters (a_i, σ_i).

4.1   Objective Functional

Consider the following optimization problem:

$$\text{find } F^* = \arg\min_F J(F), \text{ where}$$

$$J(F) = \sum_{x_i \in X_1} \mathcal{D}(Y_i, F_i) + \gamma \sum_{x_i, x_j \in \mathbf{X}} \mathcal{D}(F_i, F_j) W_{ij}. \tag{3}$$

Here $\mathcal{D}$ is a statistical distance between two distributions (such as the Wasserstein distance, Kullback-Leibler divergence, or other metrics). The first sum in the right side of (3) is aimed to reduce the dissimilarity on labeled data; the second component plays the role of a smoothing function: its minimization means that if two points x_i, x_j (either labeled or unlabeled) are similar, their labeling distribution should not be very different.

In this work, we use the Wasserstein distance w_p [6] (also known as the Kantorovich-Rubinstein distance or transportation metric) between distributions P and Q over a set D_Y as a measure of their dissimilarity:

$$w_p(P, Q) := \left(\inf_{\gamma \in \Gamma(P,Q)} \int_{D_Y \times D_Y} \rho(y_1, y_2)^p \, d\gamma(y_1, y_2) \right)^{1/p},$$

where $\Gamma(P, Q)$ is a set of all probability distributions on $D_Y \times D_Y$ with marginal distributions P and Q, ρ a distance metric, and $p \geq 1$.

It is known that for normal distributions $P_i = \mathcal{N}(a_i, \sigma_i)$, $Q_j = \mathcal{N}(a_j, \sigma_j)$ and the Euclidean metric, the w_2 distance is equal to [9]

$$w_2(P_i, Q_j) = (a_i - a_j)^2 + (\sigma_i - \sigma_j)^2.$$

We use the w_2 distance in (3) for weakly supervised regression and slightly modify the objective functional in (3) adding an L_2 regularizer:

$$J(a, \sigma) = \sum_{i=1}^{n_1} \left((y_i - a_i)^2 + (s_i - \sigma_i)^2 \right)$$

$$+ \gamma \sum_{i,j=1}^{n} \left((a_i - a_j)^2 + (\sigma_i - \sigma_j)^2 \right) W_{ij} + \beta(\|a\|^2 + \|\sigma\|^2), \tag{4}$$

where $\beta > 0$ is a regularization parameter, $a = (a_1, \ldots, a_n)^\top$, $\sigma = (\sigma_1, \ldots, \sigma_n)^\top$.

4.2   Optimal Solution

To find the optimal solution, we differentiate (4) and get:

$$\frac{\partial J}{\partial a_i} = 2(a_i - y_i) + 4\gamma \sum_{j=1}^{n} (a_i - a_j) W_{ij} + 2\beta a_i = 0, \quad i = 1, \ldots, n_1, \tag{5}$$

$$\frac{\partial J}{\partial a_i} = 4\gamma \sum_{j=1}^{n} (a_i - a_j) W_{ij} + 2\beta a_i = 0, \quad i = n_1 + 1, \ldots, n. \tag{6}$$

Denote $Y_{1,0} = (y_1, \ldots, y_{n_1}, \underbrace{0, \ldots, 0}_{n-n_1})^\top$ and let B be a diagonal matrix with elements

$$B_{ii} = \begin{cases} \beta+1, & i=1,\ldots,n_1 \\ \beta, & i=n_1+1,\ldots,n. \end{cases}$$

Combining (5), (6) and using vector-matrix notation, we finally get:

$$(B + 2\gamma L)a = Y_{1,0},$$

thus the optimal solution is

$$a^* = (B + 2\gamma L)^{-1} Y_{1,0}. \tag{7}$$

Similarly, one can obtain the optimal value of σ:

$$\sigma^* = (B + 2\gamma L)^{-1} S_{1,0}, \tag{8}$$

where $S_{1,0} = (s_1, \ldots, s_{n_1}, \underbrace{0, \ldots, 0}_{n-n_1})^\top$.

4.3   Low-Rank Similarity Matrix Representation

For large-scale problems, the dimensionality of matrices to be inverted in (7), (8) is very large and the inversion is costly. In many applications, a low-rank matrix decomposition is a useful tool for obtaining computationally efficient solutions [13]. Nyström method (see, e.g., [10]) or hierarchical low-rank matrix approximations [14,15,18–21] can be used for obtaining such a decomposition.

Let the similarity matrix be presented in the low rank form

$$W = AA^\top, \tag{9}$$

where matrix $A \in \mathbb{R}^{n \times m}$, $m \ll n$. Further, we have

$$B + 2\gamma L = B + 2\gamma D - 2\gamma AA^\top = G - 2\gamma AA^\top, \tag{10}$$

where $G = B + 2\gamma D$.

The following Woodbury matrix identity is well-known in linear algebra:

$$(S + UV)^{-1} = S^{-1} - S^{-1}U(I + VS^{-1}U)^{-1}VS^{-1}, \tag{11}$$

where $S \in \mathbb{R}^{n \times n}$ is an invertible matrix, $U \in \mathbb{R}^{n \times m}$ and $V \in \mathbb{R}^{m \times n}$.

Let $S = G$, $U = -2\gamma A$ and $V = A^\top$. One can see that

$$G^{-1} = \text{diag}\left(1/(B_{11} + 2\gamma D_{11}), \ldots, 1/(B_{nn} + 2\gamma D_{nn})\right). \tag{12}$$

From (7), (10), (11) and (12) we obtain:

$$a^* = (G^{-1} + 2\gamma G^{-1}A(I - 2\gamma A^\top G^{-1}A)^{-1}A^\top G^{-1}) Y_{1,0}. \tag{13}$$

Similarly, from (8), (10), (11) and (12) we have:

$$\sigma^* = (G^{-1} + 2\gamma G^{-1}A(I - 2\gamma A^\top G^{-1}A)^{-1}A^\top G^{-1}) S_{1,0}. \tag{14}$$

Note that in (13) and (14) one needs to invert a matrix of significantly smaller dimensionality $m \times m$ instead of $n \times n$ matrix in (7) and (8). The computational complexity of (13) and (14) can be estimated as $O(nm + m^3)$.

4.4   Co-association Matrix of Cluster Ensemble

We use a co-association matrix of cluster ensemble as a similarity matrix in (4) [5]. The co-association matrix is calculated in the process of cluster ensemble creation.

Let us consider a set of partition variants $\{P_l\}_{l=1}^r$, where $P_l = \{C_{l,1}, \ldots, C_{l,K_l}\}$, $C_{l,k} \subset \mathbf{X}$, $C_{l,k} \cap C_{l,k'} = \varnothing$ and K_l is the number of clusters in lth partition. For each partition P_l we determine matrix $H_l = (h_l(i,j))_{i,j=1}^n$ with elements indicating whether a pair x_i, x_j belong to the same cluster in lth variant or not. We have

$$h_l(i,j) = \mathbb{I}[c_l(x_i) = c_l(x_j)],$$

where $\mathbb{I}(\cdot)$ is the indicator function with $\mathbb{I}[true] = 1$, $\mathbb{I}[false] = 0$, and $c_l(x)$ is the cluster label assigned to x. The weighted averaged co-association matrix is

$$H = \sum_{l=1}^r \omega_l H_l, \tag{15}$$

where $\omega_1, \ldots, \omega_r$ are weights of ensemble elements, $\omega_l \geq 0$, $\sum \omega_l = 1$. The weights are used to assess the importance of base clustering variants [3]. They depend on the evaluation function Γ (e.g., cluster validity index) [3]: $\omega_l = \gamma_l / \sum_{l'} \gamma_{l'}$, where $\gamma_l = \Gamma(l)$ is an estimate of the clustering quality for the lth partition.

The matrix H can be considered as a pairwise similarity matrix which determines the similarity between points in a new feature space obtained with an implicit transformation of data.

It is easy to see that H admits a low-rank decomposition in the form:

$$H = RR^\top, \tag{16}$$

where $R = [R_1 R_2 \ldots R_r]$, R is a block matrix, $R_l = \sqrt{\omega_l}\, Z_l$, Z_l is $(n \times K_l)$ cluster assignment matrix for lth partition: $Z_l(i,k) = \mathbb{I}[c(x_i) = k]$, $i = 1, \ldots, n$, $k = 1, \ldots, K_l$.

As a rule, $m = \sum_l K_l \ll n$, thus (16) gives us an opportunity of saving memory by storing $(n \times m)$ sparse matrix instead of full $(n \times n)$ co-association matrix.

The Graph Laplacian matrix for H can be written in the form:

$$L = D' - H,$$

where $D' = \mathrm{diag}(D'_{11}, \ldots, D'_{nn})$, $D'_{ii} = \sum_j H(i,j)$. One can see that

$$D'_{ii} = \sum_{j=1}^n \sum_{l=1}^r \omega_l \sum_{k=1}^{K_l} Z_l(i,k) Z_l(j,k) = \sum_{l=1}^r \omega_l N_l(i), \tag{17}$$

where $N_l(i)$ is the size of the cluster which includes point x_i in lth partition variant.

Using H in the low-rank representation (16) instead of the similarity matrix W in (9), and the matrix D' defined in (17), we obtain cluster ensemble based predictions in the form given by (13), (14).

4.5 WSR-LRCM Algorithm

The basic steps of the suggested weakly supervised regression algorithm based on the low-rank representation of the co-association matrix (WSR-LRCM) are as follows.

Input:
X: dataset including both labeled, inaccurately labeled and unlabeled samples;
a_i, σ_i, $i = 1, \ldots, n_1$: uncertain input parameters for labeled and inaccurately labeled points;
r, Ω: number of runs and set of parameters for the k-means clustering (number of clusters, maximum number of iterations, parameters of the initialization process).
Output:
a^*, σ^*: predicted estimates of uncertain parameters for objects from sample **X** (including predictions for the unlabeled sample).
Steps:
1. Generate r variants of clustering partition for parameters randomly chosen from Ω; calculate weights $\omega_1, \ldots, \omega_r$.
2. Find graph Laplacian in the low-rank representation using (16) and D' in (17);
3. Calculate predicted estimates of uncertainty parameters using (13) and (14).
end.

5 Monte-Carlo Experiments

This section presents the results of numerical experiments with the proposed WSR-LRCM algorithm. The regression quality and running time are experimentally evaluated on synthetic examples with different sample sizes and noise levels and on a real example.

5.1 First Example with Artificial Data

We consider datasets generated from a mixture of two multidimensional normal distributions $\mathcal{N}(m_1, \sigma_X I)$, $\mathcal{N}(m_2, \sigma_X I)$ with equal weights; m_1, $m_2 \in \mathbb{R}^d$, $d = 8$, σ_X is a parameter. To investigate the robustness of the algorithm, we added noise to data by appending two independent features of a uniform distribution $\mathcal{U}(0, 1)$.

Let the ground truth target feature is equal to $Y = 1 + \varepsilon$ for points generated from the first component, otherwise $Y = 2 + \varepsilon$, where ε is a normally distributed random value with zero mean and variance σ_ε^2.

During Monte Carlo simulations, we generate samples of the given size n according to the specified distribution mixture. Two-thirds of the sample points are included into the training part $\mathbf{X}_{train}$, and the remaining points compose the test sample $\mathbf{X}_{test}$. In the training sample, 10% of the points selected at random from each component comprise a fully labeled sample; 20% of the sample consists of inaccurately labeled objects; the remaining part contains the unlabeled data. This partitioning mimics a typical situation in the weakly supervised learning: a small number of accurately labeled instances, medium sized uncertain labelings and a lot of unlabeled examples. To model the inaccurate labeling, we use the parameters defined in (1): $\sigma_i = \delta \cdot \sigma_Y$, where σ_Y is a standard deviation of Y over labeled data, $\delta > 0$ is a parameter.

The ensemble variants are generated by random initialization of centroids; to increase the diversity of base clusterings, we set the number of clusters in each run as $K = 2, \ldots, K_{max}$, where $K_{max} = 2 + r$, and $r = 10$ is the ensemble size. The weights of ensemble elements are the same: $\omega_l \equiv 1/r$. The regularization parameters β, γ have been estimated using grid search and cross-validation technique. In our experiments, the best results were obtained for $\beta = 0.001$ and $\gamma = 0.001$.

The quality of prediction is estimated on the test sample as the mean Wasserstein distance between the predicted according to (13), (14) and ground truth values of the parameters:

$$\text{MWD} = \frac{1}{n_{test}} \sum_{x_i \in \mathbf{X}_{test}} \left((a_i^{true} - a_i^*)^2 + \sigma_i^{*2} \right),$$

where n_{test} is the test sample size, and $a_i^{true} = y_i^{true}$ the true value of the target feature. Note that the standard Mean squared error (MSE) quality metric can be considered as a special case of MWD for accurate labeling.

We compare the suggested method with its simplified version (called WSR-RBF), which uses the standard similarity matrix evaluated with RBF kernel (2). Different values of parameter ℓ were considered, and the quasi-optimal $\ell = 1.85$ was determined. The output predictions were calculated according to (7) and (8).

To increase the statistical reliability of the results, we average the obtained estimates over 40 Monte Carlo repetitions (except for large data size $n \geq 10^5$).

Table 1 shows the results of experiments for different sample sizes and values of parameter σ_ε. Averaged values of MWD metric and execution times for the algorithms (working on dual-core Intel Core i5 processor with a clock frequency of 2.4 GHz and 8 GB RAM) are given. We use the following parameters in the data generation procedure: $m_1 = (0, \ldots, 0)^\top$, $m_2 = (10, \ldots, 10)^\top$, $\sigma_X = 3$, and $\delta = 0.1$.

The table demonstrates that WSR-LRCM produces nearly the same or better results than WSR-RBF with respect to the WMD metric. At the same time, WSR-LRCM runs much faster. For large sample sizes ($n \geq 10^5$) WSR-RBF failed due to unacceptable memory demands.

Table 1. Results of experiments with a mixture of two distributions: averaged MWD estimates and running time for two algorithms. For $n \geq 10^5$, WSR-RBF failed due to unacceptable memory demands.

n	σ_ε	WSR-LRCM		WSR-RBF	
		MWD	Time (sec)	MWD	Time (sec)
1000	0.01	0.002	0.04	0.007	0.04
	0.1	0.012	0.04	0.017	0.04
	0.25	0.065	0.04	0.070	0.04
5000	0.01	0.001	0.14	0.004	1.71
	0.1	0.011	0.14	0.014	1.72
	0.25	0.064	0.15	0.067	1.75
10000	0.01	0.001	0.33	0.002	9.40
	0.1	0.011	0.33	0.012	9.35
	0.25	0.064	0.33	0.065	9.36
10^5	0.01	0.001	6.72	–	–
10^6	0.01	0.001	89.12	–	–

In the next experiment, we compare the proposed WSR-LRCM with semi-supervised regression algorithm SSR-RBF considered in our previous work [4]. Because in semi-supervised learning, only labeled and unlabeled instances can be used, in this algorithm, we look at inaccurately labeled objects as they were unlabeled. The same RBF kernel is applied in this algorithm as in WSR-RBF. The parameters of the data generation procedure remain unchanged, $n = 1000$, $\sigma_\varepsilon = 0.1$. Table 2 shows the results of the experiments. MWD metric is utilized for quality evaluation in both cases.

Table 2. Results of experiments with WSR-LRCM and SSR-RBF algorithms. Averaged MWD estimates are calculated for different values of parameter δ.

δ	0.1	0.25	0.5
WSR-LRCM	0.012	0.017	0.038
SSR-RBF	0.051	0.051	0.051

From this table, one may conclude that additional information on uncertain labelings improves the quality of forecasting in WSR-LRCM. The parameter δ accounts for the degree of uncertainty: the larger its value, the more similar the results of weakly supervised and semi-supervised regression become.

5.2   Second Example with Real Data

In the second example, we analyse the Gas Turbine CO and NOx Emission Data Set [1,17]. This dataset includes measurements of 11 features describing working characteristics (temperature, pressure, humidity, etc.) of a gas turbine located in Turkey. The monitoring was carried out during 2011–2015. Carbon monoxide (CO) and Nitrogen oxides (NOx) are the predicted outputs. The forecasting of these harmful pollutants is necessary for controlling and reducing the emissions from power plants.

We make predictions for CO over the year 2015 (in total, 7384 observations) and use the following experiment's settings. The dataset is randomly partitioned on learning and test samples in the proportion 2:1. The volume of the accurately labeled sample is 1% of overall data; 10% of data are considered as inaccurately labeled instances; the remaining data are regarded as unlabeled samples. As in the previous example, we use the k-means clustering as the base ensemble algorithm (the number of clusters varies from 100 to $100 + r$). All other settings are the same.

As a result of forecasting, the averaged MWD for WSR-LRCM takes the value 1.85 and for SSR-RBF the value 5.18.

In order to compare WSR-LRCM with fully supervised algorithms, we calculate the standard Mean Absolute Error (MAE) using estimates of a^* defined in (13) as the predicted feature outputs:

$$\text{MAE} = \frac{1}{n_{test}} \sum_{x_i \in \mathbf{X}_{test}} |y_i^{true} - a_i^*|.$$

A Random Forest (RF) and Linear Regression (LR) methods are evaluated taking accurately labeled examples as the learning sample. The averaged MAE is 0.634 for WSR-LRCM, 0.774 for RF (with 300 trees), and 0.873 for LR. The averaged computing time is 1.99 s for WSR-LRCM, 0.35 s for RF, and 0.38 s for LR. The growth in the computing time for WSR-LRCM in this experiment can be explained by a large number of clusters (>100) as k-means parameter.

From the experiments, one may conclude that the proposed WSR-LRCM gives more accurate predictions than other compared methods in case of a small proportion of labeled sample.

Conclusion

In this work we have introduced a weakly supervised regression method using the manifold regularization technique. We have considered the case where the learning sample includes both labeled and unlabeled instances, as well as inaccurately labeled data objects. To model the uncertain labeling, we have used the normal distribution with different parameters. The measure of similarity between uncertain labelings was formulated in terms of the Wasserstein distance between probability distributions.

Two variants of the algorithm were proposed: WSR-RBF, which is based on the standard RBF kernel, and WSR-LRCM, which uses a low-rank representation of the co-association matrix of cluster ensemble. The reason for this modification is that the low-rank decomposition reduces the memory requirement and computing time. The ensemble clustering allows a better discovering of more complex data structures under noise distortions. The co-association matrix depends on the decisions of clustering algorithms and is less noise-dependent than standard similarity matrices.

The efficiency of the suggested methods was studied experimentally. Monte Carlo experiments have demonstrated a significant decrease in running time for WSR-LRCM in comparison with WSR-RBF. It has shown that taking into consideration the additional information on uncertain labelings improves the regression quality.

We plan to improve our method by using deep learning methodology (in particular, deep autoencoder) at the stage of ensemble clustering. It would be interesting to investigate different variants of hierarchical low-rank decomposition techniques. Applications of this method in various fields are also planned, especially for spatial data processing, analysis of computed tomography images and studying the relationships between single nucleotide polymorphisms in DNA sequences.

Acknowledgements. The study was carried out within the framework of the state contract of the Sobolev Institute of Mathematics (project no 0314-2019-0015). The work was partly supported by RFBR grants 19-29-01175 and 18-29-09041. A. Litvinenko was supported by funding from the Alexander von Humboldt Foundation.

References

1. UC Irvine Machine Learning Repository: Gas Turbine CO and NOx Emission Data Set, 06 April 2021. https://archive.ics.uci.edu/ml/datasets/Gas+Turbine+CO+and+NOx+Emission+Data+Set
2. Belkin, M., Niyogi, P., Sindhwani, V.: Manifold regularization: a geometric framework for learning from labeled and unlabeled examples. J. Mach. Learn. Res. **7**(85), 2399–2434 (2006). http://jmlr.org/papers/v7/belkin06a.html
3. Berikov, V.B.: Construction of an optimal collective decision in cluster analysis on the basis of an averaged co-association matrix and cluster validity indices. Pattern Recogn. Image Anal. **27**(2), 153–165 (2017). https://doi.org/10.1134/S1054661816040040
4. Berikov, V., Litvinenko, A.: Semi-supervised regression using cluster ensemble and low-rank co-association matrix decomposition under uncertainties. In: Proceedings of 3rd International Conference on Uncertainty Quantification in CSE, pp. 229–242 (2020). https://doi.org/10.7712/120219.6338.18377. https://files.eccomasproceedia.org/papers/e-books/uncecomp_2019.pdf
5. Berikov, V., Karaev, N., Tewari, A.: Semi-supervised classification with cluster ensemble. In: 2017 International Multi-Conference on Engineering, Computer and Information Sciences (SIBIRCON), pp. 245–250. IEEE, Novosibirsk (2017)
6. Bogachev, V.I., Kolesnikov, A.: The Monge-Kantorovich problem: achievements, connections, and perspectives. Russ. Math. Surv. **67**, 785–890 (2012)

7. Boongoen, T., Iam-On, N.: Cluster ensembles: a survey of approaches with recent extensions and applications. Comput. Sci. Rev. **28**, 1–25 (2018). https://doi.org/10.1016/j.cosrev.2018.01.003. https://www.sciencedirect.com/science/article/pii/S1574013717300692

8. Borisova, I.A., Zagoruiko, N.G.: Algorithm FRiS-TDR for generalized classification of the labeled, semi-labeled and unlabeled datasets. In: Aleskerov, F., Goldengorin, B., Pardalos, P.M. (eds.) Clusters, Orders, and Trees: Methods and Applications. SOIA, vol. 92, pp. 151–165. Springer, New York (2014). https://doi.org/10.1007/978-1-4939-0742-7_9

9. Delon, J., Desolneux, A.: A Wasserstein-type distance in the space of Gaussian mixture models. SIAM J. Imaging Sci. **13**(2), 936–970 (2020). https://doi.org/10.1137/19M1301047

10. Drineas, P., Mahoney, M.W., Cristianini, N.: On the Nyström method for approximating a gram matrix for improved kernel-based learning. J. Mach. Learn. Res. **6**, 2153–2175 (2005)

11. Gao, W., Wang, L., Li, Y.F., Zhou, Z.H.: Risk minimization in the presence of label noise. In: Proceedings of the AAAI Conference on Artificial Intelligence, vol. 30, no. 1, February 2016. https://ojs.aaai.org/index.php/AAAI/article/view/10293

12. Gao, W., Zhang, T., Yang, B.B., Zhou, Z.H.: On the noise estimation statistics. Artif. Intell. **293** (2021). https://doi.org/10.1016/j.artint.2021.103451

13. Grasedyck, L., Hackbusch, W.: Construction and arithmetics of $\mathcal{H}$-matrices. Computing **70**(4), 295–334 (2003). https://doi.org/10.1007/s00607-003-0019-1

14. Hackbusch, W.: A sparse matrix arithmetic based on $\mathcal{H}$-matrices. Part I: introduction to $\mathcal{H}$-matrices. Computing **62**(2), 89–108 (1999). https://doi.org/10.1007/s006070050015

15. Hackbusch, W.: Hierarchical Matrices: Algorithms and Analysis. Springer Series in Computational Mathematics, vol. 49. Springer, Heidelberg (2015). https://doi.org/10.1007/978-3-662-47324-5

16. Huang, K., Shi, Y., Zhao, F., Zhang, Z., Tu, S.: Multiple instance deep learning for weakly-supervised visual object tracking. Sig. Process. Image Commun. **84**, 115807 (2020). https://doi.org/10.1016/j.image.2020.115807

17. Kaya, H., Tüfekci, P., Uzun, E.: Predicting CO and NO_x emissions from gas turbines: novel data and a benchmark PEMS. Turk. J. Electr. Eng. Comput. Sci. **27**(6), 4783–4796 (2019)

18. Khoromskij, B.N., Litvinenko, A., Matthies, H.G.: Application of hierarchical matrices for computing the Karhunen-Loève expansion. Computing **84**(1–2), 49–67 (2009). https://doi.org/10.1007/s00607-008-0018-3

19. Litvinenko, A., Keyes, D., Khoromskaia, V., Khoromskij, B.N., Matthies, H.G.: Tucker tensor analysis of Matern functions in spatial statistics. Comput. Methods Appl. Math. (2018). https://doi.org/10.1515/cmam-2018-0022

20. Litvinenko, A., Kriemann, R., Genton, M.G., Sun, Y., Keyes, D.E.: HLIBCov: parallel hierarchical matrix approximation of large covariance matrices and likelihoods with applications in parameter identification. MethodsX **7** (2020). https://doi.org/10.1016/j.mex.2019.07.001. https://github.com/litvinen/HLIBCov.git

21. Litvinenko, A., Sun, Y., Genton, M.G., Keyes, D.E.: Likelihood approximation with hierarchical matrices for large spatial datasets. Comput. Stat. Data Anal. **137**, 115–132 (2019). https://doi.org/10.1016/j.csda.2019.02.002. https://github.com/litvinen/large_random_fields.git

22. Matérn, B.: Spatial Variation. Lecture Notes in Statistics, vol. 36, 2nd edn. Springer, Berlin (1986). https://doi.org/10.1007/978-1-4615-7892-5

23. Muhlenbach, F., Lallich, S., Zighed, D.A.: Identifying and handling mislabelled instances. J. Intell. Inf. Syst. **22**(1), 89–109 (2004). https://doi.org/10.1023/A: 1025832930864
24. Raykar, V.C., et al.: Learning from crowds. J. Mach. Learn. Res. **11**(43), 1297–1322 (2010). http://jmlr.org/papers/v11/raykar10a.html
25. Van Engelen, J.E., Hoos, H.H.: A survey on semi-supervised learning. Mach. Learn. **109**(2), 373–440 (2020). https://doi.org/10.1007/s10994-019-05855-6
26. Xiao, Y., Yin, Z., Liu, B.: A similarity-based two-view multiple instance learning method for classification. Knowl.-Based Syst. **201–202**, 105661 (2020). https://doi.org/10.1016/j.knosys.2020.105661
27. Zhou, D., Bousquet, O., Lal, T.N., Weston, J., Schölkopf, B.: Learning with local and global consistency. In: Proceedings of the 16th International Conference on Neural Information Processing Systems, NIPS 2003, pp. 321–328. MIT Press, Cambridge (2003)
28. Zhou, Z.H.: Ensemble Methods: Foundations and Algorithms. CRC Press, Boca Raton (2012)
29. Zhou, Z.H.: A brief introduction to weakly supervised learning. Natl. Sci. Rev. **5**(1), 44–53 (2017). https://doi.org/10.1093/nsr/nwx106. https://academic.oup.com/nsr/article-pdf/5/1/44/31567770/nwx106.pdf

K-Means Clustering via a Nonconvex Optimization Approach

Tatiana V. Gruzdeva$^{(\boxtimes)}$ and Anton V. Ushakov

Matrosov Institute for System Dynamics and Control Theory of SB RAS,
134 Lermontov Street, 664033 Irkutsk, Russia
{gruzdeva,aushakov}@icc.ru

Abstract. Clustering is one of the basic tasks in machine learning and data mining. Euclidean minimum-sum-of-squares clustering problem is probably the most common clustering model. It consists in finding k cluster centers or representatives so as the sum of squared Euclidean distances from a set of points and their closest centers is minimized. The problem is known to be nonconvex and NP-hard even in the planner case. In this paper, we propose a new DC programming approach to the problem. First, we cast the original nonconvex problem as a continuous optimization problem with the objective function represented as a DC function. We then devise a solution algorithm, resting upon the global optimality conditions and global search scheme for DC minimization problems proposed by A.S. Strekalovsky. We implement the developed algorithm and compare it with k-means clustering algorithms on generated datasets.

Keywords: Minimum-sum-of-squares clustering · Nonconvex optimization · DC programming · Local search · Global search scheme

1 Introduction

Clustering is one of the basic subroutines and important tasks in machine learning and data mining. It is widely applied in a large number of diverse fields, e.g. computer vision, bio- and cheminformatics, pharmacogenomics, etc. In its most simple form, clustering is to group objects or data items from a given set into non-overlapping subsets (clusters) such that each cluster consists of similar objects, while objects from different clusters are different. Since the problem of clustering is actually ill-posed, there are numerous approaches to how define clusters and how clustering should be performed. There are several ways of categorizing clustering algorithms, e.g. one often distinguishes density-based algorithms, hierarchical clustering algorithms, and statistical approaches (e.g. based on Gaussian mixture models). However, ones of the most popular clustering approaches are the so-called partition-based algorithms that fulfil clustering by solving a nonconvex optimization problem [15]. Usually, such optimization problems are to maximize separation of clusters or minimize their homogeneity

© Springer Nature Switzerland AG 2021
P. Pardalos et al. (Eds.): MOTOR 2021, LNCS 12755, pp. 462–476, 2021.
https://doi.org/10.1007/978-3-030-77876-7_31

subject to some constraints. The most popular clustering objectives express-
ing homogeneity of clusters are minimum-sum-of-stars (k-medoids), minimum
sum-of-cliques, and minimum sum-of-squares (k-means) clustering [15,22]. Note
that the last problem is probably the most famous and widely applied clustering model. Moreover, it is one of the first clustering problems formulated as a
mathematical programming problem [37].

Given a finite set $J = \{1, \ldots, m\}$ of data items, each of which is expressed
by a feature vector $a^j \in \mathbb{R}^n$, $j \in J$. The minimum-sum-of-squares clustering
problem (MSSC) is to find k cluster centers $c^i \in \mathbb{R}^n$, $i \in I = \{1, \ldots, k\}$ so as
to minimize the overall sum of squared Euclidean distances between data items
and their closest centers. Thus, the problem can be expressed as follows:

$$\min_{C \subset \mathbb{R}^n} \left\{ \sum_{j=1}^m \min_{c \in C} \|a^j - c\|^2, \ |C| = k \right\}, \tag{1}$$

where $\| \cdot \|$ is the Euclidean distance. It can be viewed as a center-based model
where one needs to find a cluster representative so as to minimize the dissim-
ilarities between data items of a cluster and their representative. The problem
MSSC is known to be NP-hard even in the plane for arbitrary k [26]. Moreover,
it is NP-hard in general dimension even for $k = 2$ [1] and NP-hard when the
dimension is a part of the input and the number of clusters is not [9]. However,
for fixed dimension n and number of clusters k, the problem can be solved in
$\mathcal{O}(m^{nk+1})$ time. Thus, the problem is challenging even for small-size problem
instances.

MSSC has widely been studied in the literature, hence there are hundreds of
heuristics and exact methods proposed to solve it. The most well-known heuristic for MSSC is k-means (also known as Lloyd's algorithm), which is probably
the most popular clustering algorithm to date due to its speed and simplicity.
Note that in machine learning literature the objective of MSCC and the problem itself are often referred to as k-means cost function and k-means problem,
respectively. Any feasible solution of MSSC is often said to be a k-means clustering. Lloyd's algorithm is an alternate heuristic that starts from an initial set of
cluster centers and iteratively repeat two steps: (i) assigning each data item to
the closest cluster centers (according to the squared Euclidean distance) - note
that data items assigned to the same center form a cluster, and (ii) computing
the new cluster centers as the means of the data items assigned to the same
clusters. Note that an iteration of the algorithm requires $\mathcal{O}(mnk)$ time, and the
algorithm has superpolynomial running time in the worst case. Other notable
traditional heuristic algorithms for MSSC are MacQueen's algorithm which is
actually an online version of Lloyd's algorithm and Hartigan-Wong's algorithm
that is an exchange type local search heuristic. The latter tries to improve an
initial partition by performing swaps of data items between clusters. If such a
swap results in a better value of the objective function, it is accepted and the
next data item is picked.

The further research on MSSC may be broadly divided into several strands.
As the k-means clustering algorithms are local search heuristics, they in general converge only to local optima. Consequently, the algorithms are heavily

dependent on the choice of initial solution (cluster centers) and have to rerun several times in order to find a quality partition. It is known that k-means algorithms (e.g. Lloyd's) can provide arbitrary bad solutions with respect to the objective value [3]. Thus, a lot of research has been focused on developing seeding procedures and obtaining theoretical guarantees on solution quality (e.g. see [29] and references therein). One of the most well-known such modifications of Lloyd's algorithm is k-means++ [3] where a seeding procedure allowing one to obtain $\Theta(\log k)$-competitive solution was proposed. Other closely related research was devoted to the development of approximation algorithms for MSSC, e.g. a PTAS proposed for a fixed number of clusters k and dimension n [28] and a linear-time PTAS [23] for a fixed k, as well as the techniques for reducing the time complexity of the traditional k-means algorithms (e.g. see [20]).

Another strand of research was focused on developing various heuristics for MSSC aimed at improving traditional k-means algorithms, e.g. by employing various local search and metaheuristic techniques. For example, modern heuristics include global k-means [4,24], j-means [16], harmonic clustering [6], I-k-means-+ [19], and plenty of metaheuristics like the scatter search [30], genetic algorithms [12], tabu search [25], memetic algorithm [27] etc.

There are also exact methods for MSSC, however the research on this direction is not so flourishing. For example, one of the first branch and bound algorithms for MSSC was proposed in [8]. In [10] a column generation algorithm that combines an interior point method and a branch and bound algorithm was developed. In the prominent paper [31] the authors proposed a branch and bound method where lower bounds for the objective value are found using a linear programming relaxation obtained with the so-called reformulation-linearization technique. The authors reported promising results on relatively large problem instances. Recently, the aforementioned column generation algorithm was improved in [2] where a new approach to finding solutions of the auxiliary problem was proposed. The authors demonstrated that their algorithm was able to find optimal solutions of instances with more than 2000 data items.

Finally, there is also a relatively small strand of literature exploiting an idea of DC decomposition of the MSSC objective function and/or DC programming [5, 7,17,21]. For example, in [5] the authors considered a DC representation of nonsmooth problem (1) and developed a DC algorithm based on this nonsmooth representation. In [17] the authors proposed an efficient DC Algorithm (DCA) for MSSC cast as a mixed integer program which is then reformulated as a continuous optimization problem using an exact penalty approach.

In this paper we develop and implement a solution algorithm for MSSC using its continuous formulation. Our approach is based on a decomposition of the objective function into the difference of two convex functions. Then, we develop a solution approach based on a special global search strategy and global optimality conditions. We report the results of computational experiments on test problem instances and compare the proposed approach with the most popular k-means clustering algorithms.

2   Problem Statement

There are several formulations for MSSC, including the nonsmooth one (1). However, the problem can also be formulated as a mixed integer program. Recall that we are given a set of data items J that must be divided into clusters. The goal is to find cluster centers such that the total sum of squared Euclidean distances between data items and their closest centers is minimized.

Let us introduce the following binary variables

$$x_{ij} = \begin{cases} 1, & \text{if data item } j \text{ is assigned to cluster } i, \\ 0, & \text{otherwise,} \end{cases} \quad i = 1, \ldots, k, \ j = 1, \ldots, m.$$

which are often referred to as assignment variables.

We also suppose that the unknown locations of k cluster centers are decision variables $y^i \in I\!R^n$, $i = 1, \ldots, k$. Obviously, we suppose that the number of data items is greater than k, otherwise the problem is trivially solved. With these notations, MSSC can be cast as the following mixed integer program:

$$\sum_{i=1}^{k} \sum_{j-1}^{m} x_{ij} \|y^i - a^j\|^2 \downarrow \min_{(x,y)}, \tag{2}$$

$$\sum_{i=1}^{k} x_{ij} = 1 \qquad\qquad \forall\, j = 1, \ldots, m; \tag{3}$$

$$x_{ij} \in \{0, 1\} \qquad\qquad \forall i = 1, \ldots, k; \ \forall j = 1, \ldots, m. \tag{4}$$

The objective function (2) minimizes the sum of squared distances and constraints (3) guarantee that each data item is assigned to exactly one cluster. One should note that for any fixed assignments x_{ij} of data items, the objective function is convex, hence according to the first order optimality conditions

$$\sum_{j=1}^{m} x_{ij}(a_l^j - y_l^i) = 0 \quad\Longrightarrow\quad y_l^i = \frac{\sum_{j=1}^{m} x_{ij} a_l^j}{\sum_{j=1}^{m} x_{ij}} \quad \forall i = 1, \ldots, k; \ l = 1, \ldots, n.$$

Thus, the optimal centers of clusters are the means (centroids) of the corresponding clusters. On the other hand, for fixed centers y^i, the assignment variables x_{ij} take binary values in the corresponding optimal solution, since data items are always assigned to the closest cluster centers. Consequently, in our approach we consider a natural relaxation of (2)–(4) where the binary constraints $x_{ij} \in \{0, 1\}$ are replaced with $x_{ij} \in [0, 1]$. The resultant problem is to minimize a nonconvex function over a convex feasible set:

$$f(x, y) = \sum_{i=1}^{k} \sum_{j=1}^{m} x_{ij} \|y^i - a^j\|^2 \downarrow \min_{(x,y)}, \quad x \in S, \ y \in I\!R^n, \tag{5}$$

where $S = \{x_{ij} \in [0, 1] : \sum_{i=1}^{k} x_{ij} = 1, \ j = 1, \ldots, m\}$.

3 DC Representations of the Objective Function

In order to apply the Global Search Theory for DC minimization developed by
A.S. Strekalovsky [33,34], we need an explicit DC representation of the nonconvex objective function.

It is well-known that DC representation is not unique. We propose to represent the objective function of the problem (5) as the following difference of two
convex functions

$$f(x,y) = g(x,y) - h(x,y), \tag{6}$$

where

$$g(x,y) = \sum_{i=1}^{k}\sum_{j=1}^{m} \left[d_1 \| \, y^j - a^i \, \|^2 + d_2 x_{ij}^2 \right],$$

$$h(x,y) = \sum_{i=1}^{k}\sum_{j=1}^{m} \left[d_1 \| \, y^j - a^i \, \|^2 + d_2 x_{ij}^2 - x_{ij}\| \, y^j - a^i \, \|^2 \right],$$

with some constants $d_1, d_2 > 0$.

It is obvious, that the function $g(\cdot)$ is convex. To prove the convexity of
function $h(\cdot)$, let us fix i and j, introduce a vector $v \in \mathbb{R}^{n+1}: \quad v_l = y_l^j, \quad l =
1,\ldots,n, \quad v_{n+1} = x_{ij}$ and consider the following ij-term in the sum of the
function $h(\cdot)$

$$\phi(v) = d_1 \| \, v-a \, \|_n^2 + d_2 v_{n+1}^2 - v_{n+1}\| \, v-a \, \|_n^2 = d_2 v_{n+1}^2 + (d_1 - v_{n+1})\sum_{l=1}^{n}(v_l - a_l)^2.$$

To form Hessian, we obtain the second-order partial derivatives of the function $\phi(\cdot)$

$$\nabla_{v_l v_l}^2 \phi = 2d_1 - 2v_{n+1}, \qquad \nabla_{v_l v_t}^2 \phi = 0,$$
$$\nabla_{v_{n+1} v_{n+1}}^2 \phi = 2d_2, \qquad \nabla_{v_l v_{n+1}}^2 \phi = -2(v_l - a_l), \qquad l,t = 1,\ldots,n.$$

After reducing Hessian to a diagonal form, we obtain the following $(n+1)\times(n+1)$
matrix

$$H(v) = \begin{pmatrix} 2(d_1 - v_{n+1}) & 0 & \ldots\, 0 & -2(v_1 - a_1) \\ 0 & 2(d_1 - v_{n+1}) & \ldots\, 0 & -2(v_2 - a_2) \\ \ldots & \ldots & \ldots\ldots & \ldots \\ 0 & 0 & \ldots\, 0 & 2d_2 - \dfrac{\| \, v - a \, \|_n^2}{d_1 - v_{n+1}} \end{pmatrix}.$$

A matrix is positive definite if all of its corner minors are positive. Since $v_{n+1} =
x_{ij} \in [0,1]$, hence $2(d_1 - v_{n+1}) > 0$ if $d_1 > 1$. In this case, all corner minors of
the matrix $H(v)$ up to n order inclusively turn out to be positive. The following
last minor

$$M_{(n+1)(n+1)} = 2^n(d_1 - v_{n+1})^n \left(2d_2 - \frac{\| \, v - a \, \|_n^2}{d_1 - v_{n+1}} \right)$$

must also be positive. Therefore, if we choose $d_2 > \dfrac{1}{2(d_1 - v_{n+1})} \max\limits_v \| v - a \|_n^2$ then the function $\phi(v)$ is convex.

With the proposed constants $d_1, d_2 \in \mathbb{R}$, the function $h(\cdot)$ in the representation (6) is convex as it is the sum of convex functions.

4   Local Search

In order to find a local solution of the problem (5), which is turned out to be the following DC minimization problem

$$f(x,y) = g(x,y) - h(x,y) \downarrow \min_{(x,y)}, \quad x \in S, \ y \in \mathbb{R}^n, \qquad (\mathcal{P})$$

we apply the well-known DC Algorithm [18, 32, 34]. It consists of linearizing, at a current point, the function $h(\cdot)$ which defines the basic nonconvexity of Problem $(\mathcal{P})$. The resultant convex approximation of the objective function $f(\cdot)$ obtained by replacing the nonconvex part with its linearization is then minimized. It is easy to see that such an approach allow finding local solutions by employing conventional convex optimization techniques.

Thus, we start with an initial point $(x^0, y^0) : \ y^0 \in \mathbb{R}^n, \ x^0 \in S$. Suppose a point $(x^s, y^s), \ x^s \in S$, is provided. Then, we find (x^{s+1}, y^{s+1}) as an approximate solution to the linearized problem

$$\Phi_s(x,y) = g(x,y) - \langle \nabla h(x^s, y^s), (x,y) \rangle \downarrow \min_{(x,y)}, \ x \subset S, \ y \in \mathbb{R}^n. \qquad (PL_s)$$

It means that the next iteration (x^{s+1}, y^{s+1}) satisfies the following inequality:

$$g(x^{s+1}, y^{s+1}) - \langle \nabla h(x^s, y^s), (x^{s+1}, y^{s+1}) \rangle$$
$$\leq \inf_{y \in \mathbb{R}^n, \ x \in S} \{ g(x,y) - \langle \nabla h(x^s, y^s), (x,y) \rangle \} + \delta_s, \qquad (7)$$

where the sequence $\{\delta_s\}$ satisfies the following conditions:

$$\delta_s \geq 0, \ s = 0, 1, 2, \ldots; \ \sum_{s=0}^{\infty} \delta_s < \infty.$$

Note that the linearized Problem (PL_s) is quadratic and convex, whereas Problem $(\mathcal{P})$ is nonconvex.

As it was suggested in [32, 34], one of the following inequalities can be employed as a stopping criterion:

$$f(x^s, y^s) - f(x^{s+1}, y^{s+1}) \leq \frac{\tau}{2},$$
$$\Phi_s(x^s, y^s) - \Phi_s(x^{s+1}, y^{s+1}) \triangleq g(x^s, y^s) - g(x^{s+1}, y^{s+1})$$
$$+ \langle \nabla h(x^s, y^s), (x^{s+1}, y^{s+1}) - (x^s, y^s) \rangle \leq \frac{\tau}{2}, \qquad (8)$$

where τ is a given accuracy.

If one of the inequalities (8) holds, it can be easily shown that the point (x^s, y^s) is a critical point to Problem $(\mathcal{P})$ with the accuracy τ and under the condition $\delta_s \leq \dfrac{\tau}{2}$. Indeed, (8) together with the inequality (7) imply that

$$
\begin{aligned}
& g(x^s, y^s) - \langle \nabla h(x^s, y^s), (x^s, y^s) \rangle \\
& \leq \frac{\tau}{2} + g(x^{s+1}, y^{s+1}) - \langle \nabla h(x^s, y^s), (x^{s+1}, y^{s+1}) \rangle \\
& \leq \inf_{x \in S} \{ g(x, y) - \langle \nabla h(x^s, y^s), (x, y) \rangle \} + \frac{\tau}{2} + \delta_s.
\end{aligned}
$$

Therefore, if $\delta_s \leq \dfrac{\tau}{2}$, the point (x^s, y^s) is a τ-solution to Problem (PL_s).

Applying the DC representation (6), we have to solve a series of the following linearized problems for MSSC:

$$
\sum_{j=1}^{m} \sum_{i=1}^{k} \left[d_1 \| y^j - a^i \|^2 + d_2 x_{ij}^2 \right] - \langle \nabla h(x^s, y^s), (x, y) \rangle \downarrow \min_{(x,y)}, x \in S, y \in \mathbb{R}^n,
$$

$$(9)$$

where

$$
\nabla_y h(x^s, y^s) = 2 \sum_{j=1}^{m} (y_l^{sj} - a_l^i)(d_1 - x_{ij}^s),
$$

$$
\nabla_x h(x^s, y^s) = 2 d_2 x_{ij}^s - \| y^{sj} - a^i \|^2,
$$

$$
i = 1, \dots, k, \quad j = 1, \dots, m, \quad l = 1, \dots, n.
$$

We denote the solution obtained by the local search method as $z = (x, y)$ ($z \in Sol(9)$).

In the next section we will show how to escape from local solutions provided by the local search method.

5   Optimality Conditions and the Global Search Scheme

Let us recall the fundamental result of the Global Search Theory for DC minimization problem.

Theorem 1 [33,34]. *Suppose that $\exists q = (\tilde{x}, \tilde{y}) : \tilde{x} \in S$ $f(q) < f(z) = \zeta$.*

Then, a point $z = (\hat{x}, \hat{y}) : \hat{x} \in S$, is a global solution to Problem $(\mathcal{P})$ if and only if

$$
\left.
\begin{aligned}
& \forall (w, \beta) \in \mathbb{R}^{k(n+m)} \times \mathbb{R} : \quad h(w) = \beta - \zeta, \\
& g(w) - \beta \geq \langle \nabla h(w), (x, y) - w \rangle \ \forall (x, y) : \ x \in S.
\end{aligned}
\right\}
$$

$$(E)$$

As we can see, the verifying condition (E) for a given w requires solving the convex program $(PL(w))$:

$$
g(x, y) - \langle \nabla h(w), (x, y) \rangle \downarrow \min, \quad x \in S,
$$

$$(10)$$

depending on 'perturbation' parameters (w, β) satisfying $h(w) = \beta + \zeta$.

According to Theorem 1, in order to determine whether a given point z is a global solution to Problem $(\mathcal{P})$, we need to solve a family of linearized problems (10) with any conventional convex optimization method.

On the other hand, we can see that if the condition (E) is violated at a given tuple $(\tilde{w}, \tilde{\beta}, u)$, $u = (u^1, u^2) : u^1 \in S, u^2 \in \mathbb{R}^n$,

$$g(u) - \tilde{\beta} < \langle \nabla h(\tilde{w}), u - \tilde{w} \rangle,$$

due to convexity of $h(\cdot)$, then we get $g(u) < \tilde{\beta} + h(u) - h(\tilde{w})$ and conclude that $z = (\hat{x}, \hat{y}) : \hat{x} \in S$ is not optimal.

Moreover, on each level $\zeta_p = f(z^p)$, $p = 1, 2, \ldots$, it is not necessary to investigate all pairs of (w, β) satisfying (E), $\zeta_p = h(w) - \beta$, but it is sufficient to discover the violation of inequality (E) only for one pair $(\tilde{w}, \tilde{\beta})$ and $u = (u^1, u^2) :$ $u^1 \in S, u^2 \in \mathbb{R}^n$.

The properties of the Optimality Conditions (E) allow developing an algorithm for solving DC minimization problems. The algorithm comprises two principal stages:

a) local search to find an approximate local minimizer z^p with the value corresponding to the objective function $\zeta_p = f(z^p)$;
b) procedures of escaping from local pits, which are based on the Optimality Conditions (E).

Global Search Scheme.

1. Run the local search method and find a local minimizer z^p in Problem $(\mathcal{P})$.
2. Choose a number $\beta : \inf(g, S) < \beta \leq \sup(g, S)$.
   For instance, $\beta_p = g(z^p)$, $\zeta_p = f(z^p) = g(z^p) - h(z^p)$.
3. Construct a finite approximation

$$R_p(\beta) = \{w^1, \ldots, w^{N_p} \mid h(w^i) = \beta - \zeta_p, \ t = 1, \ldots, N_p\}$$

of the level surface $\{h(x, y) = \beta - \zeta_p\}$ of the function $h(\cdot)$.
4. Find a δ_p-solution $\bar{u}^t$ of the following Linearized Problem:

$$g(x, y) - \langle \nabla h(w^t), (x, y) \rangle \downarrow \min_{(x,y)}, \ x \in S, \qquad (PL_t)$$

so that $g(\bar{u}^t) - \langle \nabla h(w^t), \bar{u}^t \rangle - \delta_p \leq \inf_{(x,y)} \{g(x, y) - \langle \nabla h(w^t), (x, y) \rangle\}$.
5. Starting from the point $\bar{u}^t$, find a local minimizer u^t with the local search method.
6. Choose the best point $\hat{u} : f(\hat{u}) \leq \min_{t=1,\ldots,N_p} f(u^t)$.
7. If $f(\hat{u}) < f(z^p)$, then set $z^{p+1} := \hat{u}$, $p := p + 1$ and go to Step 2.
8. Otherwise, choose a new value of β (for instance, $\beta + \Delta\beta$) and go to Step 3.

One of the principal features of the Global Search Scheme is an approximation of the level surface of the convex function $h(\cdot)$ which generates the basic nonconvexity in Problem $(\mathcal{P})$ (Step 3). There are many ways and techniques to construct the approximation. To take into account the particularities of MSSC, we construct the approximation by varying only variables y^j, $j \in J$ (cluster centers) of the function $h(\cdot)$. Thus, the approximation $R_p(\beta)$ of the level surface $\{h(\cdot) = \beta - \zeta\}$ for each pair (β, ζ_p), $\zeta_p = f(z^p)$, can be constructed by the following rule [11,13]:

$$w^{jl} = z^j + \mu_{jl} e^l, \quad j = 1, \ldots, k, \quad l = 1, \ldots, n, \tag{11}$$

where e^l is the unit vector from the Euclidean basis of $\mathbb{R}^n$.

The search of μ_{jl} is simple and, actually, analytical (i.e. it is reduced to solving the following quadratic equation of one variable μ_{jl}) for the quadratic (with fixed variable x) function $h(\hat{x}, y)$:

$$\mu_{jl}^2 \sum_{i=1}^m (d_1 - \hat{x}_{ij}) - 2\mu_{jl} \sum_{i=1}^m (d_1 - \hat{x}_{ij}) a_l^i + \gamma = 0,$$

where $\gamma = h(z^j) - \beta + \zeta_p + \sum_{i=1}^m (d_1 - \hat{x}_{ij})(a_l^i)^2$. If for some indexes $\hat{j}\hat{l}$ the discriminant turns out to be negative, then the point $w^{\hat{j}\hat{l}}$ is not included into the approximation.

Based on the presented Global Search Scheme, we developed the algorithm GSA. The results of the computational simulation is demonstrated in the next section.

6   Computational Experiments

In this section we report some computational experiments to test the proposed global search algorithm (GSA) for MSSC. We compare our approach to the most popular k-means clustering algorithms: Lloyd's algorithm (k-means) and k-means++. We implemented all the competing algorithms using C++ and run them on a PC with Intel Core i7-4790K CPU 4.0 GHz. To solve the convex (linearized) quadratic problems, we use GUROBI 9.1 solver freely available for non-commercial research. Beside the value of the objective function, we also use the following external measures to assess accuracy of a clustering: pairwise precision, recall and F-measure:

$$Precision = \frac{TP}{TP + FP}, \ Recall = \frac{TP}{TP + FN},$$

where TP are true positive pairs of data items (correctly clustered), FP—false positive pairs, and FN—false negative pairs. The pairwise F-measure is defined as harmonic mean of pairwise precision and recall:

$$F - measure = \frac{2 \cdot Precision \cdot Recall}{Precision + Recall}.$$

The initial solutions were taken at random. As the k-means algorithms are local search heuristics that are heavily dependent on initial solutions, we restart

Table 1. The testing of GSA on synthetically generated test problems

Type I

m	k	Start Obj. Val.	Best Obj. Val.	St	PL
200	16	17788.36	3068.13	9	3955
300	16	39403.38	5958.51	4	3597
400	16	25715.84	5726.85	8	4476
500	25	30378.78	6527.64	5	11606
600	25	61401.54	10199.58	6	17568

Type III

m	k	Start Obj. Val.	Best Obj. Val.	St	PL
200	16	1175.66	147.52	8	4923
300	16	1611.07	298.96	14	16723
400	16	2558.94	438.78	9	8828
500	25	4527.55	856.91	11	14947
600	25	6757.46	995.33	8	23551

k-means and k-means++ 5 times and report the best solutions found. Our k-means algorithms were set to halt when the number of iterations exceeds 500 or when the fraction of data items that changed their cluster assignment is below 0.001. Our test bed consists of two benchmark types of data sets generated as suggested in [38]. Note that our test problems are two dimensional Gaussian mixture data instances that widely used in testing solution algorithms for clustering and facility location problems [14,35,36]. For our experiments, we generated instances of Types I and III, which differ in their complexity. The problems of Type I are considered to be easier to solve and those of Type III are harder. The test problems contain from 200 to 600 points on the plane, while the numbers of clusters vary from 16 to 25 and depend on the problem size.

First, we report the computational results on the developed global search algorithm (GSA) in Table 1, where the following denotations are employed:

- m is the number of data items;
- k stands for the number of clusters;
- *Start Obj.Val.* is the value of the objective function to the problem (5) at the starting point;
- *Best Obj.Val.* stands for the value of the objective function at the solution provided by the GSA;
- *St* is the number of the local solutions passed by the GSA;
- *PL* stands for the number of Linearized Problems solved.

The results on the computational testing of GSA confirmed the difference in complexity between the two types of instances. Indeed, though the problem instances of both classes are identical in terms of number of points and clusters, GSA required solving a larger number of linearized problems (column *PL* in

Table 1) for the instances of Type III. GSA managed to improve the value of the objective function to the problem (5) by 4 to 14 times (column St in Table 1), which proves the efficiency of the procedures of escaping from local pits.

The results of computational comparison of GSA with popular k-means and k-means++ are presented in Tables 2 and 3, where we report the best objective values found by the competing algorithms as well as the values of external measures. Observe that we compare our algorithm with k-means and k-means++ with respect to solution quality only. Our approach is not competitive against these algorithms with respect to run time.

One can see that our approach obtained very competitive results, e.g. for the problems of Type III GSA found solutions which are, in general, similar to ones found by k-means and k-means++. However, for the problems of Type I our approach outperformed both competing algorithms that stuck in relatively similar local optima. For example, for problem with 400 points and 16 clusters GSA found a solution which has 23% better objective value than ones found by k-means and k-means++ (see Table 2). This behaviour of the k-means algorithms is expected, since, for the problems with relatively large number of clusters, they may require large number of reruns to find a quality partition. It is interesting to note that solution quality according to objective value does not always correlate to that according to the external measures. For example, for the problem of Type III with 400 points and 16 clusters, we can see that GSA found slightly

Table 2. Clustering results on problem instances of Type I. Our approach is compared with k-means and k-means++

Type I

m	k	Best Obj. Val.	Precision	Recall	F-meas.	Algorithm
200	16	3877.55	0.799	0.852	0.824	k-means
		3099.05	0.847	0.894	0.870	k-means++
		3068.13	0.863	0.889	0.876	GSA
300	16	6350.60	0.825	0.845	0.835	k-means
		6116.75	0.815	0.845	0.830	k-means++
		5958.51	0.840	0.885	0.862	GSA
400	16	7590.27	0.850	0.895	0.872	k-means
		7483.61	0.850	0.896	0.872	k-means++
		5726.85	0.954	0.954	0.954	GSA
500	25	8449.67	0.872	0.929	0.900	k-means
		6533.00	0.956	0.959	0.958	k-means++
		6527.64	0.957	0.960	0.958	GSA
600	25	10810.05	0.884	0.932	0.908	k-means
		10292.67	0.898	0.917	0.908	k-means++
		10199.58	0.884	0.910	0.897	GSA

Table 3. Clustering results on problem instances of Type III. Our approach is compared with k-means and k-means++.

Type III

m	k	Best Obj. Val.	Precision	Recall	F-meas.	Algorithm
200	16	168.96	0.637	0.830	0.720	k-means
		137.79	0.639	0.845	0.728	k-means++
		147.52	0.648	0.890	0.750	GSA
300	16	312.05	0.681	0.776	0.726	k-means
		298.29	0.693	0.809	0.746	k-means++
		298.96	0.669	0.791	0.725	GSA
400	16	528.80	0.594	0.803	0.683	k-means
		438.70	0.610	0.829	0.703	k-means++
		438.78	0.569	0.863	0.686	GSA
500	25	1024.93	0.648	0.766	0.702	k-means
		879.15	0.610	0.782	0.685	k-means++
		856.91	0.699	0.827	0.758	GSA
600	25	1094.19	0.671	0.822	0.738	k-means
		1008.80	0.653	0.829	0.731	k-means++
		995.33	0.786	0.838	0.811	GSA

worse solution than k-means++. However, it has slightly better Recall. This may happen due to mislabelled outliers.

7 Conclusion

In this paper we addressed the so-called minimum-sum-of-squares (k-means) clustering problem, one of the best known clustering models. This problem can be formulated as a nonconvex mathematical programming problem, i.e. a problem of minimizing a DC function over a convex set. Using the special global search scheme based on global optimality conditions by A.S. Strekalovsky, we developed an algorithm for finding quality clustering solutions. In our computational experiments we demonstrated that the proposed approach is competitive with conventional k-means heuristics and, in most cases, provides better solutions for problem instances of relatively small size.

Our further research will be focused on improving the proposed methodology to make our algorithm tractable for large-scale problem instances involving thousands of data items. Our research will also aim at adaptation of the algorithm for other clustering problems with different dissimilarity measures.

Acknowledgement. The research was funded by the Ministry of Education and Science of the Russian Federation within the framework of the project "Theoretical foundations, methods and high-performance algorithms for continuous and discrete optimization to support interdisciplinary research" (No. of state registration: 121041300065-9).

References

1. Aloise, D., Deshpande, A., Hansen, P., Popat, P.: NP-hardness of Euclidean sum-of-squares clustering. Mach. Learn. **75**, 245–248 (2009). https://doi.org/10.1007/s10994-009-5103-0
2. Aloise, D., Hansen, P., Liberti, L.: An improved column generation algorithm for minimum sum-of-squares clustering. Math. Program. **131**(1–2), 195–220 (2012). https://doi.org/10.1007/s10107-010-0349-7
3. Arthur, D., Vassilvitskii, S.: K-means++: the advantages of careful seeding. In: Proceedings of the Eighteenth Annual ACM-SIAM Symposium on Discrete Algorithms, SODA 2007, pp. 1027–1035. SIAM, Philadelphia (2007)
4. Bagirov, A.M.: Modified global k-means algorithm for minimum sum-of-squares clustering problems. Pattern Recogn. **41**(10), 3192–3199 (2008). https://doi.org/10.1016/j.patcog.2008.04.004
5. Bagirov, A.M., Taheri, S., Ugon, J.: Nonsmooth DC programming approach to the minimum sum-of-squares clustering problems. Pattern Recogn. **53**, 12–24 (2016). https://doi.org/10.1016/j.patcog.2015.11.011
6. Carrizosa, E., Alguwaizani, A., Hansen, P., Mladenović, N.: New heuristic for harmonic means clustering. J. Glob. Optim. **63**, 427–443 (2015). https://doi.org/10.1007/s10898-014-0175-1
7. Demyanov, V., Bagirov, A., Rubinov, A.: A method of truncated codifferential with application to some problems of cluster analysis. J. Glob. Optim. **23**, 63–80 (2002). https://doi.org/10.1023/A:1014075113874
8. Diehr, G.: Evaluation of a branch and bound algorithm for clustering. SIAM J. Sci. Stat. Comput. **6**(2), 268–284 (1985). https://doi.org/10.1137/0906020
9. Dolgushev, A.V., Kel'manov, A.V.: On the algorithmic complexity of a problem in cluster analysis. J. Appl. Ind. Math. **5**(2), 191–194 (2011). https://doi.org/10.1134/S1990478911020050
10. du Merle, O., Hansen, P., Jaumard, B., Mladenovic, N.: An interior point algorithm for minimum sum-of-squares clustering. SIAM J. Sci. Comput. **21**(4), 1485–1505 (1999). https://doi.org/10.1137/S1064827597328327
11. Gaudioso, M., Gruzdeva, T.V., Strekalovsky, A.S.: On numerical solving the spherical separability problem. J. Glob. Optim. **66**(1), 21–34 (2016). https://doi.org/10.1007/s10898-015-0319-y
12. Gribel, D., Vidal, T.: HG-means: a scalable hybrid genetic algorithm for minimum sum-of-squares clustering. Pattern Recogn. **88**, 569–583 (2019). https://doi.org/10.1016/j.patcog.2018.12.022
13. Gruzdeva, T.V.: On a continuous approach for the maximum weighted clique problem. J. Glob. Optim. **56**(3), 971–981 (2013). https://doi.org/10.1007/s10898-012-9885-4
14. Hansen, P., Brimberg, J., Urosević, D., Mladenović, N.: Solving large p-median clustering problems by primal-dual variable neighborhood search. Data Min. Knowl. Discov. **19**(3), 351–375 (2009). https://doi.org/10.1007/s10618-009-0135-4
15. Hansen, P., Jaumard, B.: Cluster analysis and mathematical programming. Math. Program. **79**(1–3), 191–215 (1997). https://doi.org/10.1007/BF02614317

16. Hansen, P., Mladenovići, N.: J-means: a new local search heuristic for minimum sum of squares clustering. Pattern Recogn. **34**(2), 405–413 (2001). https://doi.org/10.1016/S0031-3203(99)00216-2

17. Hoai An, L.T., Hoai Minh, L., Tao, P.D.: New and efficient DCA based algorithms for minimum sum-of-squares clustering. Pattern Recogn. **47**(1), 388–401 (2014). https://doi.org/10.1016/j.patcog.2013.07.012

18. Hoai An, L.T., Tao, P.D.: The DC (difference of convex functions) programming and DCA revisited with DC models of real world nonconvex optimization problems. Ann. Oper. Res. **133**, 23–46 (2005). https://doi.org/10.1007/s10479-004-5022-1

19. Ismkhan, H.: I-k-means-+: an iterative clustering algorithm based on an enhanced version of the k-means. Pattern Recogn. **79**, 402–413 (2018). https://doi.org/10.1016/j.patcog.2018.02.015

20. Kanungo, T., Mount, D.M., Netanyahu, N.S., Piatko, C.D., Silverman, R., Wu, A.Y.: An efficient k-means clustering algorithm: analysis and implementation. IEEE Trans. Pattern Anal. Mach. Intell. **24**(7), 881–892 (2002). https://doi.org/10.1109/TPAMI.2002.1017616

21. Karmitsa, N., Bagirov, A.M., Taheri, S.: New diagonal bundle method for clustering problems in large data sets. Eur. J. Oper. Res. **263**(2), 367–379 (2017). https://doi.org/10.1016/j.ejor.2017.06.010

22. Kel'manov, A.V., Pyatkin, A.V.: Complexity of certain problems of searching for subsets of vectors and cluster analysis. Comput. Math. Math. Phys. **49**(11), 1966–1971 (2009). https://doi.org/10.1134/S0965542509110128

23. Kumar, A., Sabharwal, Y., Sen, S.: A simple linear time $(1 + \varepsilon)$-approximation algorithm for k-means clustering in any dimensions. In: 45th Annual IEEE Symposium on Foundations of Computer Science, pp. 454–462. IEEE, New York (2004). https://doi.org/10.1109/FOCS.2004.7

24. Likas, A., Vlassis, N., Verbeek, J.J.: The global k-means clustering algorithm. Pattern Recogn. **36**(2), 451–461 (2003). https://doi.org/10.1016/S0031-3203(02)00060-2

25. Liu, Y., Yi, Z., Wu, H., Ye, M., Chen, K.: A tabu search approach for the minimum sum-of-squares clustering problem. Inf. Sci. **178**(12), 2680–2704 (2008). https://doi.org/10.1016/j.ins.2008.01.022

26. Mahajan, M., Nimbhorkar, P., Varadarajan, K.: The planar k-means problem is NP-hard. Theor. Comput. Sci. **442**, 13–21 (2012). https://doi.org/10.1016/j.tcs.2010.05.034. Special Issue on the Workshop on Algorithms and Computation (WALCOM 2009)

27. Mansueto, P., Schoen, F.: Memetic differential evolution methods for clustering problems. Pattern Recogn. **114**, 107849 (2021). https://doi.org/10.1016/j.patcog.2021.107849

28. Matoušek, J.: On approximate geometric k-clustering. Discrete Comput. Geom. **24**, 61–84 (2000). https://doi.org/10.1007/s004540010019

29. Ostrovsky, R., Rabani, Y., Schulman, L.J., Swamy, C.: The effectiveness of Lloyd-type methods for the k-means problem. J. ACM **59**(6) (2013). https://doi.org/10.1145/2395116.2395117

30. Pacheco, J.A.: A scatter search approach for the minimum sum-of-squares clustering problem. Comput. Oper. Res. **32**(5), 1325–1335 (2005). https://doi.org/10.1016/j.cor.2003.11.006

31. Sherali, H.D., Desai, J.: A global optimization RLT-based approach for solving the hard clustering problem. J. Glob. Optim. **32**, 281–306 (2005). https://doi.org/10.1007/s10898-004-2706-7

32. Strekalovsky, A.S.: On local search in d.c. optimization problems. Appl. Math. Comput. **255**, 73–83 (2015)
33. Strekalovsky, A.: On the minimization of the difference of convex functions on a feasible set. Comput. Math. Math. Phys. **43**, 380–390 (2003)
34. Strekalovsky, A.S.: On solving optimization problems with hidden nonconvex structures. In: Rassias, T.M., Floudas, C.A., Butenko, S. (eds.) Optimization in Science and Engineering, pp. 465–502. Springer, New York (2014). https://doi.org/10.1007/978-1-4939-0808-0_23
35. Ushakov, A.V., Vasilyev, I.: Near-optimal large-scale k-medoids clustering. Inf. Sci. **545**, 344–362 (2021). https://doi.org/10.1016/j.ins.2020.08.121
36. Ushakov, A.V., Vasilyev, I.L., Gruzdeva, T.V.: A computational comparison of the p-median clustering and k-means. Int. J. Artif. Intell. **13**(1), 229–242 (2015)
37. Vinod, H.D.: Integer programming and the theory of grouping. J. Am. Stat. Assoc. **64**(326), 506–519 (1969). https://doi.org/10.2307/2283635
38. Zhang, T., Ramakrishnan, R., Livny, M.: BIRCH: a new data clustering algorithm and its applications. Data Min. Knowl. Discov. **1**(2), 141–182 (1997). https://doi.org/10.1023/A:1009783824328

Machine Learning Algorithms of Relaxation Subgradient Method with Space Extension

Vladimir N. Krutikov[1], Vladimir V. Meshechkin[1], Elena S. Kagan[1], and Lev A. Kazakovtsev[2]

[1] Kemerovo State University, 6 Krasnaya Street, Kemerovo 650043, Russia
[2] Reshetnev Siberian State University of Science and Technology, prosp. Krasnoyarskiy Rabochiy 31, Krasnoyarsk 660031, Russia
levk@bk.ru

Abstract. In relaxation subgradient minimization methods, a descent direction, which is based on the subgradients obtained at the iteration, forms an obtuse angle with all subgradients in the neighborhood of the current minimum. Minimization along this direction enables us to go beyond this neighborhood and avoid method looping. To find the descent direction, we formulate a problem in a form of systems of inequalities and propose an algorithm with space extension close to the iterative least squares method for solving them. The convergence rate of the method is proportional to the valid value of the space extension parameter and limited by the characteristics of subgradient sets. Theoretical analysis of the learning algorithm with space extension enabled us to identify the components of the algorithm and alter them to use increased values of the extension parameter if possible. On this basis, we propose and substantiate a new learning method with space extension and corresponding subgradient method for nonsmooth minimization. Our computational experiment confirms their efficiency. Our approach can be used to develop new algorithms with space extension for relaxation subgradient minimization.

Keywords: Subgradient methods · Space extension · Relaxation

1 Introduction

We consider a problem of minimizing a convex, not necessarily differentiable function $f(x)$, $x \in \mathbb{R}^n$. One of the possible approaches to constructing nonsmooth optimization methods is based on smooth approximations [1–3]. For minimizing such functions, Shor [4] proposed an iterative subgradient minimization algorithm, which was further developed and summarized in [5,6].

Space extension (dilation) methods, or r-algorithms [4], are based on successive extension in specially selected directions. The method constructs a certain linear transformation which alters the metric of the space at each iteration, and

© Springer Nature Switzerland AG 2021
P. Pardalos et al. (Eds.): MOTOR 2021, LNCS 12755, pp. 477–492, 2021.
https://doi.org/10.1007/978-3-030-77876-7_32

uses the direction opposite to the subgradient in the space with transformed metric. Such a direction forms an acute angle with the direction from the given point to the point of minimum for convex functions. The first relaxation subgradient minimization methods (RSMMs) were considered in [7,8], and their versions with space extension [9,10] led to the emergence of a number of effective approaches such as the subgradient method with space extension in the subgradient direction [10] that are relaxation by distance to the extremum [6,11,12]. Using the concepts of machine learning (ML) theory [13] led to the construction of a number of effective RSMMs [14–16] and formed a theoretical basis for their development. In RSMMs, the structure of the method is defined, where the procedure for solving inequalities based on a certain learning algorithm is embedded. We present an approach to accelerating the convergence of RSMM algorithms with space extension and give an example of its effective implementation.

In the RSMM, successive approximations [7,8,14,16,17] are:

$$x_{k+1} = x_k - \gamma_k s_{k+1}, \ \gamma_k = arg \min_{\gamma} f(x_k - \gamma s_{k+1}), \tag{1}$$

where x_0 is a given starting point, k is the iteration number, γ_k is a step size, descent direction s_{k+1} is a solution of a system of inequalities on $s \in \mathbb{R}^n$ [8]:

$$(s, g) > 0, \ \forall g \in G. \tag{2}$$

Hereinafter (s, g) is a dot product of vectors. In (2), G is a set of subgradients calculated on the descent trajectory of the algorithm at a point x_k.

For smooth functions, a subgradient set consists of a single gradient vector, and subgradient methods for unconstrained problems use the same search direction as the steepest descent. Nevertheless, when following a narrow ravine, the method for solving the inequalities of the subgradient method based on the gradient values finds the direction along the ravine.

Denote $S(G)$ as a set of solutions to inequality (2), $\partial f(x)$ as a subgradient set at point x. If the function is convex, $G = \partial_\varepsilon f(x_k)$ is an ε-subgradient set at point x_k, and s_{k+1} is an arbitrary solution of the system (2), then the function will be reduced by at least ε [8] after iteration (1). Since there is no explicit assignment of ε-subgradient sets, we use subgradients $g_k \in \partial f(x_k)$ calculated on the algorithm descent trajectory and satisfying the condition

$$(s_k, g_k) \le 0. \tag{3}$$

Thus, we use vectors violating (2). According to this principle, the choice of learning vectors is also made in the perceptron method (for instance, [13,18]). A sequence of vectors $g_k \in \partial f(x_k)$, $k = 0, 1, ...$, is not predetermined, but determined by the minimization algorithm according to (1) with a built-in method for finding the vector at each iteration s_{k+1} by a ML algorithm.

Let a convex set $G \subset \mathbb{R}^n$ belong to a hyperplane, and its vector η nearest to the origin be also a hyperplane vector closest to the origin. Then a solution of the system $(s, g) = 1 \ \forall g \in G$ is also a solution of (2) [14]. It can be found as a solution to a system of equations using a sequence of vectors from G [14]:

$$(s, g_i) = 1, \ g_i \in G, \ i = 0, 1, ...k. \tag{4}$$

Based on the iterative least squares (ILS) method, paper [14] proposes a method for minimizing the quality functional with special weights w_i and quadratic regularization $F(s) = \sum_{i=1}^{k} w_i Q_i(s) + \frac{1}{2} \sum_{i=1}^{n} s_i^2$, $Q_i(s) = \frac{1}{2}(1 - (s, g_i))^2$. Taking into account the specifics of subgradient sets, a modified iterative process is proposed for solving the system of inequalities (2) based on (4):

$$s_{k+1} = s_k + \frac{H_k g_k [1 - (s_k, g_k)]}{(g_k, H_k g_k)}, \tag{5}$$

$$H_{k+1} = H_k - (1 - \frac{1}{\alpha_k^2}) \frac{H_k g_k g_k^T H_k^T}{(g_k, H_k g_k)}. \tag{6}$$

Here, $\alpha_k > 1$ is a space extension parameter, H_k is a symmetric matrix, $s_0 = 0$, $H_0 = I$.

The use of the quality functional $Q_k(s) = \frac{1}{2}(1 - (s, g))^2$ enables us to implement various gradient ML algorithms where the maximum weight is automatically given to the last observation. Taking into account the specifics of a problem, at the learning step, the gradient consistency condition $(s_{k+1}, g_k) > 0$ for the descent direction must be fulfilled, which requires specialization of the applied ML algorithm.

The convergence rate of the ML method (5)–(6) grows with an increase in the admissible value of α in (6) [14], and depends on the characteristics of the set G. In algorithm (5)–(6), we distinguish 2 stages: correction stage (5) reducing the residual between the optimal solution s^* and s_k, and extension stage (6) resulting in the increase of the residual in the extended space without exceeding its initial value, which limits the magnitude of the extension parameter. To create more efficient algorithms for solving systems of inequalities, we have to choose the directions of correction and extension so that it enables us to increase the extension parameter value.

The paper presents one of the special cases of the correction stage and extension stage implementation. It was proposed to use linear combinations of vectors g_{k-1}, g_k in transformations (5)–(6) instead of a vector g_k when it is appropriate:

$$s_{k+1} = s_k + \frac{H_k p_k [1 - (s_k, g_k)]}{(g_k, H_k p_k)}, \tag{7}$$

$$H_{k+1} = H_k - (1 - \frac{1}{\alpha_k^2}) \frac{H_k y_k y_k^T H_k^T}{(y_k, H_k y_k)}. \tag{8}$$

Here,

$$y_k = g_k - g_{k-1}, \ p_k = g_k - \frac{g_{k-1}(g_k, H_k g_{k-1})}{(g_{k-1}, H_k g_{k-1})}. \tag{9}$$

Iterations (7)–(8) are conducted under the condition:

$$(g_k, H_k g_{k-1}) \le 0. \tag{10}$$

For the proposed ML algorithm, the convergence in a finite number of iterations is proved when solving problem (2) on separable sets. Based on the proposed ML algorithm, we have developed a method for minimizing nonsmooth functions and substantiate its convergence on convex functions.

2 A Space Extension Method for Solving Systems of Inequalities on Separable Sets

If condition (10) is satisfied, our new Algorithm 1 includes iterations (7)–(8) instead of (5)–(6). The algorithm scheme does not define a method for setting parameters $\alpha > 1$ and $\alpha_y k^2$. Below, we define the dependences of the admissible space extension parameters on the set G characteristics.

Algorithm 1. Method for solving systems of inequalities

1: Set $k \leftarrow 0$, $s_0 \leftarrow 0$, $g_{-1} \leftarrow 0$, $H_0 \leftarrow I$. Set $\alpha > 1$, where α is the limit for choosing the admissible value of the parameter α_k in the case of transformations (5)-(6).
2: Find $g_k \in G$, satisfying the condition (3)
3: **If** such a vector does not exist **then**
4:    the solution $s_k \in S(G)$ is found, stop the algorithm. **End if**
5: **If** $k = 0$ and condition (10) is not satisfied **then**
6:    go to step 10. **End if**
7: Calculate the limit of the admissible value of the squared extension parameter $\alpha_y k$ for combination of transformations (7)-(8).
8: **If** $\alpha_y k^2 < \alpha^2$ **then** go to step 10
9: **else** set α_k so that $\alpha^2 \le \alpha_k^2 \le \alpha_y k^2$; perform (7)-(8); go to step 11. **End if**
10: Set α_k so that $\alpha_k^2 \le \alpha^2$ and perform transformations (5)-(6).
11: Increase k by one and go to step 2.

Let us make assumptions on the separable set G. Vector η is the shortest vector from G. Denote the length of the minimal vector of the set by $\rho = \|\eta\|$, the length of the maximal vector of the set by $R = \max_{g \in G} \|g\|$, normalized vector η by $\mu = \eta/\rho$, a vector associated with the sought solution of the systems (2) and (4), when analyzing the ML algorithm, by $s^* = \mu/\rho$, an upper bound value of the set G in the direction μ by $R_s = \max_{g \in G}(\mu, g)$, the ratio of the upper and lower bounds of the set along μ by $M = R_s/\rho$, the ratio of the minimal and maximal vectors of the set by $r = \rho/R_s = M^{-1}$, $V = \rho/R$. We use the denoted characteristics as functions of a set Q, for example, $\eta(Q)$, $r(Q)$.

Assumption 1. *Set G is convex, closed, limited $(R < \infty)$ and satisfies the separability condition, i.e. $\rho > 0$.*

Parameters ρ and R_s characterize the thickness of the set G in the direction μ, which can be formulated as a two-sided inequality:

$$\rho \le (\mu, g) \le R_s, \ \forall g \in G. \tag{11}$$

From (11), taking into account the form of s^*, we get:

$$1 \le (s^*, g) \le R_s/\rho = M, \ \forall g \in G. \tag{12}$$

Taking into account (11), R_s satisfies the constraints: $\rho \le R_s \le \|\mu\| \max_{g \in G} \|g\| \le R$. At zero thickness of the set, when $R_S = \rho$, we have the case of a flat set, and the work of the ML algorithm is reduced to solving the system (4).

Denote the algorithm approximation sequence by s_k, residual vector by $\Delta_k = s^* - s_k$. While vector s_k is not a solution to (2) for vectors g_k selected in step 2 of Algorithm 1, from (3) and (12), the inequality holds:

$$(\Delta_k, g_k) = (s^* - s_k, g_k) = (s^*, g_k) - (s_k, g_k) \geq 1. \qquad (13)$$

Denote $A_k = H_k^{-1}$. Applying the Sherman-Morrison equation to (6) and (8), we obtain a matrix transformation equation:

$$A_{k+1} = A_k + (\alpha^2 - 1)\frac{g_k g_k^T}{(g_k, H_k g_k)}. \qquad (14)$$

Similarly, for (8), we obtain

$$A_{k+1} = A_k + (\alpha_k^2 - 1)\frac{y_k y_k^T}{(y_k, H_k y_k)}. \qquad (15)$$

For a symmetric strictly positive definite matrix A, we will use the notation $A > 0$. For a matrix $A > 0$, denote a matrix $A^{1/2}$ such that $A^{1/2} > 0$ and $A^{1/2}A^{1/2} = A$. For vectors s_k and g_k of Algorithm 1 from (13), using the Schwarz inequality, we obtain:

$$1 \leq (\Delta_k, g_k)^2 = (\Delta_k, A_k^{1/2}H_k^{1/2}g_k)^2 \leq (\Delta_k, A_k\Delta_k)(g_k, H_k g_k). \qquad (16)$$

Inequality (16) is a key point in justifying the convergence rate of Algorithm 1. The choice of extension parameters should ensure that the values $(\Delta_k, A_k\Delta_k)$ do not increase when the values $(g_k, H_k g_k)$ decrease at the geometric progression speed. Successively, after a finite number of iterations, the right side of (16) will be less than one. The resulting contradiction will mean that problem (2) is solved, and our assumption that we have find a vector g_k satisfying condition (3) is false.

Denote $\tau_k = \min_{0 \leq j \leq k-1}[(g_j, H_j g_j)/(g_j, g_j)]$. With respect to the decreasing rate of the sequence $\{\tau_k\}$, the following theorem [16] was formulated.

Theorem 1. *Let a sequence $\{H_k\}$ be a transformation result (6) with $H_0 = I$, $\alpha_k = \alpha > 1$ and arbitrary $g_k \in \mathbb{R}^n$, $g_k \neq 0$, $k = 0, 1, 2, \ldots$. Then*

$$\tau_k \leq k(\alpha^2 - 1)/[n(\alpha^{2k/n} - 1)], \ k \geq 1. \qquad (17)$$

Let us show that for Algorithm 1 with fixed values of the parameter α, estimates similar to (24) are valid and get expressions for the admissible parameters α_k in (6), (8) at which the values $(\Delta_k, A_k\Delta_k)$ do not increase. To simplify the analysis, the corresponding vectors and matrices of iterations (5)–(6) and (7)–(8) are transformed by equations $\hat{s} = A_k^{1/2}s$, $\hat{g} = H_k^{1/2}g$, $\hat{A}_k = H_k^{1/2}A_k H_k^{1/2} = I$, $\hat{H}_k = A_k^{1/2}H_k A_k^{1/2} = I$. To switch to new variables, we multiply Eq. (5) on the left by $H_k^{1/2}$, Eq. (6) on the left and right by $A_k^{1/2}$, and (14) on the left and right by $H_k^{1/2}$. The transformed equations of the algorithm are:

$$\hat{s}_{k+1} = \hat{s}_k + \frac{\hat{g}_k[1 - (\hat{s}_k, \hat{g}_k)]}{(\hat{g}_k, \hat{g}_k)}, \tag{18}$$

$$\hat{H}_{k+1} = I - (1 - \frac{1}{\alpha_k^2})\frac{\hat{g}_k\hat{g}_k^T}{(\hat{g}_k, \hat{g}_k)}, \tag{19}$$

$$\hat{A}_{k+1} = I + (\alpha_k^2 - 1)\frac{\hat{g}_k\hat{g}_k^T}{(\hat{g}_k, \hat{g}_k)}. \tag{20}$$

Similarly, we transform expressions (7)–(9) and (15): $\hat{s}_{k+1} = \hat{s}_k + \frac{\hat{p}_k[1-(\hat{s}_k,\hat{g}_k)]}{\hat{g}_k,\hat{p}_k}$, $\hat{H}_{k+1} = I - (1 - \frac{1}{\alpha_k^2})\frac{\hat{y}_k\hat{y}_k^T}{(\hat{y}_k,\hat{y}_k)}$,

$$\hat{y}_k = \hat{g}_k - \hat{g}_{k-1}, \ \hat{p}_k = \hat{g}_k - \frac{\hat{g}_{k-1}(\hat{g}_k, \hat{g}_{k-1})}{(\hat{g}_{k-1}, \hat{g}_{k-1})}, \tag{21}$$

$\hat{A}_{k+1} = I + (\alpha_k^2 - 1)\frac{\hat{y}_k\hat{y}_k^T}{(\hat{y}_k,\hat{y}_k)}$.

For the residual, the equality $(\Delta_k, A_k\Delta_k) = (\hat{\Delta}_k, \hat{\Delta}_k)$ holds. From equality $(s, g) = (A_k^{1/2}s, H_k^{1/2}g) = (\hat{s}, \hat{g})$, inequalities (12) for new variables are:

$$1 \le (\hat{s}^*, \hat{g}) \le R_s/\rho = M, \ \forall \hat{g} \in \hat{G}. \tag{22}$$

Let Z be a plane formed by vectors $\tilde{g}_k$, $\tilde{g}_{k-1}$. Characteristics of set $\hat{G}$ in the plane of vectors Z are shown on Fig. 1. Here, lines W_1, W_M are projections of hyperplanes, i.e. corresponding inequality (22) boundaries for vector $\hat{s}^*$ projection defined by the normal $\tilde{g}_{k-1}$. Similarly, lines Z_1, Z_M are boundaries of inequalities (22) for vector $\hat{s}^*$ projection, defined by the normal $\tilde{g}_k$. For the indicated points on Fig. 1, we denote the segment by AB, its length by $|AB|$, and the vector with the origin at point A by $\overrightarrow{AB}$. Let ψ be the angle between vectors $\tilde{g}_k$, $\tilde{g}_{k-1}$. On Fig. 1, angle ψ is obtuse, i.e. condition (10) is satisfied: $(g_k, H_k g_{k-1}) = (\tilde{g}_k, \tilde{g}_{k-1}) \le 0$. Hence, angle φ, formed by lines W_1 and Z_1, showed on Fig. 1, is acute. Since vectors $\tilde{g}_k$, $\tilde{g}_{k-1}$ are normals for the lines W_1 and Z_1 (see Fig. 1), $\varphi = \pi - \psi$. Hence, from the angle ψ definition, $sin^2\varphi = sin^2(\psi)$,

$$cos^2\varphi = cos^2\psi = \frac{(\tilde{g}_k, \tilde{g}_{k-1})^2}{(\tilde{g}_k, \tilde{g}_k)(\tilde{g}_{k-1}, \tilde{g}_{k-1})} = \frac{(g_k, H_k g_{k-1})^2}{(g_k, H_k g_k)(g_{k-1}, H_k g_{k-1})}. \tag{23}$$

Lemma 1. *Let the values* a, b, c, β *satisfy the constraints* $a \ge a_m \ge 0$, $b > 0$, $c > 0$ *and* $0 \le \beta \le 1$, *then:*

$$\min_{\alpha,\beta}\left(\frac{(a + \beta b)^2 - \beta^2 b^2}{\beta^2 c^2}\right) = \frac{a_m^2 + 2a_m b}{c^2} = \frac{(a_m + b)^2 - b^2}{c^2}. \tag{24}$$

Proof. Proof is obtained by expanding the parentheses in (24). ∎

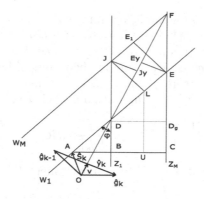

Fig. 1. The set G characteristics in the plane of vectors $\tilde{g}_k, \tilde{g}_{k-1}$.

The next lemma enables us to calculate the line segment lengths.

Lemma 2. *Let vectors p_1, p_2 and g be linked by equalities $(p_1, g) = a$, $(p_2, g) = b$, the difference of vectors $p_2 - p_1$ be collinear to the vector p, and ξ be an angle between p and g. Then:*

$$\|p_1 - p_2\|^2 = \frac{(a-b)^2}{(g,p)^2}\|p\|^2 = \frac{(a-b)^2}{\|g\|^2 cos^2\xi}. \tag{25}$$

Proof. From the collinearity of vectors $p_1 - p_2 = \gamma p$, based on equality $(p_1 - p_2, g) = (\gamma p, g) = (a - b)$, we find $\gamma = (a - b)/(p, g)$. This implies (25). ∎

Lemma 3. *As a result of (5) in step 10 of Algorithm 1, the equality holds*

$$(s_{k+1}, g_k) = (\hat{s}_{k+1}, \hat{g}_k) = 1, \tag{26}$$

and as a result of (7) in step 7, we have:

$$(s_{k+1}, g_{k-1}) = (\hat{s}_{k+1}, \hat{g}_{k-1}) = 1. \tag{27}$$

Proof. Equality (26) is established by verification. From (21), we have $(\hat{p}_k, \hat{g}_{k-1}) = 0$. Consequently, after transformation (7) in step 7, the equality $(s_k, g_{k-1}) = (\hat{s}_k, \hat{g}_{k-1}) = 1$ is preserved, which proves (27). ∎

Lemma 4. *Let a set G satisfy Assumption 1. Then the limit α of the admissible parameter value $\alpha_k \leq \alpha$ in Algorithm 1 providing inequality $(\Delta_{k+1}, A_{k+1}\Delta_{k+1}) \leq (\Delta_k, A_k\Delta_k)$ in the case of transformations (5)–(6) is given by equation*

$$\alpha^2 = M^2/(M-1)^2 = 1/(1-r)^2. \tag{28}$$

Proof. Let us examine changes in $(\hat{\Delta}_k, \hat{\Delta}_k)$ as a result of transformations (18) and (19) carried out in step 10. According to constraints (22), the admissible projection area $\hat{s}^*$ on Fig. 1 has the form of stripes between lines W_1, W_M

and Z_1, Z_M, and belongs to parallelogram $DEFJ$. Vector $\overrightarrow{OA}$ is $\hat{s}_k$ projection. Residual $\hat{\Delta}_k$, according to (18) and (19), changes only in the direction $\hat{g}_k$. As a result (18), we pass from point A to point B. According to step 2, at point A, $(\hat{s}_k, \hat{g}_k) \leq 0$, and at point B, according to the results of Lemma 3, equality (26) holds. Using Lemma 2, where $g = p = \hat{g}_k$, we get an equation for the length of the segments $|AB|$ and $|BC|$ and limitation of their lengths

$$|AB| = \frac{(1 - (\hat{s}_k, \hat{g}_k))}{\|\hat{g}_k\|} \leq \frac{1}{\|\hat{g}_k\|}, \ |BC| = \frac{(M - 1)}{\|\hat{g}_k\|}. \tag{29}$$

Denote by U a projection point $\hat{s}^*$ onto line segment BC.

$$|BU| = \beta|BC| = \beta(M - 1)/\|\hat{g}_k\|, \ 0 \leq \beta \leq 1. \tag{30}$$

As a result of transformation (18), the squared residual will decrease:

$$(\hat{\Delta}_k, \hat{\Delta}_k) - (\hat{\Delta}_{k+1}, \hat{\Delta}_{k+1}) = (|AB| + |BU|)^2 - |BU|^2. \tag{31}$$

After transformation (19) according to (20), the squared residual increases by

$$(\hat{\Delta}_{k+1}, \hat{A}_{k+1}\hat{\Delta}_{k+1}) - (\hat{\Delta}_{k+1}, \hat{\Delta}_{k+1}) = (\alpha_k^2 - 1)|BU|^2. \tag{32}$$

Let us formulate a condition that the residual does not increase in the extended space for a given location of the point U. From (31), (32), we have $(\hat{\Delta}_{k+1}, \hat{A}_{k+1}\hat{\Delta}_{k+1}) \leq (\hat{\Delta}_k, \hat{\Delta}_k)$, or

$$(\alpha_k^2 - 1)|BU|^2 \leq (|AB| + |BU|)^2 - |BU|^2. \tag{33}$$

Hence, we obtain the limitation for the space extension parameter $\alpha_k^2 - 1 \leq \frac{(|AB|+|BU|)^2-|BU|^2}{|BU|^2}$ Taking into account the possible position of points A and U, using (29), (30) and (24) of Lemma 1, setting $a = 1/\|\hat{g}_k\|$, $a_m = |AB|$, $b = |BC|$, $\beta b = |BU|$, we obtain an estimate which is valid for an arbitrary position of the point U: $\alpha_k^2 - 1 \leq \min_{|AB|,|BU|}\{\frac{(|AB|+|BU|)^2-|BU|^2}{|BU|^2}\} = \frac{M^2-(M-1)^2}{(M-1)^2} = \frac{M^2}{(M-1)^2} - 1$. From this, we obtain the estimate (28). ∎

Lemma 5. *Let set G satisfy Assumption 1. Then the limit $\alpha_y k$ for the admissible value of parameter α_k in step 7 of Algorithm 1 providing inequality $(\Delta_{k+1}, A_{k+1}\Delta_{k+1}) \leq (\Delta_k, A_k\Delta_k)$ in the case of transformations (7)–(8), is given as*

$$\alpha_{yk}^2 = \min\{\alpha_{Ek}^2, \alpha_{Jk}^2\}, \tag{34}$$

where

$$\alpha_{Ek}^2 = 1 + \frac{(2M - 1)(y_k, H_k y_k)}{(M - 1)^2(g_k, H_k g_k)sin^2\varphi}, \tag{35}$$

$$\alpha_{Jk}^2 = 1 + \frac{(y_k, H_k y_k)}{(M - 1)^2(g_k, H_k g_k)sin^2\varphi}(1 + \frac{2(M - 1)(g_k, H_k g_k)^{1/2}cos\varphi}{(g_{k-1}, H_k g_{k-1})^{1/2}}), \tag{36}$$

the value $cos^2\varphi$ is defined in(23), and $sin^2\varphi = 1 - cos^2\varphi$.

Proof. We assume that the projection $\hat{s}^*$ is on the segment EE_1 orthogonal to the line W_1. Then

$$(\hat{\Delta}_k, \hat{\Delta}_k) - (\hat{\Delta}_{k+1}, \hat{\Delta}_{k+1}) = (|AD| + |DE|)^2 - |DE|^2 = \frac{(|AB| + |BC|)^2 - |BC|^2}{\sin^2 \varphi}. \tag{37}$$

Denote a vector of minimum length from the shell of vectors $\tilde{g}_k$, $\tilde{g}_{k-1}$ by v. It possesses the property of equality of dot products $(v, \tilde{g}_{k-1}) = (v, \tilde{g}_k)$ and orthogonal to the vector $\hat{y}_k$, $(v, \hat{y}_k) = 0$. Therefore, the ray along this vector passes through the points O, D, F. The deviations of $\hat{s}^*$ from the segment DF along $\hat{y}_k$ are maximum when its projection is at points J and E. Based on (25) in Lemma 2, we find the length of the segment $|EEy| = |JJy|$. Since vector $\overrightarrow{OEy}$ is parallel to v, $(\overrightarrow{OEy}, \hat{y}_k) = 0$. Taking into account the properties of lines Z_1, Z_M, we obtain $(\overrightarrow{OEy}, \hat{y}_k) = (M - 1)$. Assuming in (25) $p = g = \hat{y}_k$, taking into account the last equalities, we find the length of the segments

$$|JJy| = |EEy| = (M - 1)/\|\hat{y}_k\|, \tag{38}$$

with extension

$$(\hat{\Delta}_{k+1}, \hat{A}_{k+1}\hat{\Delta}_{k+1}) - (\hat{\Delta}_{k+1}, \hat{\Delta}_{k+1}) = (\alpha_k^2 - 1)|EEy|^2. \tag{39}$$

Based on the condition that the residual does not increase in an extended space, similarly to (33), using (39), we have

$$(\hat{\Delta}_{k+1}, \hat{A}_{k+1}\hat{\Delta}_{k+1}) \leq (\hat{\Delta}_k, \hat{\Delta}_k), \tag{40}$$

or $((\alpha_k^2 - 1)|EEy|)^2 \leq (|AD| + |DE|)^2 - |DE|^2$. Thus,

$$\alpha_k^2 - 1 \leq \min_{AD}\{\frac{(|AD| + |DE|)^2 - |DE|^2}{|EEy|^2}\}. \tag{41}$$

According to Fig. 1, the maximum deviations along the segment DF are at the boundaries of the parallelogram $DEFJ$. In triangle DEF, the position of the projection $\hat{s}^*$ on the boundary EF will just increase the estimate (41), since the numerator of the expression will grow, and the component along $\hat{y}_k$ will decrease. An estimate similar to (41) for the position of the projection $\hat{s}^*$ on the boundary DE also leads to an increase in the estimate for the extension parameter, which is easy to show using the result of Lemma 1. Therefore, estimate (41) is the smallest estimate of the space extension parameter at the position of the projection $\hat{s}^*$ in the triangle DEF. With the projection $\hat{s}^*$ position in the triangle DEF, based on (41), using (37), (38), we obtain the estimate (35) of the admissible limit of the space extension parameter.

Segment JL is parallel to the vector $\tilde{g}_{k-1}$. Since the points of the segment are located on the lines W_1, W_M based on inequalities (12) and equality (25) in Lemma 2, we find its length $|JL| = (M - 1)/\|\tilde{g}_{k-1}\|$. Since triangle DJL is rectangular, we obtain:

$$|DL|^2 = |JL|^2 \frac{\cos^2 \varphi}{\sin^2 \varphi} = \frac{(M - 1)^2 \cos^2 \varphi}{\|\tilde{g}_{k-1}\|^2 \sin^2 \varphi}. \tag{42}$$

Taking into account the equality $|JJy| = |EEy|$, similarly to (40)–(41), using (38), (42), we get the estimate of α_k:

$$\alpha_k^2 - 1 \leq \min_{AD} \left\{ \frac{(|AD| + |DL|)^2 - |DL|^2}{|JJ_y|^2} \right\} = \min_{AD} \left\{ \frac{|AD|^2 + 2|AD||DL|}{|JJ_y|^2} \right\}$$

$$= \frac{\|\hat{y}_k\|^2}{(M-1)^2 \|\hat{g}_k\|^2 sin^2\varphi} \left(1 + \frac{2(M-1)\|\hat{g}_k\|cos\varphi}{\|\hat{g}_{k-1}\|}\right).$$

Returning to the original variables, we get the estimate of (36). Estimate (36) is the smallest estimate of the space extension parameter for the position of projection $\hat{s}^*$ in the triangle DJF, which can be shown similarly. ∎

In the next theorem, an estimate similar to (17) is obtained directly for vectors g_k generated by Algorithm 1.

Theorem 2. *Let set G satisfy Assumption 1, and the sequence $\{\pi_k = \min_{0 \leq j \leq k-1}(g_j, H_j g_j) = (g_{Jk}, H_{Jk}g_{Jk})\}$ be calculated based on the characteristics of Algorithm 1 for fixed values of the space extension parameters $\alpha_k^2 = \alpha^2$ specified in steps 7 and 10, where parameter α is specified according to (28). Then:*

$$\pi_k = (g_m, H_m g_m) \leq \frac{4R^2 k(\alpha^2 - 1)}{n[\alpha^{2k/n} - 1]}, \ k \geq 1, \tag{43}$$

where $m = arg\min_{0 \leq j \leq k-1}(g_j, H_j g_j)$.

Proof. If transformation (6) is performed at the iteration of Algorithm 1, then estimate (17), taking into account the inequality $\|g_m\|^2 \leq R_G^2$, proves (43). Let transformation (8) be performed at the m-th iteration, where y_m has part. In this case (6), based on the condition $(g_m, H_m g_{m-1}) < 0$, we get $(y_m, H_m y_m) = (g_m, H_m g_m) + (g_{m-1}, H_m g_{m-1}) - 2(g_m, H_m g_{m-1}) \geq (g_m, H_m g_m)$. This implies $\frac{(y_m, H_m y_m)}{(y_m, y_m)} \geq \frac{(g_m, H_m g_m)}{(\|g_m\| + \|g_{m-1}\|)^2} \geq \frac{(g_m, H_m g_m)}{4R_G^2}$. With (17), this proves (43) ∎

Theorem 3. *Let a set G satisfy Assumption 1, and the sequence $\{(\Delta_k, A_k \Delta_k)\}$ be calculated based on the characteristics of Algorithm 1, extension parameter α satisfy constraint (28), admissible value α_{yk}^2 be given by (34). Then:*

$$(\Delta_{k+1}, A_{k+1}\Delta_{k+1}) \leq (\Delta_k, A_k\Delta_k) \leq (\Delta_0, \Delta_0) = \rho^{-2}, \ k = 0,1,2... \tag{44}$$

Proof. According to the condition of the theorem, for Algorithm 1, all constraints on the space extension parameters are satisfied, which are necessary to fulfill the conditions of Lemmas 4, 5, in which the first of inequalities (44) is proved for each of the space extension cases in steps 7 and 10 of Algorithm 1. Continuing the chain of inequalities, we obtain the proof of (44). In (44), equality $(\Delta_0, \Delta_0) = \rho^{-2}$ follows from $s_0 = 0$ in step 1 and definition $s^* = \mu/\rho$ ∎

For fixed values of the space extension parameter with respect to the convergence of Algorithm 1, the following theorem holds.

Theorem 4. *Let the set G satisfy Assumption 1, in Algorithm 1, the values of the space extension parameters specified in steps 7 and 10 be fixed $\alpha_k^2 = \alpha^2$, and parameter α be given according to constraints (28). Then a solution to system (2) will be found by Algorithm 1 in a finite number of iterations, which does not exceed K_0 equal to the minimum integer k satisfying the inequality*

$$\frac{4kR^2(\alpha^2 - 1)}{n\rho^2[\alpha^{2k/n} - 1]} = \frac{4k(\alpha^2 - 1)}{nV^2[\alpha^{2k/n} - 1]} < 1. \tag{45}$$

In this case, until a solution is found, the inequalities hold:

$$(g_k, H_k g_k) \geq \rho^2, \tag{46}$$

$$\frac{(g_k, H_k g_k)}{(g_k, g_k)} \geq \frac{\rho^2}{R^2} = V^2. \tag{47}$$

Proof. Conditions of Theorems 3 and 4 are satisfied. Inequality (46) follows from (16) and (44), and (47) follows from (46) and definition of R. Using (16), taking into account (43), (44) and notations of Theorem 2, we get an estimate:

$$1 \leq (\Delta_m, A_m \Delta_m)(g_m, H_m g_m) \leq \frac{4kR^2(\alpha^2 - 1)}{n\rho^2[\alpha^{2k/n} - 1]} = \frac{4k(\alpha^2 - 1)}{nV^2[\alpha^{2k/n} - 1]}. \tag{48}$$

The right side of (48) decreases with increasing k and becomes less than 1 after a finite number of iterations if, as before, step 2 of Algorithm 1 allows us to find a vector $g_k \in G$ satisfying (3). Due to (16), this is impossible. Therefore, after a finite number of iterations, there are no vectors $g_k \in G$ satisfying (3), i.e. system (2) solution is found. As follows from (48), a guaranteed estimate of the number of iterations required to obtain a solution to the inequalities is given by the minimal integer $k = K_0$, at which inequality (45) does not hold ∎

For the purposes of analyzing the properties of the subgradients set obtained on the minimization method trajectory, we will study the behavior of the set characteristics depending on the degree of its perturbation. Denote by $S_\varepsilon(G) = \{z \in \mathbb{R}^n \mid \|z - x\| \leq \varepsilon, \forall x \in G\}$ neighborhood of the set G.

Lemma 6. *Let the set G satisfies the Assumption 1. Then with $0 \leq \varepsilon < \rho(G)$ the following relations will hold [14]*

$$R(S_\varepsilon(G)) \leq R(G) + \varepsilon, \ R_s(S_\varepsilon(G)) \leq R_s(G) + \varepsilon, \ \rho(S_\varepsilon(G)) \geq \rho(G) - \varepsilon, \tag{49}$$

$$r(S_\varepsilon(G)) \geq r(G) - \frac{2\varepsilon}{R_s}, \ V(S_\varepsilon(G)) \geq V(G) - \frac{2\varepsilon}{R_G}. \tag{50}$$

3   Minimization Algorithm

An iteration of the stated minimization method includes step (1). Due to exact 1-dimensional search in the subgradient set at point x_{k+1}, there is always a subgradient satisfying condition (3): $(s_{k+1}, g_{k+1}) \leq 0$. In the built-in algorithm for solving systems of inequalities, the extension parameter is set to enable us to solve the system of inequalities for combining subgradient sets in some neighborhood of a current approximation x_k. This allows the minimization algorithm to get out of the neighborhood after a finite number of iterations. In order to exclude situations of a significant subgradients set expansion, we introduce an update to the algorithm for solving systems of inequalities. To track the updates, we used a stopping criterion, formulated based on Theorem 4. Description of the minimization method is given in Algorithm 2.

Algorithm 2. $RA(\alpha)$

1: Set $x_0 \in \mathbb{R}^n$, $w_0 \leftarrow x_0$, $k, q, l \leftarrow 0$, $s_0 \leftarrow 0$, $H_0 \leftarrow I$. Set $\sigma > 0$, parameters $M > 0$, $r \leftarrow 1/M$ and the limit α for extension parameter according to equality (28). Calculate $g_0 \in \partial f(x_0)$.
2: **If** $g_k = 0$ **then** stop the algorithm. **End if**
3: **If** $(g_k, H_k g_k)/(g_k, g_k) < \sigma$ **then**
4:    update $q \leftarrow q + 1$, $w_q \leftarrow x_k$, $l \leftarrow 0$, $H_k \leftarrow I$, $s_k \leftarrow 0$. **End if**
5: **If** $l = 0$ and condition (10) is not satisfied **then**
6:    go to step 10. **End if**
7: Calculate the limit of the admissible value of the squared extension parameter α_{yk} for a combination of transformations (7), (8)
8: **If** $\alpha_{yk}^2 < \alpha^2$ **then** go to step 10
9: **else** set α_k so that $\alpha^2 \leq \alpha_k^2 \leq \alpha_{yk}^2$; perform (7), (8); go to step 11. **End if**
10: Set $\alpha_k^2 \leq \alpha^2$ and perform the transformations (5), (6).
11: Calculate a new approximation of the minimum point $x_{k+1} \leftarrow x_k - \gamma_k s_{k+1}$, $\gamma_k \leftarrow arg \min_\gamma f(x_k - \gamma s_{k+1})$.
12: Calculate subgradient $g_{k+1} \in \partial f(x_{k+1})$ from the condition $(g_{k+1}, s_{k+1}) \leq 0$.

13: Increase k and l by 1; go to step 2.

In step 12, due to exact 1-dimensional descent condition in step 12, the desired subgradient always exists, which follows from the condition for the 1-dimensional function extremum. For the sequence of approximations of the algorithm, due to the exact 1-dimensional descent in step 11, Lemma 7 holds [8].

Lemma 7. *Let function $f(x)$ be strictly convex on $\mathbb{R}^n$, set $D(x_0)$ be limited, and the sequence $\{x_k\}_{k=0}^\infty$ be such that $f(x_{k+1}) = \min_{\gamma \in [0,1]} f(x_k + \gamma(x_{k+1} - x_k))$. Then $\lim_{k \longrightarrow \infty} \|x_{k+1} - x_k\| = 0$.*

Denote $D(z) = \{x \in \mathbb{R}^n \mid f(x) \leq f(z)\}$, let x_* be a minimum point of function, x^* be limit points of the sequence $\{w_q\}$ generated by Algorithm 2. The existence of limit points of a sequence $\{w_q\}$ when the set $D(x_0)$ is bounded follows from $w_q \in D(x_0)$. Concerning the algorithm convergence, we formulate Theorem 5:

Theorem 5. *Let function $f(x)$ be strictly convex on $\mathbb{R}^n$, set $D(x_0)$ be limited, and for $x \neq x_*$,*

$$r(\partial f(x)) \geq r_0 > 0, \quad V(\partial f(x)) \geq V_0 > 0, \tag{51}$$

where parameters M, r and α of Algorithm 2 are given according to the equalities

$$M = \frac{4}{3}r_0, \ r = \frac{3}{4}r_0, \ \alpha = \frac{1}{1 - 3r_0/4}, \tag{52}$$

and parameters α_k, set in steps 7 and 10, are fixed $\alpha_k = \alpha$. If $\sigma = (3V_0/4)^2$ then any limit point of sequence $\{w_q\}$ generated by Algorithm 2 is a minimum point on $\mathbb{R}^n$.

Proof. Assume that the statement of the theorem is false: suppose that some subsequence $w_{q_s} \longrightarrow x^*$, but $x^* \neq x_*$. Then at this point the conditions (51) hold. Set $\varepsilon = \rho(\partial f(x^*))/8$. Choose $\delta > 0$, such that

$$\partial f(x) \subset S_\varepsilon(\partial f(x^*)) \equiv S_\varepsilon^* \ \forall x \in S_\delta(x^*). \tag{53}$$

Such a choice is possible due to the upper semicontinuity of the point-to-set mapping $\partial f(x)$ [8].

For the set S_ε^* characteristics according to (50), (51), taking into account the choice of ε, estimates can be obtained:

$$r(S_\varepsilon^*) \geq r(\partial f(x^*)) - \frac{2\varepsilon}{R_s(\partial f(x^*))} = \frac{3r(\partial f(x^*))}{4} \geq \frac{3r_0}{4}, \tag{54}$$

$$V(S_\varepsilon^*) \geq V(\partial f(x^*)) - \frac{2\varepsilon}{R(\partial f(x^*))} = \frac{3V(\partial f(x^*))}{4} \geq \frac{3V_0}{4}. \tag{55}$$

In the minimization algorithm, the method for solving inequalities works, and according to (52), the parameters of Algorithm 2 correspond to the characteristics of the set $G = S_\varepsilon^*$. Therefore, by virtue of definitions (53), for the method for solving inequalities, all results with the set $G = S_\varepsilon^*$ will be valid as long as the search trajectory $\{x_k\}$ is in the set $S_\delta(x^*)$. By virtue of the exact 1-dimensional search and the conditions of the theorem, the conditions of Lemma 7 are satisfied. Hence, the value $\|x_{k+1} - x_k\|$ decreases and for a sufficiently large k, due to the insignificance of changes, the sequence $\{x_k\}$ will not leave the $S_\delta(x^*)$ neighborhood for a period greater than the period between updates in step 3 of the algorithm. This period for the built-in algorithm for solving inequalities is more abundantly determined by the number of iterations required to violate inequality (47) for the characteristics of the set (54), (55).

On set S_ε^* for characteristics (54), (55), according to (43), the sequence $(g_k, H_k g_k)$ will decrease until an update occurs in step 3, taking into account the value $\sigma = (3V_0/4)^2$, under the condition $(g_k, H_k g_k)/(g_k, g_k) < (3V_0/4)^2$. An update necessarily occurs due to the validity of estimate (43) and constraints (49) for the perturbed set S_ε^*. According to Theorem 4, taking into account estimate (54) for S_ε^*, violation of (47) is fulfilled only if the sequence x_k goes beyond the set $S_\delta(x^*)$. This is a contradiction ∎

4   Computational Experiment

Tables 1, 2 show calculation results for various algorithms with space extension at fixed $\alpha^2 = 6$. Denote relaxation method with space extension in the direction of the subgradient [14] by SSM, r-algorithm [4], implemented in [14,16], by $r_{OM}(\alpha)$, r-algorithm [4] by $r(\alpha)$. Algorithm 2 was implemented with a fixed value $(RA(\alpha = const))$ for $\alpha^2 = 6$ and dynamically selected extension parameter $(RA(\alpha_k))$ for $\alpha^2 = (M/(M-1))^2 = 6$. We used 1-dimensional search [14,16] with simultaneous calculation of the function and gradient. Quadratic and piecewise linear test functions with a high degree of the level surfaces elongation, which increases with dimension are: $f_1(x) = \sum_{i=1}^{n} x_i^2 i^6$, $x_0 = (10/1,\ 10/2, ..., 10/n)$, $f_2(x) = \sum_{i=1}^{n} |x_i| i^3$, $x_0 = (10/1,\ 10/2, ..., 10/n)$. Tables 1, 2 show the number of function and subgradient calculations spent on achieving the required accuracy for the function $f(x_k) - f^* \leq \varepsilon$. The $RA(\alpha = const)$ algorithm outperforms the SSM, $r_{OM}(\alpha)$ and $r(\alpha)$ methods, and with dimension increase, its superiority grows. Therefore, the change in the directions of correction and space extension have a positive effect on the convergence rate. In the $RA(\alpha_k)$ algorithm compared to $RA(\alpha = const)$, an additional factor of convergence acceleration is involved due to an increase in the space extension parameter, which, according to the results of Tables 1, 2, led to the convergence rate increase. None of presented problems can be solved by the multistep minimization method [12].

Table 1. Function $f_1(x)$ minimization results for $\varepsilon = 10^{-10}$.

n	$RA(\alpha_k)$	$RA(\alpha = const)$	SSM	$r_{OM}(\alpha)$	$r(\alpha)$
100	1494	1834	2127	2333	2637
200	3474	3896	4585	5244	6572
300	5507	6317	7117	8480	10634
400	7690	8548	9791	11773	15058
500	9760	11510	12366	15281	19370
600	12133	13889	15537	19073	24536
700	13933	16394	18450	22500	29218
800	16492	18721	21387	26096	33473

Table 2. Function $f_2(x)$ minimization results for $\varepsilon = 10^{-4}$.

n	$RA(\alpha_k)$	$RA(\alpha = const)$	SSM	$r_{OM}(\alpha)$	$r(\alpha)$
100	2248	2714	3006	3817	4152
200	4988	6010	9939	8494	8862
300	7680	9301	11114	14050	14812
400	10625	12808	23687	19549	19392
500	13490	16656	28037	24865	25981
600	16466	20207	39703	31502	32757
700	20122	22850	44573	38796	38133
800	23016	27653	52380	44200	44287

5   Conclusion

We presented a RSMM with space extension and a ML algorithm [14,16] originating in the iterative least squares method. Taking into account the specific problem formulation of solving systems of inequalities in minimization methods, we determined the structure of the ML algorithm and altered the method of filling the structure of known ML algorithm [14]. A significant increase in the convergence rate was achieved due to altering the directions of correction and space extension in the ML algorithm, as well as by the use of the method for choosing the extension parameter at iterations. Algorithms of this type are of great practical importance due to their fast convergence with non-convex functions, e.g., when estimating the parameters of mathematical models under conditions of nonsmooth regularization [19–21].

Acknowledgment. This study was supported by the Ministry of Science and Higher Education of the Russian Federation (Project FEFE-2020-0013).

References

1. Lemarechal, C., Nemirovskii, A., Nesterov, Y.: New variants of bundle methods. Math. Program. **69**, 111–147 (1995). https://doi.org/10.1007/BF01585555
2. Richtarik, P.: Approximate level method for non smooth convex minimization. J. Optim. Theory Appl. **152**, 334–350 (2012)
3. Gasnikov, A.V., Nesterov, Y.E.: Universal method for stochastic composite optimization problems. Comput. Math. Math. Phys. **58**, 48–64 (2018)
4. Shor, N.Z.: Nondifferentiable Optimization and Polynomial Problems. Springer Science, New York (1998). https://doi.org/10.1007/978-1-4757-6015-6
5. Polyak, B.T.: The conjugate gradient method in extremal problems. USSR Comput. Math. Math. Phys. **9**(4), 94–112 (1969)
6. Polyak, B.T.: Introduction to Optimization. Optimization Software, New York (1987)
7. Lemarechal, C.: An extension of Davidon methods to non-differentiable problems. Math. Program. Study **3**, 95–109 (1975)
8. Dem'yanov, V.F., Vasil'ev, L.V.: Non-differentiable Optimization. Springer-Verlag, New York (1985)
9. Nemirovsky, A.S., Yudin, D.B.: Problem Complexity and Method Efficiency in Optimization. Wilcy, Chichester (1983)
10. Shor, N.Z.: Minimization methods for non-differentiable functions. In: Springer Series in Computational Mathematics, vol. 3. Springer, Heidelberg (1985). https://doi.org/10.1007/978-3-642-82118-9
11. Polyak, B.T.: Optimization of non-smooth composed functions. USSR Comput. Math. Math. Phys. **9**(3), 507–521 (1969)
12. Krutikov, V.N., Samoilenko, N.S., Meshechkin, V.V.: On the properties of the method of minimization for convex functions with relaxation on the distance to extremum. Autom. Remote Control **80**(1), 102–111 (2019)
13. Shalev-Shwartz, S.: Online Learning and Online Convex Optimization, Now Foundations and Trends (2012)
14. Krutikov, V.N., Petrova, T.V.: Relaxation method of minimization with space extension in the subgradient direction. Ekon. Mat. Met. **39**(1), 106–119 (2003)
15. Cao, H., Song, Y., Khan, K.A.: Convergence of subtangent-based relaxations of nonlinear programs. Processes **7**(4), 221 (2019)
16. Krutikov, V.N., Gorskaya, T.A.: A family of relaxation subgradient methods with two-rank correction of metric matrices. Ekon. Mat. Met. **45**(4), 37–80 (2009)
17. Nurminskii, E.A., Thien, D.: Method of conjugate subgradients with constrained memory. Autom. Remote Control **75**(4), 646–656 (2014)
18. Neimark, J.I.: Perceptron and pattern recognition. In: Mathematical Models in Natural Science and Engineering. Foundations of Engineering Mechanics. Springer, Heidelberg (2003). https://doi.org/10.1007/978-3-540-47878-2_27

19. Krutikov, V.N., Shkaberina, G.Sh., Zhalnin, M.N., Kazakovtsev, L.A.: New method of training two-layer sigmoidal neural networks with regularization. In: 2019 International Conference on Information Technologies (InfoTech), Varna, pp. 1–4 (2019). https://doi.org/10.1109/InfoTech.2019.8860890
20. Amini, S., Ghaernmaghami, S.: Sparse autoencoders using non-smooth regularization. In: 26th European Signal Processing Conference (EUSIPCO), Rome, pp. 2000–2004 (2018). https://doi.org/10.23919/EUSIPCO.2018.8553217
21. Tibshirani, R.J.: Regression shrinkage and selection via the lasso. J. R. Stat. Soc. Ser. B (Methodol.) **58**(1), 267–288 (1996)

Author Index

Printed in the United States
by Baker & Taylor Publisher Services